Powertrain

Michael Trzesniowski

Powertrain

Springer Vieweg

Michael Trzesniowski
Pankl Racing Systems AG
Drivetrain and Suspension Systems
Kapfenberg, Austria

ISBN 978-3-658-39884-2 ISBN 978-3-658-39885-9 (eBook)
https://doi.org/10.1007/978-3-658-39885-9

This book is a translation of the original German edition „Antrieb“ by Trzesniowski, Michael, published by Springer Fachmedien Wiesbaden GmbH in 2019. The translation was done with the help of artificial intelligence (machine translation by the service DeepL.com). A subsequent human revision was done primarily in terms of content, so that the book will read stylistically differently from a conventional translation. Springer Nature works continuously to further the development of tools for the production of books and on the related technologies to support the authors.

This Springer Vieweg imprint is published by the registered company Springer Fachmedien Wiesbaden GmbH, part of Springer Nature.
The registered company address is: Abraham-Lincoln-Str. 46, 65189 Wiesbaden, Germany

Preface

Now the manual series is already in its second edition and the family of these books is growing. While there were already five volumes at the beginning, another volume has been added for this edition. The authoritative idea that individual special volumes can go into depth without space problems has thus proven itself. Since the publication of the first edition, further findings have been added, which have found their way into the corresponding chapters, or new chapters have been added.

The fact that the contents nevertheless fit together and complement each other as if in a single book – one of the great strengths of the original book *Racing Car Technology* – is ensured by the editor in a comparable way to the way the project manager keeps an eye on the overall function in a major design project.

The racing car technology handbook series is dedicated to the racing vehicle from conception, design and calculation to operation and its (further) development.

The first volume, *Basic Course in Race Car Technology*, thus offers not only current considerations but also a historical overview of motorsport and racing operations, such as the rescue chain, and a comprehensive overview of the technology used in racing cars as a general introduction to the subject. For more than 15 years, the author has been concerned with the driving dynamics and chassis tuning of production passenger cars.

Volume two, *Complete Vehicle*, starts with the chronological design process and therefore begins with concept considerations, considers safety aspects and the design of the driver's environment, describes aerodynamic influences, and then looks at the frame and body work design.

Volume three, *Powertrain*, deals with all forms of drive systems and their energy storage, continues in the sense of load flow via start-up elements and transmission up to the side shafts. Electrical systems and electronic driving aids have also found their appropriate place in this volume.

Volume four, *Chassis*, is devoted exclusively to the decisive subsystem and its components that determine driving behaviour. Tyres and wheels, wheel-guiding parts, springs and dampers, steering, and brakes are covered.

Volume five, *Data Analysis, Tuning and Development*, deals with the phase that follows once the vehicle has been designed and built. The development and tuning of a racing vehicle requires a much different approach than its construction, and key tools – such as data acquisition and analysis, simulation, and testing – are therefore presented. The subject of data acquisition and analysis is profoundly presented by an author who is confronted with this activity on a daily basis.

For volume six, *Practical Course in Vehicle Dynamics*, authors have been recruited who have decades of experience as race engineers and race drivers on the race track. In their work, they describe the practical set-up of racing vehicles, underpin what they present with examples of calculations and thus also build a bridge to the theoretical considerations in the other volumes.

I wish all readers that they will find "their" volume in the abundance offered and that they will get essential impulses for their studies, profession and/or leisure time from reading it, be it because they are designing a vehicle, building one, operating and improving one or because they are analysing one with a thirst for knowledge.

Kapfenberg, Austria Michael Trzesniowski

Greeting

27.05.2019
EC Todsen

Second Edition: Racing Car Technology Handbook – Six Volumes

Motorsport continues to inspire. For as long as there have been cars, drivers have been pushing their racing cars to their technical and physical limits, engaging in gripping and exciting competitions. But the competition doesn't just take place on the race track. The foundation for success is laid in the development departments and design offices. In-depth knowledge of vehicle technology and development methodologies, along with thorough and timely project management, creative problem-solving skills and unconditional team play, determine victory or defeat.

Motorsport continues to be a model and guide for technological progress – be it in lightweight design, material selection or aerodynamics. Chassis and tire technology also benefit immensely, and new safety concepts are often based on experience from the race track. However, the influence of motorsport is particularly evident in the powertrain: In addition to the impressive increase in performance and efficiency of the classic internal combustion engine drive system, the key future technologies of hybrid and purely electric drive have also successfully arrived in motorsport competition and are continuing this successfully and with public appeal in a partly completely new setting. It remains very exciting to observe which attractive innovations digital networking solutions and autonomous driving systems will generate in motorsport. I recommend that young engineers in particular acquire the tools for their future careers in motorsport. What you learn in motorsport sticks. Formula Student already offers an ideal environment to start with.

I am very pleased that the book series Handbuch Rennwagentechnik has been so well received and that the second edition has been published within 2 years. This shows that the competencies addressed are clearly presented in this work and conveyed in an understandable way.

This book series has deservedly become a well-known and valued reference work among experts. The work brings students closer to the fascination of motorsport and racing enthusiast laymen to a deeper technical understanding.

I wish you much success on and off the race track!

Prof. Dr.-Ing. Peter Gutzmer

Deputy Chairman of the Executive Board and Chief Technology Officer, Schaeffler AG, Herzogenaurach, Germany

Abbreviations, Formula Symbols and Units *List of Symbols and Units*

Equations given in the text are generally quantity equations. The quantities can be used in any units, preferably in the SI units (meter-kilogram-second system). The unit of the quantity to be calculated then results from the selected units of the variables. Sometimes the numerical value equations commonly used in practice are also given. With these, the equation is only correct if it is calculated with the specified units. The unit of the result variable is therefore also given in the text.

Geometric *Points*

C to G	*Reference points, in general*
M	*Centre point*

Indices

If more than one index occurs, they are separated by a comma. The order of indices is this:

For forces, the first index indicates the location or point at which the force is applied, and the second index indicates the direction of the force, e.g. $F_{W,Z}$ ... Wheel contact force (vertical force at the tyre contact point). The vehicle fixed coordinate system used is defined in the glossary.

Additional specifications, such as front, rear and driven, follow as further indices.

0	*Zero-point position or starting point. Ambient*
1	*Before the compressor, in*
2	*After the compressor, out*
α	*Side slip*
a	*Driven, accelerating (one wheel only)*

(continued)

A	*Drive-off condition, accelerating (one axle)*
B	*Braking (one axle)*
Bt	*Battery*
C	*Coolant*
C	*Motor controller*
Ca	*Carburettor*
cl	*Clutch*
co	*Cornering*
D or d	*Axle drive (differential)*
dr	*Drag*
dyn	*Dynamic*
e (engine, motor)	*Effective*
f	*Stator*
fr	*Friction*
G	*Gearbox*
Ga	*Gas*
hyd	*Hydraulic*
i	*Inner wheel, inner*
Ic	*Inter cooler*
id	*Ideal value*
K	*Fuel*
critical	*Critical*
L	*Aerodynamic*
l	*Left, left side*
lim	*Limit*
lo	*Slipping, lock resp.*
ls	*Loss*
M	*Engine resp. motor*
m or med	*Middle, mean*
max	*Maximum permissible*
min	*Minimum*
n	*Rated value*
N	*Naturally aspirated engine*
No	*Cam*
o	*Outer wheel, outer*
Pi	*Piston*
q	*Gradient*
q	*Quick*
R	*Rolling (wheel)*
R	*Rotor*
Rd	*Rod, linkage resp.*
ref	*Reference ~*

(continued)

rs	*Right, right-hand side*
Rs	*Restrictor*
rsl	*Resulting*
s	*Slow*
Sp	*Spring*
t	*Total, nomial value resp.*
T	*Turbocharged engine*
ts	*Torsional*
V	*Overall vehicle*
v	*Valve*
W	*Wheel*
X or x	*Longitudinal direction in general*
Y or y	*Lateral direction*
Z (engine)	*Cylinder liner*
Z or z	*Vertical direction*

Distances in mm

a	*Centre distance (transmission)*
a - p	*Distances and length (in general)*
B	*Bore (diameter)*
C_R	*Dynamic rolling circumference at 60 kph*
$C_{R,dyn}$	*Dynamic rolling circumference at higher speeds*
d or D	*Diameter, in general*
h or H	*Height, in general*
L	*Total length*
l_{Rd}	*Length of conrod*
r	*Effective control arm length or force lever in general*
R	*Path radius*
r_{dyn}	*Force dynamic rolling radius of the Tyre at 60 kph*
s	*Travel or stroke, in general*
t	*(wall) thickness*

Angle in °

α	*Angle of gradient of the road*
α_f or α_r	*Slip angle of front or rear Tyre*
β	*Valve seat angle*
β	*Angle, in general*

Masses, Weights in Kg

m	*Mass, weight or load in general*
$m_{V,t}$	*Gross vehicle weight*

Forces *in* N

$F_{L,X}$	*Aerodynamic drag*
$F_{L,Z}$	*Aerodynamic downforce*
F_R	*Rolling resistance of the tyre*
F_{rsl}	*Resulting force*
$F_{V,X,A}$	*Traction force*
$F_{V,X,ex}$	*Excess force*
$F_{W,X,a}$ or $F_{W,X,A}$	*Accelerating force in the Centre of tyre contact of one wheel (a) or both wheels (A)*
$F_{W,Y}$	*Lateral force at wheel*
$F_{W,Z}$	*Vertical force at the Centre of tyre contact*

Torques and Moments in Nm

M_M	*Engine/motor torque*
T	*Torsional moment in general*

Dimensionless Key Figures

α_0	*Factor representing different load cases*
λ	*Air-fuel ratio, excess-air factor*
η	*Total efficiency of geartrain and final drive*
η_{cl}	*Efficiency of energy transformation (clutch)*
λ_a	*Charging efficiency: Mass of air corresponding to cylinder volume*
λ_{Rd}	*Stroke-to-conrod ratio*
η_e	*Effective efficiency*
Φ	*Gradation of ratio of speed*
Φ_L	*Ratio intake period to one revolution of crankshaft*
λ_l	*Volumetric efficiency*
$\mu_{W,X}$	*Coefficient of friction in longitudinal direction*
$\mu_{W,Y}$	*Coefficient of friction in lateral direction*
μ_{cl}	*Friction coefficient of clutch discs*
ν	*Bias of driving torque front/rear*
c_A	*Downforce coefficient*
c_W	*Drag coefficient*
ε	*Compression ratio*
i	*Factor representing number of strokes of engine*
i_D	*Axle ratio (final drive ratio)*
i_G	*Gearbox ratio*
j	*Number in general*
k	*Correction factor*
k_{Ca}	*Correction factor for carburettor*
k_m	*Factor representing rotating masses (allowance factor)*
k_R	*Rolling resistance coefficient*
k_{dyn}	*Dynamic amplifying factor*
k_v	*Factor for dynamic rolling circumference*
k_α	*Factor for tyre side-slip resistance*
k_Φ	*Factor for progressive ratio of speed*
L_{min}	*Air requirement*
n	*Numbering index*
Re	*Reynolds number*
S	*Longitudinal slip of tyre*
u	*Gear ratio*
z	*Number of cylinders*

Other Sizes

Δ	*Change, difference*	
ρ	*Density*	Kg/m^3
σ	*Stress*	N/mm^2
τ	*Shear stress*	N/mm^2
ρ_L	*Density of air*	kg/m^3
ω	*Circular frequency*	s^{-1}
A	*Area, cross-section area*	m^2
a_x	*Longitudinal acceleration in general*	m/s^2
b_e	*Specific fuel consumption*	kg/kWh
B	*Density of magnetic flux*	T
c_B	*Burn rate*	m/s
c_F	*Speed of flame front*	m/s
c_s	*Sonic speed*	m/s
c_T	*Speed of transport*	m/s
f	*Frequency*	Hz
g	*Acceleration due to gravity*	m/s^2
H_G	*Calorific value of mixture*	Y/m^3
H_u	*Specific calorific value*	J/kg
I	*(electric) current*	A
J	*Polar moment of inertia*	kgm^2
n	*Revolutions per minute or vibration frequency*	min^{-1}
n	*Polytropic exponent*	–
n_{krit}	*Critical rotational speed for bending, whirling speed*	min^{-1}
n_{crit}	*Critical rotational speed for torsion*	min^{-1}
P	*Power*	W
P_e	*Effective power of engine*	kW
p	*Pressure*	$Pa = N/m^2$
P_{ls}	*Power loss*	W
$p_{m,e}$	*Mean effective pressure*	bar[1]
p_0	*Ambient pressure*	bar[1]
q	*Gradient*	%
$\dot{Q}$	*Heat flow*	W
R	*Electrical resistance*	Ω
R_e	*Yield strength*	N/mm^2
R_L	*Gas constant of air*	kJ/(kgK)

(continued)

[1] 1 bar = 100 kPa. Although the valid SI unit for pressure is Pascal (Pa), the book uses the unit bar, which is more "handy" in practice.

R_m	*Ultimate tensile strength*	N/mm^2
$R_{p0,2}$	*0.2% yield strength*	N/mm^2
T	*Thermodynamic temperature*	K
t	*Time*	s
U	*(electrical) voltage*	V
v_L	*Air flow velocity*	m/s
v_V or v_X	*Longitudinal velocity*	m/s or km/h
v	*Velocity*	m/s
V	*Volume*	l = dm^3
V_c	*Compression volume*	l = dm^3
V_h	*Swept volume of one cylinder*	l = dm^3
V_H	*Swept volume of engine*	l = dm^3
v_m	*Mean piston velocity*	m/s
v_W	*Circumferential tyre velocity*	m/s
W	*Work*	J

Other Abbreviations

UT = BDC	*Bottom dead Centre*
OT = TDC	*Top dead Centre*
It	*Intake closes*
Eö	*Intake opens*
As	*Exhaust closes*
Aö	*Exhaust opens*
FVW	*Fibre composite material*

Contents

1 Combustion Engines

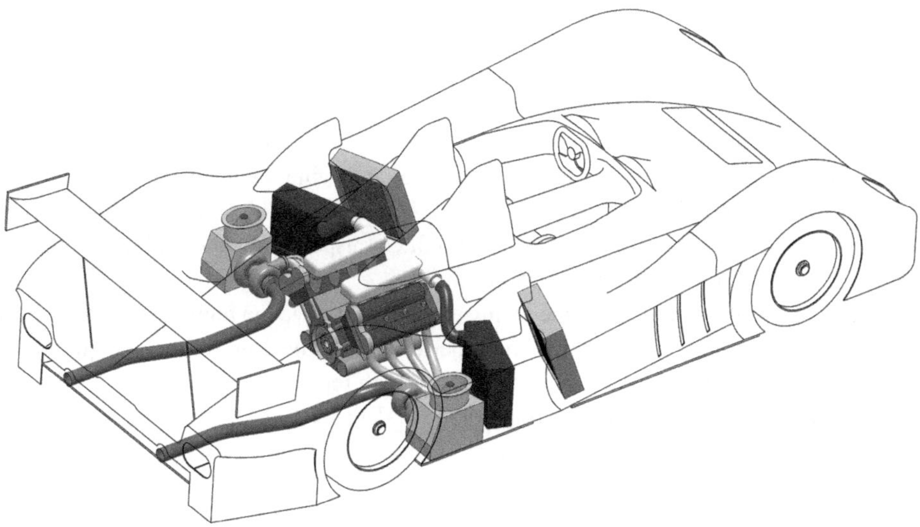

The engine is what makes a vehicle an AUTOMOBILE, i.e. capable of moving forward on its own. In addition, the combustion engine exudes a great, if not the greatest, fascination of all the individual assemblies and is a symbol of performance. In the case of racing vehicles, great attention is paid not only to its performance but also to its acoustic appearance.

M. Trzesniowski, *Powertrain*,
https://doi.org/10.1007/978-3-658-39885-9_1

1.1 Fundamentals

In general, experts do not consider the engine's contribution to the performance of racing cars to be high compared to other components (tyres, chassis).[1] Nevertheless, it is not unimportant; after all, it gives the sport its name. An engine developer[2] sums up this seemingly paradoxical situation thus: It's hard to win a race because of the engine, but easy to lose one because of the engine. The engine has to be stable and, above all, it has to have a steady power curve over the revs. This makes its behaviour predictable for the racer, who basically only uses two positions of the accelerator pedal or twist grip anyway: Idle and full load, meaning the engine is driven digitally, so to speak. In Formula 1, the full-load proportion of a lap is between 35% (Monaco) and 70% (Monza), depending on the circuit [1] Such a maximum value for engine development is also assumed for endurance races, Fig. 1.1. As engine capacity decreases, the full-load proportion increases on the same circuit. In hill climbs, however, the full-load percentage may well be less than 15%.

In terms of basic design, racing engines do not differ from series engines. Compared to engines in everyday vehicles, however, racing engines are exposed to greater longitudinal and lateral accelerations, which is important for the lubrication system, for example. The desired high performance is achieved, among other things, through high engine speed. This leads to greater inertia forces with correspondingly higher component loads, and the valve train becomes a critical system. Driver shifting errors can lead to engine destruction without electronic protection systems [3]. The service life of an engine is considerably shorter than that of a utility engine, while the power released is considerably higher. Specific outputs of up to 500 kW/litre displacement can be achieved. Corresponding values of series engines are 50–100 kW/l for supercharged gasoline engines.

From a physical point of view, the racing engine must have the highest possible values for two parameters in order to achieve high performance. Powerful torque is required for high acceleration and high power is necessary for high top speeds.

The power of an internal combustion engine is described by the following numerical equation.

$$P_e = \frac{i}{600} \cdot z \cdot n_M \cdot p_{m,e} \cdot V_h \tag{1.1}$$

P_e	Effective power, kW
i	Number of cycles, -. $i = 0.5$ for 4-strokes and $i = 1$ for 2-strokes
z	Number of cylinders, -
n_M	Speed of the engine, min^{-1}

(continued)

[1] Cf. Racing Car Technology Manual, Vol. 2 *Complete Vehicle,* Sect. 2.3.2.

[2] Luca Marmorini, former technical director of F1 engine division Toyota Motorsport.

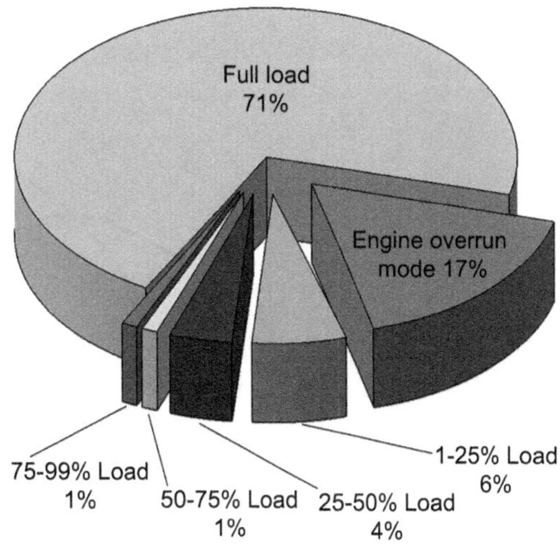

Fig. 1.1 Load shares at the 24-hour race of Le Mans, after [2]. A typical breakdown of throttle settings. The full load portion predominates as one extreme and the other follows in second place, namely operation with the throttle closed

$p_{m,e}$	Mean effective pressure, bar. Maximum values for $p_{m,e}$ approx. 12–35 bar for racing engines, between 8 and 13 bar for series-production diesel (compression ignited) engines, 7–14 bar for passenger car diesel engines
V_h	Displacement of a cylinder, l

For the torque applies:

$$M_M = \frac{100}{2\pi} \cdot i \cdot z \cdot p_{m,e} \cdot V_h$$

M_M	Engine torque, N m

The basic possibilities for increasing the power and torque of an internal combustion engine follow directly from the consideration of the equations:

Number of Strokes Theoretically, a two-stroke engine has twice the power of a four-stroke engine, all other parameters being equal. In practice, however, it does not achieve the same mean pressure. Apart from this, its use is often (and increasingly so) hindered by regulatory requirements.

Number of Cylinders
A large number of cylinders leads to a large total displacement for a given individual cylinder displacement and thus to correspondingly more power. However, a very large number of cylinders has disadvantages due to a correspondingly small individual cylinder volume with constant displacement and due to the large number of parts. However, small

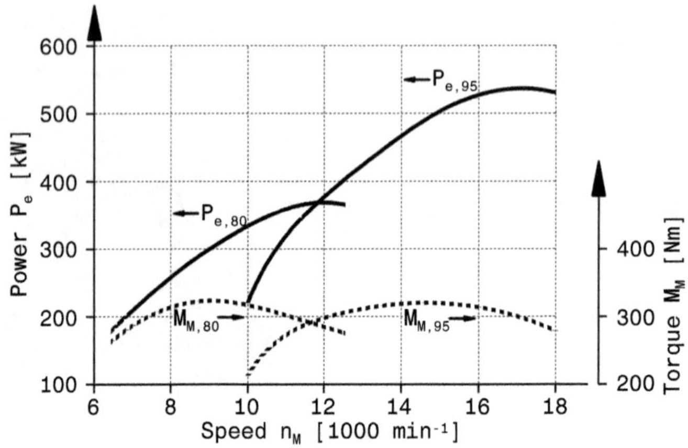

Fig. 1.2 Performance comparison of the 3.0 l naturally aspirated engines from 1980 and 1995 [4]. The increase in power was only due to the higher engine speed. The mean pressure and thus the torque remained practically unchanged

cylinder displacement volumes also allow higher speeds in principle due to the higher natural frequency of the gas-dynamic intake manifold-cylinder system.

Speed

With fixed displacement and mean pressure, the only thing left to do is to increase the engine speed in order to increase the power output, Fig. 1.2. Formula 1 naturally aspirated engines thus achieve far more than twice the rated speed of series-production Otto engines (spark ignited e.).

The limits for a speed increase primarily specify three ranges individually, i.e. if already one range can no longer be increased, the speed limit has been reached:

- Gas throughput
- Burn rate
- Component strength.

Gas Throughput

(Mixture Flow). The gas flow through the engine comes up against physical limits as the speed increases, because the valve opening areas cannot be increased at will. Once the critical pressure ratio in the valve area is reached, the gas velocity does not exceed the local speed of sound, even with a further increase in speed. The enlargement of the valve opening area is therefore an important criterion in determining the engine design. See also Sect. 1.4.1 *Cylinder Head.*

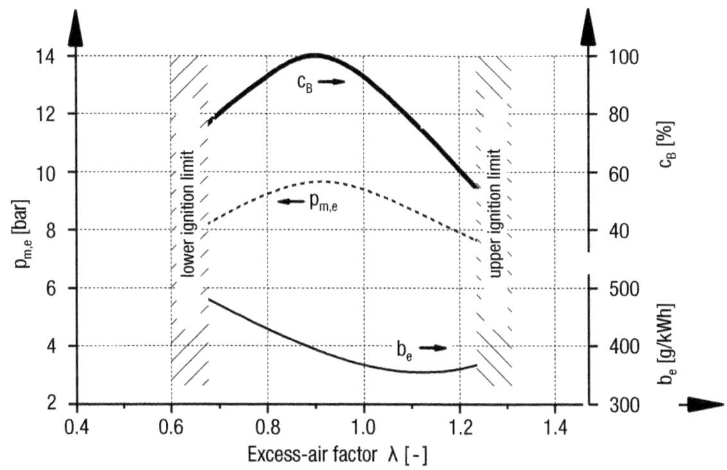

Fig. 1.3 Influence of excess-air factor (air ratio) λ on combustion rate c_B, mean pressure $p_{m,e}$ and specific fuel consumption b_e of a gasoline engine. The excess-air factor influences the combustion rate and thus the mean pressure and fuel consumption of an internal combustion engine

Combustion Rate, Piston Speed

The combustion process must be capable of completely burning the mixture supplied per working cycle as quickly as possible. The flame front speed c_F is composed of the combustion speed c_B (relative to the unburned mixture) and the transport speed c_T, with which the flame front is transported by the gas mixture's own motion: $c_F = c_B + c_T$. Engine speed and combustion chamber geometry influence the transport speed c_T . The burning speed c_B is determined by the state of the mixture, the chemical composition of the fuel and the air ratio λ, see Fig. 1.3.

The maximum mean pressure $p_{m,e}$ is reached for gasoline at an air ratio of 0.85–0.9 (excess fuel), where the greatest burning rate c_B occurs. The effective efficiency, on the other hand, is determined primarily by the perfection of combustion rather than by the burning rate. Therefore, at $\lambda = 1.1$ (10% excess air) the consumption minimum is reached. As a result of the fact that fuel consumption has also become an issue for racing engines due to social perception, the lean running capability of engines has become the focus of developers. For example, gasoline direct injection is standard in modern racing engines. The transport speed can be influenced by the intake process and the combustion chamber geometry. The design of the combustion chamber and the piston crown as well as the spark plug position thus also determine performance.

Here, the engine developer has to solve a classic compromise: stronger charge movements and thus short combustion durations are achieved with intake ports that have lower flow rates and thus limit the power output.

Figure 1.4 shows how the resulting *flame front velocity* c_F changes with speed.

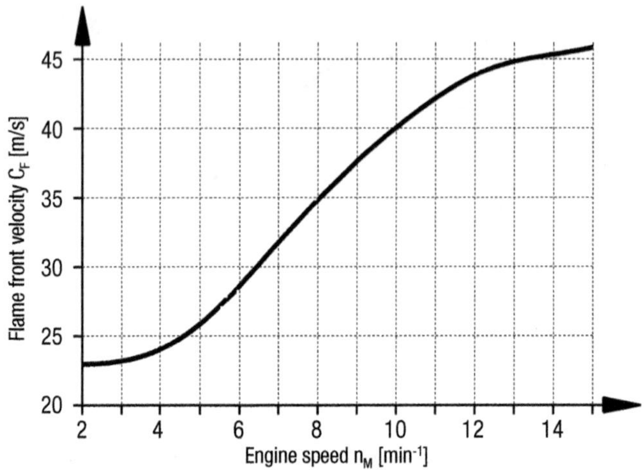

Fig. 1.4 Flame front speed c_F for 4-valve petrol engines with commercial fuel. At low speeds, the flame front speed is practically equal to the burning speed of 24–25 m/s. As the speed increases, the transport speed and thus the speed of the flame front increases. At 45 m/s the flame front speed approaches an upper limit value

If one wants to push the limit value of the speed of the flame front further up, the fuel composition must be changed. This is only possible with a few regulations. Therefore, a shortening of the flame paths in the combustion chamber is more purposeful. This is done by smaller bore or more spark plugs per cylinder.

The speed of gasoline engine combustion is therefore determined to a large extent by the existing turbulence intensity. The consequence is that the combustion duration – expressed in degrees of crank angle – is practically independent of the engine speed at constant load. Consequently, the engine speed is not the determining similarity variable in gasoline engines, but the mean piston speed v_m. This characterizes the most important tribological and fluid mechanical processes. It is observed that approximately at $v_m > 18$ m/s the pressure drops increase to such an extent that good cylinder filling is no longer guaranteed [5]. The engine life also suffers noticeably when the mean piston speed exceeds this value. From the relationship for v_m it follows that the stroke must be reduced if the speed is increased, if the known limit for v_m is not to be exceeded.

$$v_m = \frac{s \cdot n_M}{30,000} \tag{1.2}$$

Numerical value equation:	
v_m	Average piston speed, m/s
s	Stroke, mm
n_M	Engine speed, min^{-1}

Values for maximum mean piston speeds:

- $v_{m,\max} < 20$ m/s Empirical value for series-production car engines
- $v_{m,\max} = 19$ to 21 m/s for long distance engines (Le Mans etc.)
- $v_{m,\max} \approx 25$ to 27.3 m/s for Formula 1 engines.

In diesel (compression ignited) engines with direct injection, an influence of the injection pressure on the mixture formation can be determined. As fuel pressure increases, fuel atomization becomes more refined. This increases the engine's power output, assuming an adapted port design for cylinder air movement and filling, including an optimised piston bowl shape. However, the power gain decreases with increasing rail pressure due to the decreasing air ratio. Figure 1.5 shows the influence of rail pressure on engine performance using the example of the Audi TDI engines for Le Mans. Fuel injection in these turbo-charged diesel engines is provided by a common-rail system.

Component Strength

The components that set the limits of an increase in speed are the piston, the connecting rod, the crankshaft and the main bearings. Friction losses increase sharply with increasing speed and oversquare stroke-bore ratios (see appendix), i.e. $s/B < 1$, have an advantage over other designs in this respect as well. The friction power gain outweighs the higher heat losses. A smaller stroke results in a larger bore diameter for a given displacement. This allows larger valve diameters, which helps to increase the charging efficiency, especially at high speeds. With a small stroke, the oscillating and rotating inertial forces also remain smaller. High-speed engines therefore benefit several times over from oversquare stroke-bore ratios.

In addition to this consideration, other criteria that must be taken into account for an increase in speed are the mechanical (heat) resistance of components, the maintenance of lubrication and vibrations in the valve train.

Mean Effective Pressure The higher the mean effective pressure $p_{\mathrm{m,e}}$, the greater the power and torque developed by the engine. The mean effective pressure is a practical parameter for comparing engines with different displacements. It is the torque related to the total displacement. With a high mean pressure, it is also possible to achieve high power at relatively low engine speeds. This improves drivability and fuel economy. In addition, with a prescribed air restrictor, the engine speed must remain below the limiting speed if the engine is not to stall (see also Sect. 1.4.5). Low fuel consumption offers the advantage of lower starting weight and also a more favourable power-to-weight ratio throughout the race.

The effect of the gas force on the piston can also be increased by reducing the back pressure on the underside of the piston. If the pressure in the crank chamber is reduced, the

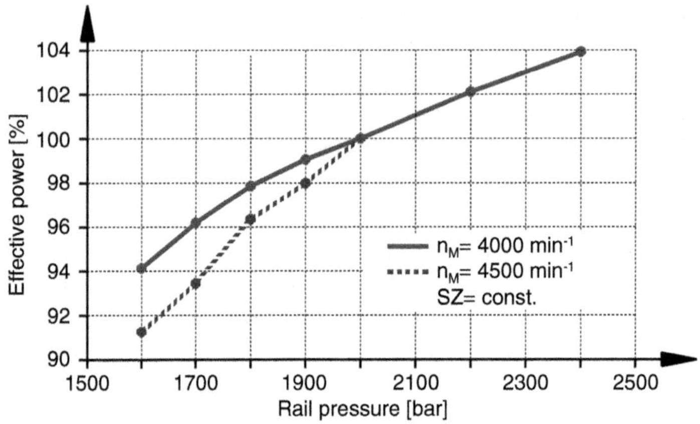

Fig. 1.5 Influence of rail pressure on the engine performance of diesel engines [6]. SZ Bosch blackening number, a measure of diesel smoke (*FSN filter smoke number*). n_M Engine speed. The engine power increases with increasing injection pressure. At the same time, the curve reflects the development of power over time. The data up to 2000 bar rail pressure comes from the first racing diesel engine R10 V12 TDI and shows its development over three years. The more recent R15 V10 TDI was already running at 2400 bar

power increases accordingly (1.1): $\Delta P_e = \frac{i}{600} \cdot z \cdot n_M \cdot \Delta p_{m,e} \cdot V_h$. If the crank chamber of a naturally aspirated engine is evacuated, the increase in mean pressure is about 1 bar (this corresponds to the ambient pressure otherwise acting on the underside of the piston). For a 3l engine, this results in a power increase of $\Delta P_e = 25$ kW at 10,000 min^{-1}. Of course, some of this is lost to the vacuum pump. In addition, the resistances of the moving crankshaft drive parts caused by air friction are almost completely reduced.

The mean pressure is largely determined by the air consumption and the mixture heating value.

$$p_{m,e} = \eta_e \cdot \lambda_a \cdot H_G \tag{1.3}$$

η_e	Effective efficiency, -. Is the ratio of the work done to the fuel energy used. Best values for η_e for racing engines up to about 0.3 (i.e. 30%), for series-production engines between 0.25 and 0.35. These values are only achieved at certain operating points.
λ_a	Air effort (mass of air corresponding to cylinder volume), -. Is the ratio of the actual fresh charge supplied to the theoretically possible, i.e. introduced without losses, charge mass in the cylinder.
H_G	Mixture heating value, kJ/m^3. Is the fuel energy related to the volume of the fresh charge.

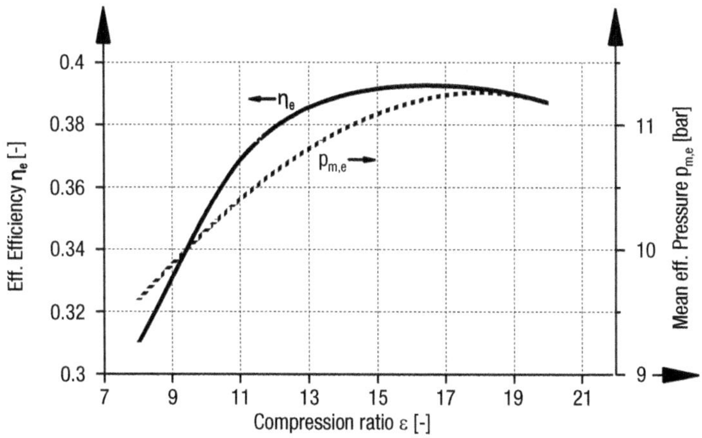

Fig. 1.6 Influence of compression ratio on mean effective pressure and effective efficiency at full load of a gasoline engine [7]. Above a compression ratio of 17:1, the efficiency drops. Due to increasing frictional forces and the effects of the shape of the combustion chamber

From this relationship, further measures follow to achieve the greatest possible power and torque:

Efficiency

The effective efficiency must be high, i.e. all losses (friction, drive of auxiliary units,. . .) must be kept low. In racing series with a prescribed standard fuel and limited fuel mass flow, this remains the only way to increase the effective engine power.

The *compression ratio* is a variable that can be easily influenced to increase the efficiency, see Fig. 1.6.

In the gasoline engine, the practically usable compression ratio is limited by knocking and glow ignition. In order to push the limit as far as possible, combustion chambers must be compact (small surface-to-volume ratio keeps wall heat losses small) and less fissured. Targeted combustion chamber roof cooling is of great value in this context. Fuel composition is also a measure of eliminating knock. The octane number is a measure of knock resistance. However, this measure can only be used if the regulations allow it. Even in Formula 1 (in the meantime) only fuels are permitted which practically correspond to the unleaded super fuel with RON 98 (research octane number) at the filling station. In this case, the octane number can only be increased within narrow limits (RON 95 to 102, [8]) by means of sophisticated blending processes, which are only possible to the required extent for fuel manufacturers.

Air Effort

(Mass of Air Corresponding to Cylinder Volume). The mass of air should be as large as possible. One way of achieving this is turbocharging. In this case, more charge is

Fig. 1.7 Charging efficiency λ_l over the engine speed n_M. The charging efficiency is made up of the following proportions: Share of scavening losses, **2** Share of flow losses, **3** Share of heating losses. The theoretical maximum value of the charging efficiency is fixed with ε/(ε–1)

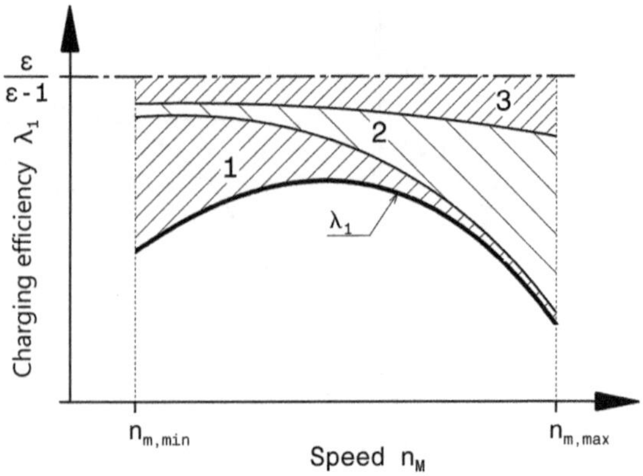

introduced into the combustion chamber than the naturally aspirated engine can theoretically achieve, i.e. the air requirement is greater than 1. The effect of turbocharging can thus also be compared with an increase in the displacement of the naturally aspirated engine. The amount of air required is also greater if the intake areas are designed with favourable flow conditions and the supercharging effects are achieved by gas-dynamic phenomena which act at least in a narrow speed range.

The charge temperature should be as low as possible. A high temperature of the fresh charge causes a smaller charge mass in the cylinder than would be possible, due to the lower density of the extended charge. Another comparative parameter for the success of the charge exchange is the degree of delivery (*charging efficiency*) λ_1. This compares the actual mass of fresh gas in the cylinder with the theoretically possible (= swept volume times air density). For 4-stroke naturally aspirated engines, the best values are in the range 0.8–0.9 and above. Turbocharged engines achieve values of 1.2–1.6. Figure 1.7 shows the basic curve of the charging efficiency over the speed. The charging efficiency is composed of the scavenging (1), flow (2) and heating (3) losses.

Throttling losses result from flow resistance in the intake system and at the valves. Heating losses result from heat exchange of the air with the intake manifold walls and the cylinder walls. Scavenging losses are the result of valve overlap and insufficient exhaust backpressure. With increasing engine speed, throttling and heating losses increase; at low engine speeds, scavenging losses predominate, so that the charging efficiency λ_l has a maximum in the medium speed range. The position of this maximum can be influenced by the choice of control times and by matching intake manifold lengths and diameters.

Mixture Heat Value The size of the mixture heat value H_G is determined by the energy chemically stored in the fuel. The fuel composition therefore also allows the performance to be influenced. However, only a few racing classes can make noticeable use of this. Dragsters are a classic example of an enormous increase in performance (and at the same time one of the few exceptions) through appropriate fuel composition. In acceleration races, exotic fuels ranging from nitromethane to di-olefins provide litres of power up to 500 kW/l.

The mixture heating value is calculated as follows:

$$\begin{aligned} H_G &= \frac{H_u \rho_L}{\lambda L_{\min}} \quad \text{for } \lambda \geq 1 \\ H_G &= \frac{H_u \rho_L}{L_{\min}} \quad \text{for } \lambda \leq 1 \\ \lambda &= \frac{m_L}{m_K \cdot L_{\min}} \end{aligned} \tag{1.4}$$

H_u	Specific calorific value of the fuel, J/kg Super gasoline $H_u = 43{,}170$ kJ/kg, diesel fuel $H_u = 42{,}500$ kJ/kg, methanol $H_u = 19{,}600$ kJ/kg.
ρ_L	Air density, kg/m^3
$L_{\min}$	Air requirement for stoichiometric combustion, kg air/kg fuel Super petrol $L_{\min} = 14.7$ kg/kg, Diesel fuel $L_{\min} = 14.5$ kg/kg, Methanol $L_{\min} = 6.4$ kg/kg
λ	excess-air factor, – (see also Annex).
m_L	Mass of air actually present in the cylinder, kg
m_K	Mass of fuel present in the cylinder, kg

Displacement The larger the displacement, the greater the maximum power. The limits of an increase in displacement are set on the one hand by the regulations, and on the other hand thermodynamic findings restrict the individual cylinder volume to a usable range (see below). Combustion chambers that are too large have the disadvantage that the flame paths become too large and the charge no longer burns through completely at high engine speeds.

In summary, Fig. 1.8 shows the specific power of different engines versus engine speed. The maximum litre power of racing diesel (compression ignited) engines is comparatively 90 kW/l.

Stability of Combustion

The differences in the combustion of individual cylinders in multi-cylinder engines should be as small as possible. Large differences become apparent, for example, through strongly deviating optimum ignition angles of individual cylinders. Such differences are caused by incorrect distribution of air and/or fuel to the individual cylinders [10]. Individual throttle devices and injection systems as well as intake manifolds have proven to be extremely effective measures to avoid this. In contrast, many production engines are equipped with distributor intake manifolds and central injection (single-point injection) for cost reasons.

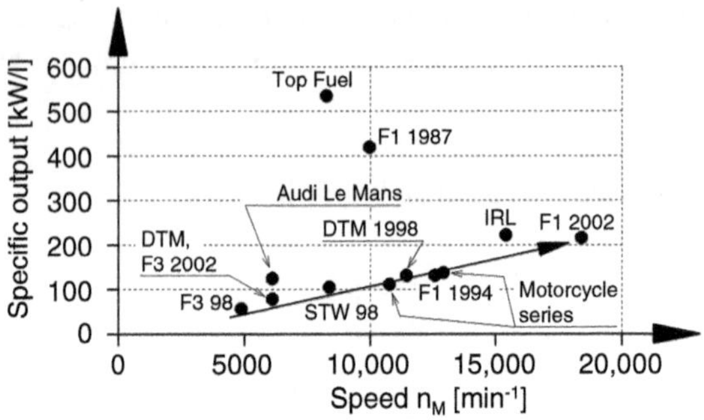

Fig. 1.8 Specific output of engines, after [9]. The plotted compensation line confirms the practical validity of (1.1). For a given mean pressure, the power is increased with the speed. A further increase is achieved by increasing the charging efficiency (e.g. supercharging: F1 1987 = Formula 1 1987) or by using a different fuel (top fuel). The most powerful engines have the dragsters (Top Fuel). The specific outputs of the turbocharged 1.5-liter Formula 1 engines of the 1980s come in second. The Indy Racing League (IRL) cars have the most powerful engines on the circuit, closely followed by the former Formula 1 naturally aspirated engines

In addition, the cyclical fluctuations of the individual cylinder combustion should be as small as possible. Cyclic fluctuations (*combustion variability*) are stochastically occurring differences in the cylinder pressure curve at constant load and speed, which, depending on their characteristics, are noticeable, among other things, in rough engine running, limited performance and poorer fuel consumption. Cyclic fluctuations are influenced by [10]:

- Charge movement at the spark plug and at the ignition timing
- Differences in the amount of air and/or fuel introduced per cylinder
- Mixing of the fuel with residual gas components
- Mixture preparation (droplet size, spray angle, *targeting*, swirl flows)
- Excessive dilution due to exhaust gas recirculation (valve overlap) or air
- Long burning time due to unfavourable combustion chamber design
- Low ignition energy or small spark plug electrode gap.

In general, the stability of combustion increases at:

- Increasing speed and load
- Higher compression ratio
- Less valve overlap
- Higher ignition energy at the electrodes
- Higher temperatures
- Lower humidity.

Other measures to increase stability are:

- Exactly the same combustion chamber shape in all cylinders.
- Equal lengths of inlet and exhaust ports and the lines connected to them.
- Active crankcase ventilation or evacuation.
- Injectors that deliver small droplet diameters and have jets oriented to minimize (intake manifold) wall wetting.
- High voltage cable of the ignition system with low resistance.
- Largest drivable spark plug electrode gap.

The desired properties, depending on the intended use, are generally achieved in designed motors by the following of the conceivable – above derived – measures [11]:

- High speed or wide speed range, whereby continuous operation at speeds of over 19,000 min^{-1}, e.g. Formula 1 until the beginning of 2009, is essentially made possible by the use of pneumatic valve actuation.
- Maximum possible throttling of the intake path.
- Performance-optimized components such as intake manifold, exhaust manifold and exhaust gas routing.
- Large valve lift and four valves per cylinder, with the valve material being mostly titanium (lower mass with about similar strength to steel).
- Improved cooling, especially of the cylinder head.
- Dry sump lubrication due to extreme acceleration of the vehicle.
- For gasoline engines, combustion chambers with as little knock as possible, which means relatively small valve angles, central plug position and moderate compression ratios and pistons with as flat a piston crown as possible.
- Due to higher thermal and mechanical loads, adaptation of the structure, the materials and the connecting elements (bolting) to the increased requirements.
- Lowest possible mass of the components used (e.g. titanium, ceramics, carbon fibre reinforced plastics). A light engine leads to a light vehicle and, with a required minimum weight, leads to the free choice of the location of additional masses to improve the vehicle balance.
- Service life adapted to racing operation (Fig. 1.9), special measures for quality assurance of the installed parts (individual testing) and routine replacement of parts after a certain mileage.

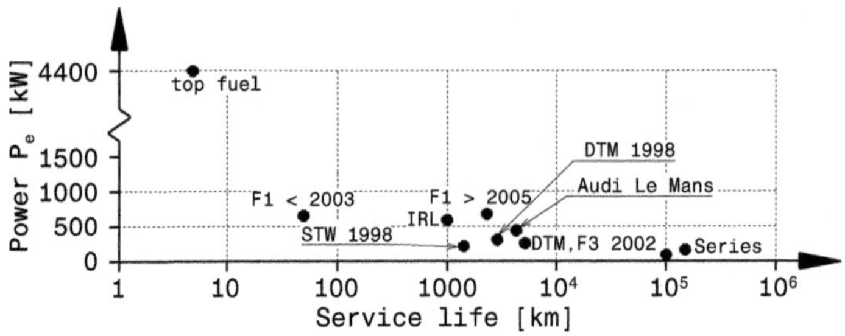

Fig. 1.9 Service life of engines, according to [9]. Although the power units of the Top Fuel Dragsters have impressive performances, they also have to be overhauled after only eight race runs (= 5 km). Conversely, this also explains why endurance power units (e.g. Audi Le Mans) are so similar to the production versions

1.2 *Choice of* Engine

When designing racing vehicles, a new engine is not always constructed, but existing engines are often used. Many vehicle manufacturers also do not have their own engine at all, but offer their vehicle without an engine. The customer subsequently buys or leases an engine from an engine manufacturer. The following are therefore some considerations for selecting suitable existing engines. However, these criteria can also be used to advantage when considering new engine design.

The main evaluation criteria are [4]:

- Engine power
- Driveability
- Engine weight: cylinder distance, bank angle, material
- Cylinder spacing: stroke-bore ratio
- Outer dimensions: Cylinder spacing, bank angle
- Center of gravity height: Bank angle, materials
- Fuel consumption
- Inflow conditions for the fresh load (air-feed conditions): Bank angle, height
- Outlet conditions for the exhaust gases: Bank angle, width
- Number of cylinders, ignition distance and firing order
- Power loss
- Vibration behaviour (mass balance and rotating masses): Cylinder spacing, bank angle, stroke/bore
- Rotational (cyclic) irregularity: Bank angle, stroke/bore, crankshaft
- Suitability for charging
- Harmony with the vehicle
- Ignition and injection system

- Auxiliary units
- Spare parts availability, service
- Potential for further development.

Engine Power
High nominal engine power is a prerequisite for high top speed. However, it says nothing about driveability and acceleration capability. In addition, the stated maximum power only comes into effect when the driver is fully on the accelerator pedal and when the rated speed is reached in the process.

Driveability
The course of an engine's torque curve and the characteristics between accelerator pedal position and engine torque output are decisive for driveability. The more powerful an engine and the lighter the vehicle, the more important it is to be able to control the engine torque sensitively. This is especially important when traction control is not allowed. With decreasing displacement, the drivability of high-revving racing engines generally improves [12].

The aim of engine development is to achieve outstanding responsiveness and very good drivability in all weather and track conditions. This applies to both sprint and endurance competitions.

When selecting an engine, in addition to the maximum value of the torque, its curve over the speed also plays a role in terms of acceleration capacity. The area under the torque curve between two speeds represents a power. This power corresponds to the rate of change of the kinetic energy of the accelerating vehicle between the speeds corresponding to the two engine speeds – if one disregards the losses in the drive train and at the tires. A curve with a larger area is therefore preferable to a curve with a smaller area but higher maximum torque in terms of better acceleration.

Engine Weight The size and design of the engine are primarily determined by the requirement for power, torque, comfort, consideration of exhaust, consumption and noise regulations, the choice of mixture preparation, the ignition system and the requirement for low maintenance and good accessibility.

The engine power or the torque in relation to the engine mass offers a decision criterion that is interesting for all vehicles. For production gasoline engines, values for naturally aspirated engines range from 0.8 to 0.9 kW/kg. 3l Formula 1 engines showed values up to 4.6 kW/kg [11]. The best values were delivered by the Formula 1 1.5 l turbo engines with 5 kW/kg.

Type
(Cylinder Arrangement). The choice of design – e.g. V-engine, W-engine, in-line engine or boxer – is not only influenced by the desired displacement requirement, but also by the space available in the vehicle.

For example, the space required by V-engines of the same size with different cylinder angles varies. At the same time, this is an example of the fact that not always machine-dynamic aspects (cylinder angle 60°,120° or 180°) are used, but the space requirement is given priority – and then solutions such as 6-cylinder V-90° engine and V8–75°, V10–67°, V10–72° and V12–65° (an even ignition distance per cylinder bank is, however, maintained) [4].

Centre of Gravity Height
The cross-sections of the different engine concepts determine not only the centre of gravity height, but also how well the engines fit into the overall vehicle system. If only the centre of gravity heights are evaluated, the flat engines perform best, Fig. 1.10. A 144° V-angle is more favourable than 180° because the exhaust system and the auxiliary units (oil pumps, centrifuges,. . .) require a higher installation position in the vehicle for this engine. Nevertheless, angles of 65° to 72° are preferred for Formula 1 vehicles because of the better installation. The unfavourable centre of gravity heights are accepted. In the meantime, the minimum centre of gravity height for Formula 1 engines is limited by the regulations to 165 mm.

The clutch size (diameter) limits the possibility of lowering the engine in the vehicle, especially in the case of a flat oil pan made possible by dry sump lubrication.

Engine Width A wide engine provides a good base for a mid-engined vehicle to bolt to the frame. Ideal for a monoposto is the shoulder width of the driver plus about 100 mm (for wall thickness and clearance). If the engine is wider, it interferes with the narrow span (air resistance) and the flow of the rear wing. If, as with the C.G. position, comparisons are made between engines with different V-angles, engines with a small fork angle perform better, as expected, Fig. 1.11.

Engine Length The engine length directly influences the wheelbase in a mid-engine concept.[3] A short engine allows the rear end of the vehicle to be kept narrow, which brings significant aerodynamic advantages for single-seaters with open wheels.[4]

---

[3]Cf. Racing Car Technology Manual, Vol. 2 *Complete Vehicle*, Chap. 2 Concept, in particular Fig. 2.15.

[4]Cf. Racing Car Technology Manual, Vol. 2 *Complete Vehicle*, Chap. 5 Aerodynamics, in particular Fig. 5.14.

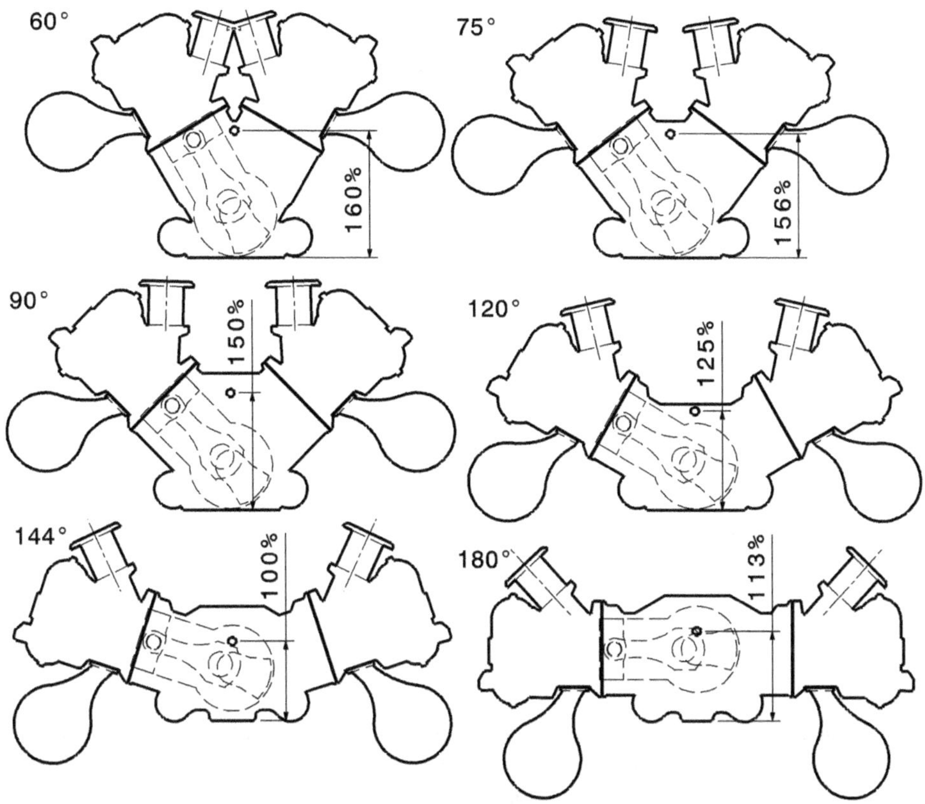

Fig. 1.10 Centre of gravity positions of different engine concepts. The basis for comparison is a V-144° engine with a centre of gravity of 100%. For reasons of comparison, the exhaust system is located on the outside in all variants. However, there are also engines in which the exhaust system is located on the inside, i.e. between the cylinder banks

Besides the pure dimensions, the symmetry around the longitudinal plane of the engine is interesting. V-engines can easily be designed symmetrically (apart from the connecting rod offset). A symmetrical design results in better utilization of the available space for engines whose crankshaft lies in the longitudinal center plane of the car, and saves transferring lines from one side of the vehicle to the other.

Stroke-to-Bore Ratio

The stroke/bore ratio determines the basic characteristics of the single stroke volume. Accordingly, a distinction is made between square designs ($s/B = 1$), undersquare or long-stroke designs ($s/B > 1$) and oversquare or short-stroke designs ($s/B < 1$). The following advantages or disadvantages of extreme designs should be considered when

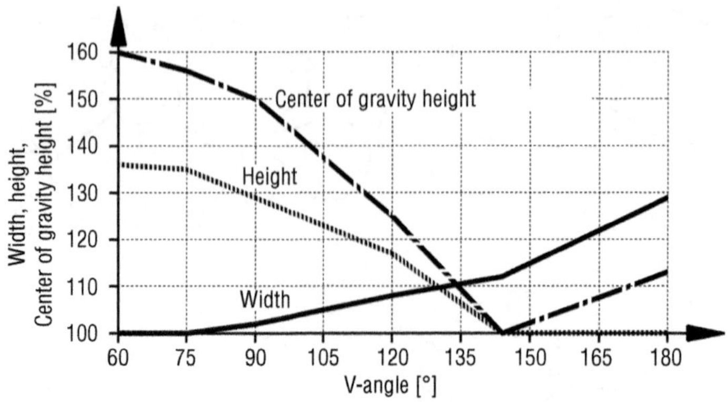

Fig. 1.11 Dimensions of different engine concepts. The values follow from the sketches of Fig. 1.10. In the V angle range 120–150° a usable compromise results

choosing a design, where the advantages of one design are the disadvantages of the other and vice versa [13].

Advantages of a long stroke:

- Engine characteristics geared to high torque
- compact combustion chamber with short combustion paths and favourable surface-to-volume ratio ensures high combustion quality
- smaller oscillating masses per cylinder unit
- smaller bore means less engine load due to throttle force.

Disadvantages of long-stroke designs:

- small bore diameter also leads to small valve cross sections. However, this only has a negative effect at high speeds.
- the large stroke leads to high average piston speeds, which increases friction losses and represents an upper speed limit
- with relatively short connecting rods, the oscillating mass forces increase strongly
- the rotating inertia forces increase in any case
- larger stroke generally increases the connecting rod skew and thus the piston side force. This in turn leads to increased piston deformation and friction.

For a final decision, the nominal speed of the engine is decisive. At moderate speeds, long-stroke engines prove to be more favourable overall (friction, mean pressure, fuel consumption). If a high speed is aimed for, short-stroke engines are to be preferred (*s/B* around 0.55).

Air Feed Conditions. The engine should be supplied with the coolest and cleanest air possible. The fewer baffles are required and the shorter the external air ducts, the better.

Fig. 1.12 Air intake on a Formula 1 car (BMW Williams FW 18, 1997 season). The combustion air flows in centrally above the driver's helmet, supported by the dynamic pressure. At 260 km/h, this amounts to about 200 mbar. With a 3.5 l naturally aspirated engine, this results in an additional output of approx. 7.3 kW [4]

The intake points are usually designed as NACA inlets or chimneys. In the latter, the back pressure creates a slight charge, Fig. 1.12.

Number of Cylinders

For a given displacement, the cylinder number results from the desired volume of a cylinder. For 4-stroke gasoline engines, approx. 300–350 cm^3 represents a favorable single cylinder volume [4]. For low fuel consumption, 450–500 cm^3 represents the optimum [11].

Keeping the mean effective pressure and the mean piston speed fixed, for a given displacement an increase in the number of cylinders (z' = higher number of cylinders) with (1.1) leads to the following remarkable result, Fig. 1.13.

Theoretically, the engine power increases with the number of cylinders for a given displacement ($P_e/V_H \propto \sqrt[3]{z}$). This is also observed in practice, but only up to the number of cylinders twelve. Above this number, the individual volumes become too small and the power output decreases.

The stroke bore ratio becomes smaller with increasing number of cylinders at constant displacement. Due to the large journal overlap, short-stroke crankshafts are more resistant to bending and torsion than long-stroke designs. Short-stroke also means a relatively large bore. This makes it possible to accommodate larger valves, which increases the air flow through the engine, Fig. 1.14.

The power loss can increase with the number of cylinders [4]. However, in [15], only an influence of the number of cylinders on the distribution of losses is found in the comparison of 4-cylinder to 6-cylinder. In this investigation, the total losses are almost independent of the number of cylinders.

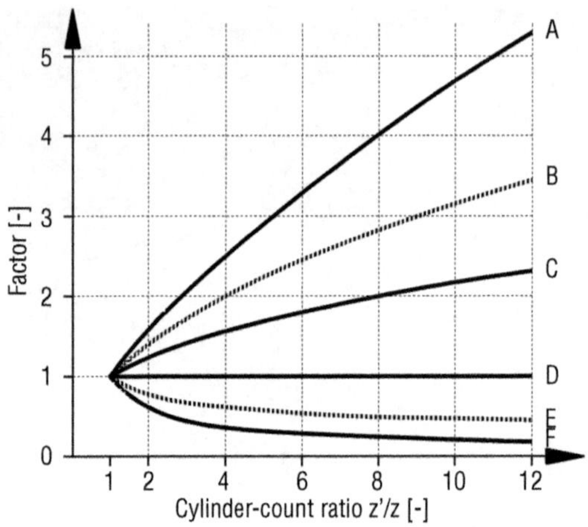

Fig. 1.13 Engine characteristics as a function of the number of cylinders, according to [14]. If the number of cylinders is increased from z to z' with the same displacement, the following characteristics change according to the curves marked A to F. (**a**) total length of the engine, (**b**) friction losses, (**c**) speed, power, total piston area, valve area, (**d**) specific bearing load, (**e**) bore, stroke, engine height, engine width, power-to-weight ratio [kg/kW], (**f**) piston area/cylinder, mass inertia, gas forces. An 8-cylinder engine delivers twice the power (course C) of a single cylinder with the same displacement, mean effective pressure and mean piston speed. Its rated speed has also been doubled, but this has no effect on the piston speed because of the correspondingly smaller stroke

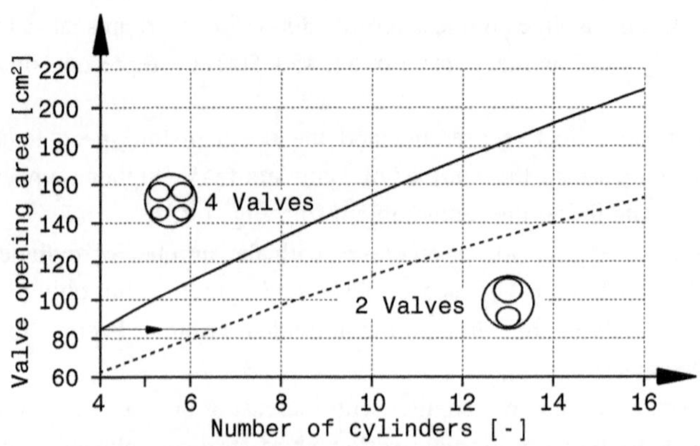

Fig. 1.14 Influence of the number of cylinders on the valve cross-sectional area. Base 3.5 l displacement, with stroke bore ratio of 0.7; valve lift 10 mm and valve diameter inlet/exhaust ratio is 1.2. As the number of cylinders increases, the valve opening area increases. It can also be seen that a 2-valve 6-cylinder engine has about the same opening area as a 4-valve 4-cylinder engine. If you want to significantly increase the area of a 4-valve engine compared to a 6-cylinder engine, the engine must have at least 5 cylinders. In this case, however, the area corresponds to that of an 8-cylinder 2-valve engine

Table 1.1 Evaluation of three cylinder numbers on 3l-V engines [4]

Criterion	V8	V10	V12	
Power	1	2	3	
Torque	2	3	2	
Single cylinder volume	2	3	2	
Number of components	3	2	1	
Dimensions	3	2	1	
Fuel consumption	3	2	1	Scoring (points):
Centre of gravity height	1	2	3	1 = bad
Vibration	1	2	3	2 = medium
Point total	16	18	16	3 = good

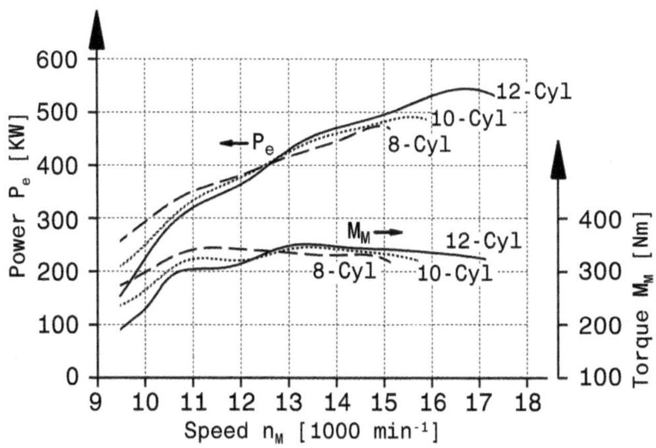

Fig. 1.15 Performance comparison of 3.0-l naturally aspirated Formula 1 engines (2nd generation 1995), after [4]. The 12-cylinder engine has an advantage in terms of performance at very high speeds, while the 8-cylinder engine has the most favourable torque curve at low speeds

In general, engines with higher numbers of cylinders are heavier. A direct comparison is hardly possible because multi-cylinder engines have a different design (e.g. V-engine) and larger displacement than engines with small numbers of cylinders (single-cylinder, in-line engine). In any case, more cylinders lead to more components.

If an unweighted evaluation is carried out for 3-litre naturally aspirated engines on three cylinder numbers in question, the V10 cylinder emerges as the best compromise, Table 1.1. Figure 1.15 illustrates this comparison on the basis of power and torque.

Cyclic Irregularity. (*Rotational Unevenness*)

The rotational irregularity is directly related to the number of cylinders and their arrangement. Another influencing factor is the ignition distance, which can be influenced by the arrangement of the crank throws. The more evenly the torque of the engine acts on the tyres, the more favourable this is. The tyres are protected and can be driven longer in a high grip area. If the torque fluctuates strongly, this has a negative effect on the circumferential and lateral force build-up of the tyre and it also degrades more quickly.

Fuel

Another consideration is the quality of the fuel. The compression depends on this and the displacement can be varied at the same power.

Racing engines are almost exclusively gasoline engines. Racing trucks and off-shore boats are an exception. Diesel (compression ignited) engines reach their maximum speed, which is limited by the combustion speed, at around 5000 min^{-1} More precisely, the ignition delay between the end of injection and the start of combustion is the cause. The ignition delay is almost constant and independent of the engine speed. As the engine speed increases, a value is reached at which the ignition delay no longer allows time for useful combustion. To increase performance, therefore, the only practical option is to increase the mean pressure. Diesel racing cars were at the start of the Indianapolis 500 (1931, 1934, and 1952) [16]. In the meantime, as in other racing classes, the regulations have been changed. In Formula 1, for example, diesel engines are banned. In touring cars and rallies, diesel-powered vehicles have been able to attract attention in more recent times. A new field of activity has emerged in endurance races of sports prototypes at Le Mans. Diesel engines have been able to compete since 2004. In 2006, a vehicle with a diesel engine was the overall winner for the first time (Audi R10). Diesel engines dominated the classic race at the Sarthe without interruption until 2014.

Both firing processes offer advantages that make it worthwhile to consider their use. Table 1.2 provides a rough overview.

Fuel consumption can therefore also be a decisive criterion for racing engines. Vehicles with lower fuel consumption need fewer pit stops with the same tank (long distance) or they can drive with a smaller tank (= lighter vehicle) (faster lap time) Two numerical values for volumetric fuel consumption of multi-track racing vehicles: 46 l/100 km (3.6 l V8 Biturbo, Otto DI, vehicle 900 kg) [17] and about 60 l/100 km (3.0 l V10 Otto naturally aspirated, vehicle 600 kg) [18]. These values occur in racing due to the corresponding driving style. The specific values are quite favourable, e.g. 258 g/kWh were achieved in Formula 1 with turbocharged 1.5-l engines [4].

Ancillaries All components of the engine periphery (water pumps, oil pumps, fuel pumps, separation systems, generators, starters . . .) should be designed and arranged modularly. This facilitates maintenance and inspection and reduces the time needed to replace parts. A high degree of integration of components also reduces the number of connection points and thus increases the reliability of the entire unit by avoiding assembly errors and leaks during operation.

Table 1.2 Comparison of basic characteristics of gasoline and diesel engines

Property	Combustion principle			
	Otto (spark ignited)	Diesel (compression ignited)		
Power	+	o		
Torque	o	+	Legend:	+ positive property, advantage
Consumption	–	+		– negative attribute, disadvantage
Costs	+	–		o medium, i.e. no pronounced Advantage or disadvantage
Exhaust emissions	+	–		

Spare Parts, Service

Racing engines have an adjusted life span, which can range from a few minutes to over 24 hours. In any case, the engines must be serviced regularly, which is usually the task of specialists. In some cases the engines are delivered sealed and the engine customer is not allowed to disassemble it. So, when choosing an engine, it will also be a consideration how good the support of the engine supplier is and what service intervals have to be kept.

1.3 Losses

The conversion of chemical energy into mechanical energy in the internal combustion engine involves losses. The lower these losses, the higher the power output for a given displacement. If individual process steps are considered, improvement measures can be developed in a more targeted manner. The quality of individual steps is described by efficiencies.

- *Volumetric efficiency* is a measure of the air flow rate per unit time through the motor.
- *Thermal efficiency* describes the efficiency with which the fuel-air mixture is formed and burned, releasing heat.
- Finally, *mechanical efficiency* is a measure of the amount of energy left over from the conversion of the heat of combustion as work done by the crankshaft.

The reduction of all these losses is part of the elementary work in the development of an engine and especially a racing engine.

Map

The indicated efficiency of an engine with external mixture formation and spark ignition drops above all in the lower map range, Fig. 1.16. The causes are to be found both in the

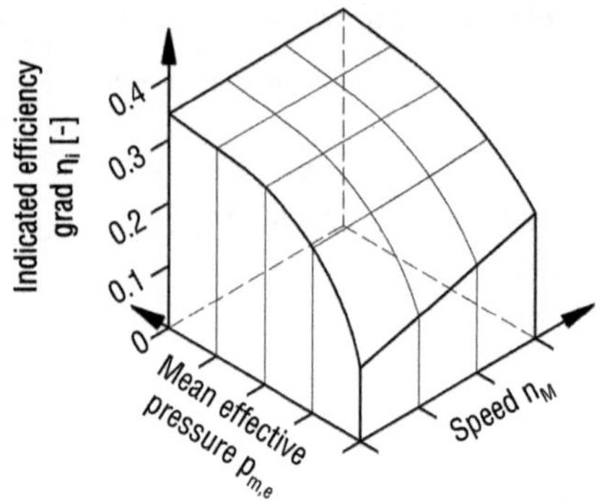

Fig. 1.16 **Curve** of the indicated efficiency of a gasoline engine with throttle control over load and speed. A gasoline engine achieves the best efficiency at high load

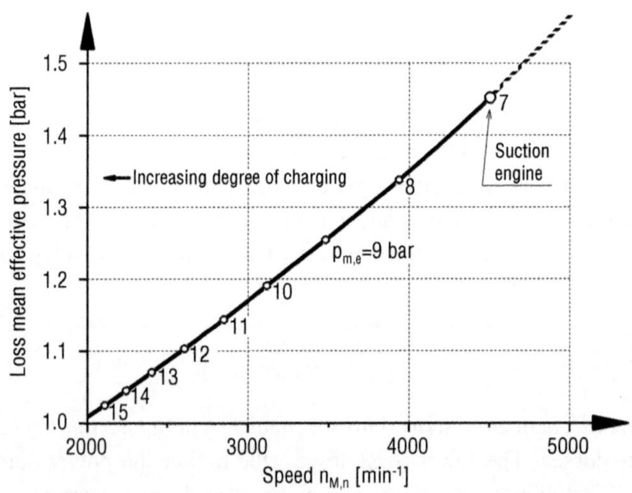

Fig. 1.17 Dependence of the mean pressure loss on the engine speed, after [19]. The values are taken from a 5-l petrol engine, $P_e = 130$ kW = const. An increasing degree of charging causes the mean effective pressure to increase and the same power can be achieved at a lower nominal speed. The losses decrease in the process. $p_{m,e}$ = mean effective pressure

quality of the combustion (too low turbulence, too low charge density) and in the most unfavourable gas exchange efficiency. The unfavourable mechanical efficiency in this map range causes a further reduction of the effective efficiency. All measures that are suitable for avoiding these lower map ranges thus improve the overall efficiency of the engine.

For example, if engine power and displacement are held fixed, a reduction in rated speed causes an increase in mean pressure. This increase in mean pressure causes an improvement in the quality ratio and the gas exchange efficiency. In addition, the friction mean pressure (mechanical losses) is reduced and thus the mechanical efficiency is improved, Fig. 1.17.

Fig. 1.18 Influence of the stroke-bore ratio on the friction losses in a 4-cylinder gasoline engine, after [20]. The mean friction pressure decreases with increasing stroke-bore ratio

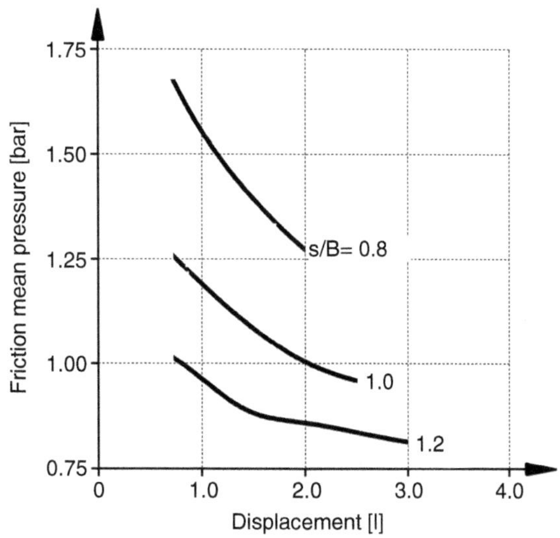

Stroke-to-Bore Ratio

The stroke-bore ratio *s/B* also has an influence. A large stroke-bore ratio results in a reduction of the friction losses, Fig. 1.18. This is mainly due to the reduction of the speed because, on the one hand, the acting inertia forces significantly determine the design of the components and, on the other hand, the shear gradient in the separating lubricating film between the friction partners decreases.

Friction Split

Towing performance is measured on the engine test bench. The oil temperature is set to approx. 120 °C for comparison purposes. When the throttle valve is closed, the drag power is greater than when it is open, because the throttle losses are greater. The absolute power loss is greater the larger the displacement and number of cylinders. Small turbocharged engines have a significantly smaller drag power than larger, multi-cylinder naturally aspirated engines.

Detailed analyses of the drag performance of engines reveal the shares of individual assemblies in the frictional power that is lost in the engine as heat and wear for the useful power. The typical breakdowns are similar for all engines, regardless of whether they are racing or production engines. The differences are found in the magnitude of the absolute values. Racing engines have much lower losses [4]. The piston group has the largest share with up to more than 40%, Fig. 1.19. The valve train also has a significant share in the lower speed range. In conventional tappet valve trains, the valve train friction consists of the friction of the camshaft bearings, the valve and tappet guides and the friction of the sliding contact in the contact area between cam and tappet. The latter has by far the largest

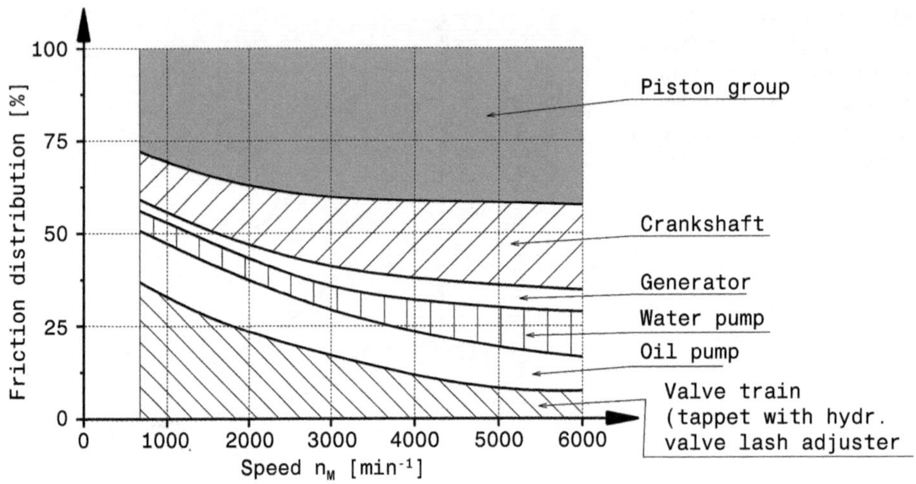

Fig. 1.19 Distribution of friction losses in the gasoline engine, according to [21]. Comparatively, Table 1.3 shows the distribution of the losses of a Formula 1 aggregrate. In addition, the torsion of the co-supporting (semi-stressed) engine during acceleration can lead to a power loss of up to 50 kW

Table 1.3 Distribution of losses of a Formula 1 3-l V10 engine [22]

Assembly	Share (%)	Assembly	Share (%)
Pistons, rings	15	Oil pumps	13
Bearing, crankshaft, camshaft	30	Pulsation	22
Pneumatic valve spring	10	Injection pump	6
Water pump	2	Fuel pump	2
		Sum =	100

share (> 50%) in the valve train. Influencing parameters are the contact force and the relative speed between cam and tappet.

Friction in the Cranktrain

The necessary sealing of the piston, its precise fit and the conversion of the reciprocating motion into a rotating one by means of the crank mechanism generate the main part of the frictional power.

Long connecting rods and piston rings with low preload and large bores with small stroke have a positive effect here. An elegant method of reducing mechanical losses is to desaxle the crankshaft. In this case, the crankshaft axis is offset towards the pressure (thrust) side in relation to the cylinder axis, which reduces the piston side force in the relevant power stroke. The offset is in the range of 10–15% of the bore. Specially coated cylinders that are as round as possible are also good. Pistons with only two rings have long been used in Formula 1 engines. When bolted to the cylinder head, the liners deform and form cloverleaf shapes or similar in cross-section, depending on the number of bolts. Careful design of the bolted joint helps to keep this liner distortion to a minimum.

Friction losses also occur in the bearing positions. The number and dimension of the bearings is important here. The fewer bearings and the smaller the diameter, the better. This is why, quite apart from the overall length, V-engines are cheaper than in-line engines. A six-cylinder V-engine, for example, has only four main bearings, whereas a four-cylinder in-line engine has five. Only the eight-cylinder V-engine has as many main bearings as the four-cylinder in-line engine. If the connecting rod bearings are included, a six-cylinder V-engine has only one bearing position [1] more than a four-cylinder in-line engine [11]. Since the force acting on the individual bearing position is smaller in the engine with the larger number of cylinders, the bearing diameters can be selected smaller. However, excessive downsizing is detrimental to both the load carrying capacity and the strength of the crankshaft [4].

Seals at the shaft passages from the oil chamber of the crankcase also cause losses. These are design-related and speed-dependent. At 12,000 min^{-1}, for example, about 3 kW are absorbed by elastomer sealing rings in a 3-litre engine [23]. In the 1980s, there were actually Formula 1 engines that used a cotton braid to seal the two crankshaft ends. For all their disadvantages, their friction losses were extremely low [23].

Friction in Valve *Train*

The heavier the valves, the higher the valve spring forces, the higher the power loss. The power loss increases with the speed. Two camshafts with four valves per cylinder have a higher power loss than a camshaft with two valves. However, these higher losses are accepted because other advantages outweigh them. With conventional materials and technologies, the valve train soon reaches its limit, which is approximately 13,500 min^{-1}. The "power explosion" in the 3.5 and 3.0 litre Formula 1 naturally aspirated engines was only possible by significantly increasing the engine speeds. Whereas the 3.0 litre Formula 1 naturally aspirated engines of the first generation already reached the end of their revving capacity at approx. 13,000 min^{-1}, the 3.0 litre naturally aspirated engines used from 1995 onwards revved up to just under 18,000 min^{-1} (cf. also Fig. 1.2). This enormous increase in engine speed was essentially achieved by two measures: Firstly, the use of titanium valves, which are considerably lighter than the steel valves used until then. Titanium valves are used on both the intake and exhaust side. Secondly, the use of pneumatic valve springs, which not only require less drive power, but also allow extreme opening and closing ramps. The greatly increased opening cross-sections of the valves are an important contribution to the higher specific output of recent years.

In relation to the maximum power, the drive power for the valve train has also been significantly reduced with these new technologies.

Losses *Due to Pumps*

The power required to drive a single or multiple water pumps is also missing from the final engine power calculation. The same applies to the oil extraction and oil pressure pumps. There are several of these in multi-cylinder racing engines because oil flying around in the crankcase impedes the free movement of the pistons, connecting rods and rotating parts at high speeds.

Today, racing engines use one suction pump per crankcase chamber, which is arranged on the side of the crankcase in line with the other pumps and is practically driven by a common shaft running through it. Thus, for example, a ten-cylinder engine has five of these pumps which return the extracted oil to the dry sump tank. Another pump – the pressure pump – draws oil from the dry sump tank, brings it up to the correct pressure and thus supplies all the important bearing and lubrication points of the engine.

If these pumps are operated according to demand, the otherwise usual compromise of designing for the maximum throughput can be dispensed with. If pumps are designed in a register design or are electrically driven, they can be set to the delivery quantity required in the current operating state independently of the engine speed. This reduces the losses that a pump oversized for the partial load range must have.

Losses Due to Pulsations (*Pumping Losses*)

A surprisingly large part of the generated power can also be extracted in the crankcase by pulsations. Pulsations occur because the pistons push the air in the crankcase back and forth with their undersides, partially compressing it and expanding it again. The determining factors here are the design of the engine and the movements performed by the pistons next to or opposite each other. In unfavourable designs, such as boxer engines (horizontally opposed), two opposing pistons always move in opposite directions. The air underneath the pistons is thus continuously compressed and expanded. The engine oil is strongly mixed with air, which reduces its lubricity and leads to particularly high oil temperatures at high engine speeds. For flat racing engines, the 180 °V design, as built by Ferrari until the turbo era of Formula 1 (1980), is more suitable. Here, opposing pistons always perform a co-rotating motion. This means that the volume of the air is no longer continuously changed, but the air is only pushed back and forth. Particularly good for oil separation and low pulsation losses are upright or slightly inclined in-line engines or also V-engines with fork angles of 60° to 90°.

The gas spring principle follows a completely different design philosophy: each crank chamber is closed in itself. The air is compressed by the downward moving piston, but the air returns the stored energy when moving upward (apart from wall heat losses). Ventilation losses at high speeds are thus avoided, but one notices a running unrest in the idle range of such engines. For racing engines this is irrelevant. The reason is the resulting force due to the difference between the cyclic fluctuations and the constant gas spring force. This leads to the perceptible torque irregularity. In sporty engines for passenger cars, a small hole in the main bearing bracket (i.e. a small ventilation opening after all) has been found as a successful remedy [24].

It is reported from other sports engines that although the gas spring principle was investigated in the development phase, large ventilation cross-sections remained in the production crankcase [25].

The rotating crankshaft also causes losses due to air friction. Closed circular crank webs are more favourable than offsets with pronounced counterweights. In the case of high-revving engines, the webs are also aerodynamically favourably shaped (see also Sect. 1.4.3 *Crankshaft Drive*).

Losses Due to Ancillaries

Losses are also caused by mechanical injection pumps and alternators. If the required injection pressure has to be provided by electric fuel pumps, the efficiency is particularly poor. Electricity must first be provided via the alternator, which is then "converted" into pressure in the fuel pump. Since both are associated with losses, a mechanical gasoline pump is preferred, which is usually driven by the free end of a camshaft. The still existing electric fuel pump is only needed for starting and slow driving. If the engine has exceeded a speed of approx. 4000 min^{-1}, the mechanical pump delivers so much fuel that the electric one can be switched off by the driver in order not to overload the weak on-board network. Once the engine has stopped, the driver must not forget to switch the pump back on.

Losses also occur in the drive of the auxiliary units. Timing belts are better than V-belts, but it is best to design the engine in such a way that the drive can come directly from the camshafts or even from existing transmission gears.

The effort to keep auxiliary units small and to dispense with electricity as an energy source as far as possible can also be seen in the starter motor. Heavy electric starters, which also require a large battery, are completely dispensed with in Formula 1 engines today. Starting is only possible in the pits. Compressed air starters are preferred, with the required energy being provided in the form of compressed air in compressed air cylinders [4].

Coatings Coatings are successfully used on tribological contact points to reduce friction and increase service life. For example, DLC (diamond like carbon) coatings are applied to piston pins, valve tappets, piston skirts and gear wheels. In addition, molybdenum sulfide, graphite and titanium nitride coatings reduce friction on the surfaces of valve stems, piston rings and bearings.

1.4 Modules

The design of an engine usually starts with a (middle) cylinder unit. The maximum displacement is specified by the regulations. The number of cylinders is determined according to the above considerations. The cylinder volume follows directly from this.

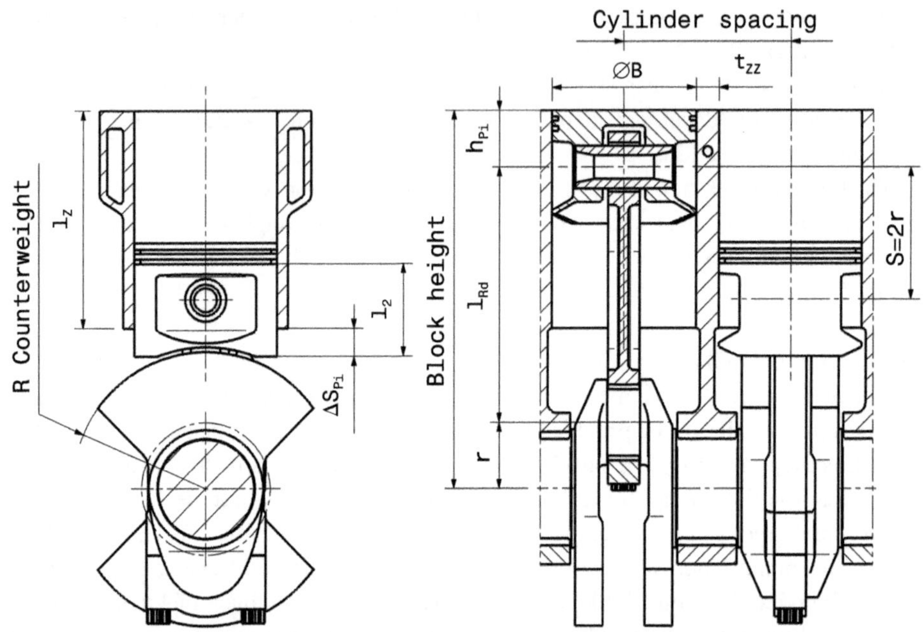

Fig. 1.20 Main dimensions of an engine. *B* Bore, *s stroke*, *r* crank radius, l_{Rd} Connecting rod *length*, h_{Pi} Compression height of the piston, l_Z *liner length*, l_2 *skirt length* of the piston, Δs_{Pi} *piston standout*, $R_{\text{counterweight}}$ *counterweight radius*, Cylinder *spacing*, *crankcase height*, t_{ZZ} *land* width between *cylinders*

The stroke-bore ratio, and thus these two variables, are chosen with consideration of the maximum piston speed, the rated speed and the combustion chamber shape. After determining other main dimensions (cylinder spacing, bank angle for V-engines, compression height of the piston), the design of the cylinder head already begins. The cylinder distance should be as small as possible, because then the engine is short and the crankshaft is stiff. This is opposed by the fact that cooling holes are to be provided in the web between the piston raceways and that the cylinder head gasket must enclose the circumferences of both adjacent combustion chambers. Valve angle, port angle, combustion chamber and valve train determine the shape of the cylinder head in addition to the cooling system. The only relevant variables that are now missing are the connecting rod length, the counterweight radius and, in the case of V-engines, the bench offset, Fig. 1.20.

The connecting rod length is determined by the rod ratio:

$$\lambda_{Rd} = \frac{r}{l_{Rd}}$$

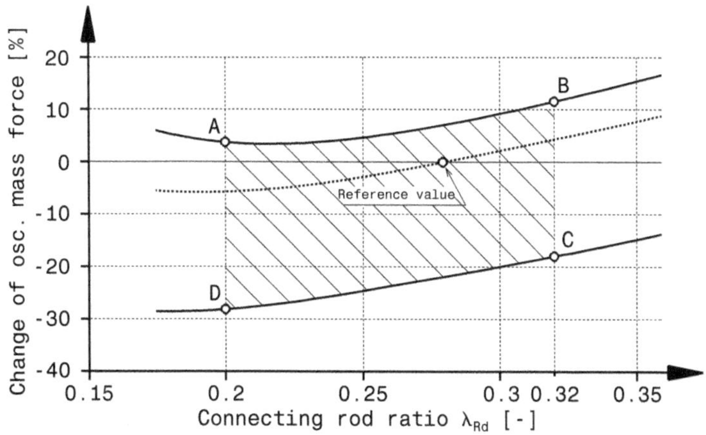

Fig. 1.21 Design range of connecting rod lengths, according to [13]. The change in mass force is related to a reference value. The design data of this value correspond to a series engine: $s/B = 1$ and $\lambda_{Rd} = 0.28$. The preferred design range is shown hatched, where point D is typical for high-speed racing engines

λ_{Rd}	(Connecting rod) rod ratio, -
R	Crank radius, i.e. half the stroke, mm
l_{Rd}	Connecting rod length (gauge), mm

Since primarily oscillating mass forces influence the running smoothness, a design range for connecting rod lengths is defined in [13] based on their magnitude, Fig. 1.21. The rod ratio is varied for different block heights. It can be seen that there is a rod ratio for which the oscillating inertia force becomes minimum. However, this is outside the useful range for racing engines.

The limits for the design range in Fig. 1.21 result from the following extrema:

Left:	below $\lambda_{Rd} = 0.2$, the bore becomes too large (internal efficiency and firing paths).
Right:	Above $\lambda_{Rd} = 0.32$, the piston diameter becomes too small. the counterweights on the crankshaft to compensate for the oscillating forces become disadvantageously large.
Above:	The crankcase height becomes too small and the connecting rod too short.
Below:	Crankcase too high, compression height of piston becomes too small.

The cornerstones of the design area represent the following interpretations:

A:	Unfavourable design: Piston diameter and compression height so large that the oscillating mass forces are unnecessarily large despite the short connecting rod.
B:	Poor design: Stroke and compression height too large.
C:	Light piston, still favourable compression height, but stroke-bore ratio so unfavourable ($s/B = 1.35$) that the average piston speed limits the speed.
D:	Typical racing engine: Light piston, low compression height (0.33B), small stroke ($s/B = 0.54$) and long connecting rod. Allows highest revs despite large bore (favourable for large valve opening).

The length of the liner l_Z is selected so that the piston does not deflect more than 15% of its shaft length l_2 at bottom dead centre. The maximum counterweight radius of the crankshaft is determined by the clearance of the piston when passing through bottom dead centre and by the size of the free crank chamber. For high revving racing engines, a cylindrical crank chamber is preferred. The counterweight radius then has its piercing point exactly on the crankshaft axis. This need not be the case with production engines. The crankcase encloses the crank mechanism and forms part of the cooling jacket. The bench offset on V-engines is determined only by the width of the connecting rod root on crankshafts without intermediate web (flying web). This in turn is dictated by the bearing shell width and the strength of the bearing cap. The smallest values for Formula 1 engines are 12 mm. The valve train can be located at the front end, at the rear end or, more rarely, in the middle of the crankshaft. The pipework (intake, exhaust system) as well as the piping (coolant, lubricant) and tubing (cooling, fuel system, air for valve spring) complete the unit. In addition to the unquestionable power and torque, the main objectives are compactness (low overall height) and low weight. For long-distance engines, tuned durability and low fuel consumption also provide competitive advantages.

A typical 10-cylinder Formula 1 engine (naturally aspirated 3-litre engine) consisted of around 3000 parts, 900 of which were moving parts. In the following, individual important assemblies and their special features will be discussed.

1.4.1 Cylinder Head

The cylinder head is the performance-determining component of an engine. It houses essential parts of the charge exchange and the cooling system. It also determines the part of the combustion chamber with the ignition device. In direct-injection engines, the cylinder head also houses the injection valve (this applies regardless of the combustion process, i.e. to both petrol and diesel (compression ignited) engines). Figure 1.22 shows a cross-section of a typical four-valve cylinder head.

The engine behaviour is primarily determined by the combustion. This in turn depends essentially on the achievable flame speed and thus the combustion duration. Combustion takes place in several phases. The first phase in the gasoline engine is initiated by the ignition spark and is called the ignition phase. It depends only on the mixture composition and lasts a certain, invariable period of time, the so-called *ignition delay*. This means that as engine speed increases and the air/fuel ratio remains fixed, there is less time for the second phase, the actual combustion. The burning time is mainly determined by the speed of flame propagation. This so-called burning speed is determined by turbulence strength and temperature course in the not yet burned mixture portion. It is greatest at an excess air of

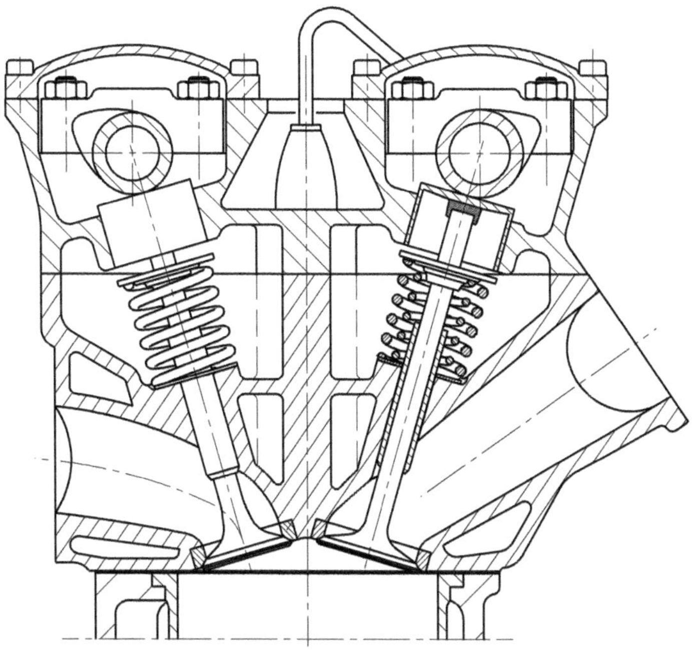

Fig. 1.22 Cylinder head Formula 1 (Ford Cosworth DFV 3.0 l V8, 1980). Cylinder head of a naturally aspirated engine with four valves per cylinder. This engine pioneered the high-revving racing engines. A small valve angle results in a narrow but relatively tall cylinder head. The combustion chamber is roof-shaped

about 10% (cf. Figure 1.3) and amounts to 20–40 m/s in normal combustion. The turbulence strength in the combustion chamber, more precisely in the flame front, can be influenced by the design of the inlet elements (e.g. swirl-generating ports), the combustion chamber shape (e.g. with turbulence-generating tear-off edges) and the utilization of the piston movement. In the latter case, the mixture is forced out of the narrowing gap between the approaching piston and cylinder head base towards the spark plug (squish flow). However, turbulence can also be generated by the flame propagation itself and the associated pressure increase. In any case, it is dependent on the engine parameters of compression, intake air temperature and engine speed.

To achieve high efficiency, short combustion times, i.e. high combustion speeds, must be aimed for. Furthermore, the correct position of the combustion process in relation to the piston movement is important. The centre of gravity of the combustion should lie between 5 and 15° crank angle after top dead centre. The position of this centre of gravity is adjusted with the ignition timing (ignition angle).[5]

[5]Engine calibration is dealt with in vol. 5 *Data analysis, tuning and development of* the series Racing Car Technology Manual, Chap. 6.

Table 1.4 Relative valve sizes

Valves/ Cylinders	Air supply	Combustion chamber	d_E/B	d_A/B	d_E/ d_A
2	Naturally aspirated	Wedge	0.43–0.46	0.35–0.37	1.25
		Hemisphere	0.48–0.5	0.41–0.43	1.10
	Supercharged		0,40–0.41	0.38–0.39	1.05
4	Naturally aspirated	Roof	0.35–0.37	0.28–0.32	1.17
	Supercharged	Roof	0.32	0.30	1.08

Legend: *B* Bore, d_E Inlet valve diameter, d_A Exhaust valve diameter

Number of Valves, Valve Angle

In the cylinder head, one part of the combustion chamber is represented, the second part is formed by the piston crown. For the shape of the combustion chamber are initially decisive the number and arrangement of the valves.

First of all, the diameter ratio of inlet to exhaust valve is interesting, Table 1.4, then it is tried to arrange valves as large as possible in the combustion chamber.

As the air consumption increases, the engine power increases (cf. considerations in Sect. 1.1), which is why the largest possible valve opening area should be aimed for. It is not without reason that numerous fundamental studies address the question of the best number of valves or combustion chamber shape, e.g [26]. The valve area is determined by the bore diameter, the roof angle of the combustion chamber and by the number of valves. The influence of the roof angle is shown in Fig. 1.23 for two- and four-valve engines. The valve areas of the two-valve engine increase greatly with the valve angle. Above 68° valve angle, the two-valve engine has even more valve area than the four-valve engine. The strengths of the four-valve engine lie at small valve angles from 0° to 40°, where it has about 25–30% more valve area than the two-valve engine. In addition, the combustion chamber shape is favourable at small valve angles. Figure 1.23 thus shows why it is not sensible to build four-valve engines with excessively large valve angles and why two-valve engines with large valve angles were preferred in the past.

An increase in the number of valves only brings the expected success to a limited extent. There are several reasons for this. First, the valve openings are circles and several can thus never completely cover the bore. In addition, a minimum land must remain in the cylinder head between the valve seat ring bores so that the valve seat rings do not come loose after pressing in, Fig. 1.24a and Table 1.5. For the same reason, there must also be a minimum land to the spark plug bore.

However, the valve openings must also have a certain distance to the cylinder, otherwise the flow resistance between the valve disc and the liner is so great that only a partial circumference of the annular valve gap is effective for the charge change. If two valves with the same name are too close together, the flow will also be obstructed.

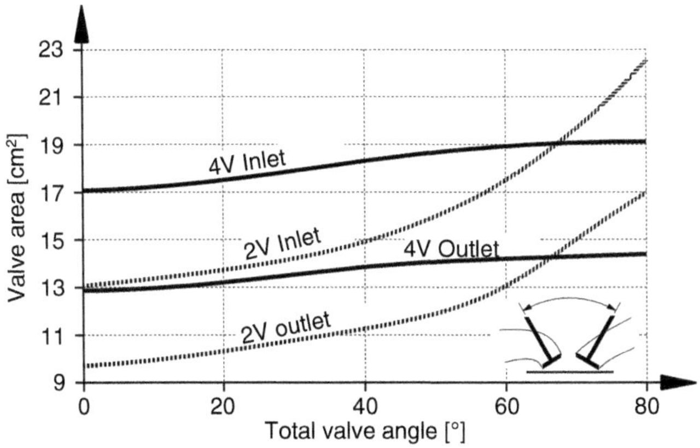

Fig. 1.23 Influence of the total valve angle on the valve surfaces [4]. The diagram is based on the following values: bore 80 mm; inlet valve diameter = 1.15 × exhaust valve diameter. However, the principle curve is the same for all usable valve size ratios. The total valve angle is the angle that includes the inlet and exhaust stems

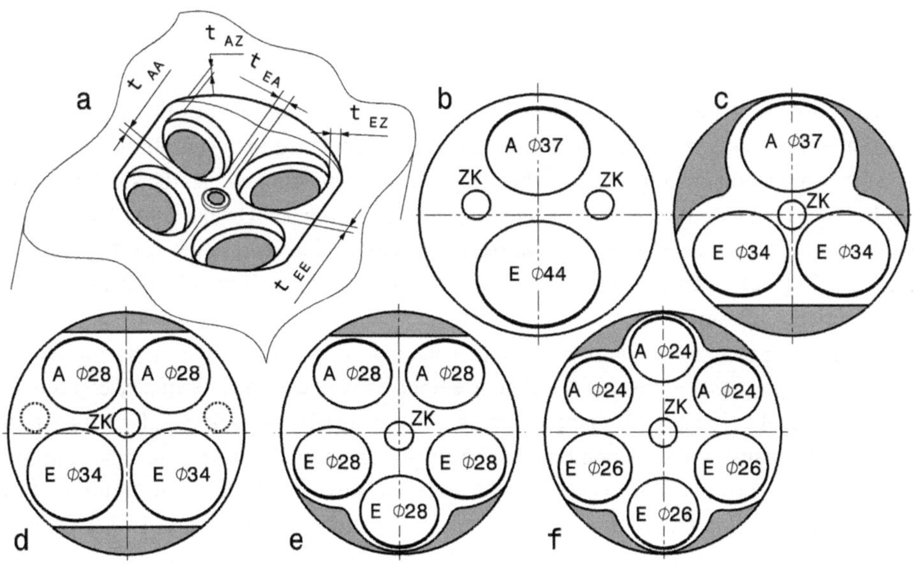

Fig. 1.24 Arrangements of valves. The listed valve diameters apply to 85 mm bore diameter. E Inlet valve. A Exhaust valve, ZK spark plug. (**a**) Definition of distances and land widths in the cylinder head, (**b**) Two-valve engine with two spark plugs, (**c**) Three-valve engine, (**d**) Four-valve engine. One or two external spark plugs possible, (**e**) Five-valve, (**f**) Six-valve

All these influences lead to the result illustrated in Fig. 1.25. Four valves per cylinder provide the largest effective opening cross-section, even though five theoretically provide a larger gap.

Table 1.5 Minimum web widths or clearances in standard cylinder heads, mm. See also Fig. 1.24

Dimension	t_{EE}	t_{AA}	t_{EA}	t_{EZ}	t_{AZ}	t_{EZk}	t_{AZk}
Value	2.5	2.5	3	1.5	1	3	3

Indexes: A Exhaust seat ring, E Intake seat ring, Z Cylinder liner, Zk Spark plug

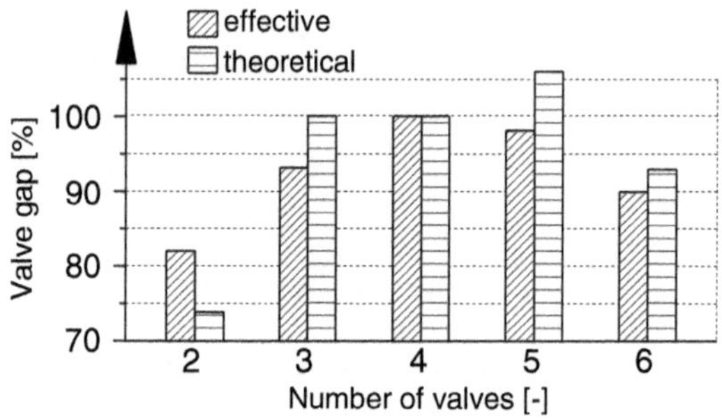

Fig. 1.25 Valve opening over the number of valves. Due to wall shielding and mutual influence, in addition to geometric constraints, the effective valve openings differ from the theoretical ones. Accordingly, four valves per cylinder provide the largest effective cross-section. Five valves bring only geometrical advantages. Six valves are theoretically and practically worse

The position of the valve in relation to the cylinder axis and the position of the port in relation to the valve axis influence the gas flow, the turbulence in the combustion chamber and the combustion chamber shape. The steeper a port is positioned, i.e. the smaller its port angle, the greater its flow [27] and thus the maximum power. On the other hand, larger valve angles result in a more favourable combustion chamber shape for high torque. Figure 1.26 gives a general overview of how the two variables ultimately affect the engine characteristics of naturally aspirated engines.

Today, with increasing compression ratios and very good knowledge of the combustion process in high-revving engines, particular emphasis is placed on a compact combustion chamber (small valve angle). The total valve angle is usually divided equally between the intake and exhaust sides. Depending on the displacement and stroke-bore ratio, the total valve angle in Formula 1 turbo engines had settled at between 22° (Renault Turbo) and 40° (BMW Turbo). For the Formula 1 naturally aspirated engines with 2.4; 3.0 or 3.5 l displacement, total valve angles between 24° and 28° are and were preferred. And the rule applies: the smaller the cylinder unit – or the larger the number of cylinders for a given displacement – the smaller the valve angle. Apart from the advantages in terms of combustion technology, cylinder heads with narrow valve angles build smaller and are therefore also lighter.

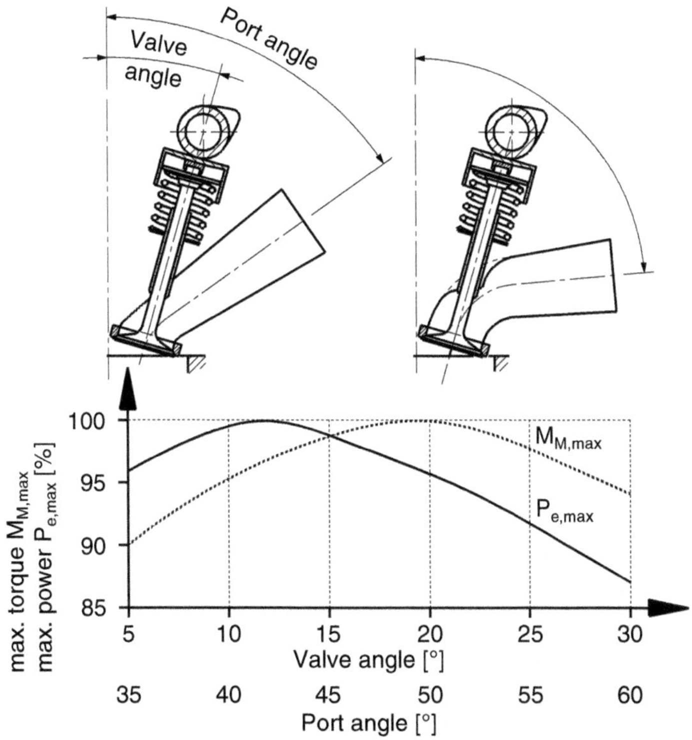

Fig. 1.26 Influence of intake valve and port angle on maximum torque and maximum power of naturally aspirated engines. The greatest torque is achieved by intake ports in the range of 50° inclination to the cylinder axis and 20–22° valve angle. Steeper valves and ducts result in the greatest power. A good compromise is provided by valves that are at an angle of 15–18° and intake ports with a port angle of about 45°

Combustion Chamber

The shape of the combustion chamber in the cylinder head is determined by the valve arrangement and the spark plug position. A distinction is made between roof, hemispherical, wedge, pan and F combustion chambers (upright valves) according to their shape.

The contribution of the combustion chamber to a high volumetric efficiency is that the fresh gas temperature at intake closure is as low as possible (because of the thermal expansion of the gas there is more mass in the cylinder) and that the residual gas content in the cylinder is as small as possible (then there is more space for fresh gas). For high thermal efficiency, the compression ratio must be as high as possible (see Fig. 1.6). The limits for this are drawn by knocking and the valve clearance in the charge change TDC, i.e. when intake and exhaust valves could touch the piston. In addition, the heat loss via the combustion chamber walls should be as small as possible. The heat transfer depends on the temperature difference between gas and wall as well as on the size of the common surface. The combustion chamber should therefore be as compact as possible.

The volume of the combustion chamber V_c follows from the targeted compression ratio:

$$V_C = \frac{V_h}{\varepsilon - 1}$$

V_c	Compression volume, mm^3
V_h	Swept volume of a cylinder, mm^3
E	Compression ratio, -

Compression ratios of gasoline engines are statically between 9:1 and 14.5:1.Formula 1 naturally aspirated engines exploit the upper limit. Turbocharged engines must stay at the lower limit. The maximum compression ratio has settled at about 13:1 even in high-revving naturally aspirated racing engines, because the more angled flame paths due to the forced flattening of the combustion chamber result in a longer burn time and the wall heat losses cancel out the thermodynamic advantage [28]. Due to their principle, diesel (compression ignited) engines compress much higher and achieve compression ratios of 17:1–21:1.

In operation, the compression ratio of racing engines sometimes changes significantly due to inertia forces. Component deformation and the consumption of bearing clearances increase the compression ratio (e.g. static $\varepsilon = 13.5{:}1$ becomes dynamic 15:1 [22]).

Even if an engine is not designed for the highest speeds, but builds up its power via high mean pressure, short combustion paths are advantageous. The combustion path is the greatest distance that the flame must travel from the spark plug to the edge of the combustion chamber. Short combustion paths allow great freedom for engine tuning. Because combustion is completed more quickly, the ignition angle can be set so that the centre of combustion lies in the favourable transmission range of the crankshaft drive.[6] If combustion is slow, the ignition angle must be set so that combustion is completed before the exhaust opens, otherwise the combustion will not do any useful work from that point on. Short combustion paths also reduce the risk of knock. Viewed from below, the spark plug position is always optimal, regardless of the valve angle, because it is practically in the middle. The situation is different from the side. With a large valve angle, the spark plug is hidden deep in the roof of the combustion chamber, i.e. not in the centre of the combustion chamber. The aim is to have the spark plugs as close as possible to the centre of gravity of the combustion chamber, because from here the combustion paths actually reach a minimum. This requirement can be better met with small valve angles. In this case, the spark plug moves closer to the place of action, because at the same time, with the smaller valve angle, the piston crown is slightly spherical at the bottom [4]. The risk of knocking is

[6] For more details see Racing Car Technology Manual, Vol. *Data Analysis, Tuning and Development*, Sect. 6.2.3 *Test Benches and Test Equipment.*

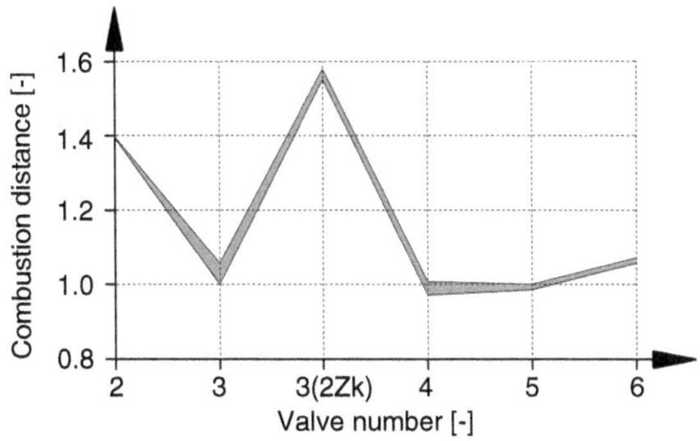

Fig. 1.27 Burning distance as a function of the number of valves, according to [26]. The different area ratio between intake and exhaust valve results in a narrow scatter band, but this does not change the essential statement. The combustion paths are short with three, four and five valves. Two spark plugs on the three-valve (3 (2 Zk)) extend the combustion path

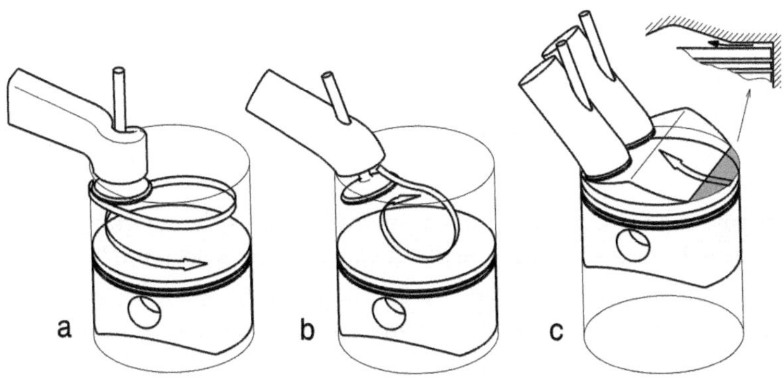

Fig. 1.28 Charge movements in the combustion chamber. (**a**) Swirl, (**b**) Tumble, (**c**) Squish flow . The charge movements are caused by the incoming charge or by the piston and shorten the burning time

reduced if the charge is in a relatively cool area of the combustion chamber at the end of combustion. This is achieved with a spark plug located close to the exhaust valves. The flame front thus propagates toward cooler intake valves. The spark plug position must also be matched with the position of the camshafts, because the plugs need a continuous shaft to the outside. The combustion paths can also be shortened with the use of multiple ignition points, Fig. 1.27.

Targeted charge movements in the combustion chamber, which are caused by the inlet port shape and arrangement, accelerate the burning through of the mixture by promoting mixing and accelerating the flame front. A distinction is made between swirl, tumble and squish flow, Fig. 1.28.

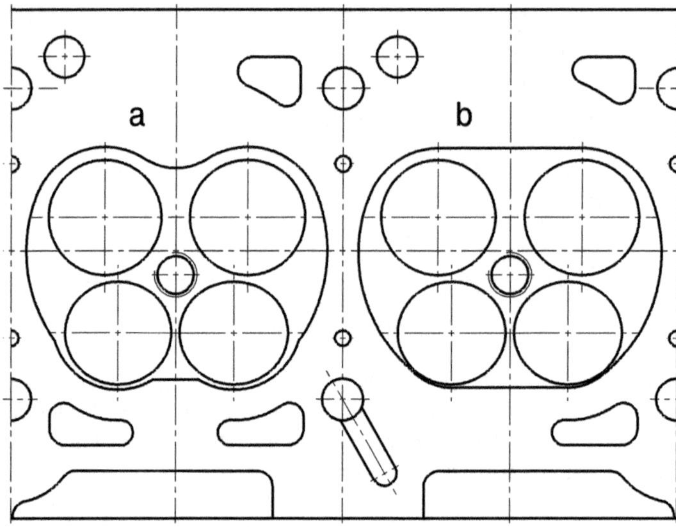

Fig. 1.29 Comparison of combustion chambers of series engine and derived racing engine in the view of the cylinder heads from below, according to [29]. (**a**) Series engine, (**b**) Racing engine. Due to turbulence at high speeds, racing engines do not have pronounced squish surfaces, but great importance is attached to the most favourable inlet and outlet conditions possible

A swirl port generates a swirl around the cylinder axis during inflow, which is still noticeable when the piston moves upwards. A similar effect is achieved by an eccentrically arranged, flat port. A cylindrical flow across the cylinder axis is called a tumble. This movement comes up at ports, where flow-parts detach at downside surface and thus prefer upper parts of the port. As a result, the flow becomes asymmetrical and the charge rotates. The flow of gas through such ports is impaired by the generation of the movement, cf. also Fig. 1.38. In another way, namely when the valves are closed and the piston approaches the combustion chamber roof, the squeeze flow is caused. For this purpose, the cylinder head is designed so that edge areas of the combustion chamber are parallel to the piston crown (squish surfaces). When the piston approaches top dead center, the mixture is forced out of the gap and generates a secondary flow. The smallest gap between the cylinder head and the piston crown is about 1 mm when the engine is at rest when cold. During operation, a few tenths of a mm remain because different thermal expansion and bearing clearances (mass force of the piston) cause a relative movement of the components. The squish flow is particularly effective at full load (in the partial load range, the cylinder filling is too low) and precisely only in the vicinity of top dead center. With increasing squish area, the combustion duration (e.g. in °CA) decreases. The squish areas amount to about 10–15% of the piston area. In Fig. 1.24, possible squish areas are shown in grey.

High-performance engines with external mixture formation manage without measures for charge movement, therefore small or no squish surfaces are provided, Fig. 1.29. In the

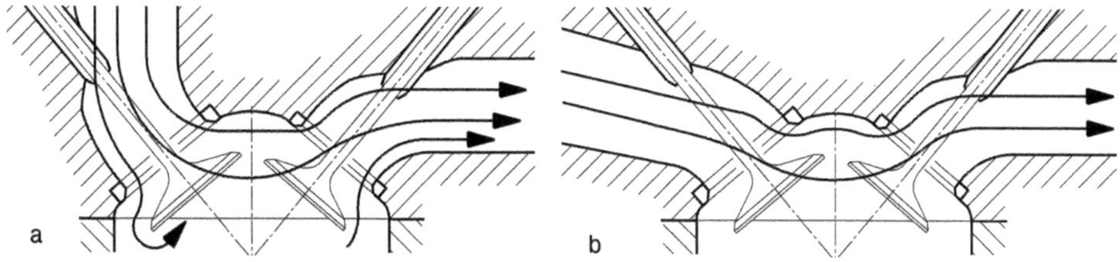

Fig. 1.30 Combustion chamber flow during charge change, after [32]. (**a**) *Downdraught intake port*, (**b**) *semi-downdraught intake port*

extreme conditions, a slight tumble movement is present at the start of combustion even without "artificial" measures [30].

For low wall heat losses, a small surface-to-volume ratio is advantageous. This is the ratio of the heated surface (combustion chamber roof) to the enclosed volume. As cylinder volume increases and stroke bore ratio increases, the surface-to-volume ratio becomes more favorable. However, the disadvantages due to heat losses for a small stroke-bore ratio are more than compensated by the possible larger valve cross sections, which is why this design is the first choice for racing engines. In general, it can be stated that for two-valve engines the *hemispherical* and for four-valve engines the *roof-shaped* combustion chamber provide the highest combustion efficiencies [31]. For high-speed engines, the four-valve roof-shaped combustion chamber is the first choice because of the larger valve cross-sections.

After passing through the valve gap, which directs the gas flow according to the seat angle of 45°, it wants to maintain its direction and spread out laterally. There must therefore be sufficient space around each open disc for the flow to pass through. In the case of four-valve heads with large valves arranged in parallel, the plates are close to the cylinder wall and their distance from each other is also so small that the flow is obstructed at these points.

In a high-performance engine, there is an intensive flow from the intake to the exhaust port during the time of the valve overlap (charge change TDC), which should be directed in such a way that the amount of waste gas that remains in the cylinder head in the normal engine (residual gas, *end gas*) is scavenged out as far as possible. The flow should be such that all residual gas in corners and niches is captured. For this reason, a downdraught inlet port appears to be advantageous, because the flow then passes under the plate of the inlet valve and also captures the part of the waste gases located there, Fig. 1.30a.

At horizontal port position however, flow runs at shortest way and leaves part of.

residual gas quantity unaffected (Fig. 1.30b). The gas exchange process at valve overlap is of decisive importance for the degree of cylinder filling. Furthermore, the valve arrangement and position of the ports are also important for the period during which the piston takes in air. The flow is least disturbed when it is in the direction of the piston that is aspirating, as is clear from flow measurements [32].

The combustion chamber must be designed for rapid energy conversion, especially at high engine speeds. The shape of the piston crown as the final element of the combustion

Fig. 1.31 Combustion chamber of a Formula 1 engine. The extremely flat combustion chamber is completely machined. A spherical clearance is fitted around each seat ring. Its shape is not ideal, but it allows a piston with low fracturing through valve pockets even at high compression

chamber often becomes a problem at high compression ratios (> 12:1). Large overlaps, valve lifts, valve diameters and large valve angles lead to highly fissured piston crowns. A typical combustion chamber for the highest compression ratios and engine speeds is shown in Fig. 1.31. The valves are not only inclined to the cylinder axis in one plane, as is common in production engines, but also form an acute angle of about 6° with each other (radial arrangement, spherical arrangement). This means that no valve is parallel to the neighboring valve. The cams of a camshaft must therefore be ground at an appropriate angle (tapered) to compensate.

The surface of the combustion chamber should be as smooth as possible and not very fissured. This enables good scavenging with little residual gas content and avoids knock nests. In production engines, the combustion chambers are generally cast. In racing engines, the combustion chambers can also be produced entirely by machining. This offers a number of advantages:

- smooth surface
- true-contour squish faces
- low dispersion of the compression ratio between the individual cylinders.

Of course, this is accompanied by disadvantages:

- increased production costs
- with large machining allowances, the advantage of directional solidification of the cylinder head base (fine cast structure due to cooled casting mould wall) is cancelled out.

Furthermore, the influence of machining is limited. Depending on the number and arrangement of valves, up to 40% or more of the combustion chamber surface is determined by the valve disc and seat ring, Fig. 1.32. In addition, there are squish surfaces and spark plug bore (s). The obvious fear that removing the homogeneous cast skin increases the susceptibility of the combustion chamber wall to cracking is not confirmed in practice.

Fig. 1.32 Combustion chamber of a racing engine. A typical cast roof combustion chamber, which is largely defined by valve surfaces and spark plug mouth. Such combustion chambers are used in Formula 3 engines or also in DTM engines

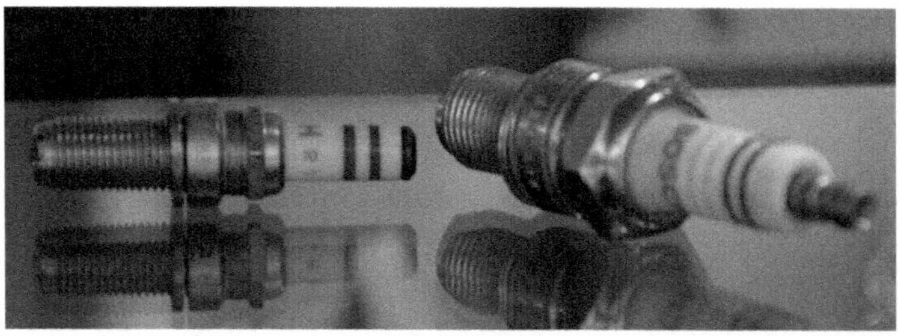

Fig. 1.33 Spark plugs. A racing spark plug (left) with M10 × 1 thread is shown for comparison with a conventional spark plug with M14 thread

Spark Plug

Spark plugs with a small thread diameter allow larger valves and disturb the combustion chamber contour less. Therefore, much smaller spark plugs are used in racing engines than in production engines. For example, plugs with M10 × 1 threads are screwed in instead of the usual M14 × 1.25 thread, Fig. 1.33. Indy car methanol engines ran with "more solid" 12-mm plugs. Some Formula 1 and LMP 1 engines are even equipped with tiny M8 × 1 plugs.

For extremely lean-running engines, as in Formula 1, the spark plugs also assume the function of a prechamber, Fig. 1.34. In this prechamber (3) there is a mixture which is easier to ignite because it is rich. Ignition takes place in the prechamber by a spark which jumps between the two electrodes (1, 2). The jets passing through spray holes into the main combustion chamber ignite the lean mixture there. Thus, air ratios $\lambda > 1.4$ can be run in the main combustion chamber. The lean mixture allows shorter conversion rates due to the higher combustion speed and leads to lower fuel consumption. The spark plug is matched to the engine. For this purpose, the prechamber volume is adapted and the position, shape and number of spray holes are determined by CFD simulations.

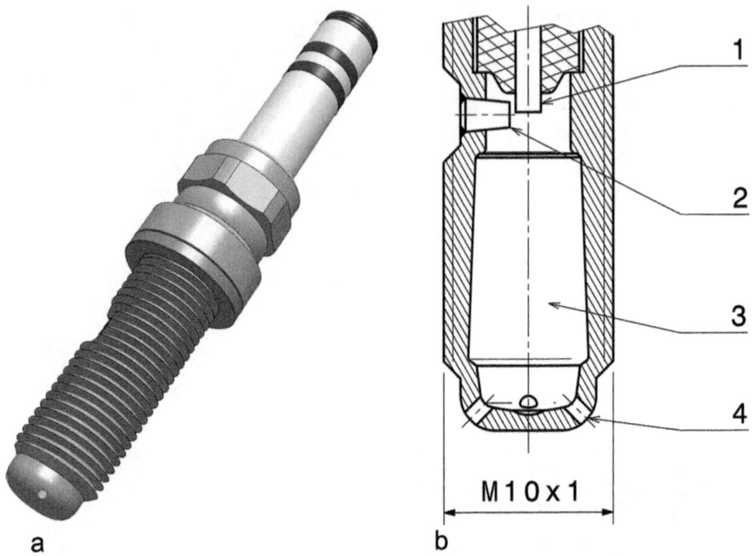

Fig. 1.34 Bosch type prechamber spark plug. (**a**) Axonometric representation, (**b**) Sectional view of the prechamber: 1 center electrode, 2 ground electrode (platinum), 3 prechamber, 4 injection hole

Port Design

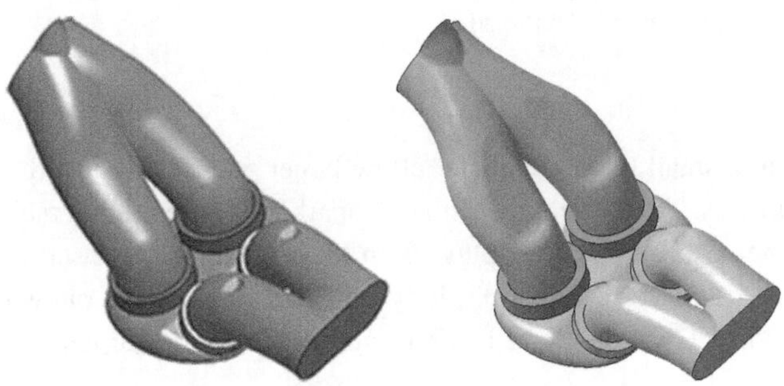

The basic position of the valves and the ports is discussed in Figs. 1.24 and 1.26. The valve position influences not only the flow rate, but also the combustion chamber shape and thus drivable pre-ignition angles, the knock tendency, etc.

As far as volumetric efficiency is concerned, there are two opposing influences acting on the flow, Fig. 1.35.

A small port diameter reduces the air flow rate due to wall friction losses. A large cross-section, in turn, reduces the flow velocity and thus reloading effects due to the inertia of the air mass. Thus there is an optimum port cross-section between these two extremes. For calculation see (1.6).

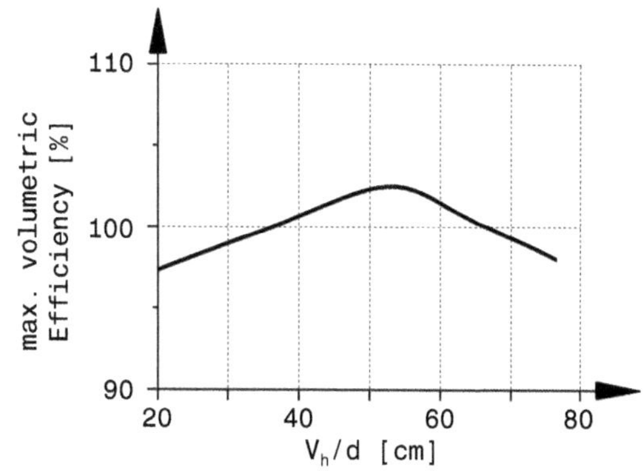

Fig. 1.35 Influence of the port diameter *d* on the flow rate (calculated) [27]. The ratio of cylinder volume V_h to port diameter *d* shows an influence on the volumetric efficiency

Intake *Port*

Inlet ducts are designed as pure filling ducts, as swirl or tumble ducts. However, swirl and tumble generation are at the expense of flow, Fig. 1.36.

High-performance engines with external mixture formation do not require any measures for charge movement; with them, the port development can be oriented purely to maximum flow. In four-valve engines, a port for generating charge movement (swirl, tumble) is often located next to a charge port. The limiting criterion for the smallest port angle is the machining for the valve spring support. A minimum wall thickness must remain in this area. The valve can be moved upwards together with the spring support, but this makes it heavier and it must be held on the cam by a stronger spring. This in turn leads to greater friction in the valve train and lower top speed. The coolant should flow around the intake port as little as possible. Because of the higher temperature of the cooling water, the fresh

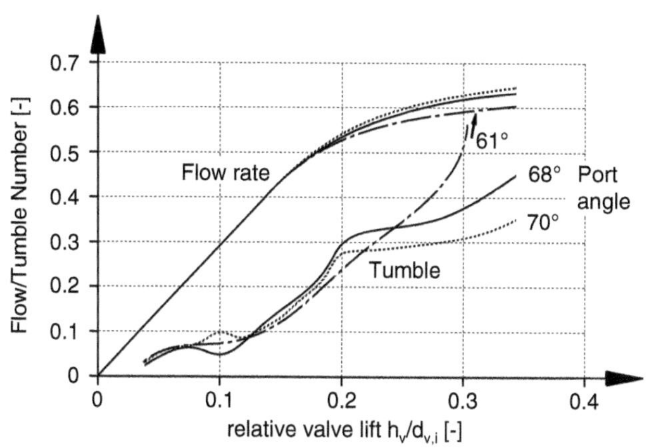

Fig. 1.36 Influence of the *port gradient* on the flow and the tumble according to [33]. The greater the flow or tumble, the greater the respective key figure

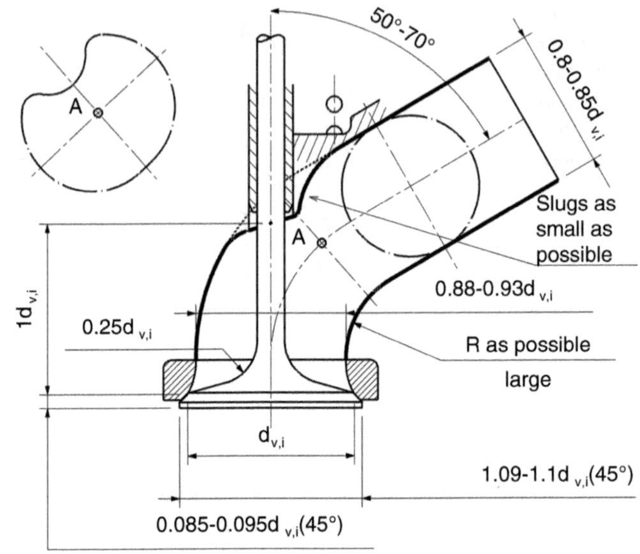

Fig. 1.37 Design recommendations inlet port for Otto (spark ignited) passenger car engine. The port is designed as a filling port. The dimensions are given as a function of the smallest valve seat diameter $d_{v,i}$. The port should be as free as possible in the cooling water jacket

gas is actually heated. In the area of the spring support, a cooling channel is dispensed with anyway, not least for reasons of space. But also in the vicinity of the combustion chamber, only the seat ring and the spark plug holder should be cooled.

The port surface should be as smooth as possible, but not polished by hand. Particularly in the case of small port diameters, cross-sectional jumps and wavy surfaces, which easily occur during manual processing, have a disturbing effect. As a result, flow may decrease even if the port wall has been polished by hand. Design recommendations for inlet ducts can be found in Figs. 1.37 and 1.38. Figure 1.39 shows a typical inlet port of a Formula 1 engine.

Valve Gap Flow

The flow cross-section at the valve is important for the charge exchange. The cross-section used by the incoming air is somewhat smaller than the geometrically available cross-section due to constriction and other gas-dynamic effects. However, the ratios can be calculated approximately and thus designs for high flow rates can be derived.

The average gas velocity is calculated from the continuity equation. The volume flow through the charge exchange ports must be exactly as large as the volume flow in the cylinder.

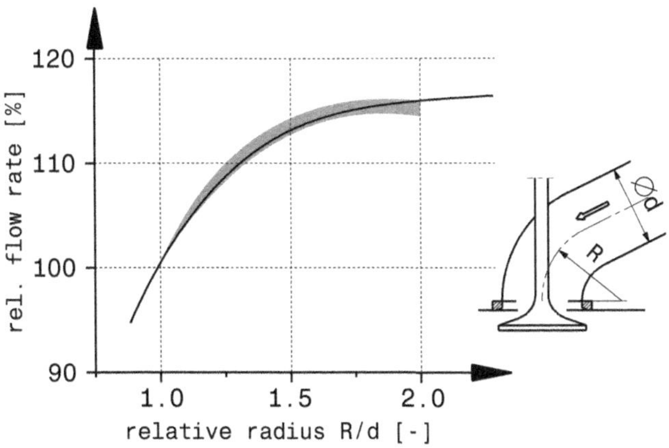

Fig. 1.38 Influence of the port bottom radius (*port curvature*) R on the flow [27]. *d* port diameter. The flow rate is related to that at $R/d = 1$. Different differential pressures result in a narrow scatter band. However, this does not change the basic statement. A ratio *R/d* of at least 2 should be aimed for

$$v_{Ga} \cdot A = v_m \cdot A_{Pi} \Rightarrow v_{Ga} = v_m \frac{A_{Pi}}{A} \qquad \text{bzw.} = v_m \frac{B^2}{d^2} \tag{1.5}$$

v_{Ga}	Mean gas velocity in the port with area A or diameter d, m/s
A_{Pi}	Piston area with bore B, m^2
v_m	Mean piston speed, m/s $v_m = 2sn$ or (1.2). In racing engines, maximum piston speeds of 20–26 m/s are achieved

With increasing gas velocity, the flow losses increase. At low velocities, the gas-dynamic effects are not pronounced enough and they do not achieve a noticeable charging effect. Thus, there exists an optimal velocity range between these two extremes.

Guide values for average gas velocities:	Inlet port $v_{Ga} \approx 70$ m/s for series engines, $v_{Ga} \approx 100$ to 130 m/s for high performance engines [35] or about 5 v_m
	Exhaust port $v_{Ga} \approx 110$ m/s for series engines

The required port cross-section can be determined from (1.5) using the specified guide values:

$$A_{erf} = A_{Pi} \frac{v_{m,n}}{v_{Ga,id}} \tag{1.6}$$

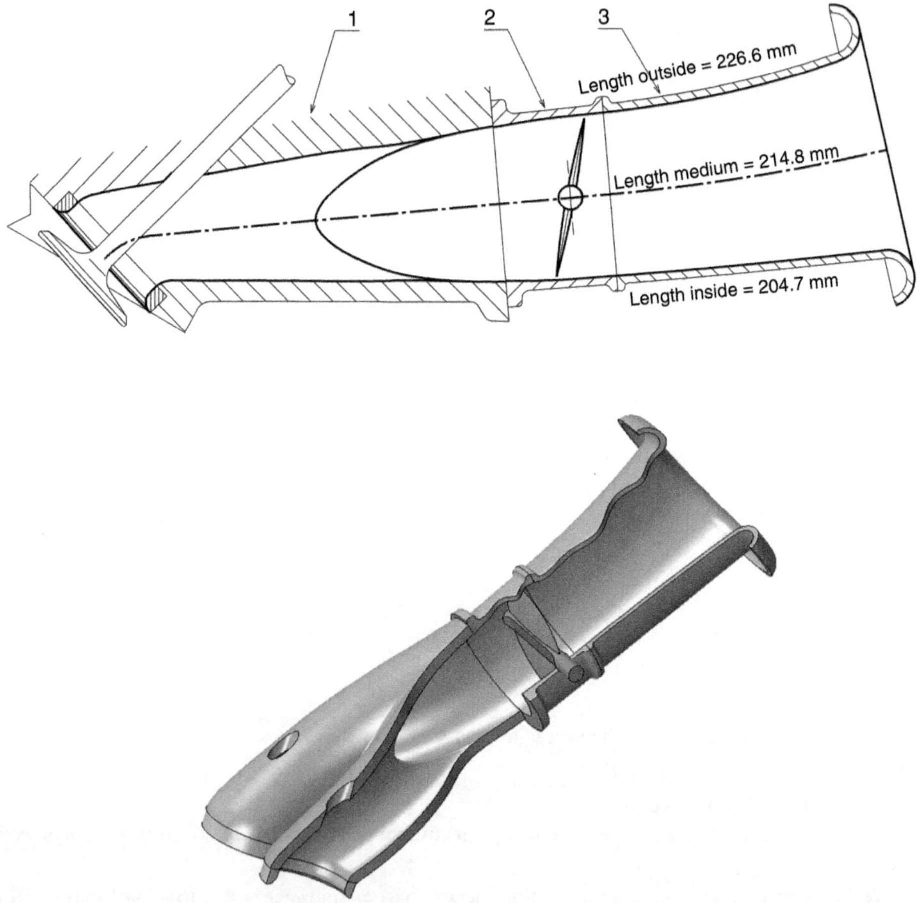

Fig. 1.39 Intake tract of a Formula 1 engine. **1** port in cylinder head, **2** throttle body, 3 intake funnel. The intake manifold is formed by parts 2 and 3 and screwed to the cylinder head. The naturally aspirated V8 engine has its rated speed at 13,500 min^{-1} with a displacement of 3.5 l. The port branches are already merged in the cylinder head to form a circular cross-section. The throttle valve is located just in front of the cylinder head flange. Its shaft is flattened in the flow area. The intake funnel is well rounded. For a one-dimensional flow calculation, the port length is assumed to be 224.8 mm, i.e. 10 mm is added to the mean port length to account for the valve disc effect [34]

A_{erf}	Required port cross-section, m^2
$v_{m,n}$	Average piston speed at nominal speed, m/s
$v_{Ga,id}$	Ideal mean gas velocity, m/s

The valve cross-section follows from the port cross-section determined in this way with the same consideration:

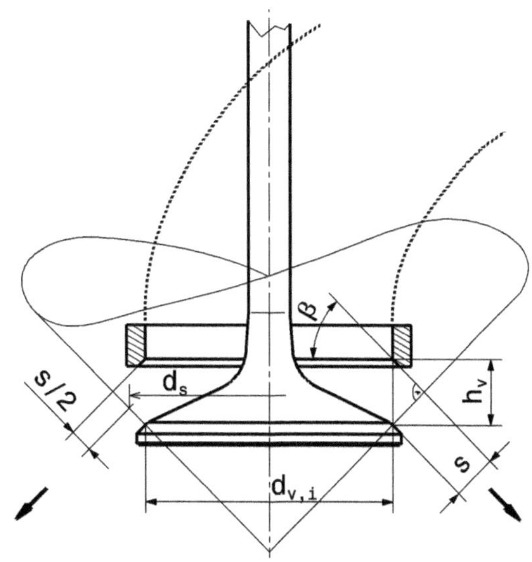

Fig. 1.40 Designations on the valve gap. The geometric opening cross-section is approximately a truncated cone surface with side length *s*. *s* valve gap, β Valve seat angle, h_v Valve lift, $d_{v,i}$ *minor valve seat diameter*(smallest valve seat diameter)

$$A_v = \frac{1}{j_v} A_{erf}$$

A_v	Free cross-section of a valve opening at max. valve lift, m^2
j_v	Number of valves per port, -

The geometric valve cross-section is calculated according to Fig. 1.40.

$$\begin{aligned} s &= h_v \cos(\beta), \quad d_s = d_{v,i} + s \cdot \sin(\beta) \\ A_v &= d_s \cdot \pi \cdot s = \pi \cdot h_v \cdot \cos(\beta) \cdot [d_{v,i} + h_v \cos(\beta) \sin(\beta)] \end{aligned} \tag{1.7}$$

s	Valve gap, m

As the valve stroke h_v increases, the opening cross-section initially increases, cf. Figure 1.36 *Flow*, but from a relative stroke of $h_v/d_{v,i} \approx 0.35$, an increase in stroke no longer brings an increase in flow because the valve cross-section no longer increases.

If the free valve cross-section is known, the valve seat diameter can be easily calculated (1.7):

$$d_{v,i} = \frac{A_v}{\pi \cdot h_{v,\max} \cos(\beta)} - h_{v,\max} \cos(\beta) \sin(\beta)$$

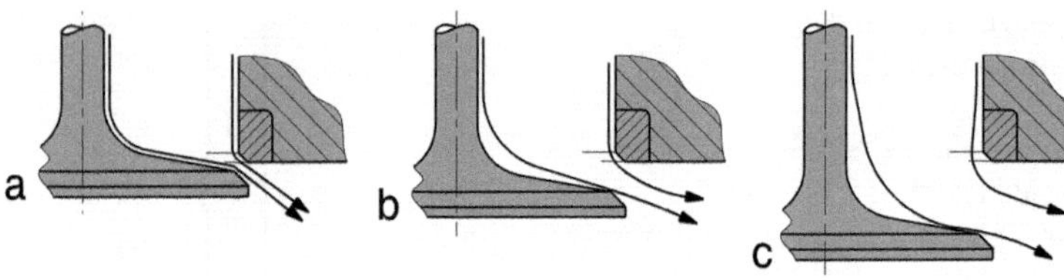

Fig. 1.41 Inlet flow in the valve gap [34]. (**a**) Small stroke: The jet fills the gap. The flow is at the valve and at the seat ring, (**b**) medium valve lift. The flow partially detaches, (**c**) large valve lift. A free jet is formed

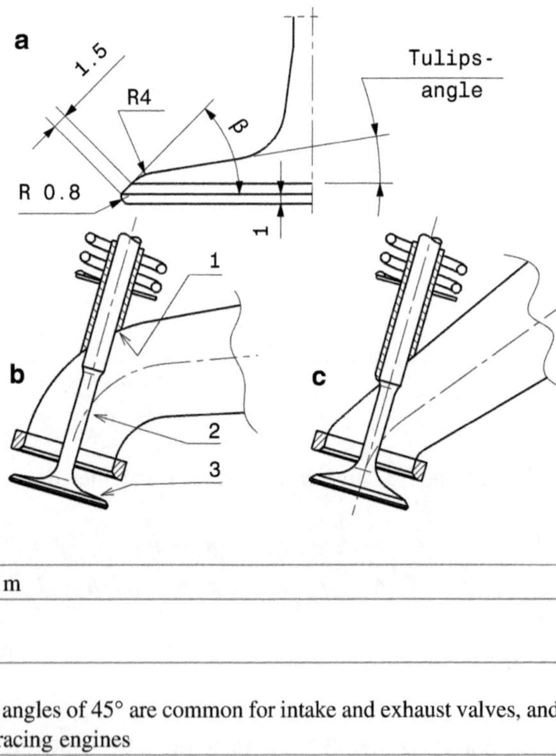

Fig. 1.42 Design of an inlet valve for high gas velocities. (**a**) typical shape of a valve head with dimensions (valve diameter about 35–40 mm). Valve seat width 1.5 mm. Further measures to improve flow (b): **1** Valve guide flush with the port, **2** Valve stem tapered in the port area, **3** Tulip angle adapted to port angle. (**b**) small tulip angle for large port angles (over 60°) [36]. (**c**) large tulip angle for small port angles (below 45°) [36]

$d_{v,i}$	Smallest valve seat diameter, m
$h_{v,\max}$	Maximum valve lift, m
β	Valve seat angle, ° In passenger car engines, seat angles of 45° are common for intake and exhaust valves, and 50–55° can also be found in racing engines

In the first approach, the exhaust port diameter and its valve diameter in the naturally aspirated engine are about 15% smaller than those of the intake port. After all, the outflow occurs at a greater pressure difference and therefore the smaller valves do not represent a flow disadvantage. In addition, the temperature load is reduced by the smaller heat-absorbing surface. For further ratios of valve sizes, see also Table 1.4.

The flow in the valve gap depends not only on the gas flow rate but also on the valve stroke, Fig. 1.41.

In order to keep the flow losses small at high gas velocities, the valve head and the area around the valve should be designed as shown in Fig. 1.42. The valve heads of racing

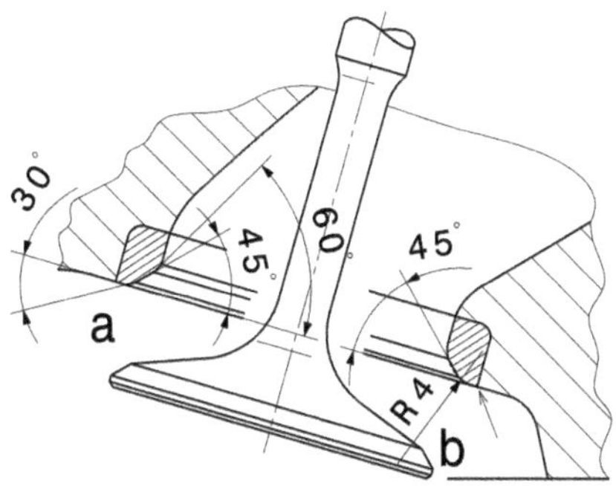

Fig. 1.43 Design of inlet valve seat rings. (**a**) Version with three chamfers to increase the flow rate [36], (**b**) Version with one rounding. The valve seat is machined to 45° after the seat ring has been pressed in. If the valve tends to backlash, the seat is changed to 50–55°

engines are flat on the combustion chamber side, i.e. they do not have a concave recess which would increase the heat-absorbing surface. Polishing the poppet also reduces the surface area and also reduces deposits.

The design of inlet valve seat rings for the highest speeds is shown in Fig. 1.43b. For comparison, a seat ring for favourable turbulence formation (flame speed) is juxtaposed, as is favourable for racing engines with medium speed (up to about 10,000 min^{-1}). In addition to the design of the seat environment, it is important in each case that the seat ring is flush with the combustion chamber wall (arrow) and does not protrude or is set back.

Materials

Seat rings are made of copper-beryllium alloys. Valve guides made of bronze or copper-beryllium alloy, valves made of steel (chrome-manganese steel), Nimonic (nickel superalloy), titanium alloys, titanium aluminide (TiAl) or ceramic (silicon nitride). Exhaust valves can also have a hollow stem with sodium filling for better heat dissipation.

Exhaust *Port*

Exhaust ports must be well cooled and their length in the cylinder head should be as small as possible (heat input). Fig. 1.44 illustrates design recommendations.

Exhaust seat rings must also be cooled well. The valve releases the majority of the absorbed heat via the seat. A much smaller part of the heat dissipation takes place via the valve guide. Seat rings of highly loaded engines are also forced-cooled with oil or water, Fig. 1.45. For example, in the Ferrari V6 1.5-liter Formula 1 engine, intake and exhaust seat rings were oil-cooled.

A direct comparison between series and racing ducts is provided by Fig. 1.46. The engines of the DTM are in fact derived from series units. Based on the existing valve angles and distances, the ducts of the racing engine are designed. The ducts are already merged in the cylinder head. However, there must still be space between the exhaust ducts for the

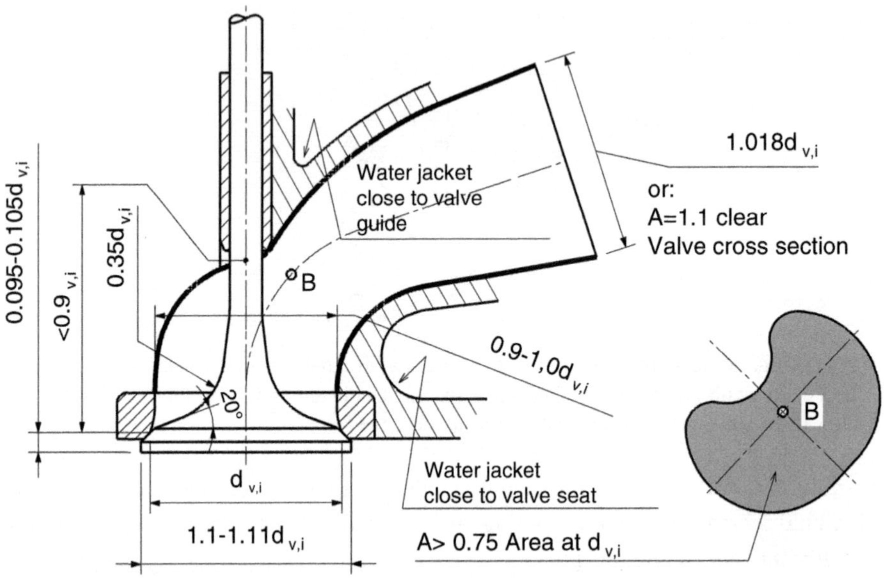

Fig. 1.44 Design recommendations for the exhaust port of a passenger car engine. The dimensions are given as a function of the smallest valve seat diameter $d_{v,i}$. It is important that the water jacket reaches critical areas well

water jacket. The gas flow is increased first by larger valve diameters compared to the series and then by adapted, straighter port routing, especially for large valve lifts.

The cooling water jacket in production engines "results", so to speak, from the space left by the combustion chamber, charge exchange ducts, screw sockets and spark plug pipes minus a wall thickness. In racing engines, the water jacket is designed specifically to meet

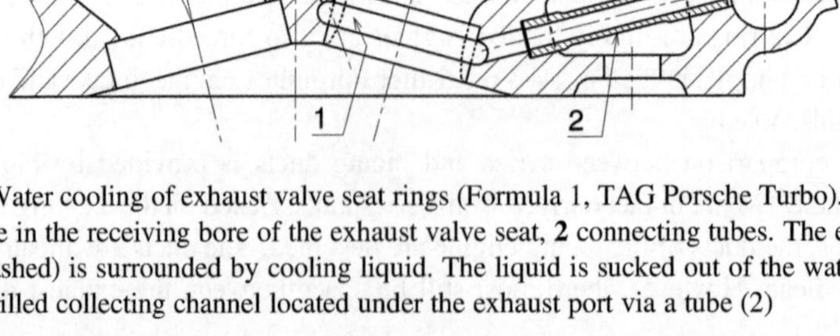

Fig. 1.45 Water cooling of exhaust valve seat rings (Formula 1, TAG Porsche Turbo). **1** circumferential groove in the receiving bore of the exhaust valve seat, **2** connecting tubes. The exhaust valve seat ring (dashed) is surrounded by cooling liquid. The liquid is sucked out of the water jacket and enters the drilled collecting channel located under the exhaust port via a tube (2)

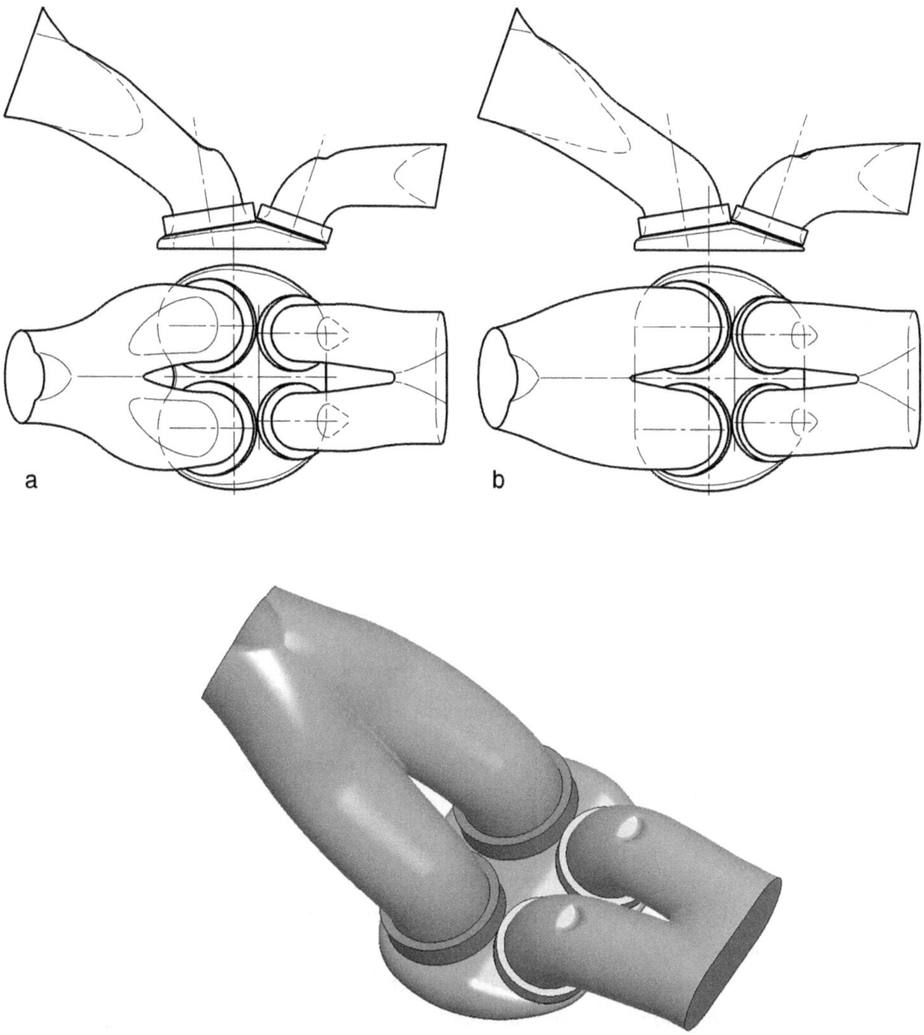

Fig. 1.46 Gas exchange ports with combustion chamber, after [29]. (**a**) series engine, (**b**) racing version derived from it (this is also shown shaded). The inlet side is on the left. The upper recess at the intake area of the intake ducts is used for the injection valve. The gas flow rate is increased compared to the standard version, especially with larger valve lifts. The associated combustion chambers are shown in Fig. 1.29

the needs. It is kept as small as possible so that the flow velocity of the water is high and dead water areas are prevented. This is achieved, for example, with two-part water jackets in which an almost horizontal partition divides the water space into a part near the combustion chamber and an area above it. In addition, holes with a diameter of approx. 5 mm guide the coolant specifically to *hot spots* (see Fig. 1.47 right cylinder head). This

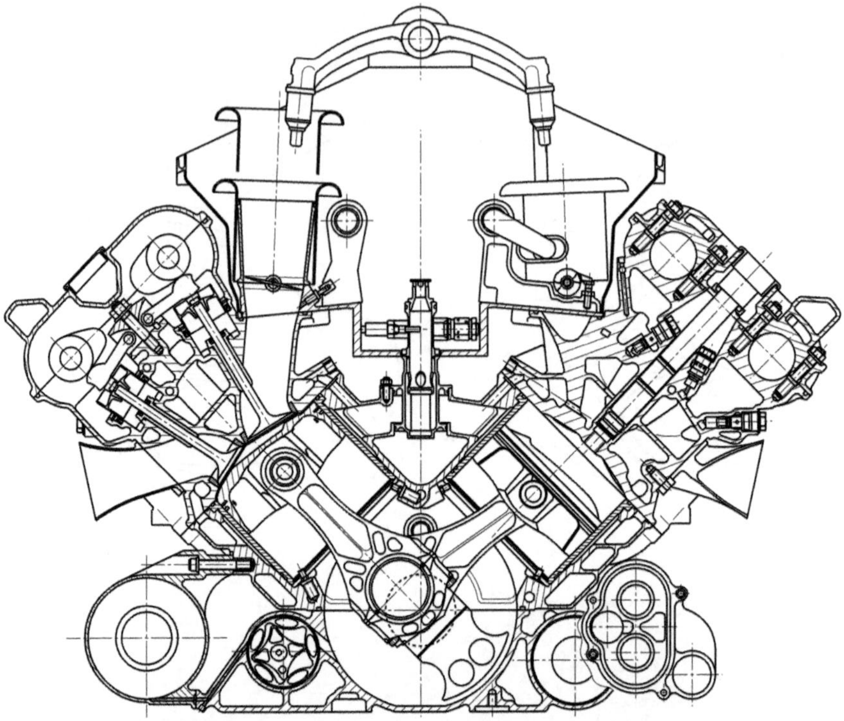

Fig. 1.47 Formula 1 3-l V10 engine (Ferrari type 049). 90° V-engine. Power 610 kW at 17,500 min^{-1}, torque 350 N m at 15,500 min^{-1}

so-called precision cooling is used, among other things, to cool the exhaust valve seat areas and spark plug seats. The combustion chamber wall is ribbed. This guides the coolant and increases the heat-dissipating surface. Constraints caused by casting technology are avoided by skeleton construction. The cylinder heads are cast partially open at the sides, i.e. with an interrupted side wall, and only before mechanical machining is the water space closed by welding in thin sheets.

In racing engines, the so-called cross-flow cooling is preferred. The coolant enters each cylinder separately on one side of the cylinder head, flows around the respective combustion chamber roof and exits again on the opposite side. In this way, each cylinder experiences the same cooling effect. This is not the case with longitudinal flow cooling, which is preferred in series engines for reasons of simplicity. In this case, the coolant flows through the cylinder head in the longitudinal direction, resulting in a correspondingly unfavourable temperature rise in the successive cylinders.

Materials
Cylinder heads are made of aluminium alloys, series parts of standard casting alloys (in order of falling strength at 250 °C): EN AC-AlSi6Cu4 (was G-AlSi6Cu4), EN AC-AlSi8Cu3 (was G-AlSi9Cu3), EN AC-AlSi7Mg0.3 (was G-AlSi7Mg), EN AC-AlSi10Mg (was G-AlSi10Mg) each T6 heat treated.

Although high-temperature alloys have unfavourable processing properties, they enable highly stressed cylinder heads: G-AlSiCu5Ni1, 5CoSbZr, G-AlCu4MgTi, G-Al2MgTi.

Particularly in the vicinity of the combustion chamber, great importance is attached to a fine-grained structure (small dendrite arm spacing). This is achieved by targeted, local cooling of the casting tool in the combustion chamber area.

Production
Cylinder heads and cylinder head covers are cast. Due to the small number of pieces, processes such as sand casting and investment casting are suitable. The gas exchange ports are represented with sand cores or merely pre-cast and partially or completely machined to the final shape.

Series parts are manufactured using gravity die casting, die casting or core packaging processes.

Cylinder Head Cover
The oil chamber of a cylinder head is closed off at the top by the cylinder head cover. If the engine is installed in the vehicle in a co-supporting manner, the hoods of V-engines usually contain engine mounts because they provide two points far away from the lower end of the engine. However, this requires the hood to have appropriate structural rigidity and tie-ins to the cylinder head. This can be achieved by integrating at least some camshaft bearings into the hood. To adjust the timing, the bearing caps closest to the drive are "conventionally" bolted to the cylinder head. Then, during assembly, the camshafts are fixed by this one bearing and subsequently the cylinder head cover can be fitted with the remaining bearing locations.

The hoods are cast from aluminium or magnesium alloys. For strength reasons, highly stressed hoods are also milled from the solid.

Minimum wall thickness of non-load-bearing areas for aluminium alloys 2 mm, for magnesium alloys 1.8 mm.

1.4.2 Valve *Train*

Poppet valves are opened and closed by a cam directly or by means of a transmission element (rocker arm, finger follower). A selection of common possibilities together with

Table 1.6 Comparison of valve train concepts

Without cam roller, with hydraulic valve lash adjustment:				
Criterion	Concept			
			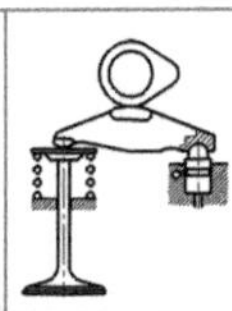	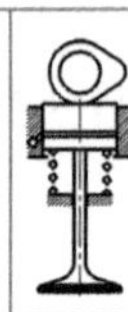
	a	b	c	d
Friction	–	–	Ø	Ø
Mass	–	+	++	+
Stiffness	–	Ø	+	++
Height	++	+	Ø	Ø
With cam roller, with hydraulic valve lash adjustment:				
Criterion	Concept			
			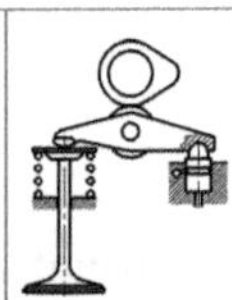	
	e	f	g	h
Friction	+	+	++	+
Mass	–	Ø	+	–
Stiffness	–	Ø	+	++
Height	++	+	Ø	–

Legend: ++ very good, + good, Ø average, – unfavourable, – very unfavourable

their most important characteristics is given in Table 1.6. One cam revolution corresponds to a complete cycle (two revolutions) of a four-stroke engine. Accordingly, a reduction of 2:1 between crankshaft and camshaft is required. The camshafts are driven from the crankshaft mostly by gears, more rarely by toothed belt or chain. In individual cases, a shaft (king shaft with bevel or crown gears) ensures synchronous transmission of the rotary motion. Toothed belts have a low mass and are easy to replace. Belt drives are often used to overcome large centre distances. Especially for high-speed motors, a pure gear drive is the optimum, Fig. 1.48. With paired gears, backlashes in the range of hundredths of a millimetre are possible. This results in precise and, above all, speed-resistant valve timing. A gear drive also makes it relatively easy to ensure several auxiliary drives.

The gears run on fixed removable axles with plain bearings or needle bearings. A removable axle can also be designed as an eccentric and thus ensure adjustability of the gear backlash. The transition to the cylinder head is a suitable location for such an intermediate gear, because the greatest tolerance jump can be expected at this point due to the cylinder head gasket.

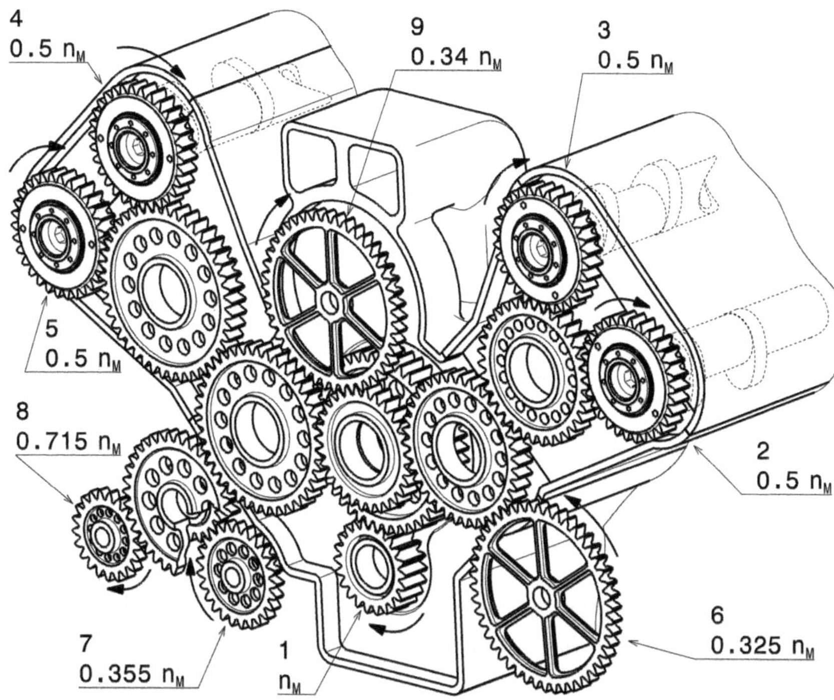

Fig. 1.48 Timing gear Ferrari 3.0 l V10 (Tipo 049). The gear train is located at the front end of the engine and, in addition to the four camshafts, also drives all the ancillary units at a speed adapted to their efficiency. **1** crankshaft ($n_{M,n}$ = 17,500 min^{-1}), **2** exhaust camshaft left, **3** intake camshaft left, **4** intake camshaft right, **5** exhaust camshaft right, **6** oil pressure pump and suction pumps, **7** suction pumps, **8** air separator, alternator and hydraulic pump, **9** water pump

Valve trains can be arranged at the front end of the crankshaft or on the flywheel side. There are also engines with centrally arranged camshaft drive. The flywheel-side extraction of the drive torque is favorable for vibration reasons, because the output takes place in a torsional vibration node of the crankshaft. However, installation and maintenance of the timing drive are more complicated than at the front end of the engine. The arrangement at the front end also has advantages for the rigidity of the engine if it is installed as a mid-mounted engine.

At extremely high speeds (over 17,000 min^{-1}) and long camshafts (such as a 10-cylinder V-engine), the torsional vibrations induced by the cam forces, in addition to possible fractures, cause the timing to fluctuate unacceptably. This is particularly noticeable in the cylinder furthest from the drive gear. To solve this problem, vibration absorbers or dampers are required in the valve train [37]. In the simplest case, an elastomeric element vulcanized into an intermediate gear is sufficient.

For racing engines, only the finger follower (concept c or g) and the bucket tappet (d) can be considered, both without hydraulic valve lash adjustment, due to their high

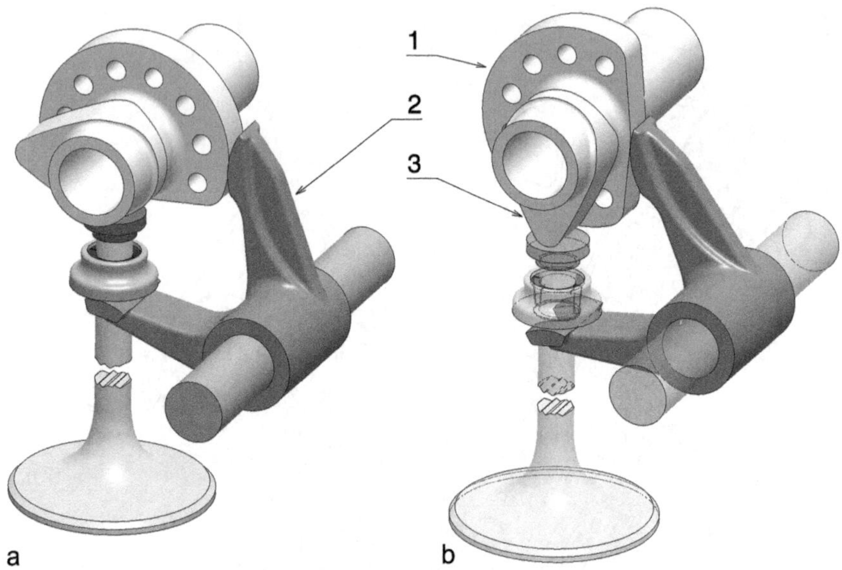

Fig. 1.49 Forced valve control with one camshaft. (**a**) Valve closed, (**b**) Valve fully open. 1 Closing *cam*, 2 Closing lever, 3 Opening cam. A cap at the end of the valve stem is the pick-up for the opening cam. Small plates are inserted between the cap and the stem to adjust the valve clearance towards the cam. To adjust the clearance to the closing lever, its axis must offer adjustment possibilities in two directions (double eccentric bearing)

rigidity and low moving masses. Cam followers allow the narrowest valve angles and thus the most compact combustion chambers. Many variable valve train systems are also based on finger follower drives. In contrast, bucket tappets offer the fullest valve lifts and thus the potential for the highest performance, but require wider valve angles and have larger oscillating masses.

A special feature is the override control (desmodromic). In this case, the valve is not only pushed open by a cam directly or via a lever, but is also pulled closed again via a second lever, which is actuated by another cam. In theory, this would mean that no valve spring would be required and that this type of actuation would be the first choice for the highest speeds, which was also the case in the past. In fact, however, springs are needed to make the system work at low rpm and at cold start. Also, a lot more parts are needed, and they have to be manufactured to tight tolerances on top of that. In the case of turbocharged engines, there is the additional complication that the intake valve may not be opened by the boost pressure under certain operating conditions. For this case, the valve spring must be dimensioned stronger than required for cold start. Figure 1.49 shows an example of a positive control system that makes do with one camshaft.

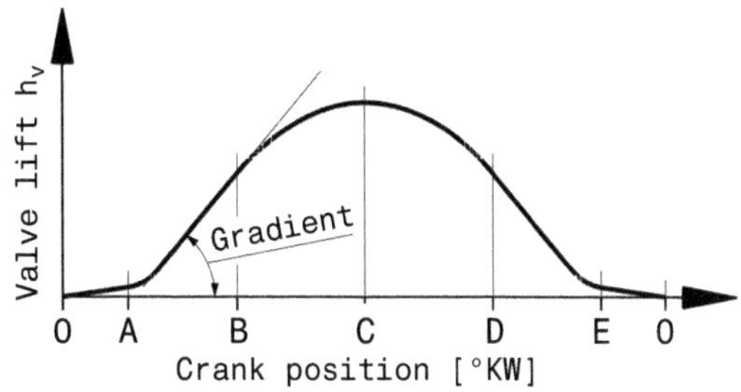

Fig. 1.50 General valve lift curve. The valve lift is plotted against the crank angle and is divided into six characteristic phases

Valve *Lift*

The valve lift curve is generated in cooperation between the cam and the pick-up or transmission element. The cam shape is determined by the shape of the pick-up. For a given valve lift curve, a pickup with a flat sliding surface produces a completely different cam shape than a pickup with a roller or circular arc surface.

A general course of the valve lift is shown in Fig. 1.50. The course begins with a start-up ramp from 0 to A, which is represented by an ascending straight line or, better, by a sine line, and on which the valve clearance is lifted. If there were no run-up, then the valve would be abruptly flung up at even a small valve clearance, the pick-up would leave the cam, and both parts would smash in. The start-up ramp must initiate a bumpless start to valve lift. Some utility engines have a very long starting ramp for safety reasons, so that larger differences in valve clearance cannot be dangerous. However, a short steep ramp is better.

After the start-up ramp, the actual valve lift begins at *A*. This is followed by a steeply rising acceleration lasting a short time up to point *B* and, after a short settling section, a slow deceleration lasting up to cam tip C. After point *B*, the valve spring already comes into action, which holds the valve or the transmission element, which wants to continue moving in a straight line, on the cam. The return movement of the valve from the cam tip must be provided by the valve spring. If the spring tension is too low, the transmission link leaves the cam contact, jumps higher and only hits the cam again later. The faster the return movement of the valve is to take place, the higher the spring force required. A valve lift curve with a large peak arc requires less spring force and offers greater speed reliability. During the closing movement, the valve follows the same symmetrical curve, is then strongly decelerated from *D* to *E* and gently touches down on its seat. Without a deceleration ramp (closing ramp), the valve would hit its seat abruptly, jump up again and extend the control time in a disadvantageous way.

The valve movement is composed of acceleration and deceleration curves. A uniform movement, which would be represented by a longer straight line, does not exist with an optimum valve lift.

The valve lift curve is judged by the area it encloses. Characteristic of a high-performance engine is a large area achieved by steep valve lift and large peak radius, but

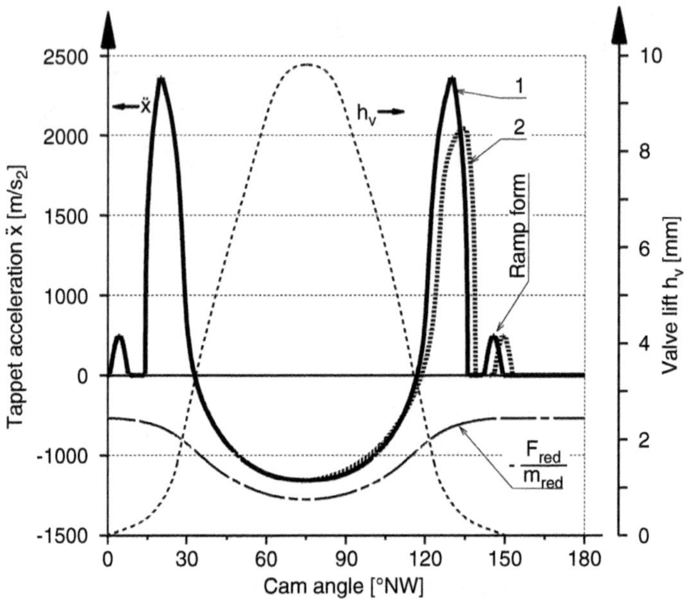

Fig. 1.51 Accelerations in the valve train. **1** inlet valve, **2** exhaust valve. The acceleration curve is influenced by the ramp shape and the cam shape. Exhaust valves can close more smoothly. The reduced acceleration F_{red} /m_{red} must be greater than the maximum deceleration at all speeds so that the transmission element does not lift off the cam

Table 1.7 Requirements for valve lifting [38]

	Opening ramp as short and steep as possible	Main cam opening phase: Fast cross-section release	Main cam closing phase: As steep as possible	Closing ramp as short and steep as possible
Inlet	Good idle quality, low idle consumption	High air expenditure	Reduce backflow (air consumption)	Reduce reverse flow, torque at low speeds
Exhaust	Minor finishing operation	Low pushing-out operation		Good idle quality, low idle consumption

the length of the timing period must not exceed a certain level. Steep valve lifts include an angle (= pitch = lift speed) up to 55° at the steepest point (for scale valve lift to crank angle (CA) = 10:1) [32]. This corresponds to 0.143 mm/°CA or 0.286 mm/°NW.

The exhaust valve should be opened quickly. In the case of "creeping" opening, the valve is heated too much by the gases flowing through it. When closing, the cam is kept flatter compared to the inlet, because only small masses flow out in this phase, Fig. 1.51. In summary, Table 1.7 summarises the most important requirements for the valve lift.

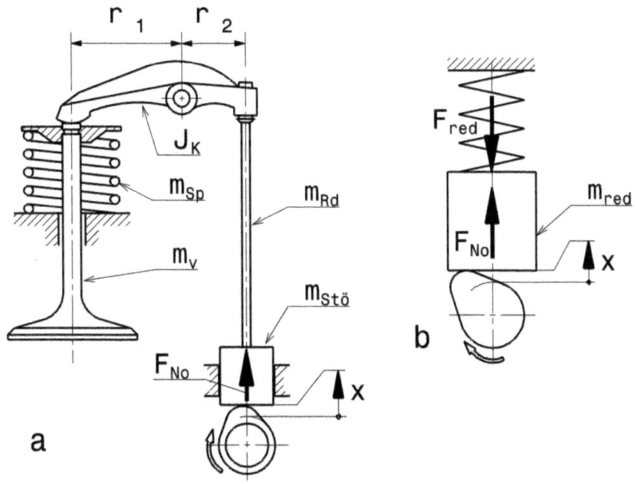

Fig. 1.52 Masses and forces in the valve train. (**a**) real system, (**b**) substitute system. The real system is summarized by "reduction" to the cam side to a simple replacement system

An equivalent system is used to determine the inertia forces and thus the required valve spring force, Fig. 1.52.

The valve spring force must ensure constant contact between the cam and the pick-up. The force F_{No} on the cam follows from the dynamic equilibrium consideration:

$$F_{No} = F_{Sp}\frac{r_2}{r_1} + \left[m_{Stö} + m_{Rd} + \frac{J_K}{r_1^2} + m_v\left(\frac{r_2}{r_1}\right)^2 + \frac{m_{Sp}}{2}\left(\frac{r_2}{r_1}\right)^2\right] \cdot \ddot{x}$$

F_{No}	Force on cam, N
F_{Sp}	Valve spring force, N
$m_{Stö}$	Mass of the tappet, kg
m_{Rd}	Mass of the push rod, kg
m_v	Valve mass, kg
m_{Sp}	Spring mass, kg
J_K	Mass moment of inertia of the rocker arm, kgm^2
r_1, r_2	Lever lengths of the rocker, m
x	Tappet stroke, m

Summarizing this equation according to the equivalent system (Fig. 1.52), we obtain the relation:

$$F_{No} = F_{red} + m_{red}\ddot{x}$$

F_{red}	Reduced valve spring force, N
m_{red}	Reduced mass of the valve train, kg
$\ddot{x}$	Tappet acceleration, m/s^2

From this follows directly a condition for the minimum spring force so that no lift-off of the pick-up ($F_{No} = 0$) is possible at a certain valve acceleration:

$$F_{red} > -m_{red}\ddot{x} \qquad \text{bzw.}\ddot{x} < -F_{red}/m_{red}$$

These ratios are shown in Fig. 1.51 as an example for a camshaft speed of a valve train.

Some design values: Ramp slope about 0.0084 mm/°NW, ramp stroke about 0.25 mm. Speed at the end of the pre-cam: For series engines max. 0.3 m/s [39], for racing engines 0.5–1 m/s. The largest acceleration (second derivative of the elevation curve) is 0.018 mm/°NW^2 [40].

Racing cams exhibit average decelerations at the cam tip of 2500–3800 m/s^2 [32]. At 4000 min^{-1} camshaft speed, this corresponds to a value of 0.0043–0.0066 mm/°NW^2. It is reported that peak values of 17,000 m/s^2 acceleration and 8000 m/s^2 deceleration were achieved by the historic Mercedes Formula 1 car W196 (first use 1954) with desmodromic valve control [41]The nominal speed of the 2.5-l two-valve engine was 8500 min^{-1}.

The correlations between the various units are provided by an analytical consideration of the valve lift curve h_v (φ):

$$\dot{h}_v = \frac{dh_v}{dt} = \frac{dh_v}{d\phi} \cdot \frac{d\phi}{dt} = h_v' \cdot \omega_{No}$$

h_v	Valve lift, mm
φ	Cam angle, rad
ω_{No}	Circular frequency of the camshaft, s^{-1}
$\dot{h}_v$	Valve lift speed, mm/s
h_v‘	Valve lift speed (first derivative of the elevation curve), mm/rad. 1 mm/° ≡ 57.29 mm/rad (1 rad = 180°/π = 57.29°)

$$\ddot{h}_v = \frac{d^2h_v}{dt^2} = \frac{d^2h_v}{d\phi^2} \cdot \omega_{No}^2$$

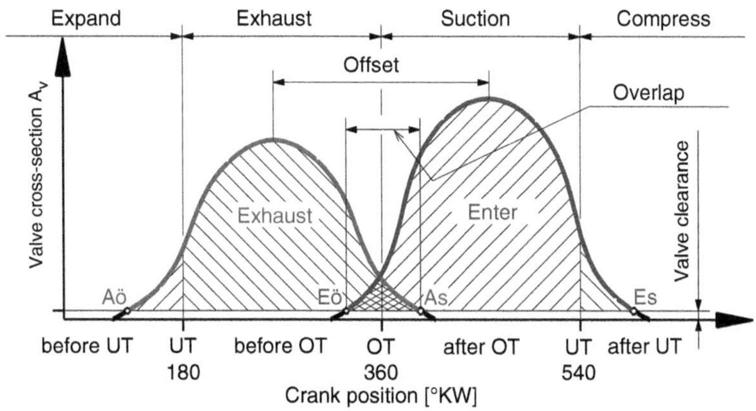

Fig. 1.53 General *timing* diagram of a four-stroke engine. The valve lifts and thus the valve openings are plotted above the crank angle. UT *bottom dead centre,* OT *top* dead *centre,* Aö *exhaust opens,* As *exhaust closes,* Eö intake opens. It intake closes. Important characteristic quantities are *offset* and *overlap*

$\ddot{h}_v$	Valve lift acceleration, mm/s^2
h_v	Valve lift acceleration (second derivative of the elevation curve), mm/rad^2. 3283 mm/rad^2 $\equiv$ 1 mm/°^{2}

Max. Compressive stress (Hertzian pressure) at the cam tip: For long-life large engines 700–800 N/mm^2. Passenger car engines run at pressures of around 1600 N/mm^2. For racing engines, values of 2100–2200 N/mm^2 are achieved.

Valve Timing

If the valve lifts of exhaust and intake are plotted against crank or cam angle (timing angle), the timing diagram is obtained. However, the control diagram also shows the valve cross-sectional area (valve opening area) as a function of the control angle, Fig. 1.53. Large valve cross-sections are aimed for to achieve a high charging efficiency and low throttling losses.

However, it is even more important to look at the valve lift curve, because it depends on how fast and how high a valve is lifted. This valve lift curve is the most important graphical representation for assessing the gas exchange processes, see section Valve *Lift* above.

The timing affects the processes of charge change and thus the engine characteristics in the following ways, [13] and [42]:

- The intake closing Es influences the filling and thus the torque characteristics much more than the other control times, see Fig. 1.54, where early Es means high torque in the lower rpm range but filling losses at higher rpm, and late Es means high rated power but filling losses at low rpm (sports engine).

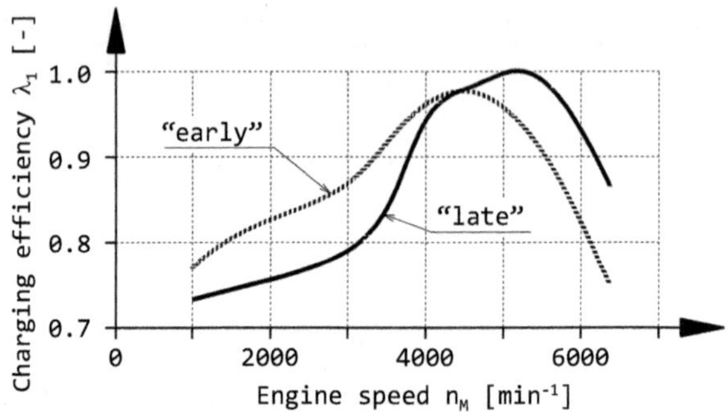

Fig. 1.54 Influence of inlet closure Es on the charging efficiency λ_l [42]. The measurement was carried out on an 8-cylinder gasoline engine with 4 valves per cylinder. By adjusting the intake camshaft by 20 °CA towards the rear, there is a clear reduction in the delivery rate in the lower speed range. On the other hand, the amount of charge increases at high engine speeds

- At low engine speeds and full load (open throttle), the mass flow at the intake valve follows the piston excitation (movement). In order to avoid charge losses (pushing back into the intake tract), the closing of the intake valve should be as close as possible to BDC. Also the valve overlap in the charge change TDC should be small to minimize the residual gas content in the fresh gas.
- At high engine speeds and full load, large valve opening areas lead to throttling and promote dynamic reloading if the intake valves are open for a sufficiently long time. This forces a late intake closure.
- At idle and part load, a late intake opening reduces the valve overlap, thus reducing the backflow of exhaust gas into the intake tract (residual gas content) (lower residual gas content leads to better energy conversion and thus fuel consumption benefits primarily through faster charge burn-through and reduced cycle fluctuations).
- A large valve overlap causes higher purging losses, which reduces the effective efficiency. However, the associated improved residual gas purging results in better cylinder filling and thus higher performance.
- Early Aö results in high losses of expansion work, but reduces the amount of push-out work required.

The control times are also recorded in simplified form as a "control diagram". This shows the points at which the valves lift from their seats and touch down again, Fig. 1.55.

Table 1.8 shows the valve timing of some utility engines as well as sports and racing engines. It should be noted that multi-cylinder engines with only one carburettor can tolerate almost no valve overlap, and that, for example, four-cylinder engines with two carburettors on Y-shaped intake pipes connected by a balancing pipe may have only very slight valve overlap. If, a higher performance is to be achieved, a separate carburettor (or a separate injector) is absolutely necessary for each cylinder. As can be seen from the table, the total timing for both the intake and exhaust valves averages 240–265° for a utility engine, up to about 320° for a sports engine, and usually between 320 and 360° for racing

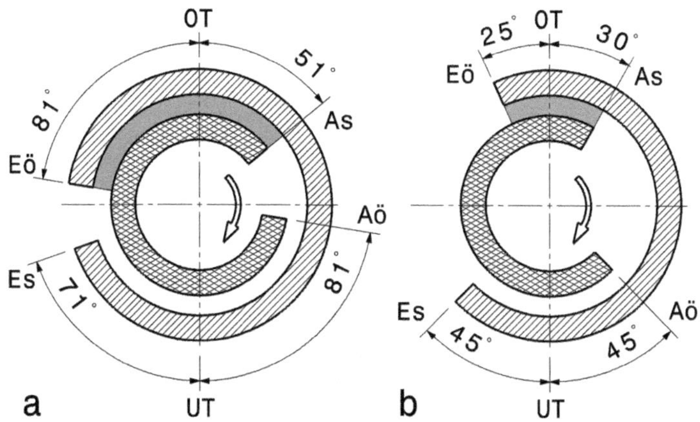

Fig. 1.55 Timing of a naturally aspirated racing engine (Porsche Formula 1, 1960s) and a production engine in °CA, after [13]. (**a**) racing engine, (**b**) series engine. The valve overlap is the gray colored area

Table 1.8 Control times of different engines in degrees of crank angle [32]

	Series engines			Sports Engines			Racing engines		
Intake opens before TDC	5	20	25	40	50	55	60	95	104
Intake closes after BDC	40	60	55	80	90	85	80	105	104
Exhaustopens before BDC	50	65	55	80	90	85	90	110	100
Exhaust closes after TDC	5	10	25	40	50	55	60	90	80

engines, but may be greater. It should be noted that cylinder heads with smaller valve cross-sections (i.e. two-valve heads) require longer timing than four-valve heads with larger valve cross-sections.

Utility engines operate with valve timings that ensure high torque and low fuel consumption and long service life even at low engine speeds; no emphasis is placed on maximum power. These relatively "tame" valve timings do not allow good cylinder filling at higher engine speeds.

Figure 1.56 compares the valve lifts of a production engine with a Formula 3 power unit derived from it. The valve stems are reduced from 7 to 6 mm diameter in the racing engine. The camshaft base circle is reduced from 34 to 30 mm so that the required valve lift is achieved without exceeding the maximum surface pressure.

Transmission Elements (*Follower*)

For racing engines, bucket tappets and rocker arms are used. To reduce friction and wear, the sliding surfaces are DLC-coated (*diamond like carbon*). These coatings are less than 5 μm thick with a hardness of over 3000 HV (Vickers hardness grades). The surface on which a DLC coating is applied must be polished, otherwise the contouring DLC coating

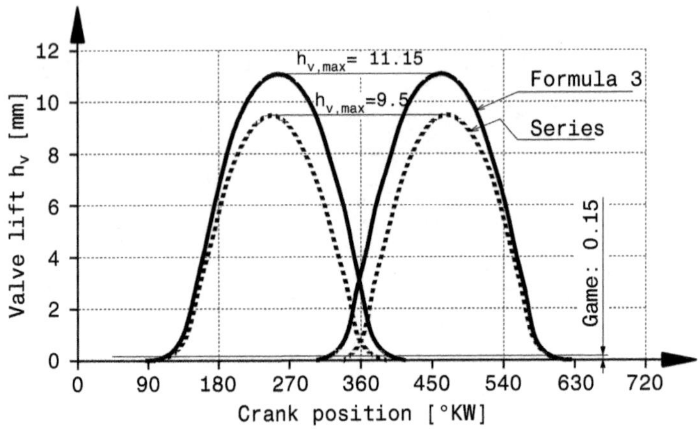

Fig. 1.56 Comparison of control times between series engine and derived racing engine, after [43]. Series engine: 2.0 l 4-valve. Racing engine: Formula 3. Valve lifts (11.15 mm compared to 9.5 mm), valve lift, offset, overlap and timing of the racing engine differ greatly from the series engine. Control times, series: Aö 60 °CA v BDC; As 32 °CA n TDC. Eö 20 °CA v TDC; Es 72 °CA n BDC. Racing engine: Aö 58 °CA v BDC; As 28 °CA n TDC. Eö 33 °CA v TDC; Es 53 °CA n BDC

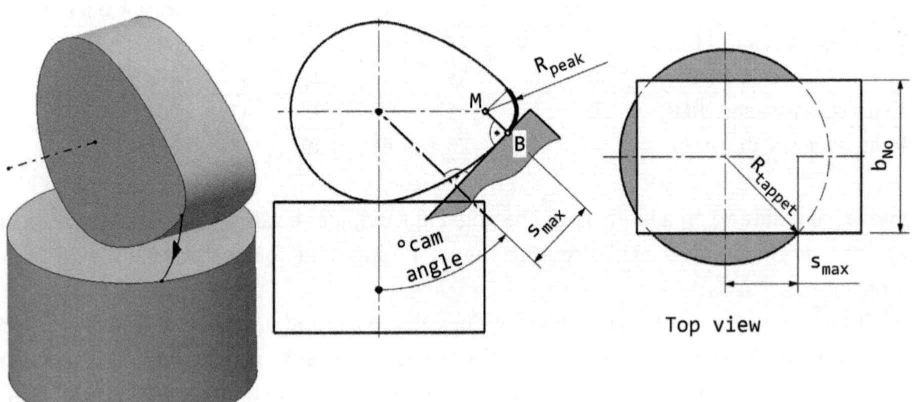

Fig. 1.57 Diameter of bucket tappet. The diameter must be so large that the cam contact point *B* always lies on the bucket bottom. The position shown is that at which the arc of a circle (R_{peak}) just touches the bucket and the greatest excursion s_{max} is achieved. Point *B* is therefore also the connection point of this circular arc to the rest of the cam flank

has an extremely abrasive effect. Coefficients of friction between DLC surfaces and steel mating surfaces are 0.1 and for both running partners DLC coated they are half of this [44].

Bucket Tappet The minimum diameter of the tappet depends on the cam shape and the cam width, Fig. 1.57.

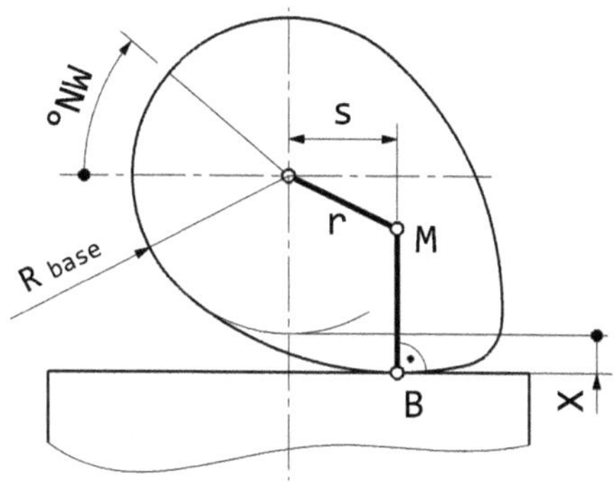

Fig. 1.58 Kinematics of the ram stroke. If the cam rotates (cam angle degrees, °NW), the tappet moves (path *x*) as soon as the cam contour leaves the base circle (radius R_{base}). The tappet movement caused by the cam can be replaced by a scotch yoke train. The crank (*r*) extends to the centre of curvature *M of* the cam contour with the respective point of contact *B*. With uniform rotation of the cam, a large excursion *s of* the point of contact *B* leads to a high stroke speed of the tappet

When the cam rotates, it touches the tappet at point B. In the picture, the cam is held in place for illustration and instead the tappet is swivelled in the opposite direction (degree cam angle °NW). In the process, the point of contact B moves out until the arc of a circle begins with the tip radius $_{\mathrm{RStip}}$ and the greatest migration s_{max} is achieved. With further rotation B remains – as long as the tip arc touches the tappet – at the same position of the bucket, because the centre of curvature M is constant. As soon as the arc runs off, other centres of curvature dictate the position of the point of contact and this moves back towards the centre of the bucket. The smallest diameter of the ram also depends on the width of the cam:

$$R_{tappet} = \sqrt{s_{\max}^2 + \left(\frac{b_{No}}{2}\right)^2}$$

R_{tappet}	Tappet radius, mm
$s_{\max}$	Largest excursion of the point of contact, mm. S. Figure 1.57
b_{No}	Cam width, mm

The tappet diameter is also important for the maximum valve stroke speed, Fig. 1.58.

For high speeds, large time cross-sections for the charge change are expedient. This requires large valve strokes for a given opening time (control time) and consequently leads to large valve lift speeds. In turn, the tappet diameters must be appropriately large for this.

Rocker Arm

(Finger Follower). Like tappets, rocker arms take over the lateral force of the cam, but have less moving mass. They can also be equipped with a roller relatively easily and thus significantly reduce friction in the valve train. In addition, a height-saving transmission ratio between cam and valve lift can be implemented. Common lever ratios range from 1.2 to 1.6. Especially with short levers, the design of the contact surface to the valve stem is critical in ensuring that no lateral force acts on the stem. The centers of curvature of the contact contour should always be on the extended valve axis when the lever is pivoted (rolling motion). In the case of a circular contour, the center should be on the valve axis at half the valve stroke.

The *valve lash* is adjusted by means of small plates or caps that are fitted to the valve stem. This saves mass compared to the convenient solution of the self-acting hydraulic balancing elements of series engines and at the same time increases the rigidity in the valve train. Levers often contain a small hole that sprays lubricating oil onto the engagement surface.

For values of maximum *surface pressure*, see section *Valve* Lift.

Valve Spring

The main task of the valve spring is to maintain contact between the cam and the pick-up at all speeds, see section Valve *lift*. Limits in the design result from the limited installation space and from the limiting speed at which lift-off occurs between the pick-up and the cam. In addition, the natural frequency of the valve train must be above the maximum speed of the camshaft (= 0.5 engine speed). To reduce friction losses, an attempt is also made to keep the spring mass as small as possible by optimizing its shape. Thus, springs with variable pitches or variable coil diameters are wound. The spring wire can have an egg-shaped profile to even out the stress distribution. Spring characteristics are also advantageously made progressive by such measures. Fig. 1.59 shows an example of a valve spring.

The speed limit of an engine with a steel valve spring is approx. 16,000 min^{-1}. Higher crankshaft speeds can be achieved with a pneumatic spring, for example. This has a strongly progressive characteristic curve with an extremely low inherent mass. The pneumatic valve spring has not yet found its way into series engines because of the high cost. Figure 1.60 contains a system overview.

The actual air spring is formed by a piston (7) running in a cylinder (9). A pressure vessel (1) with a volume of approx. 0.5–0.7 l under approx. 300 bar ensures the air supply with 10–20 bar via a differential pressure valve. The check valve (3) is only closed by the gas spring. When the air is compressed by the cam, the pressure in the cylinder (9) rises to approx. 95 bar, the temperature to approx. 300 °C. The engine may only be started when the system is completely filled. If the tank content is not sufficient for a race distance, it must be filled during a pit stop.

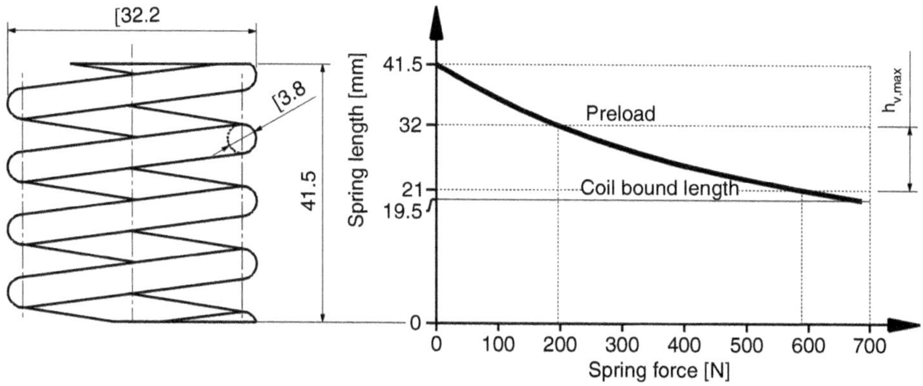

Fig. 1.59 Valve spring of a high-performance engine (Porsche 911), after [32]. This spring has a slightly progressive characteristic curve. Force at preload (9.5 mm): 196 N. Force at max. Valve stroke ($h_{v,\max} = 11$ mm): 589 N. A second spring is installed inside the motor. The total forces are then 334 resp. 961 N

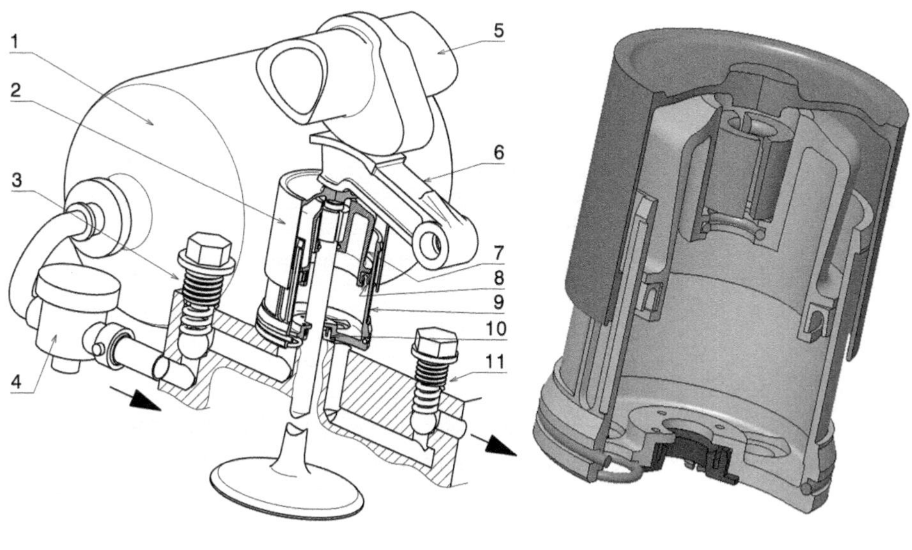

Fig. 1.60 Functional principle of a pneumatic valve spring. **1** *Pressure reservoir,* **2** *bucket tappet,* **3** Check valve *(one-way valve),* **4** *Inlet* regulator, **5** Camshaft, **6** *finger follower,* **7** Piston, **8** Piston seal, **9** Cylinder, **10** Valve *stem seal,* **11 Pressure** relief *valve*

The air spring force results from the compression of the enclosed air volume by the cam. In the case of a polytropic change of state, the following applies to the force:

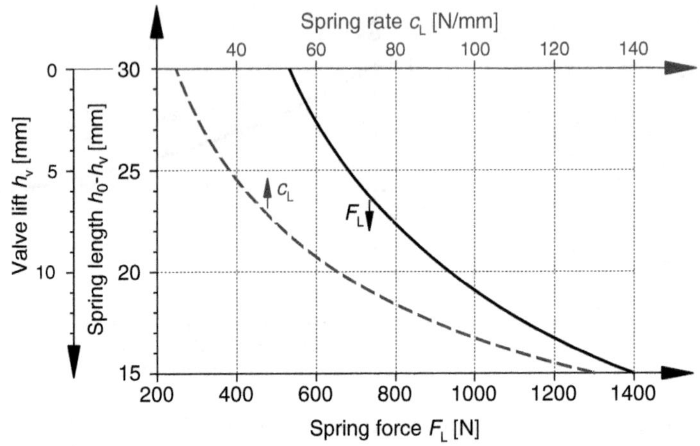

Fig. 1.61 Spring force and spring rate of a pneumatic valve spring. The effective diameter of the tappet is 22 mm, the initial pressure $p_0 = 14$ bar. The spring rate c_L increases over the valve stroke h_v, the spring characteristic is therefore progressive

$$F_L = p_0 A \frac{1}{\left(1 - \frac{h_v}{h_0}\right)^n}$$

F_L	Air spring force, N
p_0	Initial pressure in the tappet, N/mm^2
A	Effective inner surface of the tappet, mm^2
h_v	Valve lift, mm
h_0	Length of the air column in the tappet at valve lift $h_v = 0$, mm
n	Polytropic exponent, -. For air approx. 1.25–1.4

Accordingly, the air spring exhibits a strongly progressive behaviour. The spring rate is therefore not constant, but on the contrary a function of the valve stroke:

$$c_L = \frac{dF_L}{dh_v} = p_0 A \frac{n}{h_0} \left(1 - \frac{h_v}{h_0}\right)^{-n-1}$$

c_L	Spring rate of the air spring, N/mm

Figure 1.61 shows the relationship between the spring force and the spring rate and the valve stroke curve.

Cylinder *Head* Gasket

For assembly reasons, the cylinder-head gasket in production engines consists of a unit that seals all interfaces (combustion chamber, oil and water passages, and possibly secondary air ducts). These are usually metal layer beaded gaskets with vulcanized elastomer rings. In racing vehicles, individual bronze rings are used to seal the combustion chamber, which are inserted into corresponding grooves in the crankcase. The remaining sealing points are provided by O-rings. Garlock Helicoflex rings are often used in the combustion chamber area of high-speed engines. This hollow ring consists of a nickel alloy (e.g. Inconel) and is gas-filled (usually nitrogen). Under the influence of heat, the expansion of the filling gas supports the sealing effect [44].

1.4.3 Cranktrain

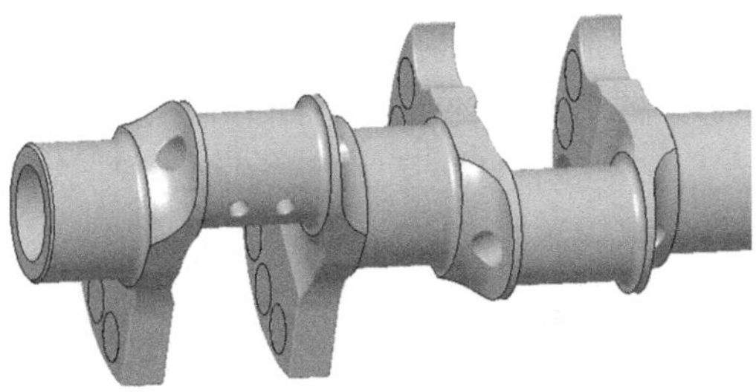

The cranktrain contains all the components that are the focus of development during a performance upgrade: The piston, the connecting rod, the crankshaft and the bearings. The limiting factor is the (hot) strength of these highly dynamically stressed components. The forces in the crank mechanism change over the crank angle and depend on the engine speed and the load, Fig. 1.62. The actual gas force doing the work is generated by the combustion pressure acting on the piston. Maximum combustion pressures in naturally aspirated Formula 1 engines are $p_{max} = 100$ bar. The maximum pressure is about 8° after the ignition TDC (0 °CA) [22]. For conventional connecting rod ratios of production engines, the ideal value is about 12–15 °CA. The speed-dependent inertia forces counteract the gas force and change their sign several times during a working cycle.

1.4.4 Crankshaft

The central component in the crankshaft drive is the crankshaft. It transmits the oscillating piston force into a rotating motion and determines the ignition sequence in multi-cylinder engines. The main dimensions of typical racing engine shafts are summarized in Table 1.9.

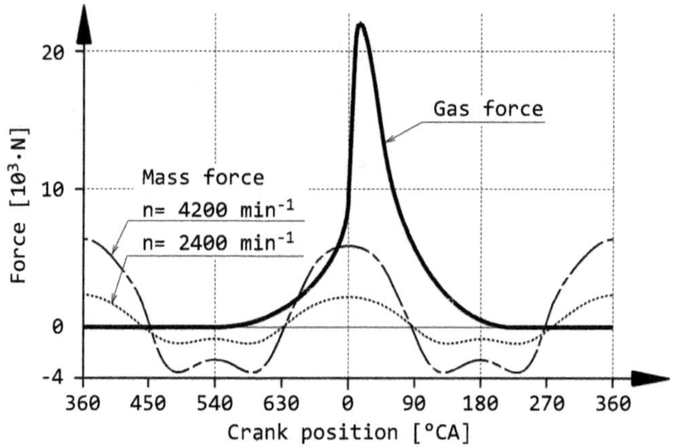

Fig. 1.62 Forces in the crankshaft drive of a four-stroke engine. The forces in the crank mechanism are highly dynamic. The gas force performs useful work. The speed-dependent inertia forces act against it

Table 1.9 Main dimensions of typical crankshafts of some racing series [45]

	World Rally Car	Indy Racing League	Formula 1
Engine type	4-cyl. series	V8	V10
Displacement, l	1.6 to 3	3.5	3.0
Stroke, mm	44	33	46
Shaft journal: Ø, mm	56	56.5	44
Width, mm	25.7	26.45	29.5
Crankpin: Ø, mm	45	46.98	36.5
Width, mm	25	47.1	39
Total length, mm	497	511	577
Counterweight radius, mm	146	152	105

The dimensions of the bearing journals influence the frictional torque as well as the torsional and bending strength. Figure 1.63 shows the influence of the most important dimensions on bearing friction. Common bearings have a width-to-diameter ratio *b*/*d* of 0.3–0.6 (max. 0.8).

For an advantageously uniform torque output, the ignition of the individual cylinders must take place at the same distance. This requires that the offsets of the crankshaft are evenly distributed over the circumference. The ideal ignition or offset distance is therefore 720°/number of cylinders for four-stroke engines (two-stroke: 360°/number of cylinders). The cranking arrangement also influences the inertia forces and moments. Crankshafts symmetrical to the center plane are preferable in this respect. Four-stroke engines have two top dead centers that can be considered as ignition TDC, therefore there are several possible ignition sequences for a given cranking arrangement. In addition to the torsional vibration

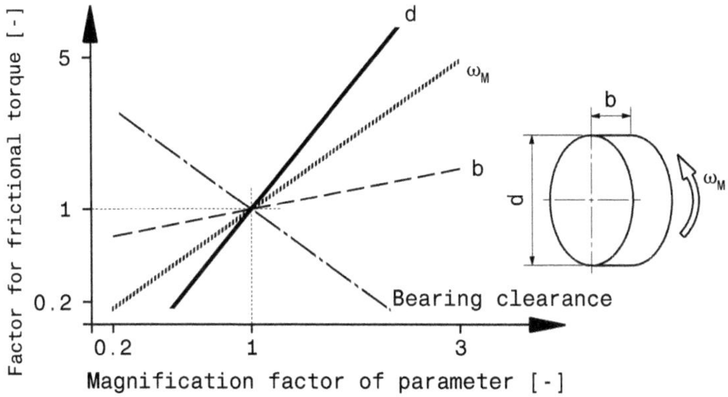

Fig. 1.63 Influence of important bearing sizes on the frictional torque. The journal diameter *d* and the bearing clearance have the greatest influence. The bearing width *b* has the smallest influence. Not shown in the diagram – because it is obvious – is the influence of the number of bearings. This increases the friction directly proportionally and is therefore also a criterion for the selection of the engine design, cf. Figure 1.64

characteristics of the shaft, another consideration in the choice of ignition sequence is the pressure vibration in the exhaust tract, which is important for exhaust turbocharging. In V-engines, equal ignition distances can be achieved if the bank angle corresponds to the ideal ignition distance. If the V-angle deviates from this, equal ignition distances can only be achieved by a crankpin offset (split pin). In this case, a pin is split into two halves offset by the difference angle between the V-angle and the ignition distance. Figure 1.64 shows crankshafts and common ignition sequences for some engine designs.

The engines are shown in the view from above with the usual cylinder numbering. The routing of the exhaust pipes and the clutch (power output side) are also entered. The engines have a uniform ignition distance as in-line engines and thus also within a bank as V-engines. Nevertheless, only in the case of the V10 engine are all ignition distances uniform. For reasons of convenience, crankshafts with offsets of 90° are therefore also used for passenger car 8-cylinder engines, which results in a uniform ignition sequence with a V-angle of 90°. The disadvantage here is the combination of the exhaust pipes, which is unfavorable for performance.

You can also see the relationship between in-line engines and double-cylinder V-engines, which use the same cranking arrangement. The number of main bearings is also interesting. In addition to the engine length, this is important for the friction power. The inline four cylinder (R4) needs the same number as the V8 engine, namely five. In addition to their compactness, V-engines also have the advantage of fewer bearing points. These can be dimensioned comparatively smaller because the forces per bearing are lower due to the higher number of cylinders.

In addition to the crank distribution, the number of counterweights is also decisive for the load on the crankshaft and its bearings, and thus on the crankcase. If each crank has a

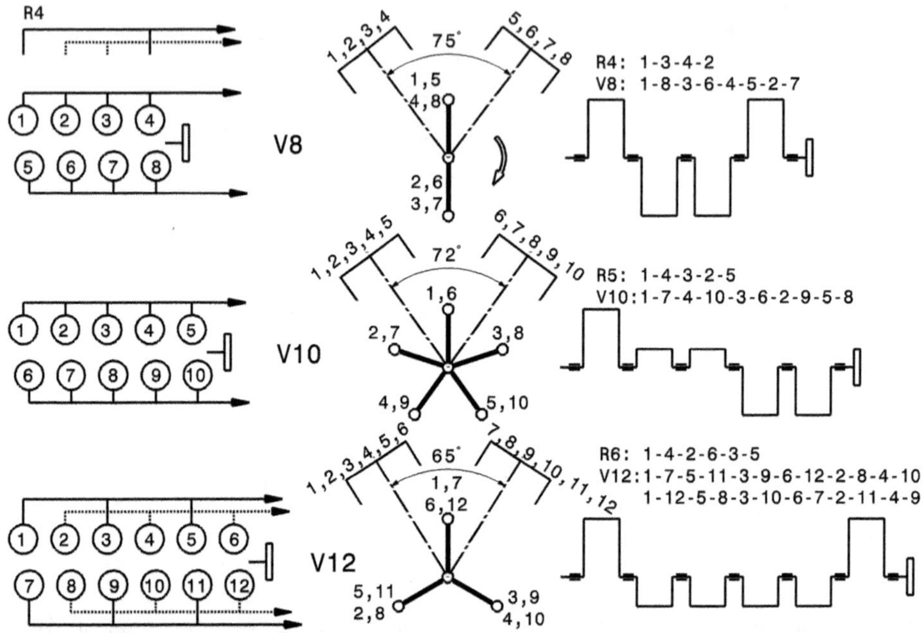

Fig. 1.64 Crankshafts and ignition sequences of some engine designs, after [4]. The crank arrangement and the direction of rotation of the crankshaft result in possible ignition sequences

pair of counterweights, this results in a well-balanced shaft, but at the same time increases the mass and the mass moment of inertia of the shaft. For racing crankshafts, a compromise is therefore sought in which the bearing load remains within the permissible limits with as few counterweights as possible.

For single-cylinder engines, 90% compensation of the oscillating mass forces is sufficient in practice when a balancer shaft is used. In karting and on racing motorcycles, balancer shafts are removed by some teams. As a direct result, damage often occurs to the starter motor. In addition, the load on the vehicle frame is greater and some drivers complain of eye flickering or visual disturbances. A balance shaft therefore compensates for its weight disadvantage in the sum of its properties.

Oil Supply

(Lubrication). The connecting rod bearings and thus also the piston pin bearings are supplied with lubricating oil via the crankshaft. The holes should open at points on the journals where the oil can escape as unhindered as possible. Areas where negative pressure occurs during a four-stroke cycle are best. At point load, the ideal orifice is about 90° forward of the power application. In fact, there will be a surface load, but its resultant can be used as a point load to define the orifice. The point of the muzzle must be well rounded. Fig. 1.65 shows some possibilities of *lubrication holes* in crankshafts.

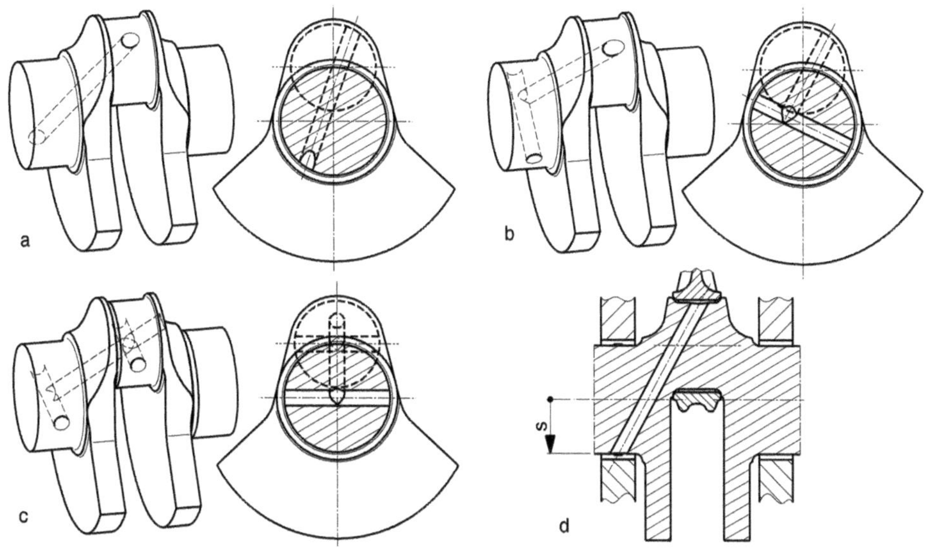

Fig. 1.65 Oil bores in crankshafts. (**a**) single bore, (**b**) additional cross bore in shaft journal, (**c**) additional cross bores in crankpin and shaft journal, (**d**) sketch for calculating the required oil pressure

In production engines, the oil passes through holes in the main bearing bracket to the main bearings, which have circumferential grooves. Via the main bearings, the oil is passed on through bores in the crankshaft. Because the shaft rotates, the oil must first be pumped against its inertia to the center of the shaft. Only from there does centrifugal force help to supply the oil. The pressure required to overcome the distance to the shaft centre therefore depends on the shaft speed, Fig. 1.65d:

$$p_{erf} = 10^{-8} \frac{1}{2} \rho_{\ddot{O}l} \cdot s^2 \cdot {\omega_M}^2$$

p_{erf}	Required oil pressure, bar
s	Distance of the oil to the shaft centre, mm
$\rho_{\ddot{O}l}$	Density of the oil, kg/dm^3. At room temperature, $\rho_{\ddot{O}l}$ is approx. 0.9 kg/dm^3
ω_M	Circular frequency of the crankshaft, s^{-1}. $\omega_M = \pi n_M/30$

At 15,000 min^{-1}, an oil pressure of approx. 8.7 bar is required for a shaft journal with a diameter of 56 mm. For high-speed engines, the *oil feed* is therefore in the middle of the shaft, Fig. 1.66.

In addition, such engines have a small stroke. This reduces the oil paths against the centrifugal force and leads to stiffer crankshafts due to large journal overlap. In addition,

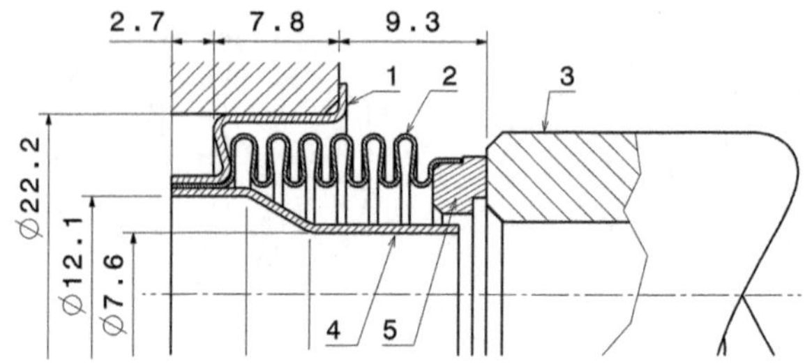

Fig. 1.66 Design of the front end of the crankshaft for oil supply, according to [4]. A mechanical seal with targeted contact pressure (approx. 42 N) ensures that no oil is lost. **1** Housing, **2** Metal bellows, **3** Crankshaft, **4** Stainless steel tube, **5** Sliding block made of resin-impregnated hard carbon

the load capacity of the main bearing shells is increased because the otherwise necessary supply grooves are eliminated. Figure 1.67 shows a crankshaft of a high-revving V6 engine. Lubricating oil is supplied only through the front end. The oil holes run primarily parallel to the shaft axis. For longer shafts, the oil is also supplied via the rear end.

Lightweight

Crankshafts are made of steel and therefore measures to avoid mass are worthwhile. First approach is to lighten the weight of the crank pin. The lighter this journal is, the less counterweight is required. This in turn reduces the overall mass of the shaft and its moment of inertia. With a smaller counterweight, its radius of motion can also be kept small and thus the crankshaft can be located low in the crankcase, which helps keep the engine center of gravity low. Fig. 1.68 shows some design variants of lightening measures.

Figure 1.68a: A deep hole drilled through the entire crankshaft removes material accordingly to reduce weight. Such a deep hole must be produced with a single-lip drill. Because its cut must not be interrupted, this step must be taken at the beginning of manufacture. The bore diameter is limited by the lifting pin. This must not be covered by the bore (arrow). The lifting pins are facilitated by two inclined bores. To reduce stress, the bottom of the hole is finished with a hemispherical cutter.

Figure 1.68b: The oil supply holes are chosen so large that they also contribute to weight reduction. The limit results from the minimum distance to the cove of the lifting journal (dimension a). These bores are also designed with a hemispherical base. With this shaft, the counterweights are bolted on. This makes it possible to use a different material with a higher density, which means that the counterweights can be made even smaller.

Figure 1.68c: The lifting pins are relieved with eccentric deep hole bores (1). In the shaft shown, the individual lifting journals are offset by 180°, therefore the counterweights are also covered by the deep-hole drilling. Lubricating oil is also supplied to the bearing points via deep-hole bores (2). In addition, not all crankings are provided with counterweights in this design, which further reduces the overall mass of the component

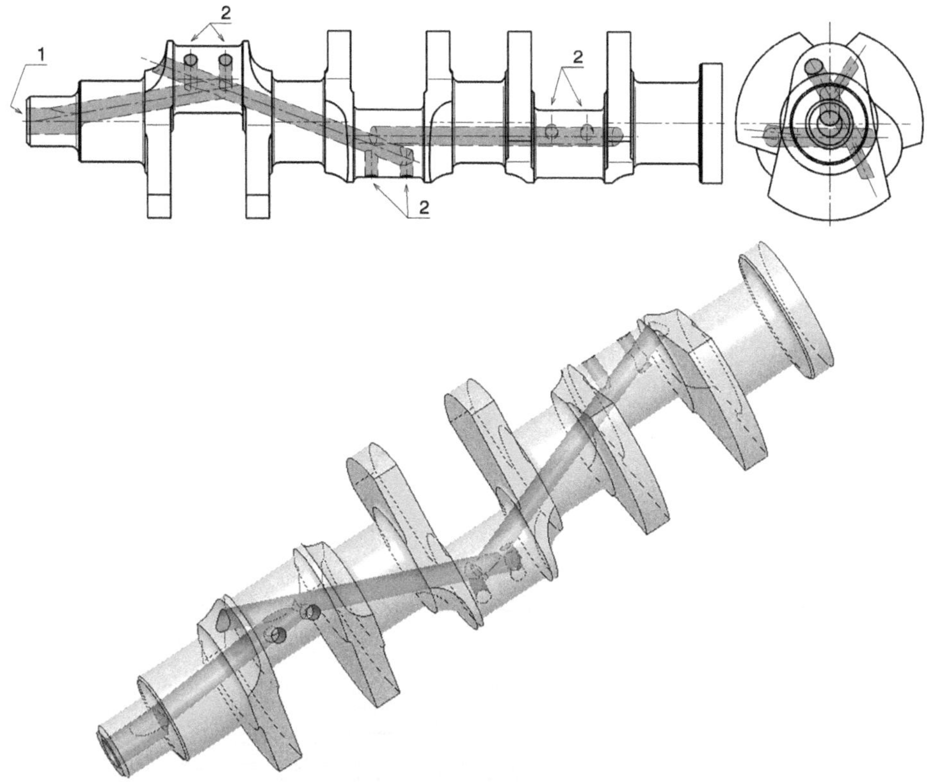

Fig. 1.67 Lubricating oil supply to the bearing points on a high-speed crankshaft of a V6 engine. The oil is fed axially into the shaft at the front and is guided to the journals with as few deflections as possible. **1** oil supply, **2** supply holes to the connecting rod bearings

Figure 1.69 shows the front end of an assembled Formula 1 crankshaft with elaborate lightening bores in the crank pin. In addition, one can see bolted-on counterweights and plugs in the bores.

Another example of a racing crankshaft is shown in Fig. 1.70. In this case, in addition to the weight, the air resistance has also been reduced. At high speeds (approx. from 10,000 min^{-1}), the air resistance (in fact, an oil-air aerosol in the crankcase generates resistance) of the rotating parts becomes noticeable. Especially the running up flank must be well rounded and chamfered. In this way, up to 30 kW of useful power can be gained in 3.0-l engines that rotate at more than 12,000 min^{1}, which would otherwise merely raise the oil temperature. An alternative way to almost completely avoid these losses is to evacuate the crankcase.

Two other special features of racing crankshafts can be seen in Fig. 1.71. To keep the counterweights of this V10 shaft small and still achieve the required mass, the webs have heavy metal plugs (tungsten, arrow) (a). The webs are not designed for aerodynamics

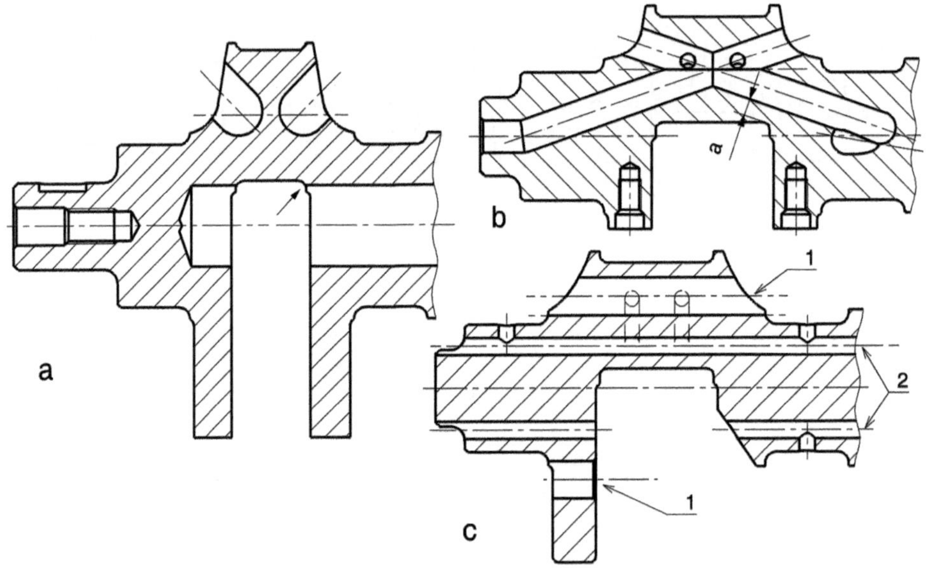

Fig. 1.68 Lightening of crankshafts, after [45]. (**a**) central deep-hole drilling, (**b**) large oil supply wells, (**c**) decentralised deep-hole drilling

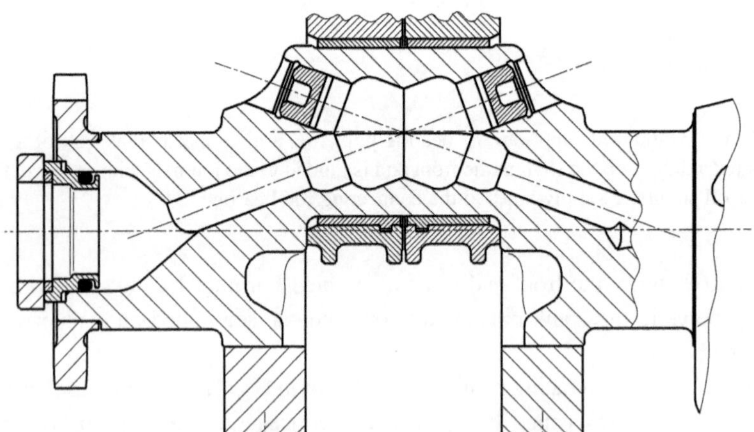

Fig. 1.69 Sectional view of the crankshaft of the BMW P82 Formula 1 engine (3.0l V10) [46]. The P82 was the engine for the 2002 season. It was developed from the predecessor P80, which had made the leap to the top in terms of performance, with the aim of also setting the best mark in terms of weight and dimensions. With 86 kg ready-to-install weight this also succeeded. Main and connecting rod bearing diameters shrank to 42 and 36 mm respectively. This and the internal machining of the crankpins saved 1 kg on the crankshaft alone

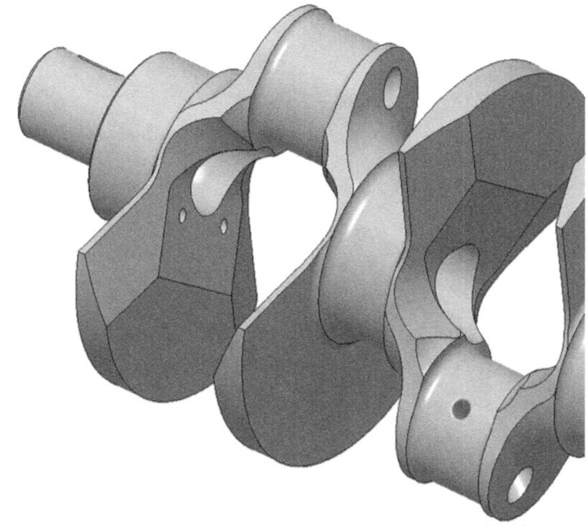

Fig. 1.70 Crankshaft of a 4-cylinder in-line engine. The shaft is lightened by a central deep hole. The webs and the counterweights are beveled to reduce air resistance

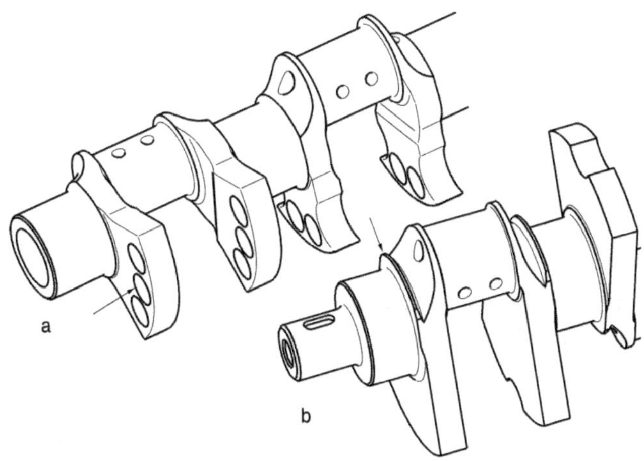

Fig. 1.71 Details on racing crankshafts. (**a**) Heavy metal plug in the counterweight (Ferrari V10 Tipo 049), (**b**) Tear-off edges on the crank web (Ford Cosworth DFV V8)

because the crankcase of this engine is evacuated. The crank webs of the V8-shaft have knife-like edges (tear-off edges, arrow), which selectively throw off the oil escaping from the main bearings (b).

Materials

Quenched and tempered steels (C45E (was Ck45), 42CrMo4), nitriding steels (31CrMoV9), micro-alloyed steels (38MnS6).

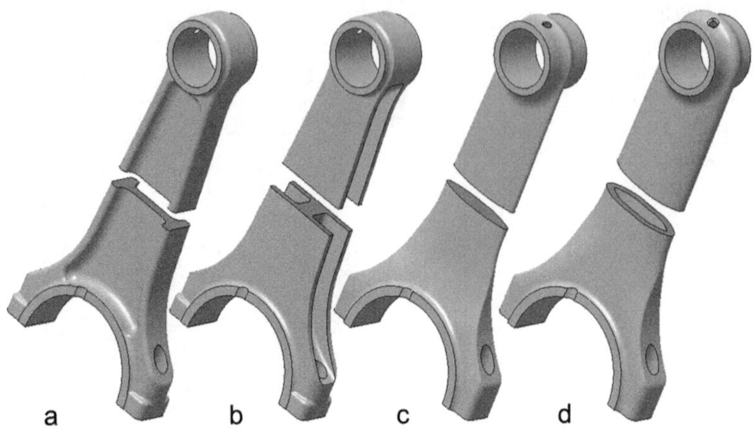

Fig. 1.72 Cross-sectional profiles of connecting rod shafts. (**a**) *I-section* (double-T), (**b**) *H-section,* (**c**) *blade section,* (**d**) hollow *section*

Production

Racing crankshafts are usually machined from solid in one piece. Forged shafts are used for engines derived from series production units. With these, the firing order of the blank can be changed compared to the series by twisting the shaft journals.

Shafts of series vehicles are forged or cast.

Flywheel

Racing engines usually have no flywheel at all, apart from the mounting plate for the clutch, which can also accommodate the starter teeth on the circumference if the engine is started with an electric starter. A smaller flywheel mass demands a higher idle speed. Unlike utility engines, racing engines therefore also stop abruptly as soon as the ignition is switched off.

This flange plate is made of steel or heat-treated aluminium.

Connecting *Rod (con Rod)*

The connecting rod connects the piston to the crankshaft. It transmits gas and inertia forces and is bent by the lateral acceleration of its own mass. The shaft is therefore designed as a flexurally rigid support. Shafts of forged or cast series connecting rods mainly have an advantageous I-section (double-T-section), Fig. 1.72. Racing connecting rods are usually milled from the solid and are then often designed as smooth-shaft connecting rods with an H-section. Short connecting rods (approx. Less than 130 mm) can also be designed as knife-edge connecting rods. They have a low air resistance and can better follow one-sided bends of crankpins. An ideal lightweight combination of high strength and low drag is offered by an oval hollow shaft. The connecting rods of high-revving Formula 1 engines typically have an I-section and are guided by the piston.

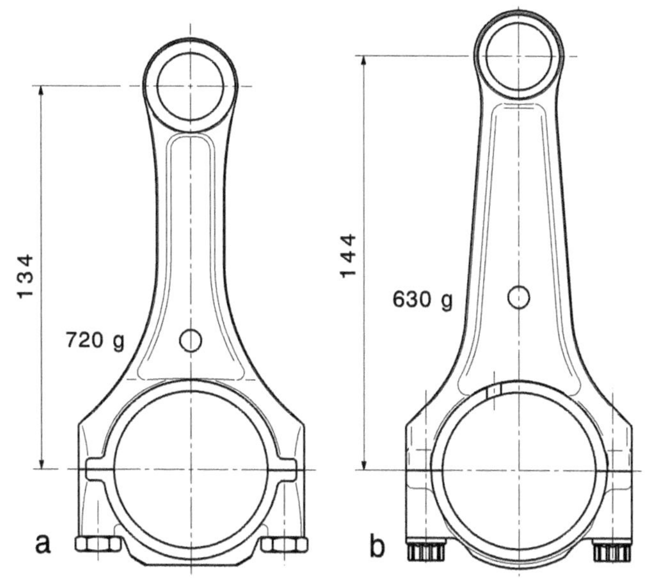

Fig. 1.73 Comparison of connecting rods of a 3.0-l petrol engine, according to [29]. A racing engine is derived from an in-line six-cylinder. While the height of the crankcase remains the same, the longer connecting rod of the racing engine results in a shortened compression height of the piston. Both connecting rods are forged, but the material of the racing connecting rod is of higher quality. This is therefore also lighter in spite of the larger gauge. (**a**) Series connecting rod, (**b**) Racing connecting rod

The greatest loads according to which a connecting rod is designed represent the gas force due to the maximum combustion pressure in the cylinder and the maximum speed in the overlap TDC, i.e. when practically no gas force counteracts the piston acceleration. Ignition pressures for race engines are in the range around 120 bar for naturally aspirated engines and 170–220 bar for turbocharged engines. A comparison between series and racing engines is shown in Fig. 1.73. A DTM engine is derived from a series engine. The racing conrod is longer and more flexurally rigid in the transverse direction, yet its total mass is lower. However, it must be mentioned that the material of the racing connecting rod is of higher quality. There are some racing series in which the connecting rod mass is regulated. But even there, there are development possibilities for the designers. For example, the installation space occupied by the connecting rod is of interest. Furthermore, an optimal ratio of rotating to oscillating connecting rod mass is sought.

The connecting rod length influences the engine height and the inertia forces. For considerations see Fig. 1.21.

The connecting rod width results from the permissible bearing load. The smallest widths of 3.0-liter Formula 1 engines are 12 mm. Series connecting rods are twice as wide.

Connecting rod feet (big end) are split for assembly reasons. One-piece connecting rods can only be used on assembled crankshafts. The bearing caps are centered to the connecting rod foot by means of pins, fitting sleeves or sawtooth profiles. Fracture-separated

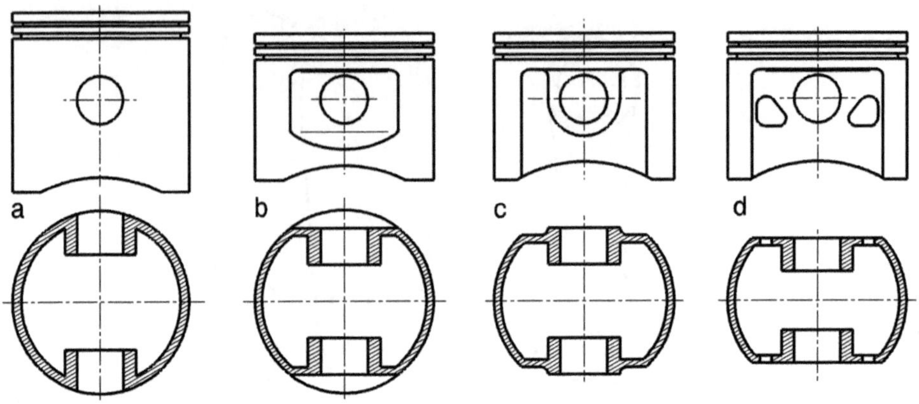

Fig. 1.74 Piston types schematic. View and cross section in hub center. (**a**) *smooth-skirt piston,* (**b**) *full* slipper piston, (**c**) *slipper piston,* (**d**) *piston with weight-saving openings*

connecting rods, as they have found their way into series production, are not used because of the material required for this. The bolting is done with high-strength bolts, e.g. made of Nimonic. The bolt axes are sometimes set slightly arrowed, deviating from the usual parallel arrangement.

Lubricating oil is supplied to the small eye either via the oil stripped from the piston or via the oil from the piston cooling nozzles, which enters via small holes in the connecting rod head, or from the large eye by means of a separate hole through the skirt. An H-section is suitable for the latter design.

Materials

Quenched and tempered steel (31CrMoV9, 42CrMo4), case-hardened steel (18CrNi8, 15CrNi6, 34CrNiMo6 V), titanium alloys (TiAl4V4). Titanium is a poor running partner for steel and must therefore be coated at the contact points (side surfaces of the eyes) or a collar bearing shell must be installed. In addition, titanium connecting rods suffer from bore expansion during operation, causing the bearing shells to become loose.

Production

Forged or machined from solid. Series conrods are cast, forged or sintered (Sinter F31).

Piston

Next to the connecting rod, the piston is the most demanding component when it comes to increasing engine speed and power. It contains part of the combustion chamber, should be as light as possible and still have high heat resistance. In addition, it needs good running properties in the liner. No wonder that pistons are among the best-kept secrets of racing engines. The choice of design, Fig. 1.74, is determined by the specific power and bore. The smooth skirt piston is only important in diesel (compression ignited) engines. In the case of

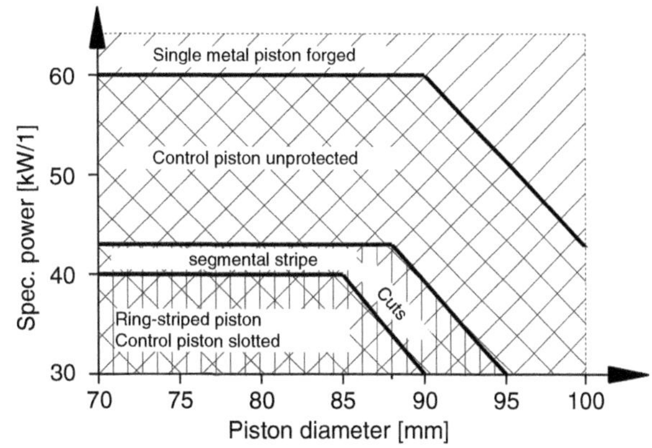

Fig. 1.75 Application limits of different piston designs. The values apply to gasoline engines with crankcase or liner made of gray cast iron

the box piston, the skirt is retracted in the hub area, thus providing the shape that gives it its name. Slipper pistons are retracted over the entire skirt area and are even more weight-optimized. Slipper pistons can have compromised straight line performance when used to their fullest extent. Which limits their use in utility engines because of the resulting noise and emissions problems.

Figure 1.75 shows that only forged pistons can meet the requirements of racing. Proven designs are the window, box and slipper pistons. Apart from these, however, racing pistons are all special designs. The compression height is very low and the piston as a whole is extremely weight-optimized. Only forged pistons are used. Weight optimization and piston cooling are decisive criteria for the design of these pistons. In Formula 1, specific powers of more than 280 kW/l are common. Speeds of more than 19,000 rpm were driven when the regulations permitted. The service life of the pistons is matched to the extreme conditions [7].

Typical dimensions (referred to bore *B*) of pistons of different engines can be compared in Table 1.10. Explanations of the dimensional designations can be found in Fig. 1.76.

The greatest influence on the piston weight is shown by the compression height [47]. Other influential areas are the eye spacing, which naturally also dictates the piston pin mass, the crown thickness and the shape of the hub support.

The pin bore area deserves high attention for pistons subjected to high loads. To relieve stresses, the bores are shaped towards the connecting rod, giving room for deflection of the pin. In addition, continuous, lateral oil pockets increase the surface pressure that can be absorbed and the oval deformation of the pin does not burst the hub.

The currently common design of racing pistons is the *box-bridged type*, Fig. 1.77. The strong ribbing in the direction of the connecting rod swivel (pressure or counterpressure side) allows low compression heights with a simultaneous reduction of the crown thickness. Figure 1.78 shows the design of the combustion chamber side of a Formula 1 piston.

Table 1.10 Piston dimensions of four-stroke engines [7, 47] Dimensional designations: See also Fig. 1.76

	Petrol engine		Diesel engine DI	
Dimension	Series	Lightweight	Series, car	Lightweight
Diameter *B*, mm	65 to 105		65 to 95	
Total length l_{Pi}/B	0.6 to 0.7		0.80 to 0.95	
Compression height h_{Pi}/B	0.30 to 0.45	0.32	0.5 to 0.6	0.47
Bolt diameter d/B	0.20 to 0.26	0.24	0.32 to 0.40	0.31
Top-land height f [mm] or f/B	2 to 8	0.04	4 to 15	0.09
1.Ring land s_1/B	0.040 to 0.055	0.045	0.05 to 0.09	0.05
Skirt length l_2/B	0.4 to 0.5	0.4	0.50 to 0.65	0.5
Boss spacing b/B	0.20 to 0.35	< 0.3	0.20 to 0.35	0.25
Crown thickness t/B or t/D[a]	0.06 to 0.10	< 0.06	0.2	0.09

[a]For diesel (compression ignited) engines

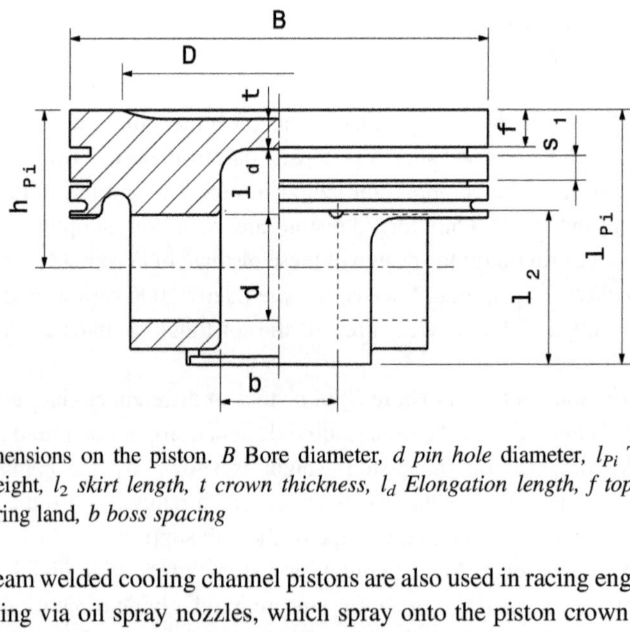

Fig. 1.76 Dimensions on the piston. *B* Bore diameter, *d pin hole* diameter, l_{Pi} Total length, h_{Pi} compression height, l_2 *skirt length, t crown thickness,* l_d *Elongation length, f top land height,* s_1 Height of first ring land, *b boss spacing*

Electron beam welded cooling channel pistons are also used in racing engines, Fig. 1.79. Injection cooling via oil spray nozzles, which spray onto the piston crown from below at the raceway end, serves to lower the piston crown temperature.

Materials

Aluminium-silicon alloys, aluminium-copper alloys and light metal composites. Silicon carbide reinforced aluminum (MMC). These light metal pistons are molybdenum or DLC coated and run in a Nikasil bore. Fiber reinforced magnesium alloys and structural carbon are promising future materials.

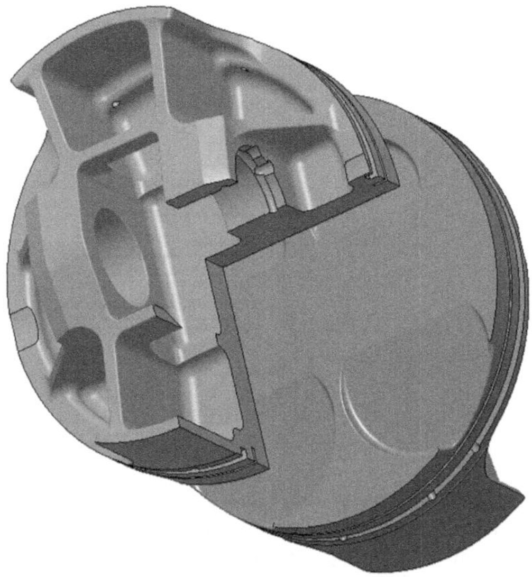

Fig. 1.77 Formula 1 pistons cut. The box-in-box piston is forged, only the web breakout between the hubs is milled out. The piston skirt is only present where it is needed, namely in the ring zone and in the pressure-counterpressure area. The location area for the piston pin is extremely short. With a bore of 95–100 mm, the piston has a mass of only 220–250 g

Fig. 1.78 Piston of a Formula 1 engine (Asiatech V10 3.0 l). Bore 91 mm, inlet valves diameter 40 mm, exhaust valves diameter 30 mm. The valve-pockets are deeply worked in and well rounded. In addition, one can also see that valves with the same name also enclose an angle, i.e. the valves are arranged radially

Since 2008, forged steel pistons have been used in diesel (compression ignited) engines at Le Mans (first Peugeot, followed by Audi in 2009). Steel pistons have greater rigidity, show less change in running clearance over temperature and significantly reduce shirt friction. They also allow the compression height to be reduced, either lengthening the connecting rod or shortening the cylinder block. The firewall height can also be reduced. The piston pin can be shortened due to the high transmittable force in the pin bore. In sum, a steel piston can match or even undercut the weight of its aluminum counterpart [6]. Steel

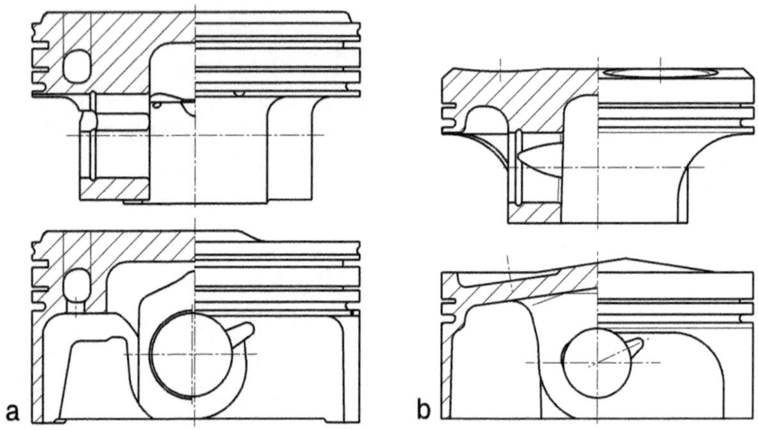

Fig. 1.79 Racing piston, according to [11]. Both pistons are basically box pistons with a narrow skirt width, resulting in a stiff piston with, however, low skirt elasticity. (**a**) Cooling channel piston, electron beam welded, (**b**) Formula 1 piston

pistons are now also regarded as a measure for further performance increases in passenger car diesel (CI) engines [48].

Production

Highly stressed pistons are forged and – if geometrically necessary (undercut) – machined. Series-produced pistons are also cast for low stresses. Short-fiber-reinforced light metals are press cast. Materials produced by powder metallurgy (e.g. RSA – Rapidly Solidified Aluminium Alloy) still suffer from permanent deformation during engine operation.

Piston Rings

Their function is to seal the piston to the raceway, to dissipate heat from the piston and to regulate the oil balance. They contribute about half of the frictional power of the piston group, which in turn accounts for about 40% of the total engine friction. The aim in engine development is therefore to use as few rings as possible with low preload and height to fulfil the required functions. Leaky rings lead, among other things, to a loss of torque or, as a result of disturbed piston lubrication, to engine damage. A standard series assembly consists of two compression rings and one oil scraper ring. Racing engines run with one ring of each type. Figure 1.80 shows some types of rings for racing engines. The L-ring (a) is fitted so that its uppermost edge coincides with that of the piston crown. It offers high flutter resistance even at high speeds due to the gas pressure acting directly behind the vertical L-leg. Two-piece compression rings (b) combine the sealing effect of two individual rings with a lower overall height and friction and have a high flutter speed. If the chamfer is at the top, only the lower edge of the ring is in contact with the cylinder wall in the depressurised state. This increases the oil wiping effect. Such one-piece rings are

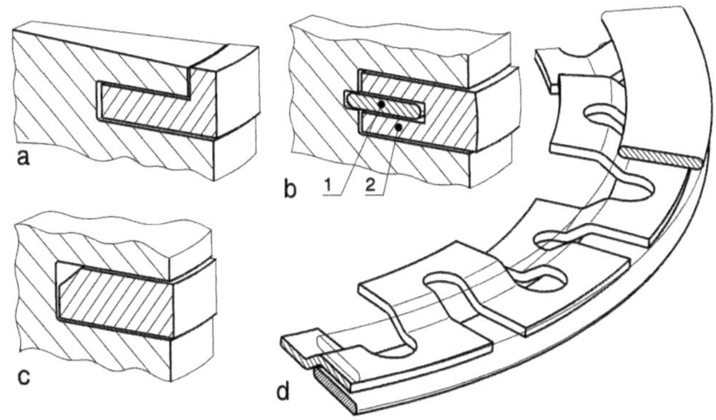

Fig. 1.80 Piston rings. (**a**) L ring, (**b**) two-piece *compression ring,* 1 sealing ring in an additional groove in the piston, 2 main ring with crowned lining. (**c**) ring *with* inside bevel. (**d**) three-piece oil scraper ring *(oil control ring).* a to c are compression rings, i.e. are inserted in the first ring groove

manufactured with minimum heights of 1 sometimes even 0.8 mm. Oil wiper rings are advantageously designed in three parts (d). Two narrow rings are kept apart by a band spring. The total height can be less than 2 mm. The wiped-off oil passes through four to eight holes at the base of the groove, through the piston wall to the inside and to the piston pin.

Materials

Spheroidal graphite iron quenched and tempered. Steels for high fracture resistance (low rings ≤1.2 mm, high speeds): Cr-Ni steel, X90CrMoV18, 67SiCr5. The running surface is provided with wear protection coatings (e.g. PVD coatings).

Production

The shape, which is decisive for the pretensioning process, is created by double-form turning steel rings are wound.

Piston Pin *(Gudgeon Pin)*

Piston pins of racing engines are floating in piston and connecting rod. In utility engines, pins are also pressed into the small connecting rod eye. Proven dimensions of piston pins are shown in Table 1.11.

The mass of piston pins can first be reduced by reducing the pin length. Further savings can be achieved by adapting the shape to the load, Fig. 1.81.

Potential investigations have also been carried out on bolts with an I-profile. The mass saving was 30% [47].

Table 1.11 Dimensions of piston pins

Dimension	Petrol engine	Diesel (compr. Ignited) engine	Racing engine
Outer diameter d_o/B	0.24 to 0.28	0.30 to 0.35	0.2 to 0.22
Inner diameter d_i/d_o	0.55 to 0.65	0.48 to 0.52	
Length l/B	0.70 to 0.75	0.70 to 0.75	0.5

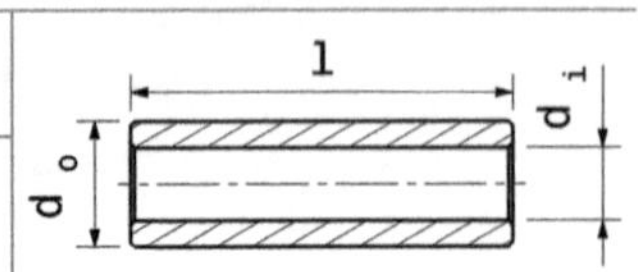

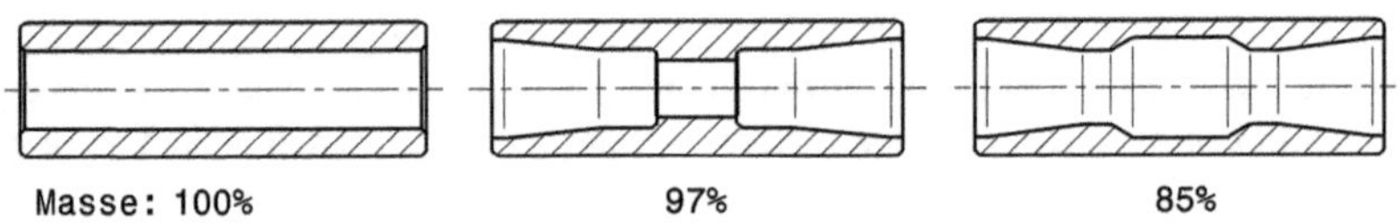

Fig. 1.81 Mass saving for piston pins, according to [47]. The bolt with dimensions 19.5 × 12 × 63 mm (100% mass) is adapted in shape to the load by changing the internal shape

Materials

Case hardening steel (16MnCr5, 15CrNi6 (DIN 73126)), nitriding steel (31CrMoV9 (DIN 73126)). Highly loaded bolts are made of ESU steel (electro-slag remelting process). Ceramic (silicon nitride Si_3N_4) allows mass savings of up to 50% compared to steel, but leads to noise problems due to low thermal expansion. However, this is only a disadvantage for utility engines.

Piston Pin Locking (*Locating Circlip*)

The usual wire ring protection can become a problem at the highest speeds (> 12,000 min^{-1}). If a fluttering or broken ring jumps out of its groove, engine damage is inevitable. Retaining rings are therefore screwed in or are specially designed, Fig. 1.82.

The nut-type retainer (a) is screwed in and its collar is caulked into a recess in the piston to prevent rotation. The wire locking ring (b) is much lighter. However, the position of the radial groove for the anti-rotation device is decisive for the speed capability. The wire end for the anti-rotation device must also be angled outwards. Standard solutions have inwardly angled ends, whose mass inertia at high speeds actually causes the ring to jump out of the groove. Teflon plugs, which are inserted into the side of the bolt hole, are another option.

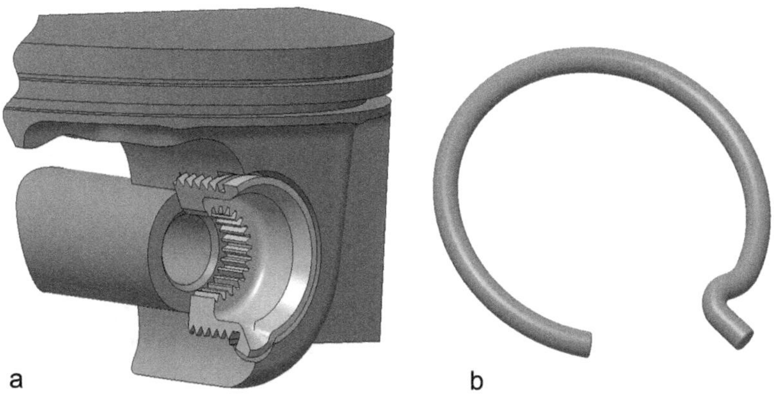

Fig. 1.82 Piston pin locks. (**a**) *nut-type retainer,* (**b**) wire snap ring (*wire circlip)*

1.4.5 Crankcase

The crankcase is the central and largest component of an engine. It houses the crankshaft and usually also accommodates the piston raceway (so-called cylinder crankcase) directly or as liners. It connects to the transmission and houses the engine mounts or bolts directly to the frame or monocoque. In the case of fully co-supporting (stressed) engines, it must also transmit a large proportion (the cylinder heads usually take on a share) of the forces and torques that occur between the axles of the vehicle. The cylinder heads are also bolted to the crankcase, as are ancillary units. In addition, part of the cooling and lubrication system is formed by the crankcase.

In production engines, the water jacket (1) usually extends to the crankcase, Fig. 1.83. The crankshaft bearing is accommodated by bulkheads (4), which are locally stiffened by ribs and webs. Gas exchange between the crank chambers of individual cylinders is facilitated by openings (3) in the bulkheads. Oil is supplied to the main bearings through holes in the bulkheads meeting the main oil gallery (2). Bearing caps hold the crankshaft via two bolts. The housing is closed at the bottom by an oil pan in which the oil supply is stored.

In racing engines, the coolant (2) only flows around a quarter to a third of the liner length, Fig. 1.84. The cylinder head sometimes has a separate cooling system with a much lower coolant temperature. The liner (1) is a separate component which is inserted from above. This ensures less distortion of the liner when screwing the cylinder head and better cooling conditions. The crankcase is smooth and circular. The lower part of the housing forms the second half and the "bearing cover". Because of the higher load, especially with V-engines, the bolting is usually done with four bolts. The bulkheads, which accommodate the crankshaft bearings, are solid or double-walled. The oil-air mixture is extracted via a plane-shaped opening (3) for each crank chamber (on V-engines the area for two connecting rods on the same crank pin).

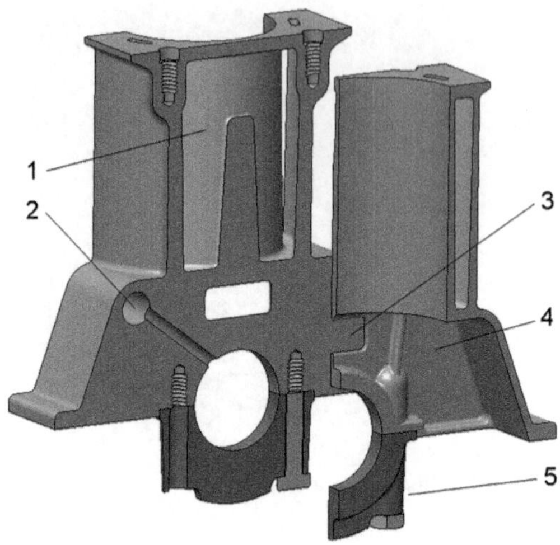

Fig. 1.83 Design features of a series crankcase. The crankcase is part of an in-line engine and directly houses the piston's raceway. The crankshaft is held from below by bearing caps. An oil sump, which also accommodates the oil supply, forms the downward seal. **1 water** jacket, **2** main oil *gallery*, **3** *passage in bulkhead*, **4** bulkhead, **5** bearing cap

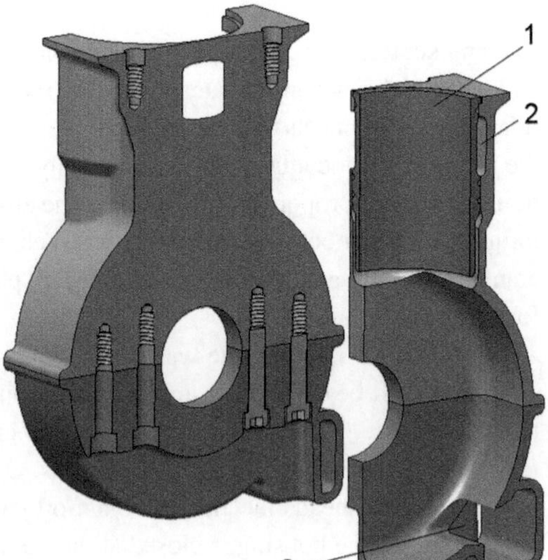

Fig. 1.84 Design features of a racing crankcase. **1** *liner* (wet), **2** water jacket, **3** *passage to suction gallery*

The piston's raceway is either placed directly by the crankcase or a separate liner is pressed or pushed in. Racing engine liners are usually made of nikasil-coated aluminium. Geometric characteristics of such liners are shown in Fig. 1.85.

The highly stressed efficient diesel (compression ignited) engines of the Audi LMP1 cars owe their low friction not least to nikasil-coated raceways, which are an integral part of the hypoeutectic Al crankcases.

On the upper side, some crankcases have flat grooves that open out at the edge of the gusset area between adjacent cylinder bores. This allows early detection of leakage of the

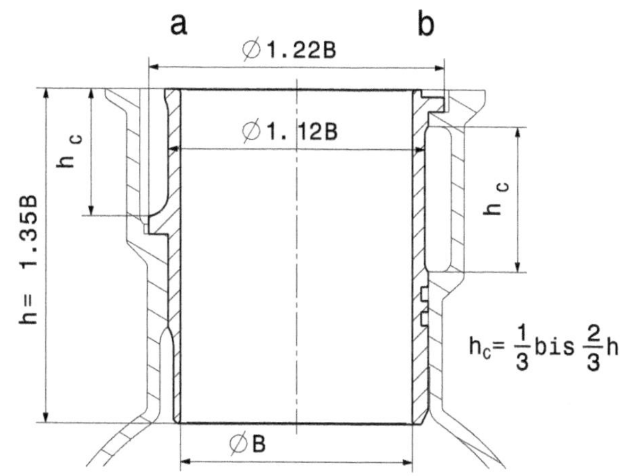

Fig. 1.85 Design features of liners. The dimensions are mean values of Formula 1 liners and are given in relation to the bore diameter B. h_C Height of the water jacket. Two different liners are shown together with the corresponding crankcase mounting hole. (**a**) *mid-stop liner* (wet standing liner), (**b**) *wet liner* (wet hanging liner)

cylinder head gaskets [23]. Figures 1.86 and 1.87 present two examples of crankcases from Formula 1.

Materials

Aluminium alloys: EN AC-AlSi8Cu3 (was AlSi9Cu3), EN AC-AlS6Cu4 (was AlSi6Cu4), EN AC-AlSi17Cu4Mg (was AlSi17Cu4Mg), EN AC-AlSi7Mg0.3 (was AlSi7Mg wa); magnesium alloys. For series engines, ferrous materials are also cast: EN-GJL-250 (was GG25), EN-GJL-300 (was GG30), GJV (was GGV, cast iron with vermicular graphite).

Production

The upper and lower parts of the crankcase are cast. Due to the small number of pieces, processes such as sand casting and investment casting are suitable. Minimum wall thicknesses are 2.5–2 mm, which not all foundries can achieve for such large components.

Series parts are manufactured using gravity die casting, die casting or core packaging processes with a minimum wall thickness of 4 mm.

Main Bearings

The service life of plain bearing shells is about 30 h for endurance races. This is sufficient for a 24-hour race including a test run [23]. In sprint competitions, the bearing shells are generally replaced after about 1200 km for safety reasons. Three-material bearings and sputter bearings are used. Some engines also have rolling bearings (cylindrical roller bearings with separable cage) and mixed variants are also chosen. For example, in the successful 12-cylinder 180° V engine of the Ferrari 312B, the first and last main bearings were ball bearings, while the others were plain bearings.

Rolling bearings are already used in some production engines, but so far for balancer shafts. Camshaft bearings are currently in the prototype stage and crankshaft bearings in the

Fig. 1.86 Crankcase of a 3.0-l V10-cylinder Formula 1 engine (Asiatech 2001). The block is cast from light metal and unites two benches at an angle of 72°. It accepts dry pressed-in liners. The clutch side (power output side) is on the left of the picture. Bore × stroke = 91 × 46.1 mm

Fig. 1.87 Titanium crankcase (Ferrari Formula 1 V12 cylinder 1995). The part is cast and machined. It includes the liners, which are inserted from above, and accommodates the upper part of the crankshaft bearings (cylinder crankcase)

concept phase. The advantages of lower friction compared to plain bearings are offset by problems in terms of acoustics and the larger installation space.

1.4.6 Intake System (Induction *System)*

The intake system must supply the engine with the required combustion air with the lowest possible losses. It is important that the air path is uniform. Cross-section jumps or offsets at separation points (intake manifold – cylinder head, etc.) must be avoided at all costs. High-performance engines have single intake manifolds throughout and no spider intake

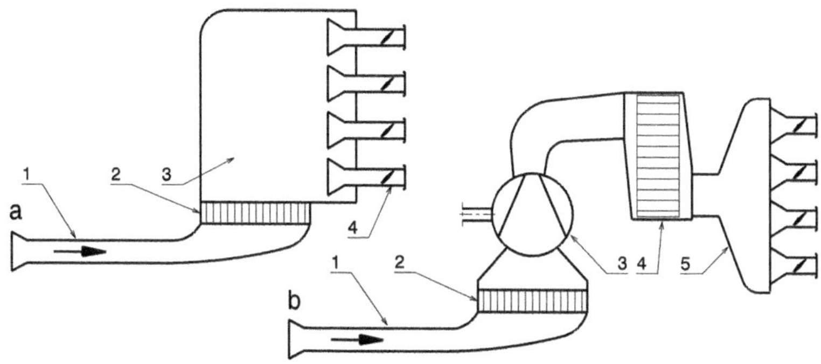

Fig. 1.88 System overview of intake systems. (**a**) for *naturally aspirated* engine, 1 raw-air intake, 2 air filter, 3 air distributor *(plenum),* 4 intake *manifold* (with throttle). (**b**) for charged engine, 1 raw-air intake, 2 *air* filter, 3 compressor, 4 *intercooler,* 5 *plenum with pipe sockets (and throttle)* The throttle element is not required for diesel (compression ignited) engines and direct-injection gasoline engines

manifolds or similar, as is the case with some commercial engines. Gas dynamic processes induced by the periodic piston movement occur in the intake system. By appropriate design of the system, resonances can be used specifically to increase the charging efficiency. The intake system is basically the same for all internal combustion engines. Differences result from the type of control (throttle element in gasoline engines with quantity control) or from supercharging, Fig. 1.88. In supercharged engines, the intake manifold consists of a large collecting tank from which short pipe connections lead to the individual cylinders.

The clean air section (downstream of the air filter) of turbocharged engines has considerably more internals than that of naturally aspirated engines. In contrast, the distribution volume, which is primarily responsible for the task that gives the engine its name, is kept simple, whereas in the naturally aspirated engine this part is elaborately developed because it has a significant influence on the engine's power curve.

The intake tract of a naturally aspirated engine on the engine side is shown in Fig. 1.39.

Air Filter

The air filter has the task of freeing the intake air from coarse impurities and dust. This reduces abrasive wear on moving parts, lowers friction and increases service life. On performance-enhanced standard engines, the standard air filter allows a sufficient amount of air to pass through in the low to medium speed range under normal driving conditions. A sports air filter is a solution if the supply of the required air volume at higher engine speeds is not possible through the original air filter. However, this is not only the case at high engine speeds, but also when the intake air has a high moisture content. The original air filter made of paper will swell due to the high water content in the air, which will hinder the intake. A cotton air filter in combination with air filter oil can eliminate this problem.

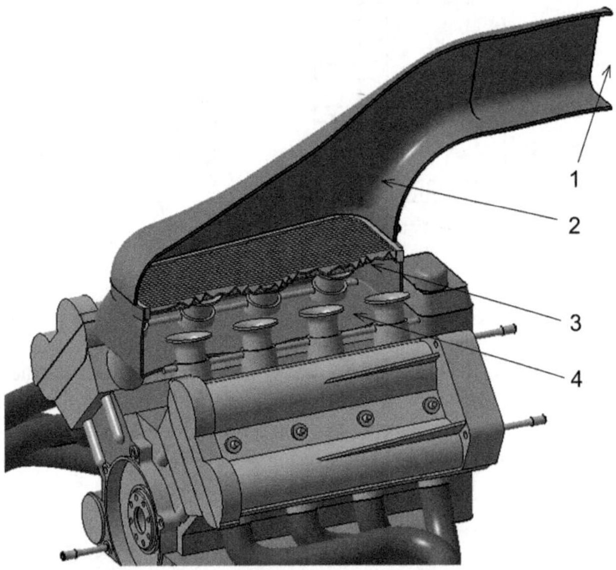

Fig. 1.89 Airbox of a formula car. **1** Air intake inside the roll bar, **2** Upper part of the airbox with diffuser and distribution volume, **3** Air filter, **4** Lower part of the airbox, attached to the engine

Airbox

The combustion air is drawn in at a convenient point on the vehicle. In single-seaters, the area above the driver's helmet and inside the roll bar is often chosen. In doing so, a slight ram effect can be exploited to increase performance. In the case of the 3-litre naturally aspirated engine of the Ferrari F1–2000, the pressure increase of 0.058 bar resulting from 350 km/h (corresponding to a 5.8% increase in air density) led to a power increase from 609 to 646 kW [37]. In closed cockpits, the air enters the vehicle through snorkels, NACA inlets or through scoops. In rally cars, the air intake is raised to roof level with a snorkel for special applications, so that the resulting bow waves cannot reach the intake tract during water crossings.

The task of airboxes (Fig. 1.89), in addition to the raw air intake, is to distribute the air evenly to the individual intake points of the cylinders and to reduce the flow velocity, which inevitably leads to an increase in pressure (diffuser effect). In addition, the lower part of the airbox keeps the heat radiation of the engine away from the intake air. In addition, the airbox usually houses the air filter, Fig. 1.90. The volume of the airbox was about 50 l for the 3-l naturally aspirated engines of Formula 1 [49].

The shape is not only interesting on the inside, but also on the outside: In formula cars, it tapers towards the rear so that the outer contour of the car is disturbed as little as possible. After all, the rear end is supposed to taper out slimly and thus make the inflow of the rear

Fig. 1.90 Air filter on a V8 engine of a formula car (Lola Zytek 3000). The air intake with the upper part of the airbox is removed. You can see the contour of the engine cover on the right side of the picture. In the foreground the intake manifolds with the flat slide valve and the fuel distribution rail

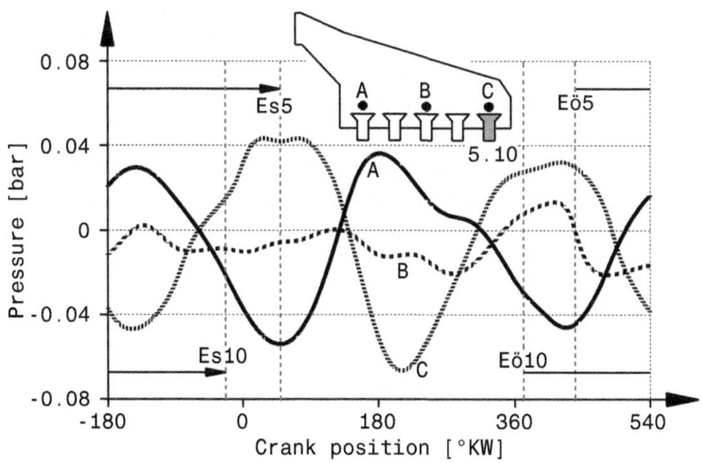

Fig. 1.91 Pressure oscillations in an airbox of a V10 cylinder engine, after [49]. The pressure curve at 15,000 min^{-1} is plotted for three positions above the crank angle. For two adjacent intake ports (cyl. 5 and 10), the timing of the intake valves is also plotted

wing more even. The pressure point of the vehicle can be shifted to the rear by a large airbox in the side view.

In fact, differences in the performance of individual cylinders also occur with airboxes. The reason lies in pressure oscillations within the air distributor, which are induced by the ignition sequence-dependent intake processes of the individual cylinders, Fig. 1.91.

Depending on their characteristics, these pressure waves lead to a preference or disadvantage of one cylinder and thus to unequal power outputs. This phenomenon is made more difficult to eliminate by the speed dependence of the pressure oscillations.

In the case of carburetor engines, it should also be noted that the ventilation holes of the float chamber also open inside the airbox, otherwise the pressure difference can become so great that the mixture composition is outside the misfire limits (ignition limits).

Airboxes are also advantageously used when an air volume limiter is prescribed, see below.

Intake Manifold

The cross section of a circular suction pipe can be determined from the following relationship [50]:

$$A_O = k_s \cdot P_{e,\max} \cdot \frac{4}{z}$$

A_O	Single pipe cross-section with circular shape, mm^2
k_s	Specific single pipe cross-section factor, mm^2/kW $k_s = 11$ to 20 mm^2/kW. The upper limit applies to high-power engines
$P_{e,max}$	Maximum effective engine power, kW
z	Number of cylinders, -

The ideal intake manifold cross-section is circular. If the cross-section is rectangular, the area must be increased to compensate for the changed friction conditions. From the conditions that both pressure drop and air mass flow rate must remain the same, it follows for the length a of the rectangle, Fig. 1.92, for single-phase turbulent gas flow ($2300 \leq Re \leq 10^5$) [50]:

$$a = \frac{d}{2}\left[\frac{\pi^7(1+\Phi)^5}{\Phi^{12}}\right]^{1/19}$$

a	Basic length of the single pipe rectangular cross section, mm
d	Diameter of the single pipe circular cross section, mm
Φ	Aspect ratio of the rectangular section, -. $\Phi = b/a$

The required increase in cross-section or base length of the rectangle as a function of aspect ratio is shown graphically in Fig. 1.92.

With a known pipe cross-section, the intake manifold length is derived from the required intake manifold volume. In order to exploit gas dynamic effects in a naturally aspirated engine, the volume should be approx. 1.5–3 times the displacement it supplies [41]. In the

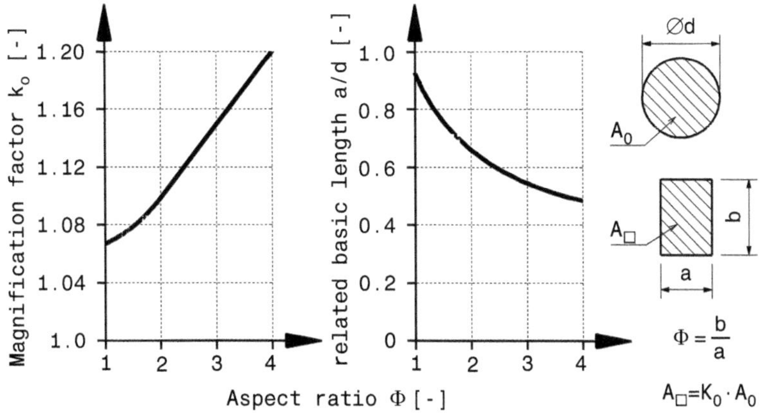

Fig. 1.92 Required increase in cross-section when changing from a circular to a rectangular intake manifold cross-section, according to [50]. From the aspect ratio Φ of the rectangle follows the necessary increase of the rectangle cross-section *A*□ or the base length *a* of the rectangle

case of the intake manifold, the same opposing gas-dynamic effects are superimposed as must also be weighed in the intake port design, cf. Figure 1.35. As the pipe diameter increases, the maximum air consumption shifts to higher engine speeds, Fig. 1.93.

A long intake manifold leads to high air consumption at low engine speeds, but also to a loss of power due to the greater wall friction associated with it. Conversely, the resonance of a short intake manifold is at high speeds and this results in high air consumption and, combined with high speed, in high power, Fig. 1.94.

The basically generally valid results of the influence of intake manifold length and diameter are described in Fig. 1.95 for a 2-l petrol engine.

For the intake manifold length, therefore, a compromise must be found between a full power curve and high power at nominal speed. Technically ideal would of course be an intake manifold with variable length and diameter. Then the optimum dimensions could be set for every load and speed. At least for the variable length, numerous designs exist. In high-end production vehicles, intake manifolds with multiple stepped lengths are used. In Formula 1, continuously variable systems were used as long as they were permitted by the regulations. Figure 1.96 illustrates the function and Fig. 1.97 presents a sectional model.

Intake Funnel Design

The design of the intake funnel is of great importance because at this point of the intake tract – apart from the air filter – the first losses can occur which can no longer be compensated. The air flows in mainly from the side (Fig. 1.98b), so an intake funnel with a curve that extends at least 90° to the funnel axis is conducive to filling. The radius of curvature should be at least ¼ of the mean flow diameter. A judicious combination of cone and cylinder sections gives the most useful performance curve, Fig. 1.98c, design (2). A

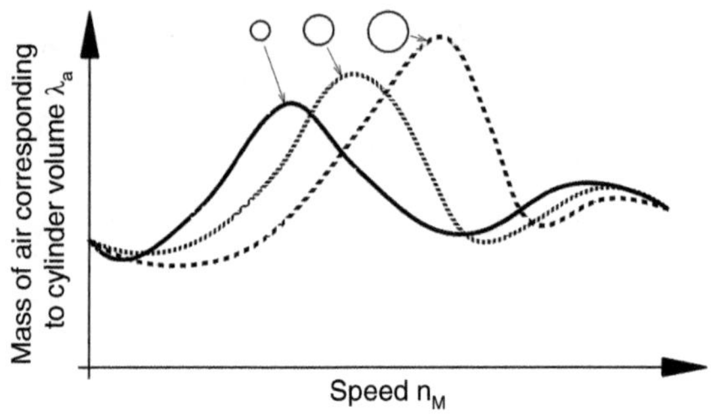

Fig. 1.93 Air consumption as a function of pipe diameter. The maximum air consumption increases with the pipe diameter, but at the same time shifts to high speeds

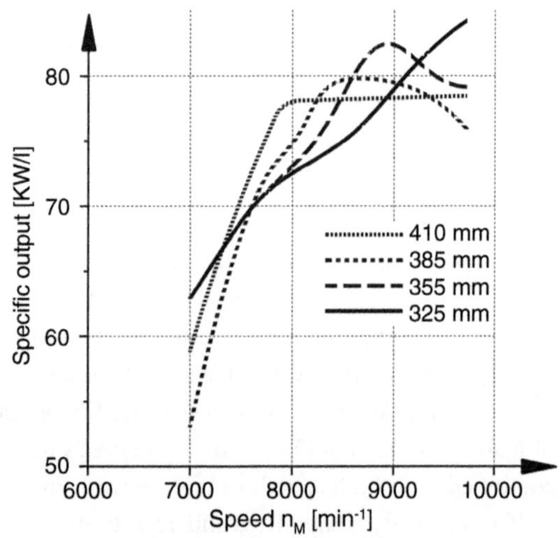

Fig. 1.94 Maximum power as a function of the suction pipe length, according to [41]. The maximum power is related to the displacement

distinctly funnel-shaped intake manifold (1) produces the highest maximum power, but the curve is very steep, i.e. a closely stepped gearbox is needed (Fig. 1.98c).

The principle of cross-sections for the intake tract of sports engines is shown in Fig. 1.99. For a nominal speed of 5500–6000 min^{-1} the length L must be about 400 mm. The air expenditure between 3000 and 6000 min^{-1} can be further increased if the length is raised above 600 mm and the intake funnel is enlarged. Due to the constriction, the air effort in the speed range 3000–5500 min^{-1} is raised. If the air consumption is to be higher above 5500 min^{-1}, a constant cross-section is more favourable.

Suction Tube with Airrestrictor

An air flow restrictor is a throttle in the intake tract prescribed by the regulations, which means that only a certain air mass can be drawn in and thus the maximum power of an internal combustion engine is also limited. For given boundary conditions such as ambient pressure and density of the ambient air, the maximum possible air mass flow is determined by the area of the narrowest cross-section A_{Rs}, Fig. 1.100. If sonic velocity occurs in the throttle cross-section (critical pressure ratio), the maximum throughput is reached:

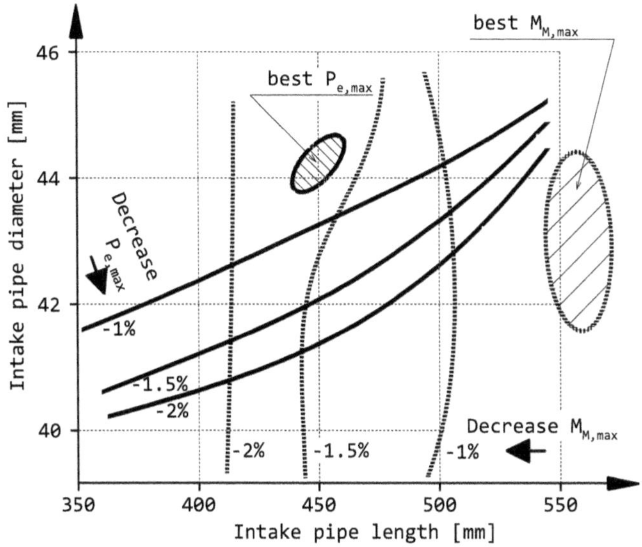

Fig. 1.95 Influence of intake manifold diameter and length on maximum torque and power, after [33]. Calculation results for a 2l- 4-valve gasoline engine. The best values for the maximum power $P_{e,\max}$ are provided by a suction pipe of medium length (450 mm) and relatively large diameter (over 44 mm). If the diameter is reduced, the maximum power decreases. It can also be seen that the power remains the same when the length decreases with the diameter. Long intake manifolds (over 550 mm) give best values of maximum torque with medium diameters (42–44 mm). The maximum torque also decreases with length

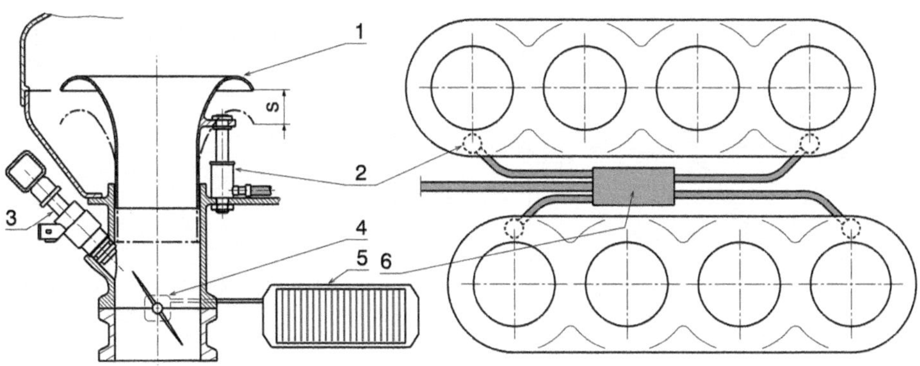

Fig. 1.96 Adjustable intake manifold length on a 3.5 l Formula 1 naturally aspirated engine. Hydraulic pistons (2) adjust the length of the intake funnels (1) and thus the natural frequency of the intake system. **1** *Intake trumpet,* **2** Hydraulic cylinder, **3** *Injection* valve, **4** *Throttle valve with* potentiometer, **5** *ECU (engine control unit),* **6** *Hydraulic* distributor

Fig. 1.97 Switch intake manifold (Ferrari Formula 1 V10 cylinder). The cutaway model shows the area of the intake manifold that accommodates the throttle valves. The intake funnels slide up and down on this. The position of the funnels is changed by the two levers visible on the left of the picture. The throttle valves are controlled by drive-by-wire unit. Note also the position of the injectors: They are located in front of the intake funnels

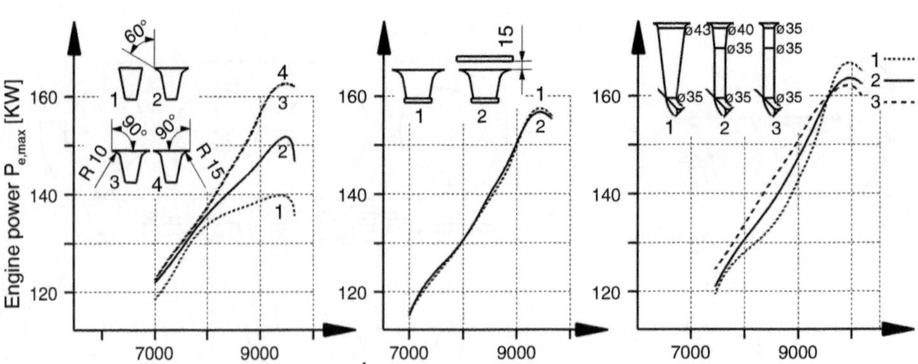

Fig. 1.98 Test results on a racing engine, after [41]. (**a**) inlet rounded with min. 0.25d to 90° to suction pipe axis (4). (**b**) Air flows in mainly from the side: The plate (2) only interferes when the annular gap area becomes smaller than the intake manifold cross-section. (**c**) Highest, but most pointed performance with version (1). Cylindrical intake manifold (3) gives favourable, linear power curve. The power curve of design (2) lies between (1) and (3) and represents a useful compromise

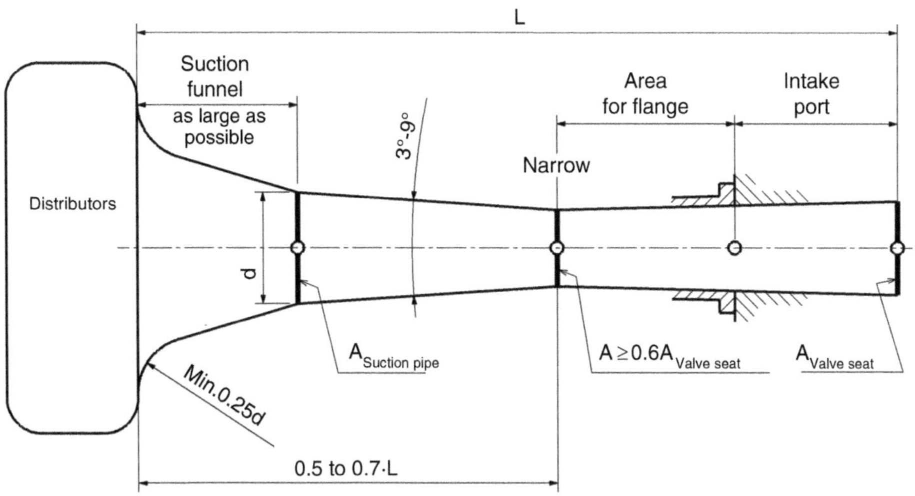

Fig. 1.99 Intake tract of sports engines, schematic. The intake tract consists of the distribution volume or airbox, intake funnel, intake manifold and intake port in the cylinder head. The cross-section A initially decreases up to a narrow point and increases again slightly from there up to the mouth in the combustion chamber

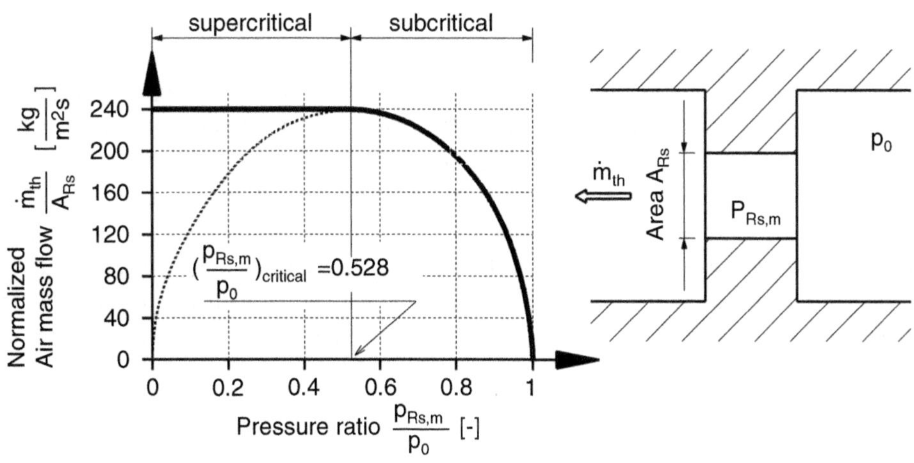

Fig. 1.100 Theoretical air mass flow through a throttle point, after [43]. If the pressure ratio is lowered, the air mass flow increases as expected. However, only up to the critical ratio. From this point on, a further pressure reduction no longer increases the mass flow rate. $p_{Rs,m}$ mean pressure in the restriction. p_0 ambient pressure. Boundary conditions: $p_0 = 1.013$ bar. $T_0 = 293.15$ K. $R_L = 287.04$ J/(kgK)

$$\dot{m}_{th,\max} = 240 A_{Rs}$$

$\dot{m}_{th,\max}$	Maximum theoretical air mass flow, kg/s
A_{Rs}	Restrictor cross-section, m^2

Based on (1.1) with (1.3) or (1.4), the maximum engine power can also be written as:

$$P_{e,th,\max} = \eta_e \cdot \dot{m}_{th,\max} \cdot \frac{H_u}{\lambda \cdot L_{\min}}, \text{ also : } P_{e,th,\max} \propto \dot{m}_{th,\max} \tag{1.8}$$

$$\text{with : } \quad \dot{m}_{th} = i \cdot \lambda_a \cdot n_M \cdot z \cdot V_h \cdot \rho_L$$

From (1.1) it follows that the critical speed at which this maximum power occurs is also given, i.e. the cylinder charge corresponds exactly to the maximum air mass flow:

$$n_{M,critical} = 60000 \frac{\dot{m}_{th,\max}}{i \cdot z \cdot V_h \cdot \lambda_a \cdot \rho_L}$$

$n_{M,crit}$	Engine speed at which the maximum power is reached with restrictor, min^{-1}
V_h	Displacement of a cylinder, l
ρ_L	Density of air, kg/m^3

Above this critical speed, the engine power cannot be increased any further because the air flow rate cannot be increased any further. Figure 1.101 shows these relationships graphically for the ideal and for a real engine.

If an air restrictor is prescribed by regulations, the inlet is designed as a Laval nozzle with the restrictor as the smallest diameter, Fig. 1.102. The airbox is aligned in the direction of travel and contains the restrictor at the air inlet. The inlet is funnel-shaped and well rounded. The actual restrictor must be 3 mm long on this vehicle with a diameter of 24 mm. Connected to this is a diffuser which has a small opening angle so that the flow does not detach.

The arrangement of such an airbox with restrictor in the vehicle can be seen in Fig. 1.103.

In [51], a so-called Dall venturi is proposed as a simple to manufacture restrictor. In this case, a conical nozzle opens directly into an equally conical diffuser and has a circumferential groove at the narrowest cross-sectional point, Fig. 1.104. This groove represents a tear-off edge for the flow and reduces the boundary layer. Compared with conventional restrictor designs, which are designed for low pressure loss with subcritical flow, the mass flow is somewhat greater over a wide range (from Mach $M = 0.6$ to 1, i.e. the speed of sound) and thus enables an increase in power in the upper speed range.

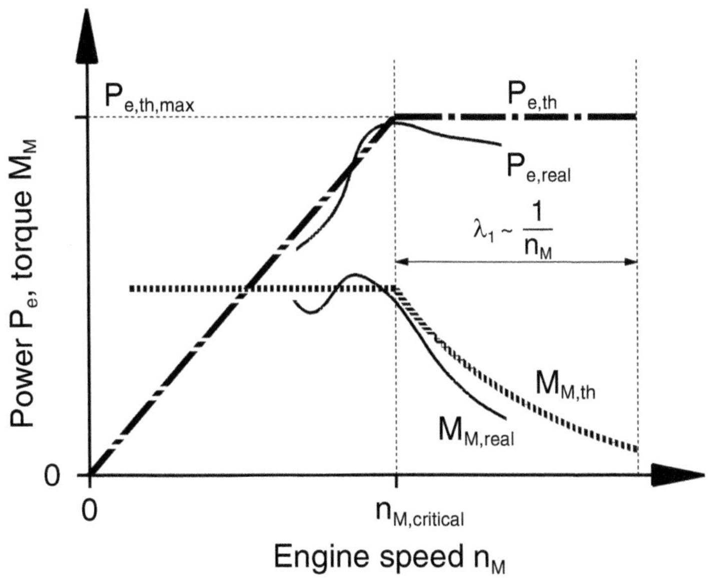

Fig. 1.101 Torque and power of an ideal and a real engine with air limiter, after [43]. The ideal engine has constant charging effieciency (degree of delivery) and effective efficiency. The power of the ideal engine $P_{e,th}$ is directly proportional to the aspirated air mass and thus to the speed. From the critical speed $nM_{,krit}$ onwards, the charging efficiency λ_1 decreases indirectly proportional to the speed

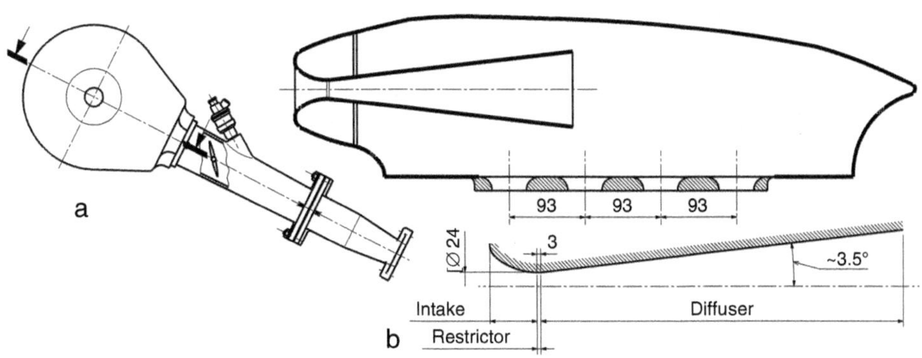

Fig. 1.102 Design of an airbox with restrictor (Opel Formula 3 2 l), according to [43]. (**a**) component, (**b**) design of the restrictor environment

Super Charging

Two types are used in principle to increase the air effort with pure gas-dynamic effects (dynamic supercharging), the oscillating tube and the resonance tube supercharging, Fig. 1.105. The effectiveness of tuned oscillating tubes increases with increasing bore-

Fig. 1.103 Airbox with restrictor on a formula car (Dallara STV2000) The restrictor supplies air to all four cylinders of the in-line engine. The restrictor is made of light metal and is laminated into the fibre-reinforced plastic airbox

stroke ratio. The effort of a 1D simulation (see appendix) therefore pays off for typical, high-revving racing engines.

Ram Pipe Supercharging The oscillating pipe effect is based on the negative pressure wave triggered by the downward moving piston, which runs in the intake pipe against the direction of flow to the collector tank and is reflected there at the open end of the pipe. The overpressure wave created in this way increases the cylinder charge by raising the pressure gradient across the inlet valve. This effect is particularly effective shortly before the inlet valves close with the piston moving upwards. Here, with the pressure wave present, the pushing of fresh charge from the combustion chamber into the intake manifold is prevented, Fig. 1.106.

The optimum oscillating tube length therefore follows from the effective inlet period within which the pressure wave must pass through the tube twice (back and forth):

$$L_1 = 30\Phi_L \frac{c_s}{n_M}$$

$$\Phi_L = \frac{Es_e - E\ddot{o}_e}{360^\circ\, KW}$$

L_1	Oscillating tube length for speed n_M, m
Φ_L	Ratio of the effective intake duration ($E\ddot{o}_e$ to Es_e) to one crankshaft revolution, -. $E\ddot{o}_e$ is about 80 °CA after TDC, Es_e is about 1/10 of the valve lift [52] For series engines Φ_L is approx.1/3–1/2, for racing engines around 1 (see also Fig. 1.55)
c_s	Speed of sound, m/s
n_M	Engine speed at which supercharging effect occurs, min^{-1}

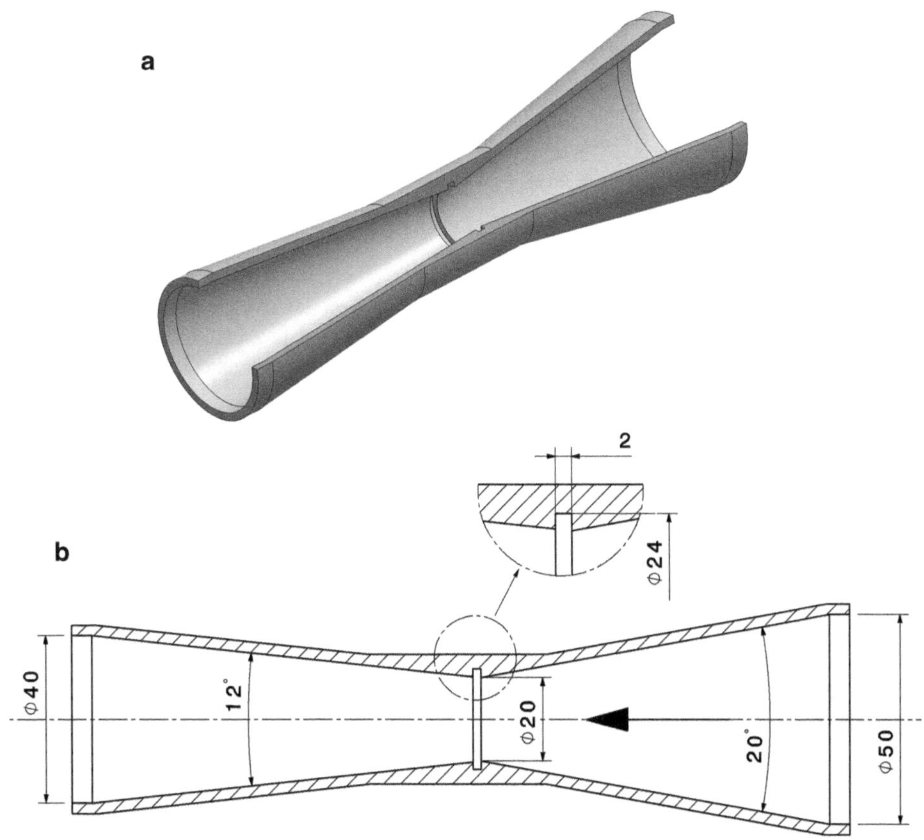

Fig. 1.104 Air volume limiter with Dall-Venturi, according to [51]. (**a**) axonometric representation (partly cut open), (**b**) drawing. The flow direction is from right to left (arrow). This restrictor was developed for the Formula SAE, i.e. the nominal diameter is 20 mm

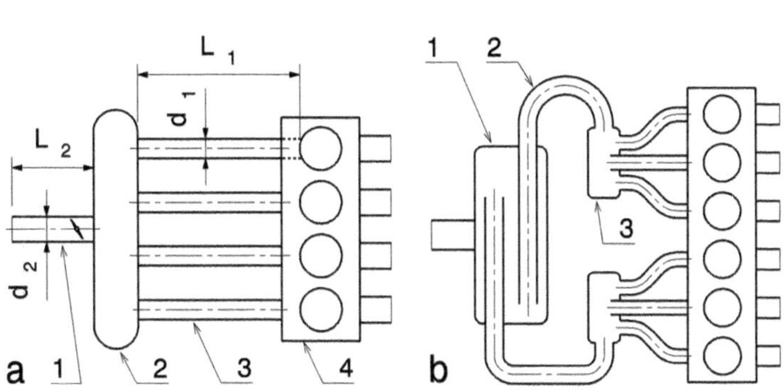

Fig. 1.105 Schematic of dynamic charging methods. (**a**) Oscillating tube supercharger, 1 Air supply tube, 2 Distribution volume, 3 Oscillating tube, 4 Engine, (**b**) Resonant tube supercharger, 1 Equalizing volume, 2 Resonant tube, 3 Resonant tank

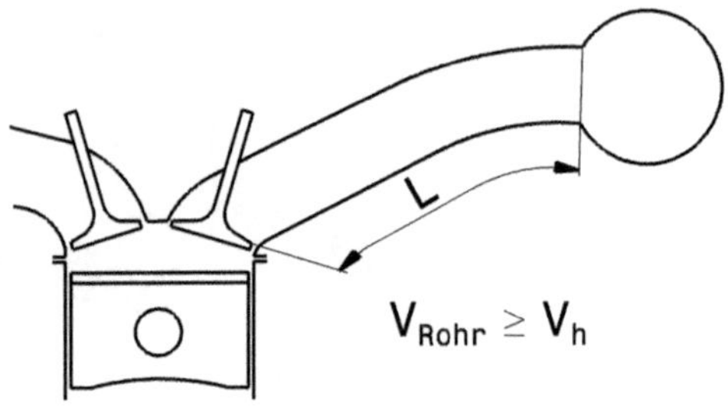

Fig. 1.106 Explanation of the ram pipe. Energy balance: The suction work of the piston is converted into kinetic energy of the gas column in front of the intake valve and this is converted into compression work of the fresh charge. The length L of the gas column extends from the valve to the intake funnel (cross-sectional jump)

A more precise statement on the tuning of the intake manifold with acoustic modeling is provided by looking at the pressure curve upstream of the intake valve (Fig. 1.107) and the well-known Helmholtz equation (inflow process into container).

According to [52], the value of $Eö_e$ is approximately 80 °CA after TDC. It_e is the point of intersection of the turning tangent of the valve lift curve with the abscissa. As an approximation, the angular value at 1/10 of the valve lift can also be used.

The optimum length L_1 of a vibrating tube for a given speed n_M can now be calculated using the following expression:

$$L_1 = \left[\frac{60}{n_M - n_{corr}} \cdot \frac{Es_e - Eö_e}{360^{\circ}} \cdot k \cdot \frac{c_s}{\pi}\right]^2 \cdot \frac{A_1}{V_h}$$

n_{corr}	Correction speed, min^{-1}. $n_{corr} = 0$ min^{-1} for open oscillating suction pipes $n_{corr} = 250$ min^{-1} for complete suction systems as in Fig. 1.105
Es_e, $Eö_e$	Crank angle at effective valve closing or opening, °CA
k	Correction factor to take account of the Helmholtz equation and the rms values of A_1, L_1, and V_h, -. $k = 0.81$
A_1	Oscillating tube cross-section, m^2. If circular, $A_1 = d_1^2\, \pi/4$
V_h	Cylinder stroke volume, m^3

Speed of sound c_s:

$$c_s = 20.02\sqrt{T_1}$$

T_1 absolute temperature of the air in the oscillating tube, K.

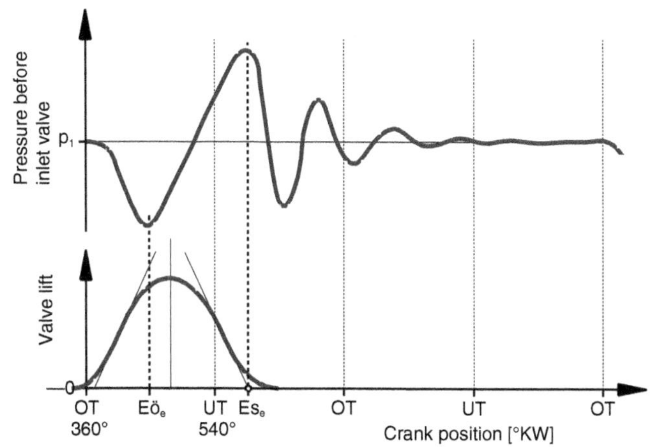

Fig. 1.107 Characteristic pressure curve upstream of the inlet valve [52]. The diagram shows the pressure curve in the intake port of a passenger car engine. In addition, the valve lift curve of the corresponding intake valve is shown. The pressure minimum occurs at the crank angle $Eö_e$. The maximum of the reflected pressure wave coincides at the angle Es_e with the closing of the inlet valve

This charging effect due to pressure oscillations in the oscillating tube can also be observed in externally charged engines [53].

The actual structural length of a vibrating tube is not exactly the theoretical length from the one-dimensional calculation, but somewhat shorter. Due to inertial effects, a vibrating gas column overhangs the touching pipe by about 60% of the inlet diameter [34]. The inlet diameter is the diameter at which the inlet torus of the intake funnel merges into the cylindrical or conical area of the intake manifold.

Resonance Induction

With this principle, the natural frequency of a tank-tube system is tuned to the desired engine speed. This method can be used particularly effectively with several cylinders that have the same firing distances. In this case, groups of cylinders are connected to a resonance container via short oscillating tubes. This container, together with a resonance tube, acts as a Helmholtz resonator with respect to the atmosphere or a compensation volume. The resonance speed for a cylinder is [7]:

$$n_M = \frac{15 \cdot c_s}{\pi} \sqrt{\frac{A_1}{L_1 \cdot (V_c + 0.5 \cdot V_h)}}$$

n_M	Speed at which resonance occurs in the suction system, min^{-1}
A_1	Cross-sectional area of suction pipe, m^2
L_1	Resonance tube length, m

(continued)

V_c	Compression volume, m^3
V_h	Displacement, m^3

Compressor

In addition to dynamic supercharging, there is also the possibility of installing a compressor in the intake tract, which significantly increases the amount of air required. When considering principle ways of increasing power, (1.3) with (1.4) provides the insight that increasing the density of air to the same extent increases the power of an engine. The density of the air is increased by a supercharger to the following extent:

$$\rho_2 = \frac{p_2}{R_L \cdot T_2} \tag{1.9}$$

ρ_2	Density of the air after the compressor, kg/m^3
p_2	Boost pressure, pa; 1 bar = 10^5 pa
R_L	Specific gas constant of air. R_L = 287 J/(kgK)
T_2	Absolute temperature after the compressor, K; x °C = (273.15 + x) K

You can also imagine the other way round that the compressor gets as much air into the cylinder as a naturally aspirated engine would only manage with x *times the* displacement. The factor x follows from the simple (actually the conditions are more complicated because of recooling etc.) comparison of the energy quantities:

$$p_1 \cdot V_1 = const. = p_2 \cdot V_2 \rightarrow \frac{p_2}{p_1} = \frac{V_1}{V_2} = x$$

V_1	Displacement of the naturally aspirated engine
V_2	Displacement of the supercharged engine
p_1	Intake manifold pressure
x	Displacement ratio

If, for example, the compressor builds up a boost pressure ratio p_2/p_1 of 2, this roughly corresponds to an increase in displacement of the naturally aspirated engine by double.

A more accurate statement for constant displacement is provided by the aforementioned extension of the expression for mean effective pressure, (1.3), with (1.4):

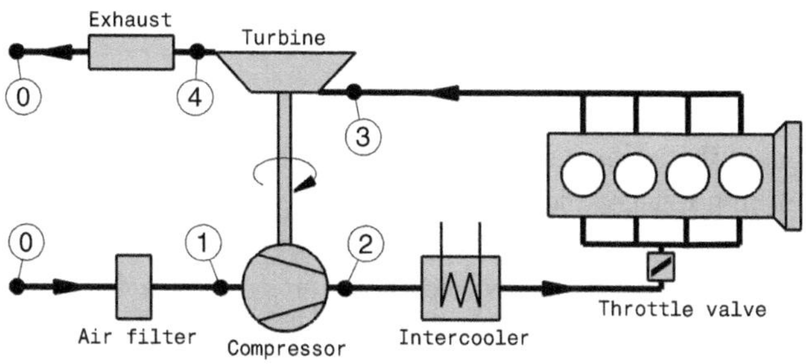

Fig. 1.108 Schematic of an exhaust gas turbocharger. Indices for state designations of air or exhaust gas: 0 Ambient state. 1,2 State before or after compressor. 3,4 State before or after turbine

$$p_{m,e} = \eta_e \cdot \lambda_a \cdot \rho_L \cdot \frac{H_u}{\lambda \cdot L_{\min}}$$

The mean effective pressure is therefore proportional to the air density ρ_L. Compression increases the density and thus increases the power of the engine to the same extent. If we assume polytropic compression and relate the mean pressures of the turbocharged engine to the naturally aspirated engine, this provides an expression for the influence of the boost pressure:

$$\frac{p_{m,e,T}}{p_{m,e,N}} = \frac{\rho_{L,t}}{\rho_{L,n}} = \frac{\rho_2}{\rho_1} = \left(\frac{p_2}{p_1}\right)^{\frac{1}{n}} \tag{1.10}$$

$p_{m,e,T}$	Mean effective pressure of the supercharged engine, bar
$p_{m,e,N}$	Mean effective pressure of the naturally aspirated engine, bar
p_2/p_1	Boost pressure ratio, -. Indices 1.2 see also Fig. 1.108
n	Polytropic exponent, -. For vehicle engine compressors, n = 1.52 to 1.62

The actual power increase with a boost pressure ratio p_2/p_1 of 2 is therefore in the range of 1.53–1.58 times. A further increase can also be achieved by recooling the compressed air. If this is cooled back to 40 °C (= 313.15 K) by charge air cooling, for example, the increase in density from 20 °C according to (1.9) and thus the increase in power is 1.87 (Fig. 1.108).

Turbocharged engines offer the advantage that smaller displacements and thus engines with smaller dimensions and lower mass can be used for the same power output (*downsizing*). This is reflected in more favorable packaging in the vehicle and lower engine friction. Compared to a naturally aspirated engine of the same power, the turbocharged engine can also be operated at a lower engine speed as an alternative to changing the displacement, which is also beneficial for mechanical efficiency (cf. Figure 1.17). Generally, a combination of both alternatives is formed as a compromise.

Depending on how the compressor is driven, a distinction is made between mechanical and exhaust gas turbocharging.

Mechanical Boost The supercharger is driven directly by the crankshaft via gears or belt drives. Although this reduces the efficiency of the engine, its mean pressure increases. Compared to a naturally aspirated engine of the same power, the supercharged engine has lower mechanical and thermal losses and thus, on balance, a better efficiency. Roots blowers, screw compressors or spiral superchargers are usually used as superchargers. More rarely, centrifugal compressors are mechanically driven by the crankshaft. With low numbers of cylinders (≤ 3), the stability of the mechanical drive can become a problem due to the rotational irregularity of the engine. A remedy was provided by an elastic intermediate element for vibration decoupling.

Pros:	Relatively simple charger units on the cold engine side Instantaneous response to load changes
Disadvantages:	Increased fuel consumption Supercharger cannot be placed anywhere on the engine because of charger drive

Exhaust Gas Turbocharging

. In exhaust gas turbocharging, a radial compressor is driven by a turbine in the exhaust tract. The engine and compressor are therefore only thermodynamically coupled. The turbine uses part of the exhaust gas energy that is otherwise dissipated to the environment and that the reciprocating engine cannot use because of the crank mechanism (incomplete expansion).

Pros:	Considerable increase of the litre capacity Richer torque curve Lower fuel consumption in comparison with a naturally aspirated engine of the same performance
Disadvantages:	Charger is installed in the hot exhaust area Low base torque at low engine speeds Delayed load absorption behaviour (*turbo lag*)

In recent years, there has been a trend in series production vehicles to reduce engine displacement while maintaining or increasing performance in order to reduce fuel consumption. This is done primarily by supercharging the engine. Some racing series have taken up this development and adapted their regulations accordingly. For example, from 2014 onwards, turbocharged V6 engines with a displacement of 1.6 l will provide the main drive in Formula 1 instead of the 2.4 l naturally aspirated V8 engines (a KERS system may

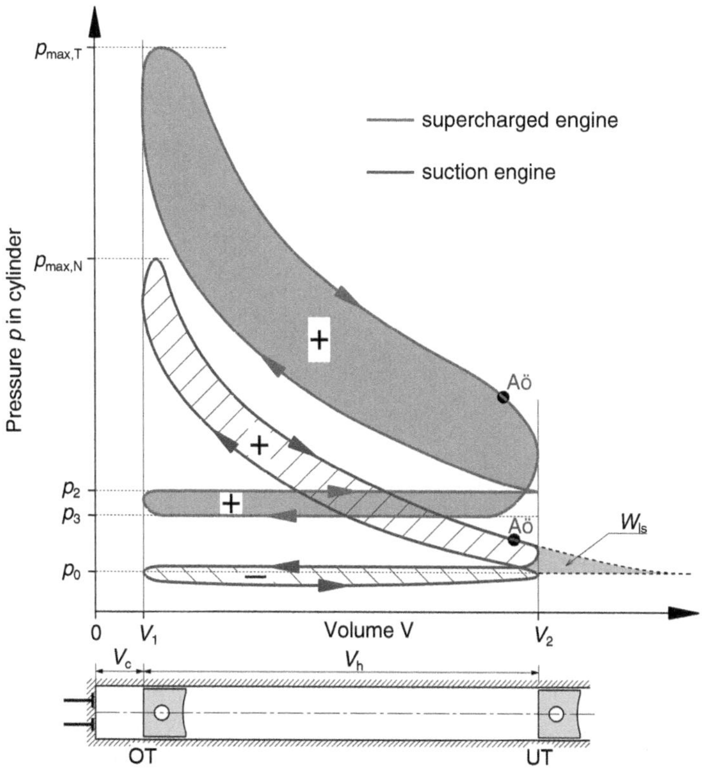

Fig. 1.109 pV diagram with and without charging. Aö Exhaust valve opens, p_0 Ambient pressure (atmospheric pressure), p_2 Boost pressure, p_3 Exhaust gas back pressure, $p_{max,N}$ Peak combustion pressure, naturally aspirated engine, $p_{max,T}$ Peak combustion pressure supercharged engine, W_{ls} Dissipated work due to incomplete expansion (exhaust energy), $V_1 = V_c$ Compression volume, V_2/V_1 Compression ratio, V_h Displacement. $V_h = V_2–V_1$. Areas circled in a clockwise direction represent useful work (+). Areas that are enclosed in an anticlockwise direction are included in the overall balance as work lost (–), e.g. the charge exchange loop in the naturally aspirated engine. For ease of comparison, both compression volumes V_1 are shown as the same size. In fact, a larger compression volume would be required for the turbocharged engine with the same displacement so that the effective compression does not become too large

serve as a further drive for a short time, see Chap. 3 *Hybrid Drives*). In addition, the maximum fuel consumption will be limited to 100 kg/h.

The indicator diagram (cylinder pressure versus volume, *pV diagram*) shows the effect of supercharging, Fig. 1.109. A continuous curve represents the entire working cycle of a 4-stroke engine. Arrows illustrate the bypass sense of the cycle. The naturally aspirated engine generates a slight negative pressure relative to the atmosphere p_0 during intake (piston at TDC) and compresses during the subsequent upward movement of the piston until top dead centre (TDC) is reached. Combustion causes the pressure to rise beyond this to the peak combustion pressure $p_{max,N}$. The piston is displaced towards bottom dead centre

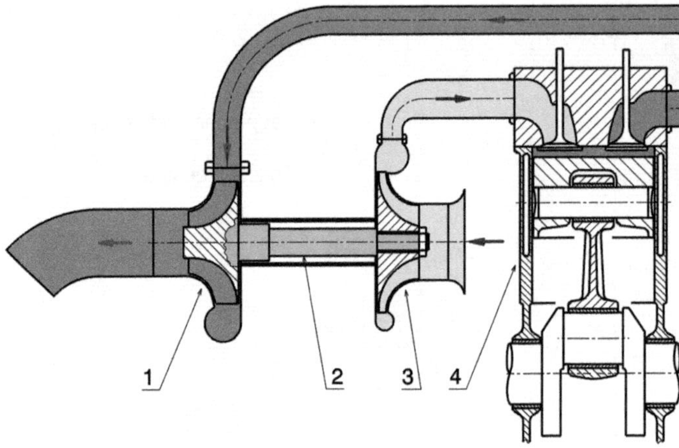

Fig. 1.110 Principle of exhaust gas turbocharging. 1 Turbine, 2 Connecting shaft, 3 Compressor, 4 *engine*. The connecting shaft together with the mounted compressor and turbine wheel is called the running gear

(BDC) by the pressure and the cylinder pressure drops due to the end of combustion and the increase in volume. When the exhaust valve (Aö) opens, however, the gas has not yet expanded to ambient pressure p_0 and the exhaust gas exits at a positive pressure relative to ambient. The cycle is completed and the process starts again. You can see a triangular area that would be usable if the exhaust gas were expanded to ambient pressure. However, this exhaust gas energy W_{ls} is lost in a conventional crank mechanism (loss due to incomplete expansion). With an exhaust gas turbine, however, part of the exhaust gas energy can be fed into the cycle.

In the turbocharged engine, the pressure level of the cycle is higher. For the intake stroke, the compressor provides a boost pressure p_2 that is higher than the ambient pressure p_0. The subsequent high-pressure loop basically differs from that of the naturally aspirated engine only in the higher pressure level and the enclosed area. The peak combustion pressure $p_{\mathrm{max,T}}$ is also significantly higher than that of the naturally aspirated engine. After opening the exhaust valve (Aö), the exhaust gas reaches the turbine. The turbine is subjected to the exhaust backpressure p_3. If this is below the boost pressure p_2 (positive scavenging gradient, as in the diagram), the piston also performs positive work during the charge change. Through the connection of the exhaust gas turbine with the compressor, part of the otherwise lost exhaust gas energy is thus used for the cycle. A negative purging gradient ($p_2 < p_3$) must be avoided. In this case, high residual gas contents reduce the cylinder filling with fresh gas and thus have the effect of reducing performance and increasing fuel consumption.

The operating principle of exhaust gas turbocharging is outlined in Fig. 1.110. The fresh air is drawn in by the compressor (3), compressed and fed to the combustion chamber of the engine (4). The exhaust gas from the combustion chamber consists of the air mass plus the fuel mass and feeds the turbine (1), which directly drives the compressor via a connecting shaft (2). After expansion in the turbine, the exhaust gas flows into the open air.

A schematic of an exhaust gas turbocharger with the most important state designations is shown in Fig. 1.108. The designations are given as an index for formula symbols, e.g. p_1, T_1, etc.

Even though the exhaust gas turbocharger is not mechanically coupled to the engine, the two turbomachines are connected. The turbine power and the compressor power are the same in steady-state operation (free-running condition). Thus, for high overall efficiency, the turbine must be matched to the compressor. Usually, an operating point is selected in the compressor map. This results in a certain rotational speed of the rotor. The turbine wheel diameter is designed in such a way that the turbine operates with the highest efficiency at this speed. The tuning is carried out with the aid of characteristic diagrams. These represent the gas throughput versus the pressure ratio. So that these diagrams can be used in a generally valid manner, i.e. independently of inlet temperature and inlet pressure, the variables are standardised, i.e. set in a relationship with reference values. Figures 1.111 and 1.112 each show an example of a characteristic diagram for a compressor and a turbine.

The usable map range of a compressor is framed by the surge (pumping) limit, the maximum speed and the stuffing limit, Fig. 1.111. If the volume flows are too small and the pressure ratios are too high, the air flows backwards through the compressor until a stable pressure ratio is established again and the air is conveyed back to the compressor outlet. The resulting increase in boost pressure causes the flow to break away from the compressor blades again and the process repeats. The periodic noise gives this phenomenon the name "pumping". When the air at the compressor inlet has reached the speed of sound, no further increase in mass flow rate is possible: the stuffing limit has been reached. The maximum speed of the impeller is dictated by the strength of the blades.

The throughput behaviour of a 4-stroke engine is plotted in Fig. 1.111b as so-called sip lines. As a volume displacement engine, the throughput of a reciprocating engine is primarily dependent on the engine speed (cf. 1.8). As the engine speed increases, more volume is displaced. At constant engine speed n_M the volume flow $\dot{V}_1$ increases only slightly and linearly with increasing pressure ratio. In the compressor map, the matching of the compressor with the engine can thus be entered. In the example a full load absorption line is entered. Starting at low engine speeds, the compressor ramps up close to the surge limit and, once a certain boost pressure is reached, it is kept constant by a control system. The full load operating line runs at higher engine speeds in the range of high compressor efficiencies. The full load line shows the entire engine operating range in the compressor diagram: At full load, the operating points are at and at part load below the entered operating line.

Regulation

The turbine is designed so that sufficient boost pressure is available even at low loads and engine speeds. It is therefore adapted to a smaller exhaust gas flow and not to the maximum exhaust gas throughput of the engine. Due to its small size, it has a low mass moment of

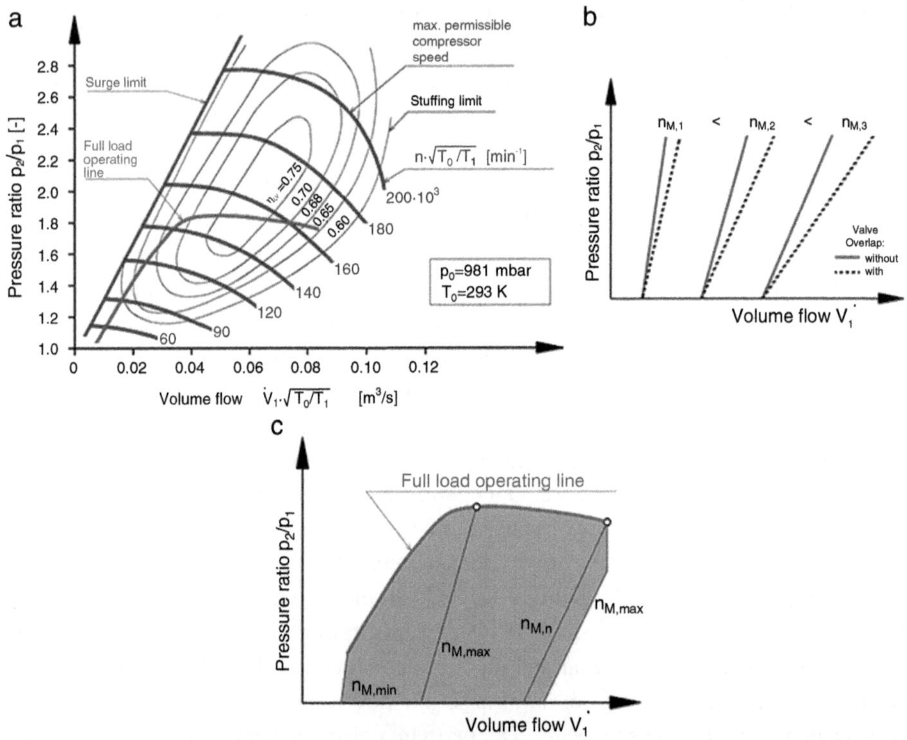

Fig. 1.111 Compressor map with engine operating line [54]. (**a**) Compressor map with full load line. The course of the engine's full-load operating curve results from the control of the turbocharger. When a certain boost pressure is reached, part of the exhaust gas volume is routed past the turbine and the boost pressure remains approximately constant. $\eta_{s,v}$ isentropic compressor efficiency. (**b**) Sip lines (absorb curve) of a 4-stroke engine at different engine speeds n_M. (**c**) Engine operating range. $n_{M,\min}$ lowest full load speed, $nM_{,\mathrm{Mmax}}$ speed at maximum torque, $n_{M,\mathrm{n}}$ speed at rated power, $n_{\mathrm{M,max}}$ maximum speed

inertia and accelerates faster, so it responds more quickly to load changes. At high engine speeds, however, this would lead to destruction of the engine due to excessive boost pressures, or the speed limit of the turbocharger would be exceeded. Therefore, it is necessary to limit the boost pressure. This is done either on the pressure side of the compressor or on the inlet side of the turbine. On the pressure side, a pressure relief valve can blow off the excess air (*pop-off valve*) and thus limit the boost pressure. However, this method is energetically clumsy because part of the compressor work is lost to the engine. A more favorable method is to regulate the turbine. The following methods are in use:

- Self-regulating boost pressure control via a boost pressure-controlled bypass valve (boost pressure limitation).

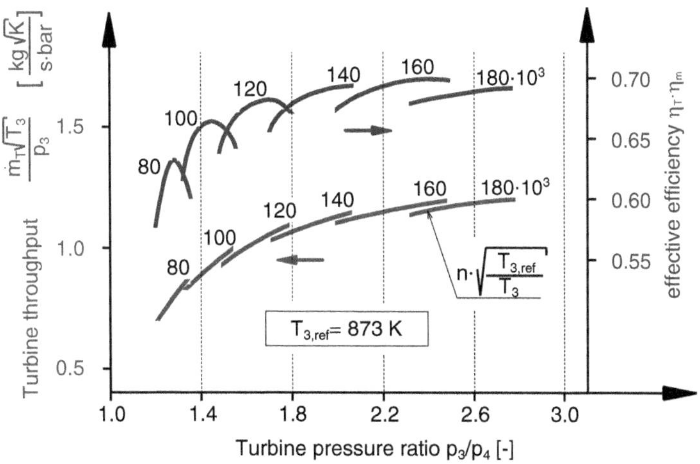

Fig. 1.112 Turbine map of an exhaust gas turbocharger [54]. The turbine speed n normalized over temperatures is given in [min^{-1}]. The higher the pressure ratio, the greater the turbine throughput and its speed. The greatest efficiencies are achieved at the highest speeds. Indices: T turbine, 3 before turbine, 4 after turbine, ref. reference value

- Electronic boost pressure control.
- Adjustable turbine geometry (VTG).

Figure 1.113 shows the schematic of an electronic control system with which the boost pressure can not only be limited but also adapted to the operating state of the engine in the partial load range. In principle, the system works like those with the volume flow to the turbine limited by the boost pressure. In such self-regulating systems, there is only the connecting line from the discharge pipe after the compressor (7) to the bypass valve (1). As soon as the boost pressure reaches a certain value, the spring-loaded valve (1) opens due to the pressure acting on a diaphragm and a partial flow of the exhaust gas bypasses the turbine (8). The spring force – and thus the boost pressure – can be adjusted via an adjusting screw (a). The electronic system shown is extended by the cycle valve (6). This valve can vary the control pressure for the boost pressure control valve (1). To do this, it is connected not only to the pressure pipe downstream of the compressor (7), but also to its intake side. The spring of the boost pressure control valve (1) is designed to be considerably softer than that of the simple boost pressure limiting systems, so that it can open even at lower boost pressures. The cycle valve is designed as a two-way valve whose central cross-section can be changed by the engine control unit (5) and thus the control pressure in the control valve (1). For this purpose, the control unit processes signals from the throttle valve position (4), the boost pressure p_2, the charge air temperature T_2, the engine speed n_M, the coolant temperature T_C and from the knock sensor in gasoline engines. This also makes it possible to briefly increase the boost pressure, as may be needed for overtaking during heavy acceleration (overboost, push-to-pass system).

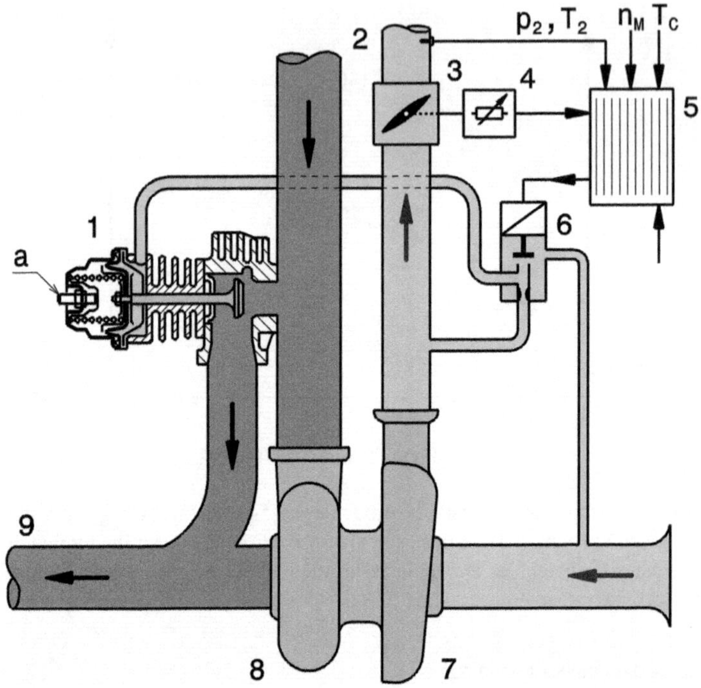

Fig. 1.113 Electronic boost pressure control. 1 blow-off valve (*wastegate*), a *adjustment spring* preload, 2 to engine, 3 throttle valve, 4 potentiometer, 5 *engine* control unit (*engine ECU*), 6 *triggered valve*, 7 compressor, 8 turbine, 9 exhaust system

In racing engines of the 1980s, the then purely mechanical control valve (6) could be adjusted via a rotary knob ("steam wheel") in the cockpit and thus the maximum boost pressure could be set by the driver.

For cost reasons, the blow-off valve is located directly in the turbine housing in series engines. For racing engines, however, the arrangement as shown in the illustration is preferable. The blow-off valve can be placed at a suitable location and the bypassed exhaust gas flow does not interfere with the turbine outlet.

With the boost pressure control described, part of the exhaust energy is lost for turbocharging when bypassing the turbine. With an adjustable inlet to the turbine, the entire exhaust gas flow can be used and the efficiency increased. Such a variable turbine geometry (VTG; *variable nozzle turbine VNT*) is shown in Fig. 1.114. Rotatably arranged guide vanes (4) change the free inlet cross-section from the volute casing to the turbine wheel. Between two extreme positions – open (b) and closed (c) – intermediate positions can be approached continuously. For this purpose, the adjusting ring (1) is turned, causing the blade levers (3) to rotate with the bearing journals of the blades. The vane journals are seated on the bearing ring (2) which seals the inlet area towards the bearing housing. The adjusting ring is operated electronically, for example, by a pressure box such as the bypass

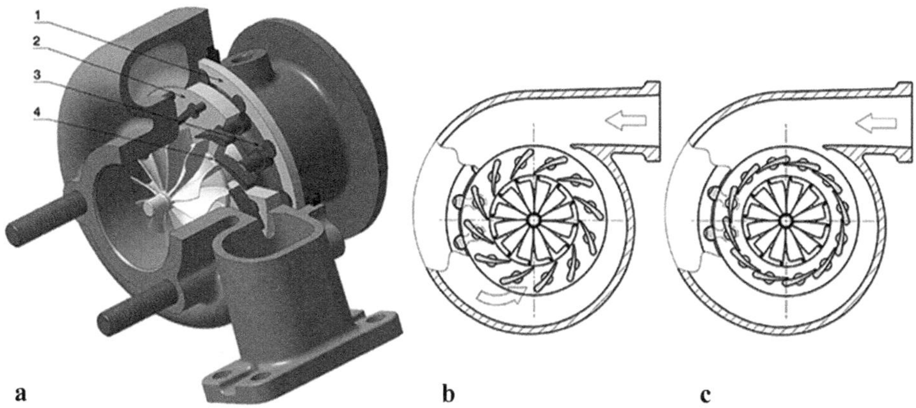

Fig. 1.114 Design and operating principle of a variable turbine geometry (only turbine and bearing housing shown). (**a**) Axonometric representation, partially sectioned, (**b**) Position of blades open, (**c**) Position of blades closed. 1 Adjustment ring *(actuator ring)*, 2 *vanes pivot carrier*, 3 *vane lever*, 4 Guide vane

valve, using only a connecting rod. At low engine speeds, the inlet cross-section is reduced. As a result, the pressure at the turbine inlet increases and the power of the turbine increases with the pressure ratio thus increased. In addition, the guide vanes create the desired angle of attack to the turbine wheel and increase turbine efficiency. The engine also responds more quickly to load changes. With increasing exhaust gas throughput, i.e. at higher engine speeds, the inlet cross-section is increased. As a result, the exhaust backpressure does not become too high and the scavenging gradient remains positive. VTG turbochargers are available for gasoline and diesel (compression ignited) engines, although the design for the spark-ignition engine was realized later because the higher exhaust gas temperatures require more heat-resistant materials.

Figure 1.115 shows the effect of VTG control in driving mode in comparison to wastegate control. A higher boost pressure is already noticeable in coasting mode due to the improved flow to the turbine wheel. During subsequent acceleration, the boost pressure also reaches the limit specified by the regulations earlier.

Response

The delay time from depressing the accelerator pedal to the noticeable build-up of torque characterizes the engine's response behavior. For a racing engine in particular, a fast and clean transition from the overrun to the acceleration phase is important. Especially in adverse road conditions (rain) and high-powered cars. In general, there are a number of measures that can be taken to eliminate or at least reduce the annoying disadvantage of the sluggish response of a turbo engine:

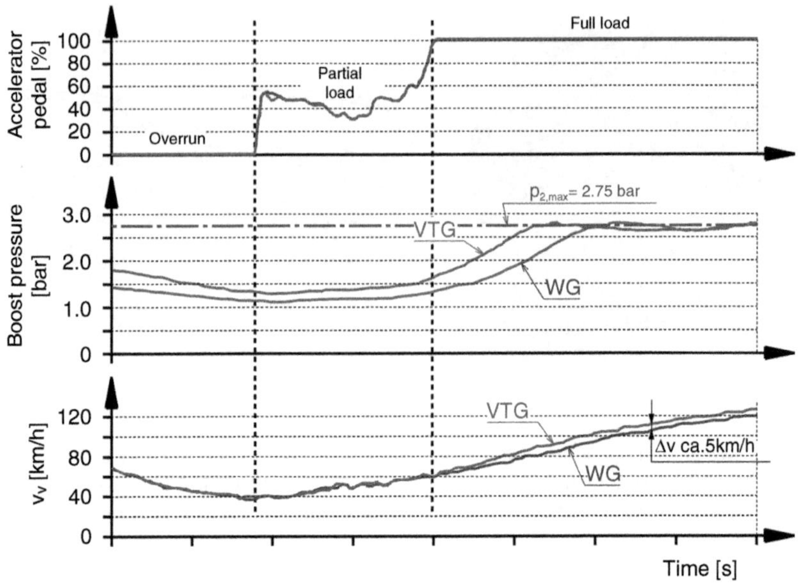

Fig. 1.115 Comparison of VTG charger and wastegate control in dynamic behaviour, Audi R15 TDI [6]. v_V Vehicle speed, $p_{2,\max}$ boost pressure limited by ACO regulations, VTG exhaust gas turbocharger with VTG control, WG Exhaust gas turbocharger with wastegate control. The response of the VTG supercharger is significantly better, which in this case results in a speed advantage of about 5 km/h

- Boost pressure control.
- Two superchargers (especially on V-engines), instead of one, or a smaller supercharger with lower mass moment of inertia.
- Flow-favourable and throttled air routing on the inlet side as well as corresponding design of the exhaust gas routing.
- Volume of the air duct (especially the air collector) downstream of the compressor as small as possible.
- Intercooling with high efficiency.
- High geometric compression of the engine.
- Variable turbine geometry.
- Rolling bearing mounted common shaft turbine – compressor (radial bearing); axial bearing omitted due to spindle bearing.
- Electric assistance of the common shaft turbine – compressor (euATL) or electric additional compressor (eBooster).
- Short-term additional air injection onto the compressor wheel by means of an additional compressor.
- Anti-Lag System (ALS).
- Multi-stage charging.

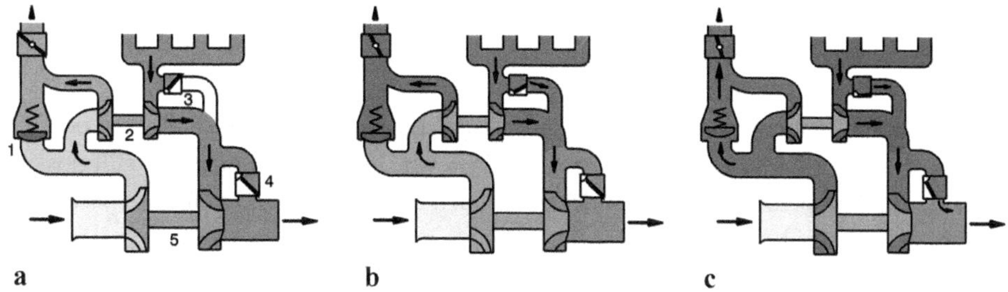

Fig. 1.116 Schematic of a register charging system, Audi 3.0 l V6 TDI Biturbo [55]. (**a**) two-stage for lower speed range. Control via high pressure VTG. (**b**) two-stage for medium speed range. Controlled via turbine switching valve. **c** single stage for high speed range. Regulated via low pressure wastegate. 1 Compressor bypass valve, 2 High pressure turbocharger with VTG turbine, 3 Turbine changeover valve, 4 Wastegate low pressure turbine, 5 Low pressure turbocharger

Anti-Lag-System

(ALS, also onomatopoeically "Bang-bang"). Especially in turbocharged rally engines, where the restrictor has a sensitive influence on the compressor behavior and the turbochargers are large, the following measure is often taken to improve the response from the sliding mode. A bypass valve directs the air compressed in front of the closed throttle valve directly into the exhaust tract in front of the turbine in sliding mode. At the same time, the engine management system reduces the ignition angle (approx. 40 °CA or more later, i.e. about 20° after TDC!) and enriches the mixture. This causes the combustion to be delayed until the exhaust and, due to the additional air introduced, the remaining fuel burns in front of the turbine or sometimes visibly later as a lambent flame at the tailpipe. The turbine is thus kept up to speed and the supercharger accelerates away from a higher boost pressure when the throttle is opened. The disadvantages are obvious: the exhaust system is exposed (for a short time) to higher temperatures and pressures, which significantly reduce its service life. It also increases fuel consumption.

Register Charge

The structure and the different operating modes of a multi-stage turbocharger are illustrated in Fig. 1.116. Two compressors and the associated turbines are connected in series. The fresh air is pre-compressed by the low-pressure compressor (5, blue) in the entire map range. In the high-pressure compressor (2, blue), a further pressure increase of the air mass flow takes place before it is directed to the combustion chamber via the throttle valve and air collector. Depending on the pressure ratio upstream and downstream of the high-pressure compressor, a bypass valve (1) opens automatically. If the compressor work of the low-pressure stage alone is sufficient for the desired boost pressure, this valve is open. At low engine speeds, the entire exhaust gas flow in the high-pressure turbine (2, red) is expanded. When this reaches its speed limit, the turbine bypass valve (3) is opened and the greater part of the exhaust gas flow is routed directly to the low-pressure turbine (5, red).

The low-pressure turbine has a wastegate (4), which is used to regulate the boost pressure in the event of high exhaust gas flow rates.

Due to the comparatively small high-pressure turbocharger, outstanding response is achieved even at the lowest engine speeds. However, high engine power is still possible because the low-pressure system is designed for the correspondingly large air or exhaust gas throughput [55].

Position of the Throttle Valve in Petrol Engines

Throttle valves on the intake side of the compressor lead to a low-weight and simple control of the engine. In addition, the dynamic behavior is better, because the pressure drop downstream of the throttle valve keeps the supercharger speed higher in sliding mode. However, this same negative pressure results in the need for special seals for the shaft outlet from the compressor housing to prevent lubricating oil from being drawn into the fresh air to the engine. Production engines have the throttle behind the compressor on the discharge side. This keeps the volume of the air collector small and the transient behaviour is thus favourably influenced. A bypass valve is installed to prevent the air trapped in front of the throttle valve from bringing the compressor to its surge limit and slowing it down. This is a spring-loaded valve in the compressor housing that directs the air from the compression duct back to the suction side. Individual throttle valves for each cylinder, common on naturally aspirated race engines, are extremely rare on turbocharged units. The low supercharged Champ Car engines (see appendix "CART") had some. V-engines like to use one throttle body to control each bank as a compromise. Each bank is thus supplied by its own air collector, at the inlet of which sits the throttle valve. Single throttle bodies on supercharged engines result in a high actuation force because the springs must be strong for reliable closing. This can become a problem for drivers in endurance races [56].

Versions

In principle, series turbochargers do not differ from those for racing. The differences lie in the details with the aim of improving the response and efficiency as well as reducing the weight, Fig. 1.117 and Table 1.12.

If a restrictor is specified, it is placed at the compressor inlet. This increases the pressure drop in the compressor due to the pressure drop in the air restrictor and the thrust bearing must absorb greater forces. In this case, it covers the entire shaft circumference instead of the otherwise usual 270° design.

Pressure Wave Supercharging

A crankshaft-driven cell rotor (Comprex supercharger) enables direct energy exchange between exhaust gas and intake air. This system results in high pressure increases at low engine speeds and responds quickly to load changes. However, the rotor must be positioned

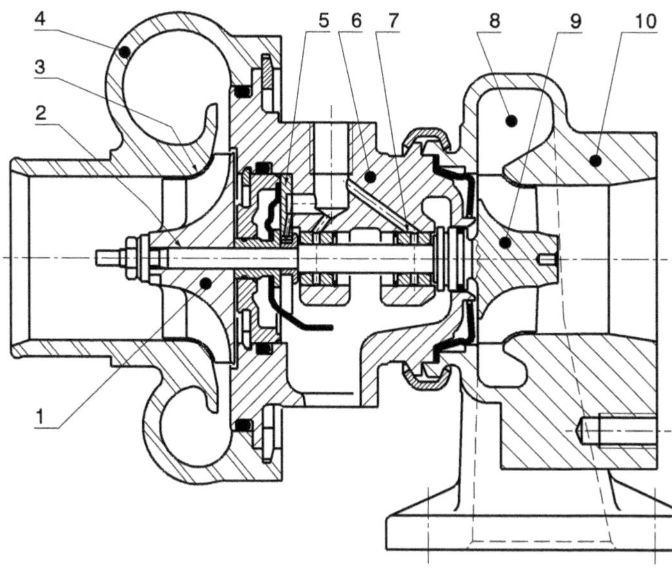

Fig. 1.117 Comparison turbocharger series – Formula 1. For explanations see Table 1.12. 1 *Compressor* wheel, 2 *Compressor wheel – shaft joint,* 3 Gap dimension, 4 Compressor *housing,* 5 Thrust *bearing,* 6 Bearing *housing,* 7 Bearing, 8 Spiral duct, 9 Turbine wheel, 10 Turbine housing

on the engine like a mechanical supercharger, and increased exhaust and scavenging air volumes result.

Intercooling

(Charge Air Cooling). The compression process raises the temperature of the air, thus cancelling out part of the increase in density. For a further increase in performance, the combustion air must therefore be recooled. This is done in air-to-air or water-to-air heat exchangers, which are arranged between the compressor and the distribution volume, Fig. 1.108.

The temperature T_2 after the compressor is a function of the boost pressure ratio and the output temperature T_1:

$$T_2 = T_1 \cdot \left(\frac{p_2}{p_1}\right)^{\frac{n-1}{n}}$$

T_1, T_2	Temperature of the air before or after compressor, K
p_1, p_2	Pressure before or after compressor, bar
n	Polytropic exponent, -.For vehicle engine compressors, n = 1.52 to 1.62

Table 1.12 Comparison of turbocharger design series to Formula 1 of the 1980s, according to [4]

Pos.	Designation	Standard version	Formula 1	Reason or measure for F 1
1	Compressor wheel	Aluminum alloy investment casting	Wrought aluminium alloy	Higher strength
2	Joint	Friction-locked	Positive-locking	Transmit higher torque
3	Gap dimensions	0.2 to 0.3 mm	0.1 to 0.15 mm	Manual assembly required
4	Compressor housing	Die-cast aluminium	Aluminium sand casting	One-piece design is more streamlined
5	Axial bearing	Hydrodynamic slide bearing	Double load capacity or none	High loading pressures. No axial bearing necessary with adjusted spindle bearings
6	Bearing housing	Water and oil cooling	Oil cooling only	Sufficient for short service life
7	Bearing	Plain bearing	Needle roller bearing, spindle bearing	Less friction; faster running up of the running gear and thus faster response
8	Spiral duct	Monoflow	Dual-flow	Pulse charging with separation of the exhaust gas routing up to the turbine inlet
9	Turbine wheel	Nickel investment castings	Heat-resistant nickel special alloy, ceramic	Higher exhaust gas temperatures, lower mass moment of inertia
10	Turbine housing	Ductile cast iron	Lightened construction	
–	Mass	6.5 kg	5.5 kg	Reduced wall thicknesses, ceramic wheels

At a boost pressure ratio of 2, for example, the temperature rises from 20 °C (293.15 K) to 98 to 110 °C (371 to 382 K).

The increase in density due to a reduction in temperature describes (1.9). From this follows directly from (1.10) the increase in mean pressure of the turbocharged engine compared to the naturally aspirated engine through charge air cooling:

$$\frac{p_{m,e,T}}{p_{m,e,N}} = \frac{\rho_{2'}}{\rho_1} = \frac{\rho_{2'}}{\rho_1} \cdot \frac{T_1}{T_{2'}}$$

$p_{m,e,T}$	Mean effective pressure of the supercharged engine, bar
$p_{m,e,N}$	Mean effective pressure of the naturally aspirated engine, bar
$\rho_1, \rho_{2'}$	Density of air before compressor or after intercooler, kg/m^3

(continued)

$p_{2'}$	Pressure after intercooler, bar; where $p_{2'} \approx p_2$
$T_{2'}$	Temperature of the air after the intercooler, K

The heat flow to be dissipated by the charge air cooler is:

$$\dot{Q} = \dot{m}_L \cdot c_{p,L} \cdot (T_{2'} - T_2)$$

$\dot{Q}$	Heat flow dissipated by the charge air cooler, W
$\dot{m}_L$	Mass air flow supplied to the engine, kg/s. see also (1.8)
$c_{p,L}$	Specific heat capacity of air at constant pressure, J/(kgK). c_P is temperature dependent. For dry air, c_p (20 °C) = 1004.6 J/(kgK) and c_p (100 °C) = 1010.4 J/(kgK)

For an initial dimensioning of the size of an air-cooled charge air cooler, the following rule of thumb can be used:

$$V_{Ic} = k_{Ic} \cdot P_e$$

V_{Ic}	Block volume of the charge air cooler, dm^3
k_{IC}	Size coefficient, dm^3/kW. k_{IC} = 0.08 to 0.11 dm^3/kW
P_e	Effective engine power, kW

Typical block depths are between 30 and 100 mm.

Throttle *Device*

In a gasoline engine with external mixture formation (intake manifold injection, carburetor), the mixture supply to the engine is influenced for load control (quantity control, in contrast to quality control in diesel (compression ignited) engines and direct-injection gasoline engines in certain operating ranges). In this case, the intake is throttled at partial load, but not at full load. Of the conceivable possibilities for influencing the mixture quantity to the combustion chamber, only a few have ultimately prevailed: Throttle valves, flat slide valves and rotary slide valves.

In production engines, the throttle bodies are located in the mixture former (carburetor or central injection) or in the intake manifold or intake tract. The closer the throttle valve is located to the engine, the more spontaneously the engine responds to load changes. For passenger car engines, therefore, the rule of thumb is that the volume after the throttle valve should not be more than 3.5 times the displacement it supplies. For competition engines, however, even this value is extremely high [28]. Therefore, in racing engines, a throttle body is used for each cylinder. In this way, the throttle bodies can be arranged directly at the engine intake. The disadvantage of this design is the higher actuating force In carburetor engines, a large valve overlap with a single carburetor has a disruptive effect, and therefore high-performance engines have separate carburetors for each cylinder. Injection valves are also arranged individually for each cylinder.

In turbocharged engines, the throttle element can be arranged upstream or downstream of the compressor. The most common arrangement today is between the compressor and the engine. For more details, see the section on *supercharging*.

In order to improve drivability, the throttle valves of both cylinder banks of the extremely powerful V-engines of Formula 1 are not opened simultaneously but – which is possible due to e-throttle actuators – with up to 30% difference, i.e. one bank "lags" behind the other. Only from approx. 70% pedal travel do all flaps open synchronously. So that one cylinder bank is not overloaded, the banks swap their roles each time the throttle valves close completely [37].

Throttle Valve

The throttle valve is simple in design, but even when fully open it interferes with the mixture flow and reduces the free flow cross-section by the projected flap shaft area. This influence can be reduced by relocating the throttle valves to the outside of the intake funnels. Here, the intake cross-section is already so large that hardly any disturbance of the intake flow occurs. This arrangement is used on successful Formula 1 engines, for example. If this arrangement is not possible, at least the shaft in the intake manifold area can be flattened.

Although the simultaneous control of several throttle valves is more complex compared to the flat slide valve, it is still the simplest option for engine control with the lowest operating forces. There is hardly any possibility of jamming and any dirt that may be sucked in can impair the function of the butterfly valves. A problem arises when a common throttle shaft is used through several intake manifold arms. Due to the different thermal expansion of the shaft and the intake manifold, butterfly valves that are fixed on the shaft come to rest laterally on a bearing point at high temperatures. Throttle valves do not require a housing such as the flat or rotary valve, but sit practically free in the intake duct and the fuel injected outside cannot settle in the free spaces or gaps of the housing and thus lead to engine lean. This supports a spontaneous response of the engine. Just as the throttle valve's law of opening supports the engine's responsiveness. This releases a lot of intake cross-section even at a small opening angle. With large throttle diameters, this can become a disadvantage: The angle encoder potentiometers for the electronic engine control must be high-resolution so that no problems occur in the lower part-load range due to uneven running. For better dosage of the air volume at low flow rates, the throttle valve can also be equipped with a progression (ball zone).

If space permits, the throttle shaft should be parallel to the duct partition of bifurcated intake ducts. With this arrangement, the throttle valve supports a beneficial vortex formation in the part-load range, which shows a similar tendency as a duct cut-off.

Where space is limited, e.g. with small cylinder clearances, an oval design of the throttle valves is a solution.

Flat Slide

Flat slides are mostly used for gasoline injection. With a slide valve, several intake manifolds can be controlled exactly evenly. However, the distance between cylinders must be relatively large so that the slide can completely close circular passages. At full load, the mixture flow is not impeded. To ensure that the slide valve remains smooth-running and thus satisfactorily dosed for the driver even when the engine is in sliding mode (e.g. when braking for a corner), where large pressure differences occur between upstream and downstream of the slide valve and thus cause a large resulting downforce, it is mounted on anti-friction bearings. This pressure difference is approx. 0.2 bar for a naturally aspirated engine (ambient pressure upstream and approx. 0.8 bar downstream of the slider). This results in a normal force of approx. 100 N per cylinder for a 3-l engine [4]. Simple, sliding slides can cause the driver to feel that they are stuck due to this friction caused by low pressure. This phenomenon is even more pronounced in turbocharged engines when the slide is located after the compressor. The compressor is still pumping air at the beginning of the slide operation due to its inertia, which makes the pressure difference with the intake area after the slide even greater. When using an exhaust gas turbocharger, throttle valves are preferred not least for this reason.

If the slide is arranged longitudinally in the vehicle, the actuation is set up so that it moves backwards when opened [23].

Flat valves require their own housing, Fig. 1.118. Therefore their system weight is higher than that of comparable butterfly valves.

Rotary Valve

Rotary valves are an interesting compromise between flat valves and throttle valves. Like flat valves, rotary valves have the advantage of fully clearing the intake cross-section at full load. Like throttle valves, they are pivoted and easy to operate, Fig. 1.119. They are lighter and smaller than the flat slide valve, but heavier than the throttle valve. In addition, rotary valves in dual-flow ducts allow different opening cross-sections in the part-load range, which means that, for example, a slight swirl flow can be produced in the combustion chamber by the different admission of the two port branches. Double rotary valves are also used in Formula 1 engines. The two rollers touch each other in the middle of the intake manifold and rotate in opposite directions. The reason for this solution is the smaller diameter of the vanes: this allows the intake manifold to be shorter, which is necessary for a higher engine speed.

Mixture *Formation*

The mixture formation in racing gasoline engines was primarily by intake manifold injection. One of the first successful direct injection engines was the Audi 3.6-l V8 FSI

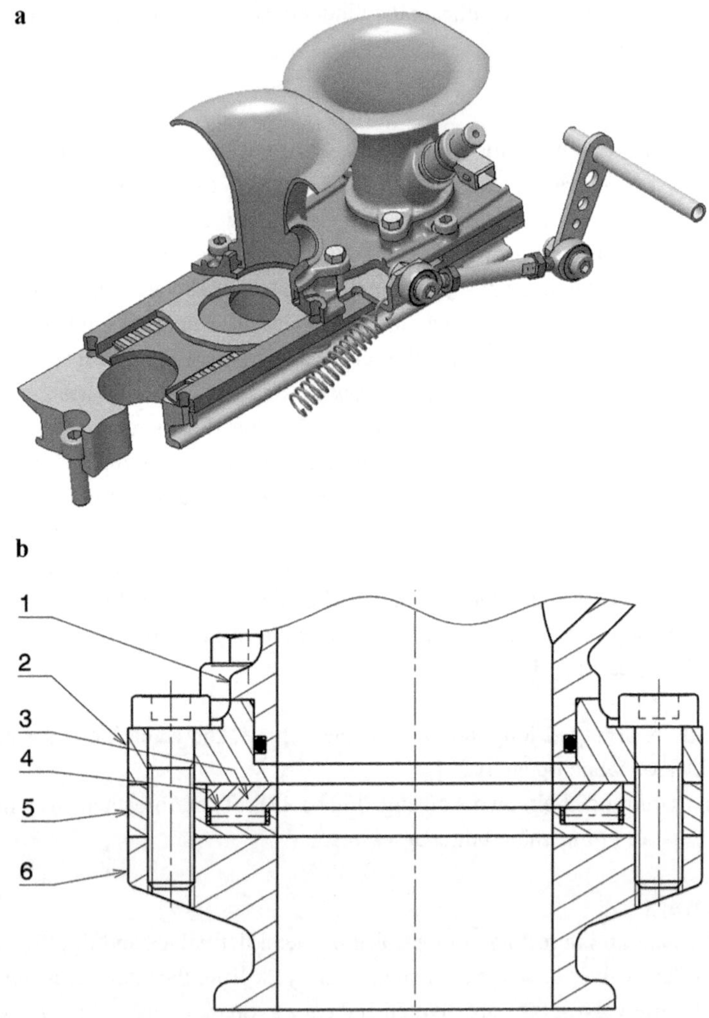

Fig. 1.118 Flat slide of a multi-cylinder engine. (**a**) Axonometric representation. (**b**) Sectional view. **1** Intake *trumpet,* **2** Housing *cover,* **3** Flat slide, **4 Roller bearing** (needle roller flat cage), **5** Bearing housing, **6** Intake *pipe resp.* cylinder head . The slider is made of steel and is mounted on roller bearings. The rollers need a running surface with sufficient hardness, therefore the lower part of the housing (5) is also made of steel. The injection valve is located in front of the slide in the intake funnel. This brings performance advantages due to more time for mixture formation. The disadvantage is that fuel can splash onto the closed slide valve and the housing must therefore be tight all round

BiTurbo at Le Mans 2000. In the meantime, the engines of many racing series inject directly into the combustion chamber. Diesel (compression ignited) engines run on the road as well as on the race track with air-distributing direct injection.

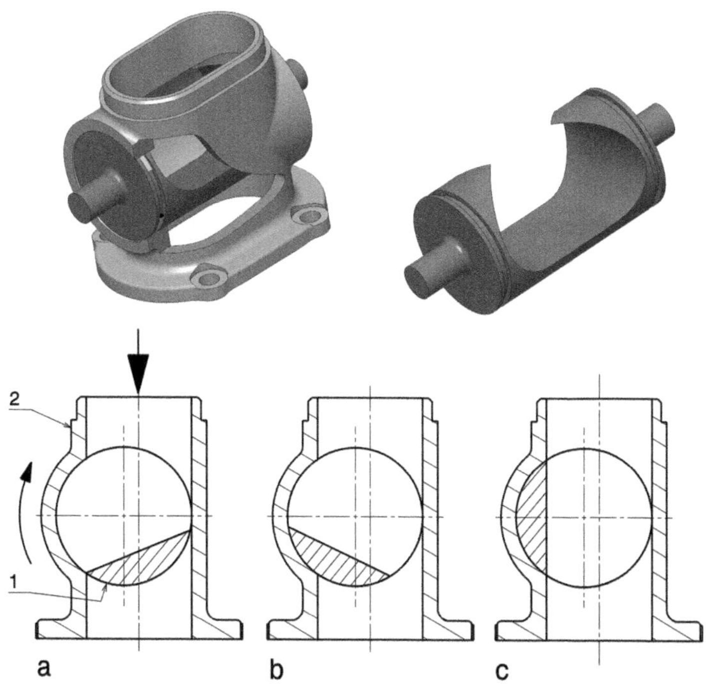

Fig. 1.119 Suction connection with rotary vane. The upper illustration shows the complete arrangement (with suction port partially removed) and the rotary vane alone. The rotary vane (1) is inserted laterally into the locating bore of the inlet connection (2). Three important positions of the throttling element are shown in the sectional view. (**a**) idle, (**b**) *part load,* (**c**) full load

With intake manifold injection, the position of the injectors is decisive for the engine behaviour. The injectors are located as far away as possible from the combustion chamber. This allows high engine speeds because there is more time for mixture formation. In addition, the fresh air can be cooled well by the evaporating fuel on the path remaining to the valves. Because cooler air means [4]:

- More power, because with the cooler air more air mass and thus more oxygen gets into the combustion chamber
- Later start of knocking because the cooler fresh charge is less prone to spontaneous ignition
- A lower thermal stress
- Less fuel consumption because more pre-ignition can be used.

Inside closed airboxes, injectors are even placed outside the intake funnel centrally at the mouth (e.g. Figures 1.97 and 1.47).

With open intake funnels, however, the injection nozzles cannot be moved further forward as desired, since otherwise fuel would leak out. A shock wave reflected by the closed valves travels outwards against the intake direction, entraining fuel droplets in the process. If these droplets are not sucked in again the next time the intake valves are opened, they are lost for energy conversion and thus increase fuel consumption.

***Choice* of *Carburettor* Size**

Despite all the possibilities of influencing engine behaviour that electronic fuel injection offers, the carburettor has at least one advantage. It is easier to maintain and repair. This advantage means that it can still successfully hold its own in some racing classes – preferably off-road motorcycles.

The selection of the nominal carburetor size for a carburetor is based on the air mass flow rate that the engine or the supplied cylinder can handle [41]:

$$d_{Ca,n} = k_{Ca}\sqrt{V_h \cdot n_{\max}}$$

$d_{Ca,n}$	Carburetor core diameter, mm $d_{Ca,n}$ must not be confused with the venturi diameter
$d_{\min}$	Venturi diameter. $d_{\min} \approx 0.8\ d_{Ca,n}$
V_h	Displacement of the single cylinder, l
$n_{\max}$	Maximum speed, $\min^{-1.}$ $n_{\max} \approx 1.1\ n_n$
	n_n: Speed at maximum power
k_{Ca}	Correction factor, -
	For 1, 2 and 4 cyl. engine: 0.82 (fa. Solex) 0.8 to 0.9 (fa. Weber)
	6-cylinder engine: 1.0; 8-cylinder engine: 1.15
	Note: In practice, twin carburettors are used for four-cylinder engines and above

The average gas velocity in the carburetor venturi v_{Ga} should be about 100 m/s (limits 90 to 110 m/s) at rated speed. The control of the cross section can thus be obtained from the continuity equation:

$$v_{Ga} = \frac{A_{Pi} \cdot v_m}{A_{Venturi}}$$

v_{Ga}	Gas velocity, m/s
A_{Venturi}	Clear cross-section of the venturi, mm^2
A_{Pi}	Piston area, mm^2
v_m	Mean piston velocity, m/s. S. (1.2)

The calculation of the gas velocity can also be determined based on the inlet timing (e.g. for fuel injection engine) [41]:

$$v_{Ga} = \frac{V_h}{A_{Suction}} \cdot i \frac{n_M}{60} \cdot \frac{720^\circ}{\beta_{Es} - \beta_{E\ddot{o}}}$$

V_h	Stroke volume single cylinder, cm^3
$A_{Suction}$	(Smallest) intake cross-section, cm^2
n_M	Engine speed, min^{-1}
i	Factor for number of cycles, -. $i = 0.5$ for four-cycle process
$\beta_{E\ddot{o}}, \beta_{Es}$	Crank angle at inlet opening or at inlet closing, °CA

Material and Production

Suction systems are cast from aluminium or magnesium alloys, and can also be welded from semi-finished products. Particularly lightweight systems can be laminated from CFRP.

1.4.7 Exhaust System

The exhaust system not only discharges the burnt air/fuel mixture to the environment at a suitable point, but also accommodates one (or two) lambda probe(s), a catalytic converter, a soot particle filter and one (or more) silencer(s), depending on the combustion process.

Exhaust Systems for Ideal Pressure Curve

However, the exhaust can transport much more than just exhaust gas. For racing engines in particular, the exhaust system and thus also the silencer are important for the torque curve. An internal combustion engine works as an air pump, so to speak, and just as gas dynamics influence cylinder filling on the intake side, they do the same on the exhaust side. In this respect, the exhaust system is partly responsible for the so-called scavenging gradient from the intake to the exhaust side. If the exhaust acts as a flow brake, the cylinder is poorly drained and the residual gas content remaining in the cylinder impedes the subsequent filling with fresh gas. Conversely, if the scavenging gradient is too strong, fresh gas will also flow into the exhaust tract before exhaust closure. This amount of energy is lost for combustion and further increases fuel consumption and emissions.

The ideal exhaust provides a vacuum after the exhaust valve when the exhaust is open and prevents fresh gas loss or even compresses the fresh charge somewhat when the exhaust is closed by overpressure.

The effect of gas-dynamic effects depends, among other things, on the temperature and the engine speed. The difficulty in designing an exhaust system is to extend its beneficial effect over a wide speed range. The exhaust systems of modern high-performance two-strokes – with or without silencers – consist of five sections: Exhaust pipe, diffuser, exhaust chamber, countercone and tailpipe. Figure 1.120 illustrates the effect for one engine speed. When the exhaust port is opened, a pressure wave travels into the exhaust at the speed of sound. The speed of sound at these conditions is about 450–550 m/s (so it increases with temperature and thus accommodates the speed characteristic). The diffuser

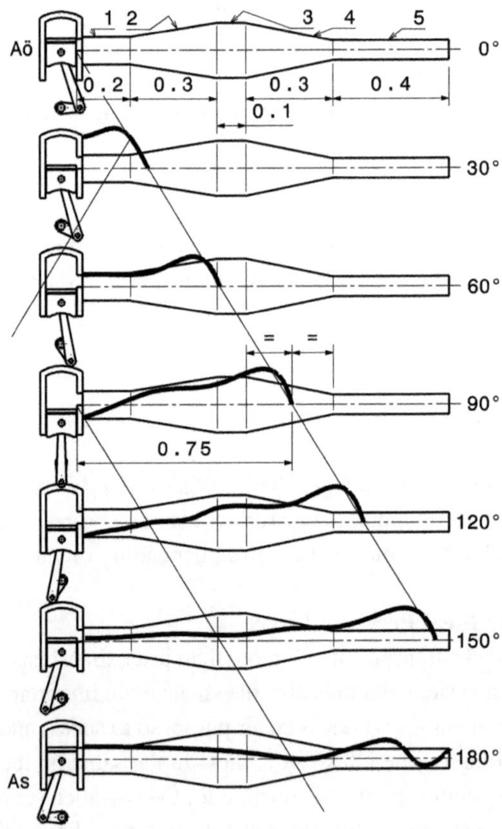

Fig. 1.120 Pressure curve in the exhaust of a 2-stroke engine (calculated), according to [41]. Structure of the exhaust: **1** exhaust pipe, **2** diffuser, **3** exhaust chamber, **4** opposite *taper,* **5** *tailpipe.* Calculated pressure curve in an exhaust bulb for a speed of 10,000 min^{-1} and a (temperature-related) sound velocity of 500 m/sec

At 30° crank angle after exhaust opening, the exhaust pressure wave has penetrated approx. 5 cm into the diffuser, with a reflected, returning vacuum wave lowering the total pressure. After 60° the pressure wave reaches the diffuser end, the pressure level has dropped considerably. – At 90°, the pressure wave is in the middle of the mating cone, where an overpressure wave is reflected, which later swells in the diffuser (which is flowed through in reverse) and causes a backflow into the cylinder just before exhaust closure. In the meantime, the original pressure wave has reached the tailpipe (at approx. 110°) and left it again (at approx. 160°); here, further reflections occur, at the start of the tailpipe a pressure increase, but when the pressure wave exits into the open air, a strong negative pressure wave returns. In order for this to complete the entire 1.3 m long "return path" and still act in time, the engine speed would have to drop below 5700 min^{-1}.

thereby reduces the flow resistance compared to a cylindrical pipe and due to its cross-sectional expansion it acts like a pipe end (reflection at the free end): A pressure wave is reflected as a vacuum wave and travels back to the cylinder. The same thing happens at the actual exhaust end at a later time, when the continuing pressure wave reaches the pipe end and is reflected as a negative pressure wave. In the diffuser, the intensity of the pressure wave decreases as the cross-section increases. The counter-cone acts like an obstacle (reflection at the fixed end) and a pressure wave is reflected as such. As it travels back to the cylinder, it is amplified as the cross-section decreases in the diffuser, which now acts like a nozzle due to the reverse direction. If the lengths fit, the pressure wave reaches the exhaust port when the piston passes over it again to close it.

As with the intake manifolds, a longer exhaust pipe causes effective resonances at lower speeds, a shorter one correspondingly at higher speeds. If the pipe already has a slight taper, the effect is similar to a shortening. Short pipes – no longer than 5–6 times the initial diameter, which in turn should provide a cross-section of 130–175% of the exhaust port – are suitable for high-speed racing two-stroke engines [41].

The diffuser also allows the occurrence of the maximum mean pressure to be shifted to higher or lower speeds. Its length should be about 30–40% of the total system. The cross-section of its mouth and thus that of the (usually adjoining) cylindrical chamber should, according to empirical values, have 3.4–4.5 times the initial area of the exhaust pipe.

Through the length of the cylindrical chamber follows a volume, which also affects the resonance speed. In the following counter cone the cross section is reduced to the width of the tail pipe. The length of the countercone should be 25–40% of the diffuser length. The tailpipe cross-section should be 21–38% of the initial area of the exhaust pipe [41].

It applies:

- smooth pipes lead to a (small) backwater, and no suction effect
- Diffusers do not create backpressure, but strong suction effect
- Baffle walls or counter cones only lead to strong backpressure
- all changes in cross-section have the same but graduated effect.

A long, narrow tailpipe generally allows for higher rpm and therefore higher power, but may cause thermal problems such as overheated piston crowns.

The length of the entire exhaust system to the centre of the mating cone (0.75 m in Fig. 1.120) can also be determined for two-strokes using the following relationship [57]:

$$l_{res} = \frac{(\beta_{As} - \beta_{A\ddot{o}}) \cdot c_s}{12 n_{M,res}}$$

l_{res}	Length of the exhaust system to the middle of the counter cone, m
$\beta_{A\ddot{o}}, \beta_{As}$	Crank angle at exhaust valve opening or exhaust closure, °CA
c_s	Speed of sound, m/s
$n_{M,res}$	Engine speed at which maximum mean pressure occurs, min^{-1}

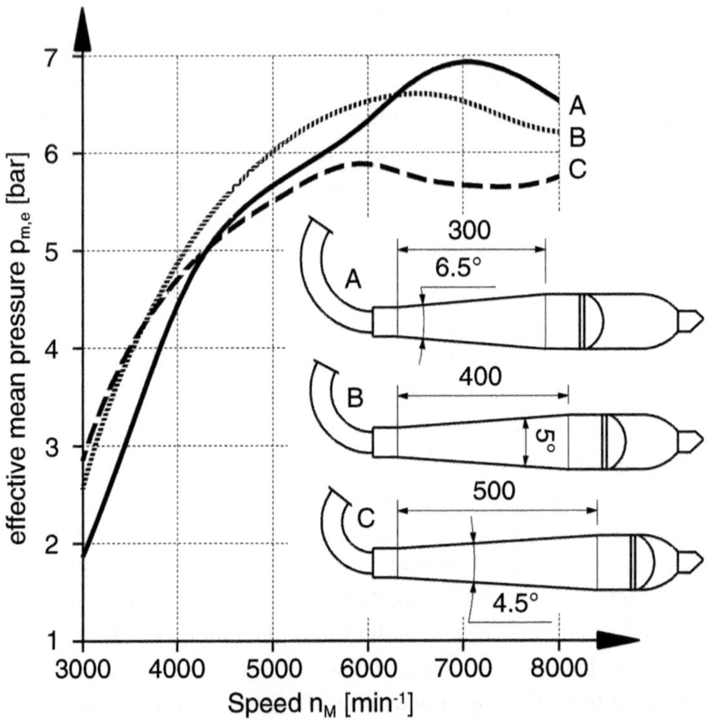

Fig. 1.121 Influence of the exhaust design on the mean pressure, after [41]. Three different exhaust systems were measured for 50-cm^3 two-stroke engines (Fichtel & Sachs). The diffuser was varied. The corresponding exhaust pipes compensate the volume in front of the baffle. The shortest exhaust leads to the highest maximum mean pressure, but also at high revs. The fullest curve offers the version *B*

Figure 1.121 shows the effects of different diffuser designs on the mean effective pressure. It seems obvious to at least partially circumvent the necessary design compromises by shifting the length of the diffuser (analogous to the intake manifold) or at least the position of the baffle as a function of engine speed. In fact, various systems using such a principle have been brought to production maturity.

On multi-cylinder naturally aspirated engines, these gas-dynamic effects of the individual pipe are supplemented by the effect of merging with the exhaust pipes of neighbouring cylinders. With a skilful arrangement of the merging according to the cylinder arrangement and above all the ignition sequence, the mean pressure curve of the engine can be changed to advantage.

The effects of possible exhaust pipe designs for a four-cylinder engine are shown in Fig. 1.122. The fan-shaped manifold (d), which was common in the past, performs worst in comparison. The best is a combination of cylinders 1 and 4 and cylinders 2 and 3 (a). With the usual firing sequence of a four-cylinder engine with a flat crankshaft (cf. Figure 1.64),

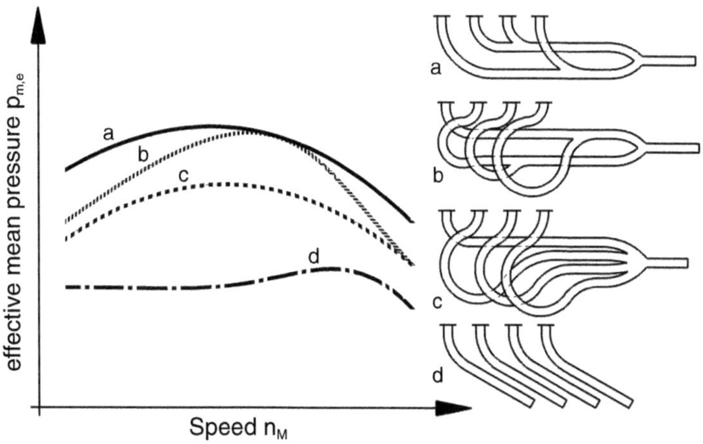

Fig. 1.122 Influence of the order of (primary) merging on the mean pressure of a 4-cylinder engine (schematic). The best design is a (merging 1-4 and 2-3), which leads to an improvement of the scavenging gradient with the usual firing sequence 1-4-3-2. The pressure wave of the cylinder firing 180° before is reflected as a vacuum wave

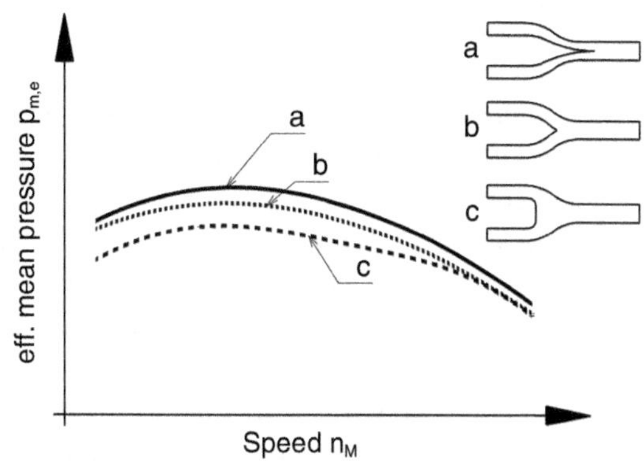

Fig. 1.123 Influence of the design of the secondary manifold on the mean pressure of a 4-cylinder engine (schematic). (**a**) Tubes tapered, (**b**) Tubes with constant cross-section, (**c**) Extension

the shock wave of the previously firing cylinder reflected as a vacuum wave improves the gas exchange.

If the ignition distance is uniform, then on eight- and ten-cylinder engines the cylinders of one bank are joined with tubes of equal length (cf. Figure 1.64). On twelve-cylinder engines, three pipes each from cylinders on one bank side are combined with the same ignition distance. The remaining four exhaust pipes are either routed as tailpipes to the muzzle or they are combined into one tailpipe on each side of the engine.

However, not only the sequence, but also the type of merging shows an influence on the mean pressure curve. Figure 1.123 shows the influence of the secondary merging, i.e. downstream of the second merging. If the individual pipes are tapered before merging (a), the gas velocity increases and the pressure decreases accordingly. This has a positive

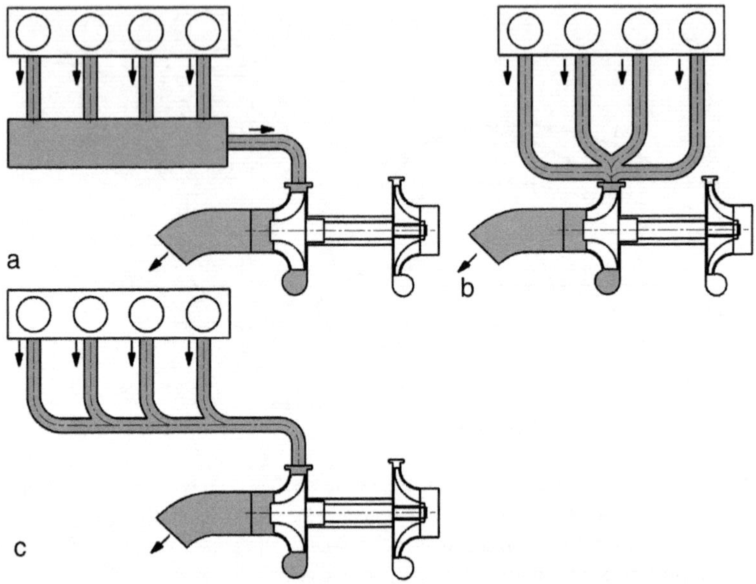

Fig. 1.124 Exhaust pipe routing for turbocharged engines (schematic). (**a**) *Ram turbocharging,* (**b**) *Pulse turbocharging,* (**c**) Compromise for passenger cars

effect on the mean pressure curve. However, when merging individual cylinders (primary merging), it is better to increase the pressure, i.e. in this area the tubes should become wider before merging.

The pipe routing for turbocharged engines is initially based on the basic process. The thermal energy of the exhaust gas can be used by collecting the exhaust gas in a tank and feeding it from there to the exhaust gas turbine (ram charging, Fig. 1.124a). This results in more even ratios between input and output for the exhaust gas turbine, but the system reacts less quickly to load changes. For the highly dynamic vehicle drives, the use of the kinetic energy of the exhaust gas flow is preferred. In this case, each exhaust line is individually routed to the turbine along the shortest possible path (b). On four-cylinder passenger car engines, a mixture of both options is often used for space reasons (c). The exhaust pipes are brought together before the turbine, but as late as possible. This way, at least part of the pulse effect is added to the turbine.

In the case of pulse charging, the exhaust pipes are often tapered. This increases the exhaust gas velocity upstream of the turbine.

Tailpipe

The length of the tailpipe depends not only on purely gas-dynamic criteria, but also on the available space in the vehicle and on any position of the muzzle required by the regulations. In addition, a silencer is added in most cases.

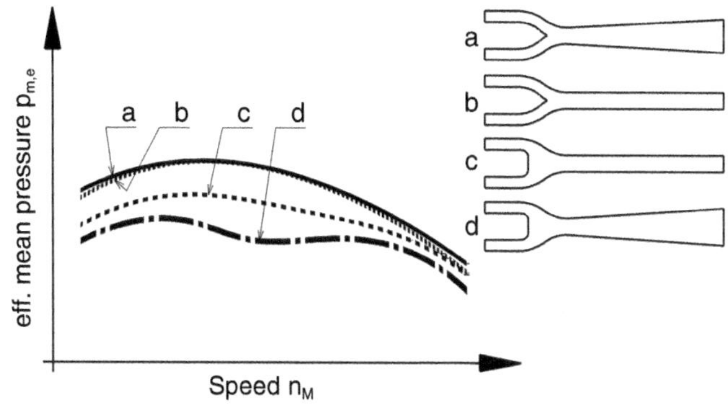

Fig. 1.125 Influence of the tailpipe on the mean pressure of a 4-cylinder engine (schematic)). (**a**) Pipes with constant cross-section on tapered tailpipe, (**b**) Pipes with constant cross-section on cylindrical tailpipe, (**c**) Extended mating on cylindrical tailpipe, (**d**) Extended mating on tapered tailpipe

The tailpipe should be at least as long as the exhaust path to it, i.e. from the exhaust valve to the secondary manifold. Short tailpipes result in high peak power at high rpm, longer pipes lift the torque in the lower rpm range, but cap the final power. The tailpipe can be cylindrical or flared. Combinations with different mergers are conceivable for this purpose. The effects on the mean pressure curve of a four-cylinder engine are shown schematically in Fig. 1.125. If the exhaust pipes are joined with a constant cross-section, practically no difference can be measured between the conical and cylindrical tailpipes (a, b). The situation is different, however, when the pipes are brought together with a wider cross-section. In this case, the combination with a cylindrical tailpipe (c) brings advantages in the torque curve.

The regulations of the individual racing classes generally specify the measurement regulations for determining noise. The tailpipe of the exhaust system is advantageously designed so that it points away from the measuring microphone.

A complete exhaust system of a cylinder bank is shown in Fig. 1.126. Figure 1.127 shows an example of the tailpipes passing through the bodywork.

Connectors and *Mounts*

The exhaust system experiences large temperature differences and thus large differences in length between operation and standstill. Rigid fittings can become problematic because they can loosen after a few heating cycles. In this context, straight, rigid tube guides also have the disadvantage that tubes or their holders can crack if expansion is impeded by temperature stress. This can be remedied by pipes that contain compensating bends and holders or fittings that have large expansion reserves, Fig. 1.128.

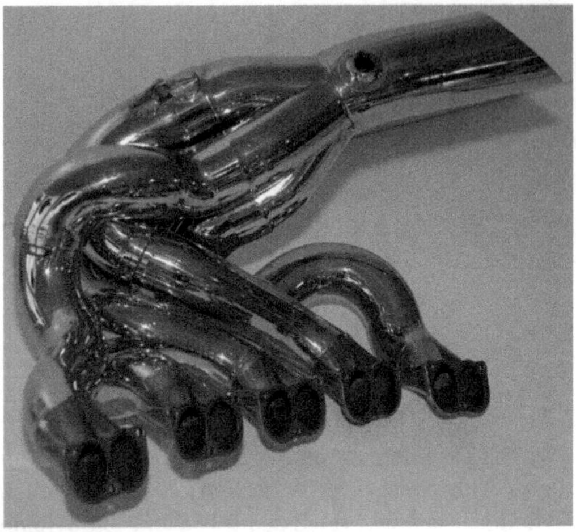

Fig. 1.126 Exhaust system (Ferrari Formula 1) for naturally aspirated V10 cylinder engine. The individual exhaust pipes are the same length until they join. The flange pattern on the cylinder head reflects the bifurcated exhaust ducts. The tailpipe carries a lambda probe and discharges into the atmosphere without a silencer

Fig. 1.127 Mouth of the tailpipe (Ferrari Formula 1). The extremely high-revving engine needs a short tailpipe. The usual integration into the diffuser area of the underbody was less and less successful with the cars getting lower and lower from season to season and the increase of temperature-sensitive CFRP. Finally this solution was born: The tailpipe points upwards and opens at the top of the engine cover

Tension springs can also wrap around a section of tubing to provide an elastic tether, Fig. 1.129.

Figure 1.130 shows the pipe assembly and the housing of the exhaust system of a V8 engine.

Noise Level Measurement (Noise *Test*)

In many racing series, the maximum noise level is limited and is checked during technical scrutineering. Although the combustion engine is the main source of exhaust and intake noise, it is also the source of crankshaft movement. The latter cause large surfaces such as

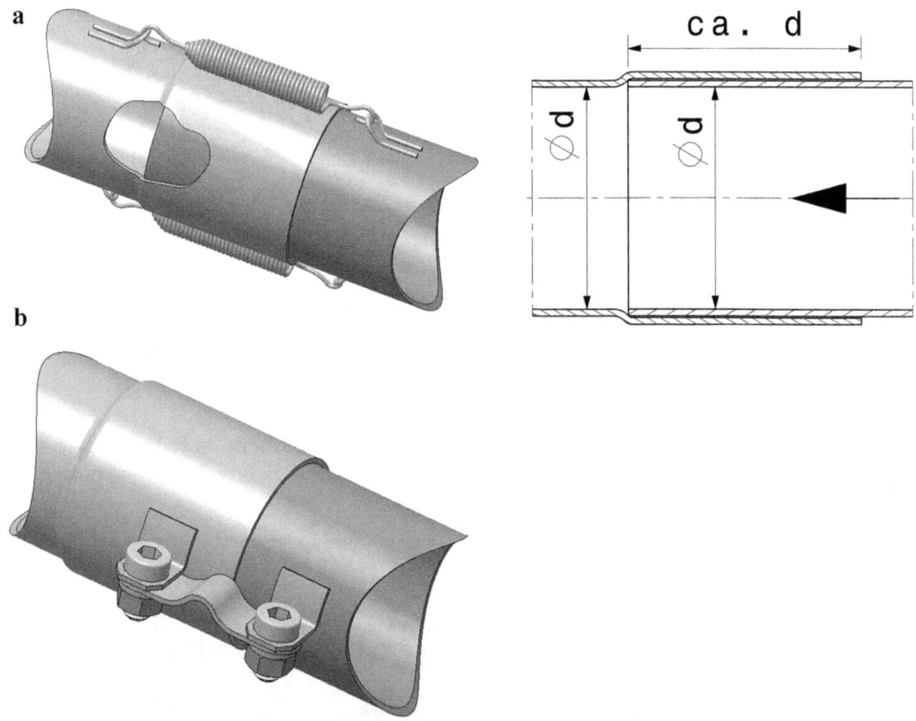

Fig. 1.128 Connections of exhaust pipes. (**a**) Plug-in connection with tension springs. (**b**) Plug-in connection with elastic strap. A cylindrical tube is inserted into a flared tube. Sealing is achieved by line contact at the end face. The direction of flow (to the left in the picture) is determined by the funnel shape of the expansion. The joining length is in the order of magnitude of the pipe diameter. This simple connection does not require any additional seal and can be easily loosened and rejoined. With this type of connection, pipes can be connected to each other or a silencer can be connected to the pre-pipe

wings, end plates, underbody, splitter, etc. to vibrate, so that these can sometimes become the main problem during technical acceptance. In addition, sound waves are also reflected off large, smooth surfaces, further increasing noise levels. "Plugging" the tailpipe of the exhaust (dB-killer) has no success in such cases. Apart from the fact that engine performance suffers because the residual gas content in the combustion chamber increases. Instead, wing elements and large planar elements must be braced with cables and/or their resonant frequencies must be shifted to other areas by means of stiffeners. The contribution of the tailpipe mouth can be mitigated by relocating the tailpipe to a more favorable area, such as on the vehicle flank facing outward. In any case, the tailpipe mouth should not be located between the end plates of the rear wing.

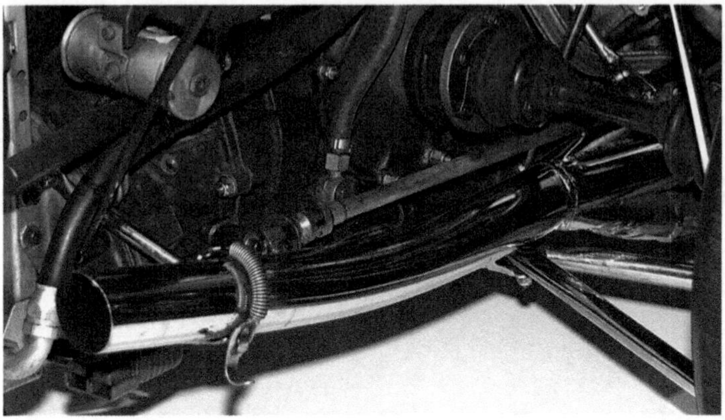

Fig. 1.129 Mounting of a tail pipe. The tail pipe is pressed against its fork-shaped receptacle with a tension spring. The connection is elastic, allows longitudinal sliding and can be opened quickly

Fig. 1.130 Exhaust system of a Monoposto. The exhaust pipes of this V8 engine are combined on one bank side like those of an inline four cylinder: The pipes of the two outer cylinders and the two inner ones are each joined together. Then the two remaining pipes are united to form a tailpipe. The tailpipe is routed to the rear of the silencer. The system adapts to the shape of the vehicle. You can easily see the bottle neck shape of the car on the underbody by the course of the strip for the engine cover

Materials

Exhaust systems are made of steel, stainless steel (1.4301, 1.4828) or titanium alloys, depending on the targeted service life and budget. The wall thickness is 1 mm, in some cases even 0.5 mm. Sections close to the engine must be able to withstand temperatures around 1000 °C. If components are further downstream from the cylinder head flange, such

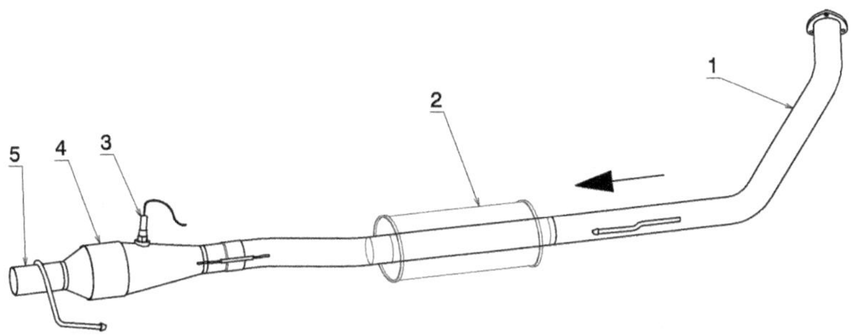

Fig. 1.131 Exhaust system of a touring car. *1 header tube*, *2* silencer (if mandatory), *3 lambda sensor*, *4* catalyst, *5 tail pipe*. The diffuser piece in front of the catalytic converter ensures that the cross-section of the cell is evenly charged with exhaust gas. This reduces the exhaust gas back pressure and increases the efficiency of the exhaust gas aftertreatment

as muffler pots, these can also be made of aluminum alloys or fiber-reinforced plastics. Other materials include Inconel (nickel-chromium-iron superalloy), which is the first choice for Formula 1 teams, and Nimonic (nickel-chromium-cobalt alloy, high temperature resistant, high creep strength i.e. low creep), which is occasionally used in Formula 1.

Titanium is not yet used in series exhaust systems for passenger cars. However, production plants are already being developed for this purpose. One argument against the long-term use of titanium is its tendency to creep (increase in plastic strain over time despite constant load).

Catalyst [58]

Racing engines are trimmed for performance and drivability. Exhaust gas detoxification – as prescribed by law for passenger car engines – is not an issue. Many compromises, which are the order of the day when tuning production engines, do not have to be made in the first place. In certain racing series, however, detoxifying catalytic converters are mandatory. These are basically the same design as their production counterpart and also work on the same chemical principle, although lambda one (excess-air factor, air ratio $\lambda = 1$, see Appendix) does not have to be run continuously. This is important for a racing engine, because the greatest power can be achieved in the rich mixture range. The racing catalytic converters also have a number of other special features. First of all, the catalytic converter is installed at the end of the exhaust system, i.e. directly in front of the tailpipe, Fig. 1.131. In a passenger car, on the other hand, the catalytic converter is located as close as possible to the exhaust manifold so that it reaches its effective temperature quickly after a cold start. Furthermore, the racing version is designed for low weight and low exhaust backpressure. The catalytic converter is constructed as a wound catalytic converter from perforated stainless steel foil and has a cell density of 100 cpsi (*cells per square inch*), Fig. 1.132. Typical passenger car cell densities are between 400 and 1000 cpsi. Wound catalysts are

Fig. 1.132 Motorsport PE catalytic converters made of coiled, perforated stainless steel foil, type HJS Emission Technology. These elements are welded directly into the exhaust pipe (can be seen in the picture on the right)

mechanically much more robust than the ceramic monolithic carriers (i.e. consisting of one piece) that are often installed in passenger car exhaust systems.

1.4.8 Lubricating Oil Supply

Reliable lubrication of high-revving racing engines is difficult to ensure, but essential for the reliability of the unit. After all, the main tasks of oil are not only lubrication (= separation of sliding surfaces, i.e. reduction of friction and wear) but also cooling. In order to keep the weight of the engine low, it is also tried to keep the oil quantity small. In principle, there is *wet sump* and *dry* sump *lubrication*, depending on the design of the oil sump. Dry sump lubrication ensures the oil supply even under extreme driving conditions and also enables air separation and reduces pan losses from the crankshaft and connecting rod. Figure 1.133 provides an overview of the individual components and their integration into the overall system.

The oil-air mixture from the crank chambers is sucked off by the suction pump (8) via the line (7) and conveyed via the cyclone filter (6) through the line (5) to the dry sump tank (2). In this tank the oil is separated from the air and the calmed oil reaches the inlet of the pressure pump (8) via the line (3). This presses the oil through the filter (9) via the oil cooler (11) and the line (12) into the main oil gallery of the engine. This is also where the pressure is taken, which is displayed in the cockpit (13). The driver is also informed about the temperature of the oil (14). Crankcase ventilation is provided by a pipe (1) connecting the cylinder head cover to the dry sump tank. In the engine, the crankcase forms a unit with the cylinder head cover via the timing chain shaft and oil return ducts, and the overall system is thus closed. In production vehicles, the design of the crankcase ventilation is much more

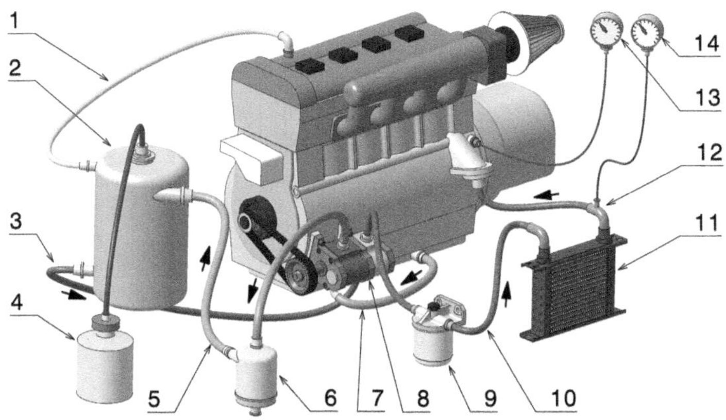

Fig. 1.133 Simplified representation of a dry sump lubrication system. 1 Breather hose, 2 Dry sump *oil tank*, 3 Pressure-*oil* supply, 4 *catch can*, 5 Return from *scavenge* pump, 6 Cyclone with *screen type filter*, 7 Supply suction pump (*scavenge in)*, 8 *Pump assembly*, 9 Oil filter, 10 Pressure-oil return (*pressure oil out)*, 11 Oil heat exchanger, 12 Pressure *oil* to engine, 13 *oil-pressure gauge*, 14 Oil-temperature gauge

complex, because throttles and non-return valves have to ensure that the leakage gases (blow-by) are safely returned to the intake tract (and thus to the combustion chamber) of the engine. And that both at partial and full load.

Figure 1.134 shows a diagram of an oil supply system. Suction pumps (1) draw the oil-air mixture from the oil sump or from the individual crank chambers. The separation into individual crank chambers prevents lubricating oil from sloshing to the rear or front when the engine is installed longitudinally during acceleration or braking and the pump cannot extract the oil. On the same drive shaft as the suction pumps sits a centrifuge (2) which separates oil and air. The system must be closed, so the centrifuge and dry sump tank are vented into an overflow tank (8). In the following oil cooler (3) the heat can then be dissipated more efficiently than with foaming oil. The dry sump tank (4) stores the oil supply and allows the remaining air to be released from the oil. The oil pressure pump (5) generates the delivery pressure. Before the oil reaches the actual lubrication points in the engine, it is cleaned of dirt particles by a filter (6). The most important supply points are main bearings, connecting rod bearings, spray nozzles for piston cooling, gear drive for camshafts and auxiliary units, valve actuation but also possible auxiliary units such as exhaust gas turbochargers or screw compressors. The arrangement of the oil cooler in the delivery circuit of the suction pumps has the advantage that the cooler is located in the low-pressure area of the system and is less sensitive to leaks – caused by stone chips, for example.

The location of the extraction point(s) in the crankcase or oil sump is of great importance. Although the suction pumps are dimensioned 20–30% larger than the pressure pumps because they suck in an oil-air aerosol, it is nevertheless important that the oil is removed from the sump before it is caught by rotating parts and "distributed" in the

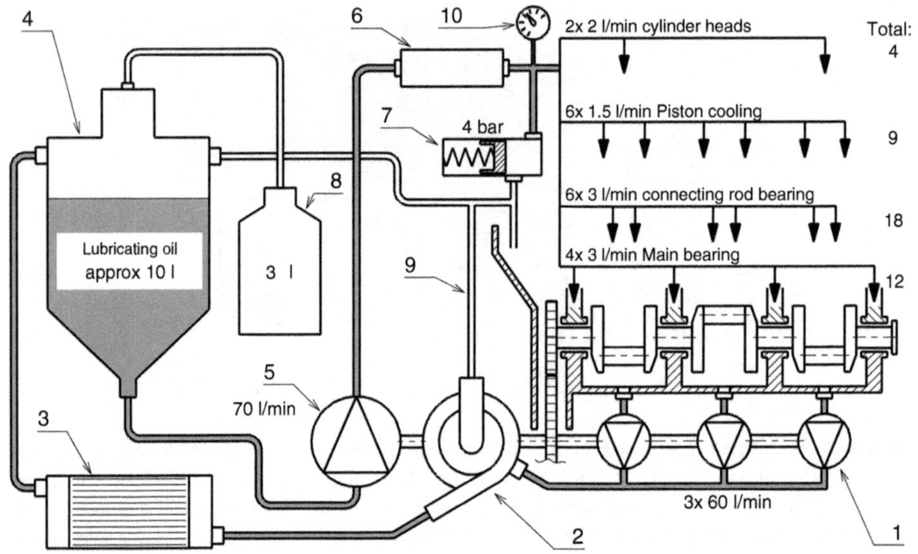

Fig. 1.134 Schematic of dry sump lubrication V6 -2.5 l, according to [4]. One bilge pump operates per crank chamber. The oil is fed to the heat exchanger only after the air separator. The total oil volume is approx. 15 l. The volume flows entered apply to an engine speed of 11,000 min^{-1}. **1** Suction pumps (bilge pumps, *scavenge pump),* **2 Oil** separator (centrifuge), **3 Oil** cooler, **4** *Dry sump oil tank,* **5** Pressure pump, **6 Oil** filter, **7 Pressure** regulator *valve,* **8** *catch tank* (Overflow tank), **9** *breather line,* **10 Oil pressure** indicator

crankcase. If the engine has only one pump, the suction line can be designed with two suction ports. A valve provides inertia-controlled closure of the part of the line on the inside of the crank, Fig. 1.135. Such a device can also be used to operate a wet sump lubrication system in a racing vehicle, as shown in the example of Formula BMW/ADAC with the K1100 motorcycle engine.

Appropriate design of the oil sump or the lower part of the crankcase can ensure reliable extraction of the oil and still keep the crankcase volume small. Suction pumps with a large flow rate can practically evacuate the crankcase, which virtually eliminates the losses of the crankshaft drive due to air and oil resistance, especially at the highest speeds. In high-speed passenger car engines, a *windage tray* is used, Fig. 1.136a. It separates the oil chamber from the crank chamber and has plane-like deflectors which are, for example, notched directly from a metal sheet. This variant also permits wet sump lubrication. The sump can be made shallower if two lateral oil drain channels collect the oil collected by deflectors, Fig. 1.136b. About 20 mm clearance of the bottom from the connecting rod violin is generally sufficient. With oil collection ducts, two suction points at the front and rear engine ends are sufficient. A side exhaust allows the lowest overall height, Fig. 1.136c. With evacuation of the crank chamber, the housing contour can be extremely close to the connecting rod locus or the counterweight path, Fig. 1.136d.

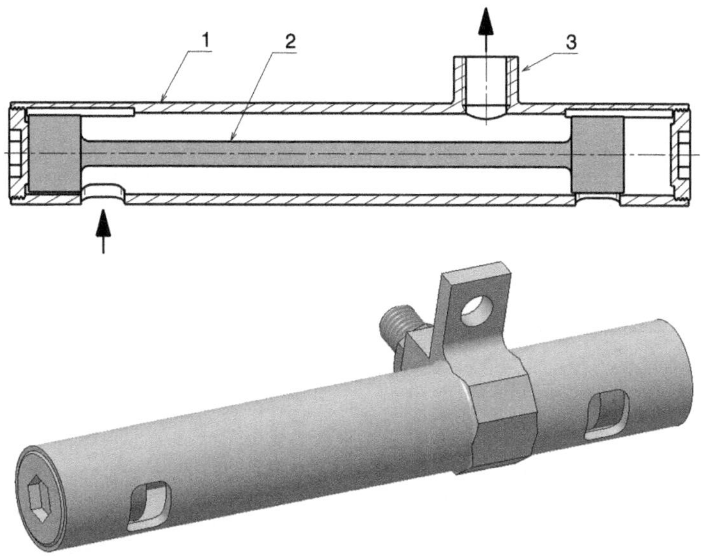

Fig. 1.135 Oil extraction valve. **1** Suction pipe, **2** Slide valve, **3** Connection to oil pump suction side. The unit is installed in the oil sump at right angles to the direction of travel. When cornering, the slider is moved to the left or right by the transverse acceleration and thus closes the side through which air could also be sucked in (in section, this is the right opening). Relief grooves on the upper side prevent the slider from stopping if oil has accumulated on its front side

Figure 1.137 illustrates the design of oil collection spaces using axonometric representations.

Pumps

The oil pressure pump generates a delivery pressure of 4–7 bar. It is driven directly from the crankshaft or from the rear shaft end (e.g. via an Oldham coupling) of another ancillary unit (coolant pump, suction pump,. . .). If racing engines are derived from a production engine, larger pumps are used which are mounted on the outside of the engine, Fig. 1.138. In thoroughly designed racing engines, the auxiliary units are often combined in separate units, Fig. 1.139.

The flow velocity in pipes and ducts should be 0.5–1 m/s.

The main types of pumps used are: Gear pump, trochoidal pump and gerotor pump.

In addition to the required flow rate, the efficiency of the pump is important. The power loss can be up to 4% of that of the total engine at certain operating points [7].

An example of a typical gear pump unit is shown in Fig. 1.140. The unit comprises one pressure and four suction pumps in a housing (5) which is inserted into a receptacle in the crankcase (4). They are driven by a common shaft (1) through a gear wheel (9) which is incorporated in the gear train of the engine. The gears (10) seated on the drive shaft are

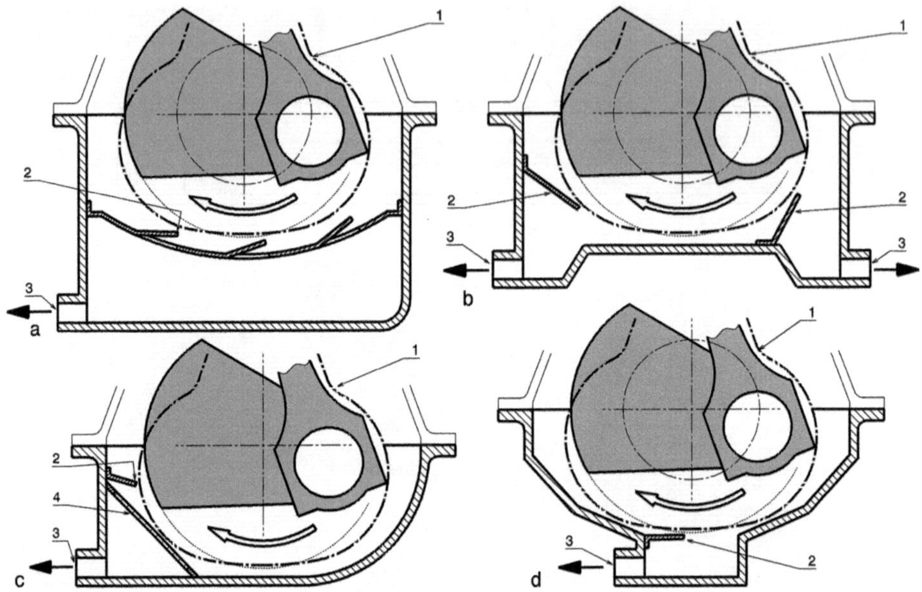

Fig. 1.136 Design of the oil pan or crankcase lower part. (**a**) oil planer, (**b**) lateral oil collecting channels, (**c**) lateral strainer, (**d**) central oil collecting channel. **1** *conrod locus,* **2** *scraper,* **3** *Oil pick up,* **4** Metal *screen*

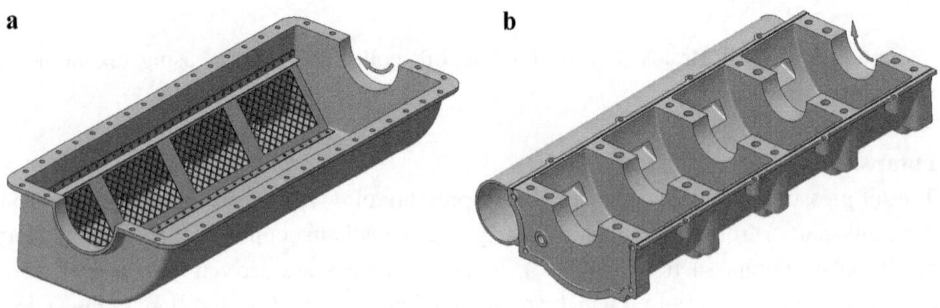

Fig. 1.137 Views of two oil collection chambers. (**a**) Oil pan with metal screen, cf. Figure 1.136c, (**b**) Crankcase bottom section of a high-revving V8 engine. The oil-air mixture is extracted tangentially from each crankcase chamber on the upstream side. For this purpose, individual suction pumps (gerotor pumps) are located on the outside in a common housing, which are driven by a common shaft

entrained via an insertion spring. The meshing idler gears (8) run on an axle (2). All pump wheels have basically the same teeth, but differ in width. This allows the flow rate to be adjusted. The delivery rate of suction pumps is generally designed to be 25–30% greater than that of pressure pumps because they do not transport pure oil but an oil-air mixture. Between the individual chambers, intermediate discs (6) separate the pumps. All inlet ducts

Fig. 1.138 External oil pump (Formula Renault). The engine is derived from a passenger car series engine. The wet sump lubrication is converted to dry sump lubrication. The external oil pump sits on the outside at the side of the flat cast oil pan and is driven via a shaft from the auxiliary unit drive at the front of the engine. The shaft is connected to the pump by a claw coupling. On the left side of the picture you can also see the oil cooler, which is installed between the engine block and the oil filter and is cooled with water

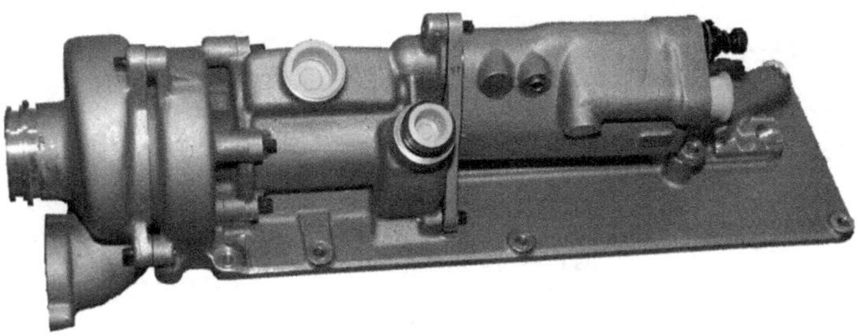

Fig. 1.139 Auxiliary unit of a racing engine. The water pump is located on the left, followed by the oil pressure and suction pumps on a common shaft. The unit is bolted to the outside of the lower part of the crankcase and is driven by the engine

are located on the crankcase side. The pressure pump delivers the oil to a pressure oil port (11). The suction pumps have separate inlets but deliver to a common outlet chamber which opens into a circular opening in the pump cover (3). The pump gears are offset on the drive shaft by an angle so that the oil displacement area moves toward the pump outlet. The pumps are self-lubricating. Small connecting holes from the relief pockets of the pressure chambers to the bearing points serve this purpose (see also section A-A). The axle (2) is thus also supplied with oil via a recirculating groove, which it in turn makes available to the idler wheels via small transverse bores. The pump unit is pre-assembled except for the drive gear (9) and then inserted and screwed into the mounting of the crankcase. Then the

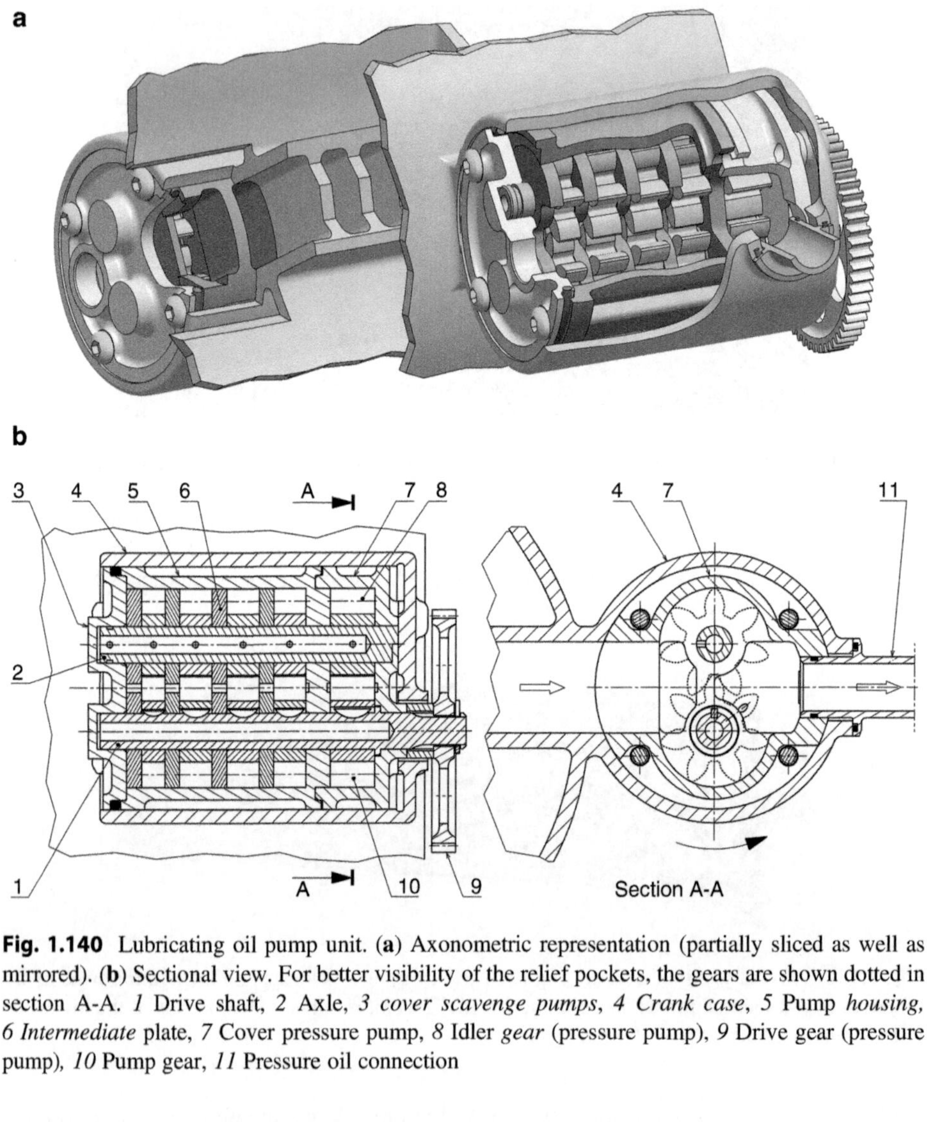

Fig. 1.140 Lubricating oil pump unit. (**a**) Axonometric representation (partially sliced as well as mirrored). (**b**) Sectional view. For better visibility of the relief pockets, the gears are shown dotted in section A-A. *1* Drive shaft, *2* Axle, *3 cover scavenge pumps*, *4 Crank case*, *5* Pump *housing*, *6 Intermediate* plate, *7* Cover pressure pump, *8* Idler *gear* (pressure pump), *9* Drive gear (pressure pump), *10* Pump gear, *11* Pressure oil connection

pressure oil connection (11) and the connection (not shown) from the suction pump outlet are plugged into the cover (3).

The maximum velocities in suction lines should not exceed 2.5 to max. 4 m/s. In pressure lines, a maximum of 5–7 m/s is used as a design basis for line cross-sections.

Dry Sump Oil Tank

The dry sump tank (Fig. 1.141) stores the oil supply and separates the returning oil from the remaining air that has remained dissolved in the oil despite the air separator. It may also be

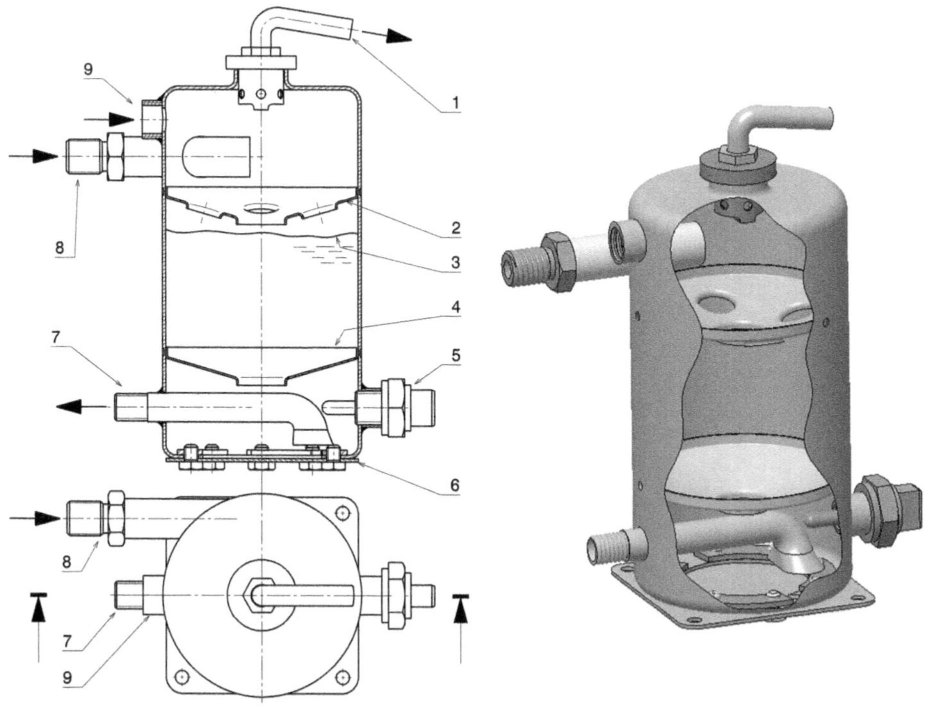

Fig. 1.141 Dry sump tank. **1** Vent (to overflow tank), **2** *Top baffle plate,* **3** Oil *level running,* **4** *Bottom baffle plate,* **5** Temperature sensor, **6** Bottom *plate,* **7** *pick up line to pressure pump,* **8** Return *from scavenge* pump, **9** From engine *breather*

provided with cooling fins for further heat removal from the oil. The dry sump tank sits as close to the engine as possible, but should also be covered by cooling air.

In operation, the level is such that a calming space remains above the oil level in which the air can separate from the oil. The return oil enters tangentially at the upper end of the vessel and thus wets the vessel wall as a thin film. This facilitates the separation of the air. The air exits the vessel through a vent line. Perforated baffles inside keep the movement of the oil level within limits, even under extreme driving conditions. Suction is near the bottom of the tank, well away from the point of entry. The suction funnel is fitted with a coarse-mesh screen as a filter. The free cross-section of this screen should be larger than ten times the suction line cross-section.

The housings are designed in two parts (base plate, central joint, . . .) so that they can be cleaned inside. Dry sump tanks are often integrated into the *bell housing*, e.g. on single-seaters and sports prototypes, Fig. 1.142.

Some tanks are equipped with an electrical *oil preheater*. This facilitates the "cold start" of the engine and prevents lubrication problems caused by too viscous oil.

Fig. 1.142 Arrangement of the lubrication system in a formula car (Arrows A10 B Megatron). You can see the rear axle with a mighty final drive (driving direction to the right). The dry sump tank sits in the clutch housing and cannot be seen. The black expansion tank is mounted on top of it. The oil cooler sits at the rear of the gearbox and dissipates heat to the air flowing through it. Its connections are provided with temperature measuring strips

Table 1.13 Capacity of dry sump oil tanks

Engine stroke volume [l]	Filling quantity [l]
Up to 2.5	5
To 4	7
Until 6	10

The tank must be dimensioned large enough to ensure lubrication even despite oil consumption during the race. Comparatively, the lubricating oil consumption of the Ford Cosworth DFV 3 l V8 engine was about 3.4 l per 320 race kilometres [23]. The engine oil is changed between practice and race. As a result, each Formula 1 car needs 160 l of lubricating oil per race weekend with about 2200 km.

Different volumes are required depending on the size of the motor. Table 1.13 can be used for reference values.

1.4.9 Cooling *System*

The cooling system dissipates part of the heat generated during combustion. This is necessary so that the heat resistance of the components is not exceeded (cylinder head, piston, liner,. . .). In the case of the petrol engine, the risk of knocking and glow ignition is

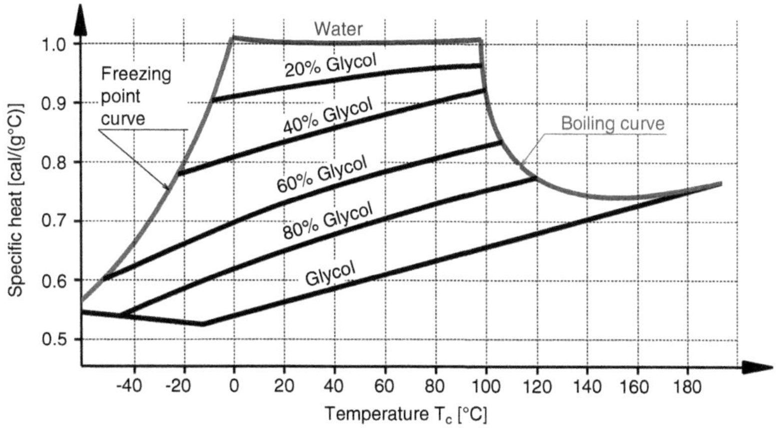

Fig. 1.143 Dependence of the specific heat on the temperature and the water-glycol mixture [11]. Details of the glycol admixtures in mass percent. Adding glycol lowers the freezing point, but at the same time reduces the storable heat of the coolant

also reduced. Moreover, the density of the fresh charge is higher at low temperatures and therefore the effective power of the engine. Faults in the system (starting with a leaking hose) quickly lead to total engine failure. The cooling system of production and rally cars also includes the heating system for the interior. A mixture of water and additives is usually used as coolant, sometimes pure water is also used, see also Fig. 1.143.

In principle, there are several ways of routing the coolant through the engine (coolant routing), Fig. 1.144. Conventional longitudinal flow (a): The coolant flows longitudinally through the cylinder block and passes into the cylinder head through openings in the cylinder deck at each cylinder. Essentially, longitudinal flow also forms in the cylinder head, but in the opposite direction to the block. How much fluid crosses over where must be determined and tuned by simulation and testing.

Forced flow in longitudinal direction (b): The cooling medium flows through the block longitudinally, the transfer to the head occurs only at one point at the output end of the engine. The cooling medium must therefore also flow through the cylinder head longitudinally and in the opposite direction.

Primary flow through cylinder head: This is a variation of the above concepts. Here, the cooling medium flows from the heat exchanger first into the cylinder head and then into the block. In this way, the combustion chamber area is cooled more strongly and the liner area, which is surrounded by the fluid already heated by the head, has a higher temperature. The latter is advantageous in terms of low piston and ring friction. Under extreme conditions, fuel also condenses less easily on the running surface of the warmer liner.

Cross flow block and head (c): The coolant flows through each cylinder at right angles to the crankshaft axis. In the process, the fluid enters the cylinder block from a manifold and

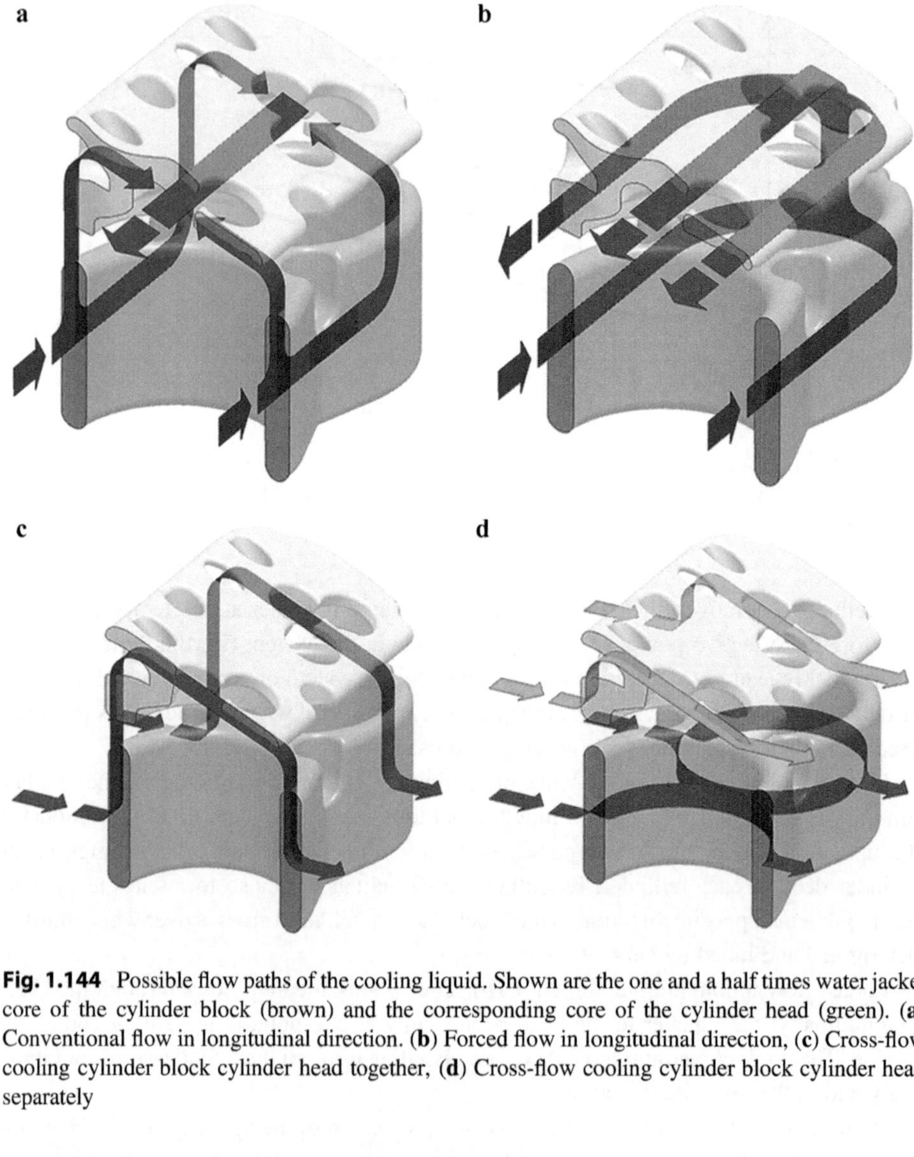

Fig. 1.144 Possible flow paths of the cooling liquid. Shown are the one and a half times water jacket core of the cylinder block (brown) and the corresponding core of the cylinder head (green). (**a**) Conventional flow in longitudinal direction. (**b**) Forced flow in longitudinal direction, (**c**) Cross-flow cooling cylinder block cylinder head together, (**d**) Cross-flow cooling cylinder block cylinder head separately

then passes into the cylinder head on the exhaust or intake port side. A manifold on the opposite side of the block discharges the coolant again.

Cross flow head and block separately (d): In this variant, the flow is the same as in (c), but the coolant for block and head is routed separately. On which side the cooling medium enters depends, among other things, on the installation space (position of the heat exchanger), the core production and the pressure loss or flow cross-section (nozzle-shaped better). Another important consideration is the heat flow. If the cooling medium has previously flowed around the exhaust port, it will reach the inlet side heated up and, in the worst case, heat the incoming air. Apart from the fact that the cooling jacket will therefore only be placed around the inlet seat rings, the coolant flow from the inlet to the exhaust side appears to be more favourable in terms of heat dissipation.

Cross-flow head alone: A compromise of the variants presented. Longitudinal flow prevails in the cylinder block and cross flow forms in the cylinder head. Longitudinal manifolds on both sides of the head force a crossflow from the exhaust or intake side of each cylinder across the combustion chamber roof to the other side. The supply manifold is fed from the block and the opposite manifold feeds the coolant back to the block.

With cross-flow cooling, the water pump can be designed smaller because it has to overcome fewer losses.

The cooling system is operated as a closed system, i.e. no part of the coolant is discharged to the environment, but in the event of excessive overpressure into an expansion tank or an overflow tank. Cooling liquid is sucked back in from this during cooling, Fig. 1.145.

The water pump (1) delivers the coolant into the cylinder block. This flows around the liners from the front to the rear. A partial flow is simultaneously directed to the cylinder head. These flows are calibrated via different holes in the cylinder head gasket or in the cylinder head deck. The cylinder head is flowed through transversely from the intake to the exhaust side. The pure longitudinal flow as found on utility engines cannot be used on highly loaded engines. This is because the first liner is the coolest and the last the hottest in this method. Conversely, due to the further flow in the cylinder head from the rear to the front, the last combustion chamber is the coolest and the first the hottest. In Formula 1 engines, they go one step further and cool the block and head completely separately. This allows the combustion chamber to run relatively cool (about 60 °C water temperature) and the liners at the usual temperatures (about 90–110 °C water temperature). Fuel would precipitate as condensate on cooler liners, thus worsening combustion and impairing lubrication.

The coolant leaves the engine in the area of the cylinder head at as high a point as possible so that air is carried along. The air is separated in the following vent tank (3) and fed to the expansion tank (6). In the heat exchanger (4), the cooling liquid gives off heat to the environment. An oil cooler (5) can be arranged in the return flow to the water pump suction side. All vent lines converge in the expansion tank. The tank is closed with a cover containing a double valve. If the system pressure rises above a limit value, the poppet valve opens and a partial quantity flows into the overflow tank. If the pressure in the system drops, e.g. during subsequent cooling, a vacuum could develop which would cause the pipes and heat exchanger to collapse. A vacuum valve in the cover prevents this by opening and sucking the missing water quantity back in from the overflow tank.

In the case of two heat exchangers, the basic scheme looks the same, Fig. 1.146. However, in order to ensure an even thermal load in highly loaded V-engines, each cylinder bank is usually supplied by its own water pump.

The system is under overpressure (approx. 1.3–1.5 bar; maximum values up to 5 bar), which is set with the overflow valve of the expansion tank. Due to the higher system

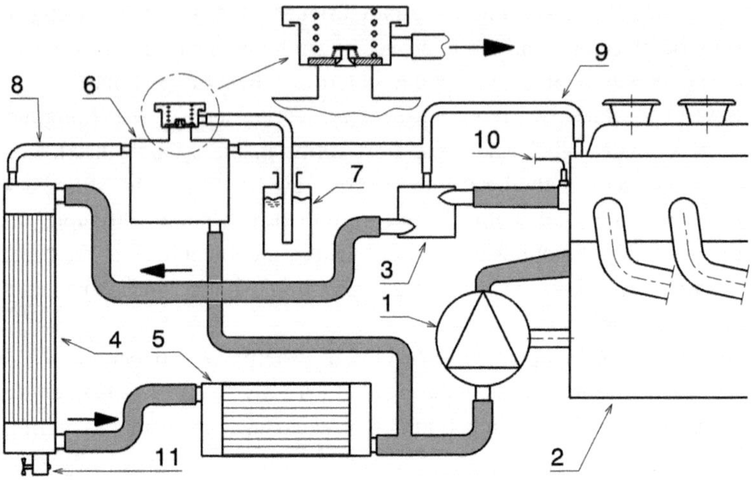

Fig. 1.145 Cooling system schematic. **1** Water pump, **2** Engine, **3** *swirl pot,* **4** Heat exchanger, **5** Oil cooler, **6** *header tank,* detail shows connection pipe with cover double valve, **7** *catch tank* (overflow tank), **8** *top radiator bleed,* **9** *engine bleed,* **10** *connection temperature indicator* (remote thermometer), **11** Drain plug

pressure, the water only boils at higher temperatures (above 120 °C). The coolant temperature should be kept at about 90 °C. In warmer countries (e.g. South Africa) this target value is exceeded due to the higher ambient temperature.

A thermostatic valve that bypasses the radiator while the engine is warming up is not needed on most race cars.

As a rough rule of thumb, the amount of heat to be dissipated per time is approximately equal to the power at the crankshaft. The heat transported by the water depends on its volume flow (water pump) and the temperature difference (heat exchanger). The temperature difference between the coolant inlet and outlet should be as low as possible, especially in the case of highly loaded engines (about 5 °C. For service engines up to 10 °C). Higher temperature differences allow a reduction of the delivery volume of the water pump, but lead to higher component stresses.

Higher system temperatures make it possible to reduce the size of heat exchangers. This simplifies the packaging (arrangement of all components in the car), the amount of coolant is smaller and the aerodynamics of the overall vehicle can be further improved due to the smaller radiator frontal area. However, the engine and its system components must deliver acceptable performance with sufficient reliability despite the higher temperature level. Among other things, more heat must be dissipated through the lubricating oil. Some Formula 1 teams have successfully taken this (laborious) step.

For average flow velocities in suction lines 2.5 m/s and in pressure lines 5 m/s is recommended. The diameter of designed connection lines is 25–32 mm, depending on the heat output to be dissipated. The pressure loss in the cylinder head water jacket, which

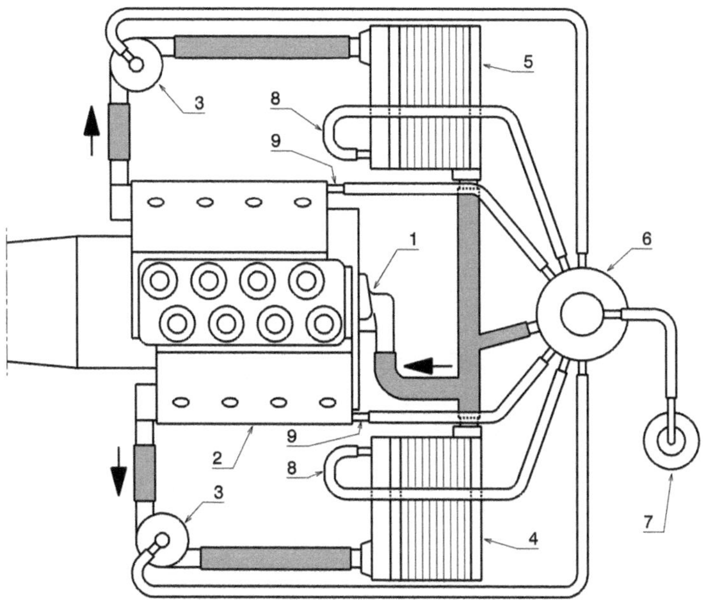

Fig. 1.146 Coolant circuit with two heat exchangers (top view). **1** Water pump, **2** V-engine, **3** *swirl pot* (breather tank), **4** *radiator right hand bank,* **5** *radiator left hand bank,* **6** *header tank* (epansion tank), **7** *catch tank* (overflow tank), **8** *top radiator bleed,* **9** *engine bleed*

the water pump must overcome in addition to other resistances, is in the range of 250–450 mbar.

Air in the system interferes considerably with heat dissipation – air bubbles act as an insulator and can cause local boiling if they accumulate. A vent tank in the return, usually after the exit from the cylinder head, separates a large part of any air that may be present by means of a centrifugal effect, Fig. 1.147. The water is forced into a helical movement by the arrangement of the pipe sockets, during which the lighter air remains near the screw axis and can leave the tank from there upwards.

Coolant Pump

The coolant pump is either driven directly by the combustion engine or by its own electric motor. Directly driven pumps are placed in the belt drive at the front of the unit or driven by gears and shafts. In the latter case, they can be placed on the side of the crankcase in a module with the oil pumps, Fig. 1.139. Electrically driven pumps offer the greatest freedom of accommodation. Further advantages result from the electric drive. The speed is variable and can be adapted to the needs of the combustion engine (load, ambient temperature, operating condition,. . .) independently of the engine. In sprint competitions, the required pump power is taken from the battery and thus does not occur as power loss during propulsion. Fig. 1.148 shows an example.

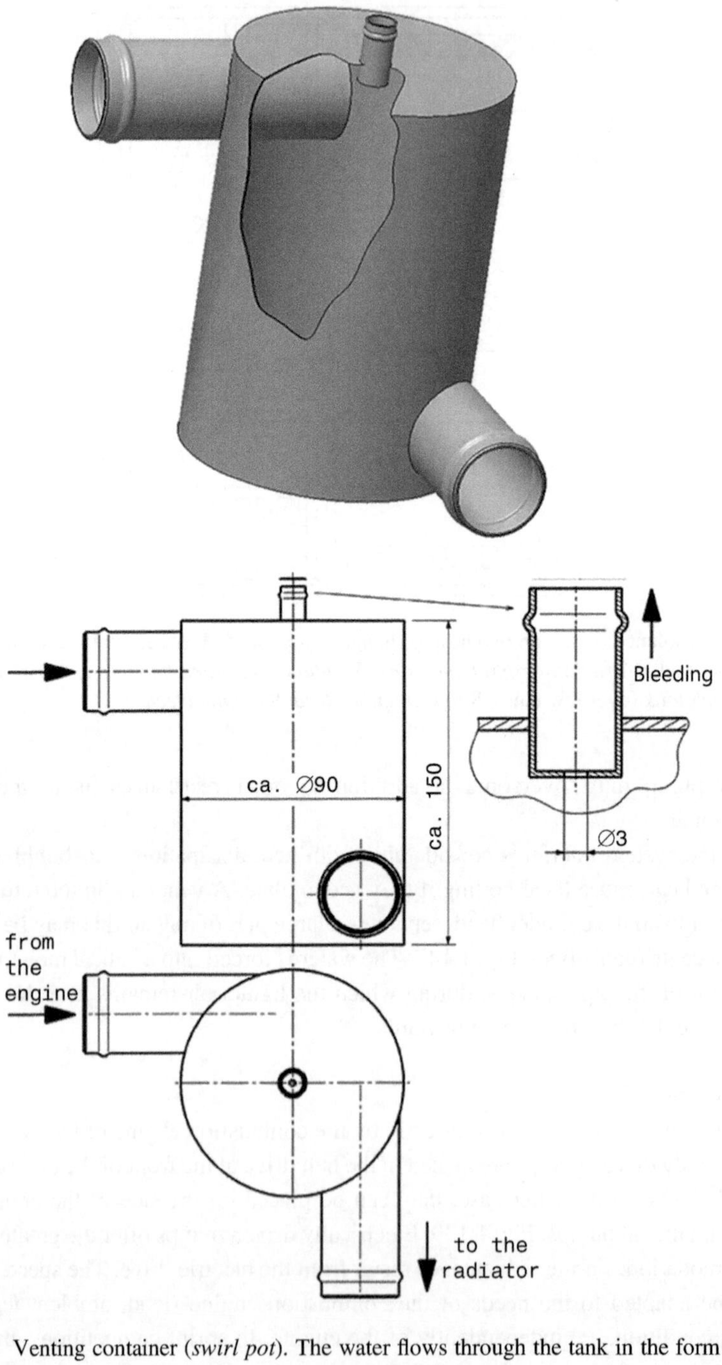

Fig. 1.147 Venting container (*swirl pot*). The water flows through the tank in the form of a helix. Due to a centrifugal effect, existing air tends to remain in the centre, from where it can escape. The vent is connected to the expansion tank

Fig. 1.148 Electric coolant pump, type Bosch ECP 160 [59]. The radial pump is driven by a brushless electric motor (BLDC motor) and controlled via pulse width modulation (PWM microcontroller on the left in the picture). With dimensions of 105 × 90 × 95 mm, the pump delivers 160 l/min at 1.5 bar. At the same time, it consumes 690 W of power. The inlet (top) and outlet (side) connections are made via Wiggins terminals (see Fig. 1.152). The electric motor is air-cooled. The air enters through the black, cylindrical (in the picture central) connection piece and leaves the motor housing through cooling slots

Heat Exchanger

Heat exchangers consist of a block with fine channels (the heat exchanger itself), the radiator network between these channels and two water boxes which divide the coolant evenly between the channels on the inlet side and collect it again on the opposite side, Figs. 1.149 and 1.150. The water boxes also accommodate the connections for the coolant hoses. The side panels reinforce the block and with their help heat exchangers are fixed in the vehicle by supporting fasteners (tabs, holes,. . .). In the case of production vehicles, water boxes are made of fibreglass-reinforced polyamide and mounted on the radiator block with a gasket via flanging. The individual solutions for racing vehicles are made of aluminium sheet and welded or brazed onto industrially produced radiator blocks. Heat exchanger blocks are made of aluminum. Copper was used as the base material in the past because of its better thermal conductivity. However, this advantage was bought with the disadvantage of a higher mass. Heat exchangers and joints must be leak-proof and able to withstand operating pressures of up to 5 bar. The functionality must not be affected by the usual stresses in racing (acceleration forces, vibrations, dirt, stone chipping, etc.). Heat exchangers should not be painted because paint acts as an insulator rather than having a beneficial effect on heat radiation.

The preferred locations for heat exchangers are the vehicle nose and side boxes next to the cockpit, Fig. 1.151. The disadvantage of the nose arrangement is the long pipe runs and the problem of hot water on open vehicles in the event of a frontal collision. An arrangement of charge air and water coolers close to the engine keeps the line volumes small,

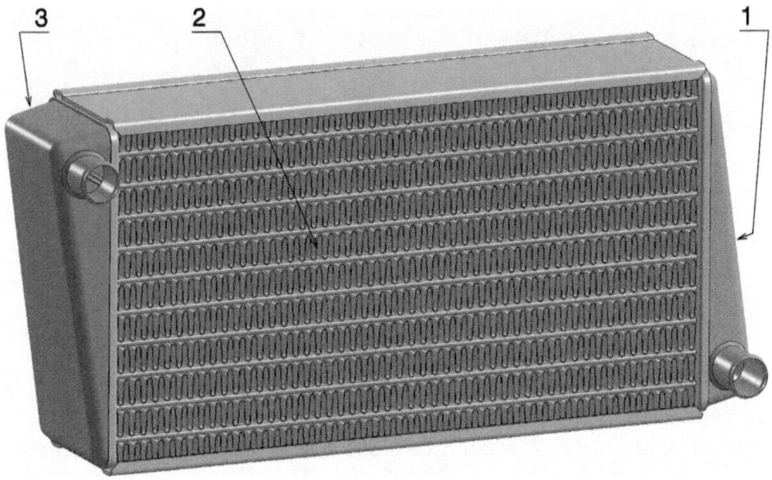

Fig. 1.149 Heat exchanger in cross-flow arrangement. The three main components are: 1 *in cap* (water box inlet), 2 radiator *core*, 3 *out cap* (water box outlet)

Fig. 1.150 Heat exchanger detail (partly cut open). Basic structure of a heat exchanger. 1 *cap* (water box), 2 Flat *tubes*, 3 *end plate*, 4 Fin (*louvers*), 5 *Side plate*. Flat tubes and fins form the cooler matrix. The cooling liquid (red arrows) flows through the flat tubes (2). The heat is transferred to the air, which flows through gaps formed by fins (blue arrows). The fins can also have turbulence-generating slits or similar to intensify the heat transfer

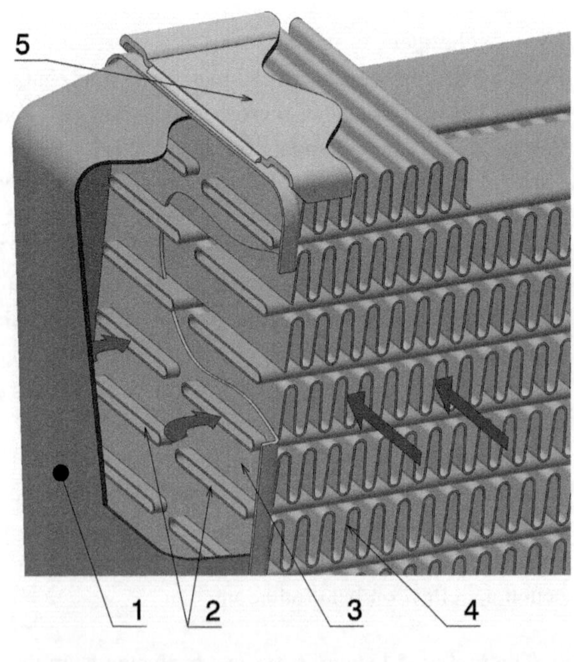

reduces flow losses and facilitates the desired modular design, which makes maintenance easier.

Coolers are flowed through from top to bottom or from one side to the other (cross-flow). With cross-flow coolers, the coolant ideally enters at the bottom and exits at the top.

Fig. 1.151 Heat exchanger (Reynard/BAR). The heat exchanger is located in the left side box of a single-seater. The shape of the side box can be seen in the contour of the underbody. The radiator is aligned at an angle to the longitudinal axis of the vehicle and thus does not widen the frontal area of the vehicle. An electric preheater is housed in the connection line in the foreground

In general, the connections should not only be at the bottom or top, but in such a way that the coolant flows through the cooler "diagonally" and thus no air pockets remain. Venting of the cooler should still be provided. The performance of a cross-flow cooler can be increased by placing both connections on the same side. Between the connections there is a separating plate in the distribution volume called water box, which forces a cross flow. The coolant then flows through the width of the radiator twice. A cooler of this design also manages with a lower installation height.

The air inlet to the radiators is positioned so as to cause little disturbance to the overall airflow around the vehicle. The cross-sectional area of the air outlet after the heat exchangers must be larger than that of the inlet. Also on the outlet side, try to positively influence the overall airflow or use a negative pressure area to lift the airflow through the radiator. Further design advice on heat exchanger shafts can be found in the literature on race car aerodynamics.[7]

Hose fittings as well as connections are a possible cause of failure and should therefore be carried out with particular care. There are many quick-release metal fittings with O-ring seals that are of interest to endurance racing vehicles where quick repair is essential (e.g. Figure 1.152). Normally, beaded tubes and adherence to minimum dimensions are sufficient for a solid hose connection, Fig. 1.153.

A (suitable fabric-reinforced) hose has the advantage of allowing movements between the connection points and compensating for tolerances between assemblies that are further away (motor, heat exchanger, tank). In contrast, short hoses and pipelines can, in the worst

[7]E.g. Racing Car Technology Manual, Vol. 2 *Complete Vehicle*, Section 5.6 *Heat Dissipation and Ventilation*.

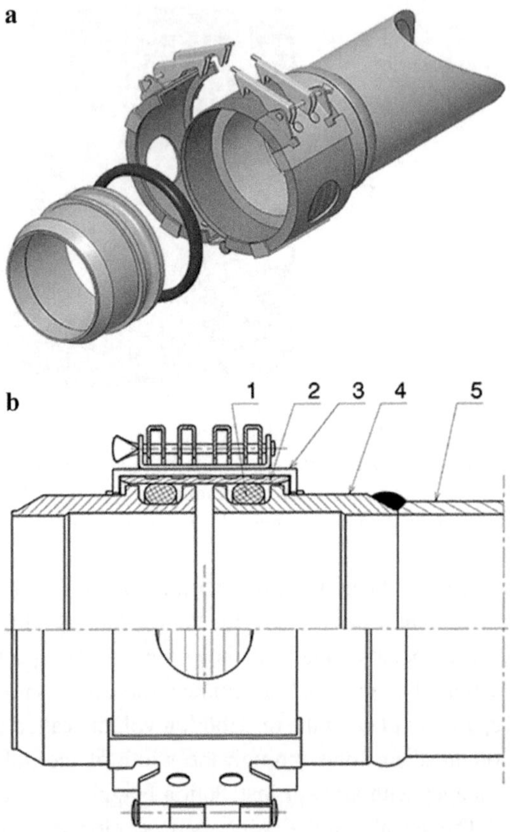

Fig. 1.152 Quick release (Wiggins system). (**a**) Axonometric representation, (**b**) Half-section representation. **1** *O-ring*, **2** Inner *sleeve*, **3** Clamp *(coupling)*, **4** *Welding* flange, **5** Tube. The two weld-on flanges each carry an O-ring (1) which creates the seal to the inner ring (2) during assembly. The flanges (4) are held in the axial direction by an articulated clamp (3). The clamp can be closed with one hand and allows easy controllability of the connection. This locking system allows an axial compensation in the range of 6 mm and an angular deviation of up to 4°

case, lead to the tearing off of connecting pieces due to the transfer of the engine movements.

In return, aluminium tubes are lighter than hoses of the same internal diameter and, due to their rigidity, hold the installation position in the vehicle much better. This increases reliability in tight spaces and with moving and hot parts in the immediate vicinity (shafts, pulleys, exhaust pipes, etc.).

1.5 Special Features of Racing Engines

Basically, there is no difference between production and racing engines. But due to the fact that the race engines are fully developed and mostly designed for a specific use, there are some noticeable differences in detail.

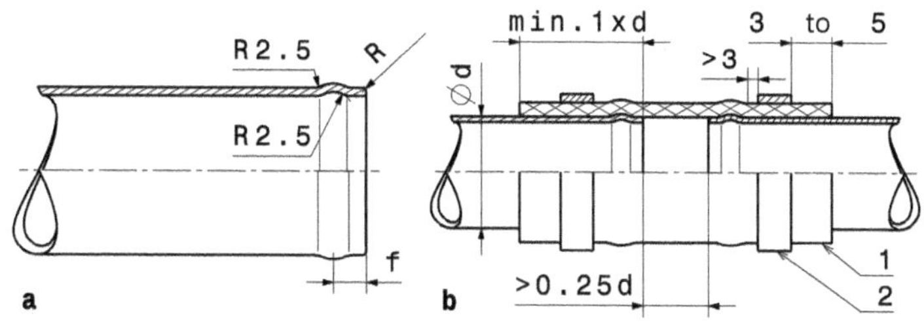

Fig. 1.153 Favourable design of pipe sockets and hose connections (tw. according to DIN 71550). (**a**) Pipe socket, (**b**) Hose connection. 1 Hose, 2 Clamp. The tube socket has a *bead* to hold the hose and a rounding to facilitate assembly

1.5.1 Engine Start

Extremely high-bred engines cannot be started like series engines. For Formula 1 engines, the actual start-up is preceded by a check lasting about one hour. Before starting, the entire engine is preheated using external equipment. Either the coolant is heated or at least the fuel metering device is preheated. The cooling system is pressurized by means of a suitable device in order to locate leaks before the fairing parts are fitted to the vehicle [23].

The fuel metering device is set to rich (see appendix: air ratio) and the electric fuel pump is switched on. The crankshaft is usually made to rotate by external devices (compressed air starter via transmission input shaft, electric starter). Once the engine is running under its own power, a mechanically operated fuel pump provides system pressure in some engines. Initially, the engine is run at about 2000 min^{-1} until oil and coolant temperatures have risen above 30 °C. Only then is the speed increased and the engine is started. Only then is the engine speed increased and the mixture leaned. The car may only leave the pit when the coolant temperatures have reached 70 °C [23].

Shut Down

Before switching off the ignition, the engine is revved up again at idle for a few seconds so that the lubricating oil is pumped out of the engine into the dry sump tank. This ensures a lossless detection of the oil level.

Transport

Pneumatic valve springs must be pressurized with a container. Otherwise the valves will slip into the combustion chamber under their own weight and damage the piston crown.

Engine *Suspension*

Vibration-isolated suspension via elastic mounts, as is standard in production vehicles, is rarely found in racing vehicles. Decoupling elements take up installation space, as do components that are given room to move by such a mounting. They represent an additional

Fig. 1.154 Front view of an engine (Ferrari Formula 1 V10 cylinder). A V-engine offers the possibility to provide distant mounting points at the crankcase (here two at the bottom) and cylinder heads (here two each at the top cover). At these points the aggregate is bolted to the monocoque

Fig. 1.155 Rear view of an engine (Ferrari Formula 1 V10 cylinder). The clutch housing and the gearbox housing are bolted to a co-supporting engine and all forces and torques are transmitted from the rear axle to the frame via this connection. On each cylinder head cover there is a mounting hole for the clutch housing, which is advantageously far away from the holes in the crankcase

mass and, in addition, elastic elements lead to energy losses as well as indirectness due to additional movement space. In general, the engine is bolted directly to the frame, if it is not even a supporting element of the vehicle, as is often the case with single-seaters, Figs. 1.154 and 1.155.

Ancillaries

In racing engines, attempts are made to integrate as many ancillary components as possible into existing castings (cylinder head, crankcase). This eliminates brackets, lines and sealing points – all potential sources of error. In addition, this integral design leads to a lighter overall system. The disadvantage that the cast parts are more expensive to produce and the overall unit is inflexible and therefore cannot simply be used for vehicles with a different design only applies to series-production vehicles, where a basic engine is used once longitudinally, once transversely, once with a manual gearbox, once with an automatic gearbox in different model series.

1.6 Fuels, Coolants and Lubricants

Common operating fluids for engines are fuel, lubricant and coolant.

1.6.1 Fuels

In the case of gasoline fuels, their boiling point and knock resistance are outstanding properties for engine operation. Knock resistance is the ability of gasoline to prevent unwanted ignition (i.e. not initiated by the spark plug). Prolonged knocking combustion leads to damage and consequent destruction of spark plugs, pistons, cylinder head gasket and valves. One measure of knock resistance is the octane rating. The higher the value, the more resistant the fuel is to knocking.

The boiling curve, i.e. the evaporation fractions above the temperature, has a great influence on the engine behaviour. Low-boiling fuel components are favourable for starting. However, too many of them lead to the formation of vapour bubbles in the fuel system, especially in summer. Low-boiling components are better in this respect. However, they can condense on the cold cylinder running surface and severely disrupt piston lubrication. Fuel components that evaporate in the medium temperature range result in good drivability. Conversely, a lack of components in the medium range is noticeable by "jerking" during acceleration. When the engine is hot-started, the demands on the fuel are exactly the opposite. By their very nature, ambient temperatures are not always the same, which is why there is winter-grade fuel and summer-grade fuel.

In the case of racing fuels, internal cooling due to high heat of vaporization (for increased filling), high combustion rate (for high engine speeds) and large calorific value (for high energy input) are added to the properties mentioned. The amount of energy bound up in the fuel is of particular interest because it has a direct effect on engine performance. (1.3) gives this relationship in simple form. If one wants to compare fuels with regard to their energy content, the mixture calorific value H_G (1.4) is therefore the decisive variable and not the specific calorific value H_u. The data in Table 1.14 may illustrate this, in that the quotient H_u/L_{min} which is essential for the mixture calorific value is given. Nitromethane

Table 1.14 Energy comparison of fuels, according to [7, 60]

Characteristic value	Nitromethane	Methanol	Ethanol	Iso-octane
Sum formula	CH_3NO_2	CH_3OH	C_2H_6O	C_8H_{18}
O_2 content, % (mol/mol)	52.5	49.9	34.8	0
Calorific value H_u, kJ/kg	11,300	19,900	26,800	44,300
Air requirement L_{min}, kg/kg	1.7:1	6.45:1	9:1	15.1:1
Specific energy H_u/L_{min}, kJ/kg	6647	3085	2978	2934
Heat of evaporation, MJ/kg	0.56	1.17	0.93	0.27

has a much lower calorific value than iso-octane, but the amount of energy to be released in the combustion chamber with the required air is more than double, because nitromethane requires much less air for stoichiometric combustion. The utilization of twice the amount of energy naturally also results in a correspondingly greater thermal and mechanical load on the engine.

The minimum air requirement is also important: the lower the stoichiometric fuel/air ratio, the more fuel can be introduced into the combustion chamber. The drivable misfire (ignition) limits permit the richest possible mixture. The higher the ratio of combustion products to reaction partners, the greater the combustion pressure and thus the possibility of power output [60].

A high combustion rate is advantageous for racing fuels in any case. With an adapted ignition map, the knock limit can be shifted and the centre of gravity of the energy conversion can be moved to the area in which the crankshaft drive has the highest mechanical efficiency (depending on the connecting rod ratio 8–15 °CA after TDC).

Methanol is used in some racing series (e.g. IRL until 2006 and ice speedway). It has a much lower air requirement than common petrol fuels (Table 1.14), which is why the mixture formation must be adapted to methanol. Maximum performance is achieved at a fuel/air ratio of 4:1 ($\lambda = 0.62$). Materials (especially gaskets and injection system parts) must also be specifically selected because of its corrosive or swelling behavior. In terms of the internal engine, the high heat of vaporization proves to be an advantage: the combustion chamber temperature is lower than with gasoline and thus the cylinder filling is higher. The compression ratio can also be selected higher than with gasoline. Values of up to 16:1 have been published [60]. The cold start behaviour of methanol is sometimes problematic due to its low vapour pressure, which is why hydrocarbons are added. This incidentally increases the drivability of carburetor engines. With fuel injection, the transition behavior of the engine proves less problematic. Alcohol is also highly hygroscopic (absorbs moisture from the air). That's why many teams flush the engine with gasoline at the end of race day, running it until the fuel system is methanol-free.

Unlike other fuels, a methanol fire can be extinguished with water. A disadvantage in this context, however, is that the flame is hardly visible in sunlight during the day.

Table 1.15 Fuel specifications for gasoline engines, according to [61]

Parameter		A	B	C
Composition, %	Toluene	30	60	84
	n-Heptane	4	9.5	16
	Iso-octane	66	30.5	0
Research Octane rating ROZ		101.6	101.9	101.8
Engine Octane Number MOZ		94.2	91.2	90.0
Density (at 15 °C), kg/dm^3		0.747	0.799	0.840
Specific calorific value H_u, kJ/kg		43,124	41,931	41,102
Air requirement L_{min}, kg/kg		14.5:1	14.0:1	13.7:1
Vapour pressure according to Reid, bar		0.151	0.138	0.118
Boiling point, °C	Start	96.0	98.5	100.0
	10%	97.5	100.5	105.0
	50%	98.5	102.0	106.0
	90%	100.0	102.0	108.0
	Final boiling point	123.5	108.0	116.0

Another alcohol-based racing fuel is ethanol. Its properties lie roughly between methanol and gasoline. The greatest performance is achieved with a fuel/air ratio of 7:1 ($\lambda = 0.78$).

For top-fuel dragsters, the fuels are hardly restricted by the regulations. Nitromethane-based fuels are the first choice here. The combustion works over a wide excess-air factor range. Nitromethane even burns without atmospheric oxygen, its own content is sufficient. Combustion takes place at a very high speed and thus places an enormous load on the engine. However, the engines are run very rich and to avoid knocking or early ignition with low compression and late ignition angle. Nitromethane is not used pure but mixed with methanol, so the race itself takes only a few seconds. Nitromethane is corrosive and attacks aluminium and magnesium, the fuel system must therefore be drained and cleaned when the vehicle is shut down.

At 36% by weight, nitrous oxide (laughing gas, molecular formula N_2O) contains more oxygen than air (23% by weight) and is therefore attractive as an oxygen supplier in racing engines. The specific energy of any fuel can be increased with it. The compound is stable and readily available. However, it is stored as a gas in pressure vessels and thus requires some care. The voluminous nature of its storage makes its use interesting only for short term operation. In the short term, a power increase of about 20% can be achieved [60].

Table 1.15 shows some specifications of fuels studied in 1988 during the performance development of the Honda V6 1.5-l turbo engine for Formula 1. The best results were obtained with fuel C: 742 kW power at 12,000 min^{-1} and 664 N m torque at 9700 min^{-1}. The fuel has a high toluene content and a density already in the range of diesel fuels. The high density means that the fuel has more energy content in absolute terms than the other two despite its lower specific calorific value.

As special products, all racing fuels have one disadvantage in common: they are harmful to health, difficult to obtain and expensive. This is one of the reasons why many racing series – including Formula 1 – prescribe fuels that correspond to commercially available petrol or diesel at the filling station.

In the course of the ten-year development (2000–2010) of the BMW Formula 1 engines, 1% more power and 2% reduction in consumption were achieved with the aid of the fuel, despite the tightly regulated fuel. In the meantime, the development potential here has probably been largely exhausted [46].

In general, race fuels available for purchase differ from their gas station counterparts in their narrower compositional spread, which is also reflected in a narrower range of density. However, this composition is of limited duration once the transport cask is opened because of the different boiling ranges of the constituents. The cask should only be opened for removal and unused fuel should be stored back in the tight cask. This also prevents moisture from being absorbed from the ambient air.

1.6.2 Lubricants

The tasks of the lubricant are manifold. First of all, it must keep friction between sliding partners low and ideally prevent mechanical contact between them. The lubricant reduces wear and transports contamination and wear particles away from the lubrication point. Another task is to dissipate heat and seal gaps. For most of these functions, not only the supply but above all the removal of the lubricant is decisive.

In engines, liquid lubricants, i.e. oils, perform these tasks.

One of the requirements for lubricating oils is that they should be as thin as possible at low temperatures and still have sufficient toughness at high temperatures. The choice of viscosity or its range for multigrade oils depends on the running clearances in the engine, its performance and the typical operating temperatures. Tighter running clearances call for a thinner viscosity oil and vice versa. As a rule of thumb, oil pressure at idle should be between 1.7 and 2 bar when the engine is warm. The influence of power and temperature on the required viscosity is shown in Table 1.16. For use in service engines, the evaporation loss should be as low as possible so that oil consumption remains low. In addition, contamination particles and combustion products should remain in suspension so that the oil change intervals can be kept long.

Oils for racing engines must be optimized for the respective application. Fully synthetic oils are used for the high-revving Formula 1 engines. They are characterized by the highest shear and temperature resistance, with simultaneous low frictional resistance. By the turn of the millennium, 15–20 kW of engine power alone could be gained through oil development [37]. In the meantime, additional power of 25–30 kW has been reported [46]. The oils are very low viscosity and need high foam damping capacity, due to the high engine speeds and the corresponding oil movements in the dry sump tank and in the engine. Synthetic oils have the advantage over mineral oils of a greater specific heat capacity and a higher

Table 1.16 Choice of SAE viscosity grade of synthetic racing engine oils, according to [62]

		Maximum engine power, kW					
		<100	150–300	450–600	750–900	1000–1200	1350–1500
Oil temperature, °C	150	10W-30	10W-40	10W-40	15W-50	15W-50	15W-50
	130	5W-20	10W-30	10W-30	10W-40	15W-50	15W-50
	100	0W-20	5W-20	5W-20	10W-30	10W-40	10W-40
	80	0W-10	0W-20	0W-20	5W-20	10W-30	10W-30

temperature resistance. Their cooling effect is therefore greater and they remain stable up to approx. 160 °C. Conventional oils should not exceed 115 °C.

Racing engines are operated very rich (see appendix: air ratio) at least at full load because of the resulting higher engine power (this does not quite apply to long-distance engines, which are operated around $\lambda = 1$ because of the importance of favourable fuel consumption). For the oil, this means increased risk of fuel dilution. The dirt solubility of the oil must therefore be higher than for gasoline. Cold startability is not an issue for circuit engines. The service life plays a minor role, but naturally a greater one in endurance races. Oil consumption and also refilling must be taken into account [7].

If engines are operated with methanol, not only the mixture builder but also the oil must be adapted to the fuel. Residue formation on pistons and combustion chamber surfaces is lower than with gasoline operation, which is why the content of detergents (oil cleaners) and dispersants (dirt solvents) can be lower. On the other hand, corrosion and wear protection must be more pronounced, because corrosion and cylinder wear increase in cold operation [7].

1.6.3 Coolant

Depending on the regulations, the coolant can be pure water or a mixture of water and a protective agent. If pure water is prescribed, then this is so that escaping coolant (leak, accident, boiling over,...) does not pollute the track. Aggregates can act like oil to tires. The protective agent added to the water prevents or delays corrosion in the heat exchanger and in the engine, especially if different materials are used in the cooling system that are far apart in the electrochemical voltage series. Common metals with which water can come into contact are: Aluminum, cast iron, copper, brass, steel and tin solder. In the case of utility engines and other vehicles used in winter, there is an additional task for the protective agent, namely antifreeze protection. Polyhydric alcohols are used as antifreeze. For Central European climatic conditions, the water is mixed with antifreeze in a ratio of about 1:1. The usual glycols have the advantage of raising the boiling point of the coolant in comparison with water. At the 50% concentration stated, the boiling point is 109 °C. However, their influence on the specific heat capacity of the coolant is disadvantageous. This decreases with increasing glycol admixture, the ability to dissipate engine heat thus decreases, Fig. 1.143. Therefore, only as much glycol should be added as required by the ambient conditions.

1.7 Examples of Engines

German Touring Car Masters (DTM)

The basic design of the engines is narrowly defined: V8 naturally aspirated gasoline engine with max. 4 l displacement. The power is limited by two 28 mm restrictors. The engine block must have a bank angle of 90° and be made of aluminium. The aluminium cylinder head may contain a maximum of four valves per cylinder and these must be actuated by tappets and coil springs. Valves must be made of steel or titanium. The camshafts may only be driven by chain or toothed belt. Variable systems are prohibited for valve timing, as well as in the intake and exhaust systems.

The engines must weigh at least 165 kg when dry but with auxiliary power units and with all parts attached to the engine except the exhaust manifold.

The mileage of the engines is in addition to the friction reduction in the foreground of the development, the engines are sealed the entire season and may only be opened to adjust the valve clearance. The durability is over 5000 km.

Typical maximum values of the engine characteristics are 340 kW power at 6750 min^{-1} and 510 N m at 5250 min^{-1}.

Formula 3

The engines are based on series units that are installed in a unit vehicle. Durable, cost-effective engines are a regulatory goal, which is why air volume limiters are used. As a result, the engines last for one racing season with revisions at intervals of 2500 km [63]. The engines may have a maximum of four cylinders and 2 l displacement. Supercharging and two-stroke processes are not permitted. The entire intake system including mixture builder must fit into a cuboid of 1000 mm length and 500 mm width and height. The engines will be run on commercial fuel. The exhaust systems shall have silencers and catalytic converters.

The 4-cylinder in-line Otto (spark ignited) engine from Opel has a square design with 86 mm bore and stroke, Fig. 1.156. Despite the restrictor, the 4-valve engine achieves a mean effective pressure of 16.1 bar. This corresponds to a torque of 256 N m at 4600 min^{-1}. The maximum power of 129 kW is achieved at 5000 min^{-1}. The cylinder block is made of grey cast iron. The two-ring pistons run directly on the honed housing surfaces (integral

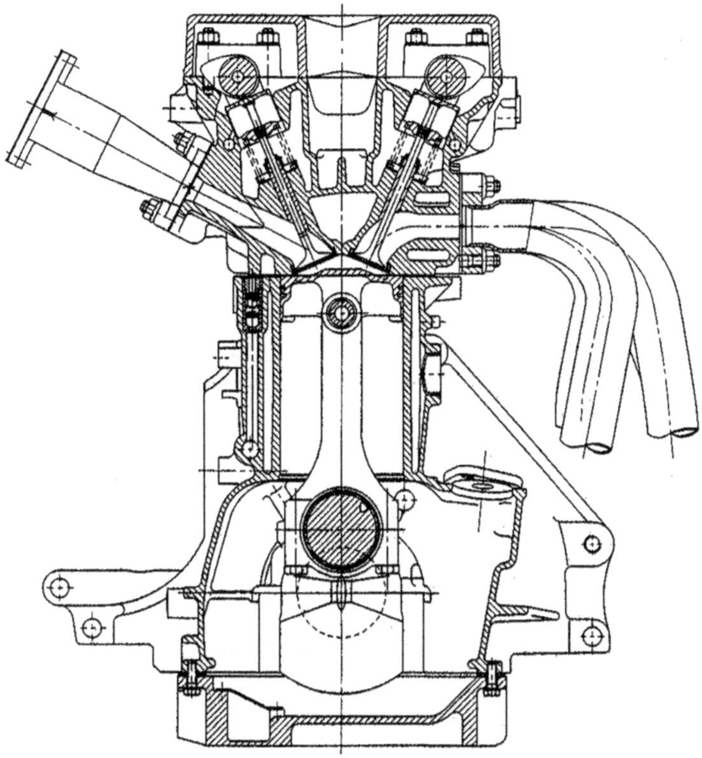

Fig. 1.156 Cross-section Formula 3 engine (Opel 1992 season) [43]. 4 cylinder 2.0 l petrol engine. Power: 129 kW/5000 min^{-1}. Torque: 256 N m/4600 min^{-1}. An air restrictor with ∅24 mm was prescribed by the regulations at that time. The intake manifold together with the airbox for this engine is shown in Fig. 1.102

liner). The valves in the aluminium cylinder head are actuated by two overhead camshafts and steel springs.

The engine does not have a generator. The on-board battery alone covers the energy requirements over a race distance. The exhaust system features a four-in-one combination. The common exhaust pipe leads into a silencer. The catalytic converter forms the end.

The aggregate weighs 96 kg.

Formula Renault

The engine is based on a passenger car production unit (Renault Clio RS 2000). Above all, the engine mounting and the lubrication system were adapted for installation in a monoposto, Fig. 1.157. It is an in-line four-cylinder petrol engine. From 82.7 mm bore and 93 mm stroke a displacement of 2 l is obtained. Four valves per cylinder are actuated by two overhead camshafts and steel springs. Using fuel with RON 98 and a compression ratio of 11.2:1, the engine produces 133 kW of power at 6300 min^{-1} and 213 N m of torque at

Fig. 1.157 Engine of a Formula Renault (right side)
The in-line four-cylinder engine is bolted to the rear wall of the monocoque via a crossbeam on the cylinder head and via lateral screw bosses on the cast, flat oil sump (not visible in the picture). It is installed in the vehicle in a co-supporting manner and thus mediates between the transmission, which carries the rear wheel suspension, and the monocoque

5300 min^{-1}. The exhaust gases are after-treated in a catalytic converter. An overhaul of the engine is due every 6000 km, which is about three times a year for racing teams.

Formula 1

Engine design is now extremely restricted by the regulations.

Until 2012, they had to be 2.4-liter naturally aspirated engines with eight cylinders. The usable rev band was very narrow. The power output was around 580 kW (about 20 kW more in qualifying). The maximum engine speed was fixed by regulations at 19,000 min^{-1}. Bore × stroke resulted from the targeted ratios of 95–100 × about 40 mm.

The engine mass was 85–90 kg for the 3l V10 engines and was specified by a minimum weight of 95 kg for the V8s. Even the centre of gravity height of the power unit was now in the rule book. It had to be at least 165 mm. In the time before that, the distribution of the mass was a declared development goal of the engine team. There should be as little mass as possible in the upper area of the engine, at the lower end it didn't matter insofar as ballast masses were already placed in the vehicle in that area.

Since 2013, efficiency has also been a declared development goal in Formula 1. Thus, the drive no longer consists of an internal combustion engine alone, but this, in contrast, works together with electrical machines (so-called MGU – *motor-generator unit*) and energy storage units, Fig. 1.158. The internal combustion engines (3) are 1.6-l V6 exhaust gas turbocharged and equipped with gasoline direct injection. However, the exhaust turbine (9) not only drives the compressor wheel (6), but also the MGU-H (*motor-generator unit heat*) (7), which is an electric machine that can use the exhaust enthalpy in regenerative

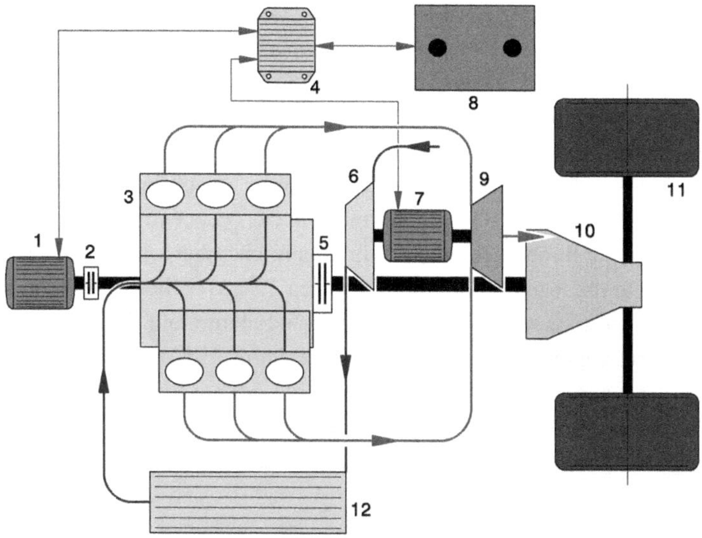

Fig. 1.158 Formula1 drive unit schematic. For reasons of clarity, the gearbox is far away from the clutch. 1 MGU-K, 2 Clutch for MGU-K, 3 V6 engine, 4 *Electronic control unit*, 5 *Clutch drive line*, 6 *Compressor*, 7 MGU-H, 8 Battery, 9 Turbine, 10 Transmission, 11 Driving wheels 12 *Intercooler.* Unlike in this illustration, the MGU-H can sit not only between but also outside the compressor and turbine. In the 2014 F1 W05 Hybrid, Mercedes went so far for the first time as to place the compressor in front of the engine and the turbine behind it (*split turbo*), which has a number of advantages (compressor on cooler side near the air intake, short lines to the intercooler and short rear end and thus smaller yaw moment of inertia of the car)

operation, is located on the same shaft. If the compressor work is not needed in a driving condition, electric current can be generated and a battery (8) can be charged. In this case, however, the MGU-H also replaces a wastegate for boost pressure control because the energy it absorbs is no longer available to the compressor. Conversely, the MGU-H can also drive the compressor if required (eBoost function) by operating as a motor and receiving energy from the battery. Another electric machine (MGU-K (1), *motor-generator unit kinetic*) represents the energy converter of the ERS system (see Chap. 3 *Hybrid Drives*). A control unit (4) decides which electric machine is used in which operating mode (motor, generator) based on the driving condition (accelerating, braking, overtaking, qualifying, emergency running), the driver's request and the battery charge level.

The maximum volume flow of the fuel is limited to 100 kg/h. A maximum of 100 kg of fuel may be consumed per race. The maximum permitted speed of the combustion engine is 15,000 min^{-1}, that of the MGU-H 125,000 min^{-1}. At this speed, the output of the MGU-H is between 60 and 80 kW. The MGU-K provides a maximum of 120 kW and 200 N m. The combustion engine provides approx. 450 kW. The max. System power is thus 630–650 kW. The boost pressure is not limited by the regulations. The teams go here up to about 3.5 bar (absolute). To achieve extremely high thermal efficiencies ($\eta_T \leq 0.45$) with a spec.

Fuel consumption below 200 g/kWh (tw. up to 170 g/kWh), the engines are operated extremely lean (approx. $\lambda = 1.2$) [64]. For comparison, current passenger car gasoline engines are between 340 and 280 g/kWh at their best point. The racing engines achieve this lean running capability by using a pre-chamber ignition system, see also Fig. 1.34. The pre-injection nozzle and spark plug sit on a central pre-chamber that opens into the combustion chamber. A small quantity of fuel (approx. 3% of the total injection quantity) is injected approx. 60° before TDC and ignited in rich concentration approx. 22° before TDC. The jets passing through small holes in the pre-chamber wall into the main combustion chamber ignite the lean mixture there between 12 and 5° before TDC [65].

The aluminium pistons run in nickel-silver coated aluminium liners.

The mean effectivepressures at rated power are about 13.4 bar and at maximum torque about 15 bar. The combustion pressure reaches its peak of 100 bar about 8 °CA after ignition TDC. With conventional connecting rod ratios, the ideal value is approx. 12–15° after ignition TDC. The combustion pressure curve exhibits extreme stochastic fluctuations over the crank angle.

The valves are opened 15–17 mm wide. The large strokes in conjunction with the high valve accelerations required are achieved by combining finger followers with bucket tappets.

The compression ratio is statically 13.4–13.8:1. At high engine speeds this value increases to approx. 15:1 due to the sum of elastic deformation of the crankshaft, connecting rod, etc. and by consuming the bearing clearances. This leads to a tight combustion chamber.

The throttle element is not controlled directly mechanically but, as in modern passenger car drives, via an drive-by-wire actuator.

There are about 10 l of oil circulating in the lubrication system, and the total amount of oil is completely circulated every 15 s [22].

When the number of engines for the teams was still unlimited by the regulations, an overhaul was common about every 400 km. Replacement intervals of other individual parts were: Valves 800 km, connecting rods 1200 km, cylinder heads 2500 km; cylinder block, crankshaft and camshaft 3500 km [66]. The change in regulations for 2004 raised the overhaul interval to about 800 km. A further change from 2005 required engines to endure two race weekends i.e. approximately 1500 km of racing.

Pistons, piston rings and bearings are generally replaced at every overhaul.

The fuel used is practically equivalent to superfuel from the filling station (RON 102) in the sense of the regulations.

Figure 1.47 shows the cross-section of a typical naturally aspirated Formula 1 engine. It is still a 10-cylinder engine, but basically there is no difference to the later V8 engines. The cylinder units had even remained the same because the displacement had also been reduced in line with the number of cylinders from 3 to 2.4 litres. A geroter pump sucks the oil-air mixture out of the cylindrical crankcase and feeds it to the centrifuge (bottom left). The crankshaft has an extremely short stroke of 41.4 mm and heavy metal plugs in the counterweights. The low pistons run in large bore (96 mm) wet liners and are cooled from below with several piston spray nozzles per cylinder. The coolant is distributed across the V-space and supplied to critical locations in the cylinder head (spark plug seat, exhaust valve seat rings) by means of targeted holes. The water jacket of the head is in two parts and thus initially keeps the coolant close to the combustion chamber. The four valves per cylinder are held on the rocker arms by pneumatic valve springs (left cylinder head in the picture). The pressure in the spring chambers is controlled by screw-in valves (right cylinder head in the picture). The lubricating oil runs back into lateral longitudinal ducts of the crankcase via external pipes (not visible in the sectional view). The length of the intake manifold is variable via adjustable intake funnels. This is now no longer permitted by the regulations. The injection valves are located outside the intake funnels. The engine is controlled by a throttle valve. The intake funnels are located inside a common airbox, the lower part of which is shown in section. Oil ducts are also cast onto the crankcase (e.g. visible on the far right under the water duct).

Le Mans

For the famous 24-hour race, the same regulations apply as in the American ALMS (*American Le Mans Series*). With an exempted number of cylinders, naturally aspirated engines may have a maximum displacement of six liters and turbocharged engines a maximum of four liters. The standard fuel for petrol engines is Eurosuper (RON 98). Diesel (compression ignited) engines must be operated with standard fuel that complies with the European standard and contains approx. 25% GTL (gas to liquid). Depending on the vehicle weight classes (675 and 900 kg), displacement and supercharging method, air volume limiters with specific diameters and prescribed charging pressures ensure approximate equality of opportunity.

An example of the power units developed according to these regulations is that of the extremely successful Audi R8. This LMP 900 vehicle is powered by a V8 engine with 3.6 l and 4 valves per cylinder. With cylinder dimensions of 85 mm bore and 79.3 mm stroke, it is a smooth short-stroke engine. The compressors of the two turbochargers draw in the air via an air restrictor of 30.7 mm diameter each and are allowed to compress the air to a maximum of 1.67 bar. Nevertheless, the compression ratio of 12.2:1 is relatively high. The engine reaches its maximum power of 405 kW at a speed like a large series engine. The highest torque specified is 700 N m. The usable speed range is between 3500 and 7800 min^{-1}. The maximum speed is thus comparatively low and underlines its influence on the durability of an engine. A Le Mans race alone covers about 5100 km, depending on the weather. Training and qualifying are added to this, so that some engines have covered over 10,000 race kilometres.

The cylinder block has a bank angle of 90° and, like the two identical (!) cylinder heads, is made of aluminium. The pistons, fitted with three rings, run on a nikasil-coated running surface directly in the crankcase (integral liner). The crankshaft has 180° crankpin offset ("flat shaft"). However, the internal mass balance is so good that vibrations caused by second-order inertia forces are not damaging to the engine, as is usually the case with this design.

The strengths of this long-distance engine lie in its good drivability and low fuel consumption (under 50 l/100 km). The gasoline direct injection (FSI) system used plays a not inconsiderable role in both of these.

Furthermore, many functions are integrated into the large cast parts of the cylinder block and head. This eliminates many weak points such as external oil and water lines and the necessary connections.

The total mass, including ancillary units and turbochargers, is 175 kg, which also shows that reliability was a top priority during development.

Since 2004, diesel (compression ignited) engines have also been allowed to compete in the LMP1 class. In 2006, a vehicle with a diesel engine clinched overall victory for the first time: Audi R10 powered by a V12 TDI diesel engine. The engine exploits the maximum permissible displacement of 5.5 litres. The basic structure of the cylinder block is made of aluminium with a 90° bank angle. Four valves per cylinder are actuated by two overhead camshafts and coil springs. The diesel fuel is injected via a common rail system at approx. 2000 bar. Two turbochargers, each with a 39.9 mm diameter air restrictor at the compressor inlet, deliver the combustion air. The boost pressure is limited by the regulations to 2.94 bar.

The maximum power is reached with 480 kW between 4500 and 5500 min^{-1} and the max. Torque with 1100 N m. The driven speed range is between 3000 and 5000 min^{-1}. However, the engine delivers usable torque practically from idle speed.

The engine is operated with a particle filter and does not emit any visible smoke, even in racing mode.

The successor to this engine, the Audi V10 TDI, is no less successful in the endurance classic at the Sarthe, Fig. 1.159. The engine's external dimensions were reduced while its displacement remained the same and its weight was reduced by 12%. The regulations also reduced the restrictor diameter and the permissible boost pressure. The 90° bank angle was retained despite the number of cylinders being reduced to 10. The engine powering the Audi R15 TDI uses steel pistons because the aluminum piston of the V12 was at its load limit despite the fiber-reinforced bowl rim. The steel piston also made it possible to reduce the block height along with the resulting weight and installation space savings. The engine has few external connections. Most of the ducts and lines are integrated in the castings. The centrally horizontally split aluminium crankcase consists of an upper section with Nikasil raceways and a ladder frame lower section (bedplate), which also forms the flat bottom end of the engine. Auxiliary units, such as oil and water pumps, are of modular design and are located on the side of the bedplate so that they can be easily replaced. The gear drive of the auxiliary units and the camshafts is located on the front side of the engine, i.e. opposite the power output side. The engine is fully integrated into the chassis. The cylinder head covers, milled from the solid, accommodate mounting eyes for this purpose. The two exhaust gas turbines are controlled via variable turbine geometry (VTG) [6].

Fig. 1.159 Audi V10 TDI [Audi Sport, 2011]. V10-cylinder diesel (compression ignited) engine with common rail injection Exhaust gas turbocharging controlled via VTG. At the front is the dry sump tank with only 10 l volume. The two turbochargers are located close to the engine on both sides. The compressor inlet with restrictor is clearly visible on the left turbocharger

World Rally Car (WRC)

The regulations are relatively permissive compared to others. The vehicles run with turbo and naturally aspirated engines with a maximum number of cylinders of eight. The displacement of naturally aspirated engines may not exceed 3 l with two valves per cylinder and 2.5 l with four valves. If the engine is turbocharged, the displacement limit is 2.5 l and the output is limited by a restrictor at the compressor inlet. The airflow-limiting inner diameter is 34 mm for gasoline engines and 35 mm for diesel (compression ignited) engines.

Common engine configurations consist of an in-line four-cylinder turbocharged at 3.5 bar. An automatic recirculation system maintains the boost pressure with an electric compressor even when the throttle is closed, thus improving the dynamics of the throttle response (bang-bang system). The maximum output of 220 kW is delivered at 5500 min^{-1}. The highest torque is 480 N m and is released at 4000 min^{-1}. The engines are designed for approx. 19 operating hours.

NASCAR

For NASCAR cars, naturally aspirated V8 engines are used, the essential basic components of which should come from series production. Only one engine may be used per race (including practice and qualifying). With a displacement of 5.86 l, the engines achieve a power output of 645 kW, which was limited to 540 kW from 2015. The engine speeds are between 4500 and 9500 min^{-1} depending on the track and the restrictor (which is required on some tracks and is fitted between the mixture former and the intake manifold) Maximum torque of around 700 N m is reached at 7200 min^{-1} The cylinder heads are made of an aluminium alloy. The two valves per cylinder are actuated by a bottom-mounted camshaft, pushrods and rocker arms. The valves are closed by steel springs. Other spring types are not permitted. Until 2011, mixture formation was handled by a four-barrel carburetor supplied with 110 octane unleaded fuel. Since 2012, the mixture is formed via gasoline injection. The cylinder block is made of cast iron. It has the classic North American design with a camshaft located centrally in the V-space (so-called *small block* for 5.8 l displacement). The block itself forms the raceway of the aluminum piston. The crankshaft has the 90° crankpin offset typical of passenger cars. The compression ratio is limited by the regulations to 12:1. The engines are designed for a service life of one race, which is about 1300 km.

IndyCar Series
(formerly IRL). The concept parameters are severely restricted by the regulations of this American racing series for the engine designers. The idea behind this is an advanced, but cost-effective engine. It had to be a 90° naturally aspirated V8 engine with a maximum displacement of 3.5 l until 2011. The bore had to be 93 mm. The engines featured a "flat" crankshaft, so ran at 180° journal offset. The dry engine mass had to be at least 128 kg without radiator, clutch, engine control unit, ignition box and filter. Since 2012, 2.2-liter V6 turbocharged engines (single or biturbo) have powered the single-seater cars. Maximum permitted bore 95 mm. Minimum mass is now 112.5 kg. The permitted maximum speed of 12,000 min^{-1} is so strictly adhered to by the uniform rev limiter that the teams install their own, smoother-acting one upstream. The official IndyCar governor simply switches off the ignition completely when the limit is reached and the engine immediately goes into overrun mode. In contrast, the custom limiter shuts down individual cylinders as the rev limit is approached, allowing the driver to approach the limit with modulated engine power.

The combustion air reaches the combustion chamber via four valves per cylinder, which are operated by two overhead camshafts. The valve springs must be made of metal. Methanol has been used as a fuel since the 1960s for safety reasons (easy extinguishing). In the meantime, ethanol (E85) is used. It is injected electronically controlled directly. The power is between 410 and 520 kW, depending on the distance, which is reached at the maximum permitted speed. The high compression ratio of 15:1 is made possible by the alcohol fuel. The crankcase and cylinder heads are made of aluminium alloys. The pistons move in dry cast iron liners.

It is driven on oval courses and the average speed is correspondingly high at around 355 km/h.

References

1. Alten, H., Illien, M.: Demands on Formula One Engines and Subsequent Development Strategies. SAE Paper 2002-01-3359. SAE International, Warrendale (2002)
2. Ulrich, W.: Audi- der Sieger von Le Mans, Vortrag der ÖVK Vortragsreihe, Wien (Mai 2004)
3. Indra, F.: Grande complication, der Opel Calibra der ITC-Saison 1996 in. Automobil Revue Nr. **50** (1996)
4. Hack, I.: Formel 1 Motoren, Leistung am Limit, 2. Aufl. Motorbuch, Stuttgart (1997)
5. Soltic, P.: Part-Load Optimized SI Engine Systems. Zürich, Eidgenössische Technische Hochschule, Dissertation Nr. 13942 (2000)
6. Baretzky, U. et al.: Der V10 TDI für die 24 h von Le Mans, Beitrag zum Wiener Motorensymposium. VDI Reihe 12 Nr. 735 Band 2, VDI, Düsseldorf (2011)
7. Van Basshuysen, S. (Hrsg.): Handbuch Verbrennungsmotor, 3. Aufl. Vieweg, Wiesbaden (2002)
8. Internetseite der FIA. http://www.fia.com/sport/Regulations/f1regs.html. Accessed on 12 Dec 2005
9. Dolt, R. et al.: Der Opel DTM-V8-Motor – Entwicklungsschwerpunkt Verbrennungsanalyse. In: Krappel, A. (Hrsg.) Rennsport und Serie – Gemeinsamkeiten und gegenseitige Beeinflussung, 1. Aufl., S. 163–175. expert, Renningen (2003)
10. Atkins, R.D.: An Introduction to Engine Testing and Development. SAE International, Warrendale (2009)
11. Van Basshuysen, S. (Hrsg.): Lexikon Motorentechnik, 1. Aufl. Vieweg/GWV Fachverlage, Wiesbaden (2004)
12. Life at the Limit: Sonderheft Race Engine Technology. High Power Media, Wedmore (2007)
13. Köhler, E.: Verbrennungsmotoren, 1. Aufl. Vieweg, Wiesbaden (1998)
14. Krautter, W.: Why Multicylinder Motorcycle engines? SAE Paper 690748, 1969 [Design of Racing and High Performance Engines, SAE PT-53]
15. Pischinger, F., Esch, H.J.: Einfluss der Zylinderzahl auf die Reibungsverluste von Personenwagenmotoren. MTZ. **42**(12), 525 (1981)
16. Hack, G.: Der schnelle Diesel. Alles über Diesel-Autos, 3. Aufl. Motorbuch, Stuttgart (1987)
17. Appel, W.: Development of the Chassis for the R8. AutoTechnol **3,** 56 ff, Vieweg, Wiesbaden (2003)
18. Andorka, C.-P., Kräling, F.: Formel 1, das Milliardenspiel. Copress, München (2002)
19. Bosch.: Kraftfahrtechnisches Taschenbuch, 22. Aufl. (1995)
20. Ebel, B.: Reibungsverluste von Pkw-Ottomotoren. MTZ. **54**(6), 294 (1993)
21. Hinz, R., Schwaderlapp, M.: Leichtbau im System Zylinderkopf. In: Oetting (Hrsg.) Leichtbau im Antriebsstrang, S. 162–173. expert, Renningen (1996)
22. Schöggl, P.: Aktuelle Trends und Methoden in der Rennfahrzeugentwicklung. Vortrag im Rahmen der ÖVK Vortragsreihe. Graz, 22. Jan. (2003)
23. Incandela, S.: The Anatomy & Development of the Formula One Racing Car from 1975, S. 2. Haynes, Sparkford (1984)
24. Eichler, F. et al.: Der Antriebsstrang des Mercedes SLS AMG, Beitrag zum Wiener Motorensymposium. VDI Fortschritt-Bericht Reihe 12 Nr. 716 Band 2, S. 97–124 (2010)
25. Kerkau, M. et al.: Hocheffiziente Performance – der Antrieb des neuen Porsche 911 Turbo, Beitrag zum Wiener Motorensymposium. VDI Fortschritt-Bericht Reihe 12 Nr. 716 Band 1, S. 41–72 (2010)
26. Schäfer, F., Barte, S., Bulla, M.: Geometrische Zusammenhänge an Zylinderköpfen. MTZ. **58**(7/8), 384–391 (1997)

27. Iguchi, S., Hirose, K., Matsuda, Y.: The lean-burn engine: recent developments by Toyota. Die Evolutionäre Weiterentwicklung des Automobils **1**, 1 ff (1995. Herausgeber: ÖVK, H.P. Lenz. Wien. Eurotax 1995)
28. Bucknel, J.: Powertrain general. In: Royce, S. et al. (Hrsg.) Learn & Compete, a Primer for Formula SAE, Formula Student and Formula Hybrid Teams, 1. Aufl. Racecar Graphic Limited, London (2012)
29. Indra, F., Tholl, M.: Der 3,0-l-Opel-Rennmotor für die Internationale Deutsche Tourenwagenmeisterschaft. MTZ. **52**(9), 454 ff (1991)
30. Alten, H.: Mercedes-Illmor, Zehn Jahre Entwicklung am V10-Formel-1-Motor. MTZ. **7**(8), 522–533 (2005)
31. Yagi, S., Ishizuya, A., Fujii, I.: Research and Development of High-Speed, High-Performance, Small Displacement Honda Engines. SAE Paper 700122, 1970 [Design of Racing and High Performance Engines, SAE PT-53]
32. Apfelbeck, L.: Wege zum Hochleistungs-Viertaktmotor, 6. Aufl. Motorbuch, Stuttgart (1983)
33. Brandstetter, W., et al.: Der 2,0-l 16V-Motor für den neuen Ford Escort RS 2000. MTZ. **52**(10), 502 (1991)
34. Winterbone, D.E., Pearson, R.J.: Design Techniques for Engine Manifolds. Wave Action Methods for IC Engines, 1. Aufl. Professional Engineering Publishing Limited, London (1999)
35. Stoffregen, J.: Motorradtechnik. Grundlagen und Konzepte von Motor, Antrieb und Fahrwerk, 4. Aufl. Vieweg, Wiesbaden (2001)
36. Burgess, P., Gollan, D.: Praxishandbuch Zylinderköpfe – Technik, Tuning, Modifikationen, 1. Aufl. Heel, Königswinter (2005)
37. Wright, P.: Ferrari Formula 1.Under the Skin of the Championship-Winning F1-2000, 1. Aufl. David Bull Publishing, Phoenix (2003)
38. Flierl, R., Oehling, K.-H., Hösl, J.: Ventiltriebauslegung moderner Motoren. MTZ. **9**, 462–469 (1993)
39. Urlaub, A.: Verbrennungsmotoren, Band 3 Konstruktion. Springer, Berlin (1989)
40. Scussel, A.J.: The Ford D.O.H.C. Competition Engine SAE-Paper 640252 (1964)
41. Hütten, H.: Schnelle Motoren seziert und frisiert, 10. Aufl. Motorbuch, Stuttgart (1994)
42. Merker, G.P., Kessen, U.: Technische Verbrennung, Verbrennungsmotoren. B.G. Teubner, Stuttgart/Leipzig (1999)
43. Indra, F., Grebe, D.: Der Formel-3-Rennmotor von Opel. MTZ. **54**(11), 576 ff (1993)
44. Bamsey, I.: V10 Formula One Engine Technology, 1. Aufl. Racecar Graphic Ltd, London (2005)
45. Lenger, B.: Statement of requirement to produce crankshafts for racing engines. Diplomarbeit an der FH Joanneum, Graz (2000)
46. Theissen, M. et al.: 10 Jahre BMW Formel-1-Motoren, Beitrag zum Wiener Motorensymposium. VDI Reihe 12 Nr. 716 Band 2, VDI, Düsseldorf (2010)
47. Kamp, H.: Leichte Kolben für Pkw Otto- und Dieselmotoren. In: Oetting (Hrsg.) Leichtbau im Antriebsstrang, S. 1–26. expert, Renningen (1996)
48. Ottlicky, E., et al.: Steel pistons for passenger car diesel engines. ATZautotechnol. **5**, 38–42 (2011)
49. Maynes, B.D.J., et al.: Virtual engineering of formula 1 engines and airboxes. AutoTechnol. **4**, 46–50 (2003)
50. Lenz, H.P.: Gemischbildung bei Ottomotoren aus der Reihe List (Hrsg.): Die Verbrennungskraftmaschine Band 6. Springer, Wien (1990)
51. Watson, H. et al.: Optimizing the Design of the Air Flow Orifice or Restrictor for Race Car Applications. SAE Paper 2007-01-3553
52. Bamer, F.: Saugrohre zur Erhöhung von Drehmoment und Leistung. In: SDP-Technik 1985, S. 28–41. Steyr-Daimler-Puch AG, Wien (1985)

53. Stehlig, J., et al.: Längenvariables Luftansaugmodul für turboaufgeladene Motoren. MTZ. **01**, 48–55 (2012)
54. Mayer, M.: Abgasturbolader: Sinnvolle Nutzung der Abgasenergie, 3. Aufl. Moderne Industrie, Landsberg/Lech (1994)
55. Bauder, R. et al.: Der neue High Performance Diesel von Audi, der 3,0-l-V6-TDI Biturbo, Beitrag zum Wiener Motorensymposium. VDI Reihe 12 Nr. 735 Band 1, VDI, Düsseldorf (2011)
56. Bamsey, I.: Bentley at Le Mans, 1. Aufl. Racecar Graphic Ltd, London (2004)
57. Blodig, G.: Einfluss der Rennmotorenentwicklung auf die Brennraumgestaltung bei Serienmotoren. In: Krappel, A. (Hrsg.) Rennsport und Serie – Gemeinsamkeiten und gegenseitige Beeinflussung, 1. Aufl., S. 63–76. expert, Renningen (2003)
58. N.N: Broschüre: Motorsport Lieferprogramm 2015. HJS Emission Technology, Menden (2014)
59. http://www.bosch-motorsport.de/content/downloads/Products/33518712459.html#. Accessed on 06 July 2018
60. Owen, K., Coley, T.: Automotive Fuels Reference Book, 2. Aufl. Society of Automotive Engineers SAE Inc., Warrendale (1995)
61. Yutaka Otobe et al.: Honda Formula One Turbo-Charged V-6 1,5L Engine. SAE-Paper 890877 (1989)
62. N.N.: Driven Racing Oil. Broschüre, Huntersville. www.drivenracingoil.com (2014). Accessed on 06 Jan 2015
63. Katalog der Automobil Revue: Büchler Grafino AG, Bern (2002)
64. www.springerprofessional.de, Westerhoff M.: Ferrari bestätigt Magerbrennverfahren in der Formel 1. Accessed on 11 May 2016
65. Brooke, L.: Pushing the ICE forward, gradually. Autom. Eng. (SAE), 24–27 (2016)
66. Ludvigsen, K.: Mercedes Benz Renn- und Sportwagen, 1. Aufl. Motorbuch, Stuttgart (1999)

Electric Drives

2

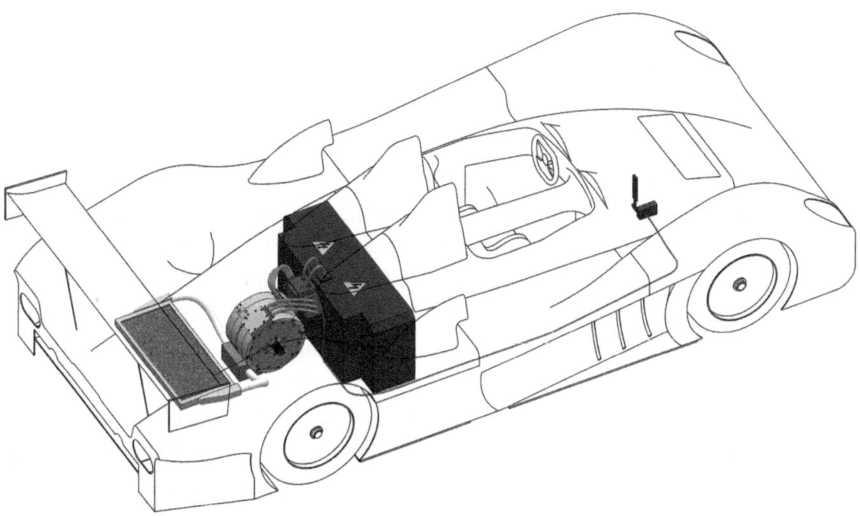

Driving a vehicle with an electric motor is not an idea of our time. It was merely brought up again by the public debate on reducing the greenhouse gas CO_2. The first road vehicle to exceed the 100 km/h mark in a record-breaking journey was a torpedo-like single-seater in 1899. The record-breaking car was driven by two 25 kW electric motors on the rear axle. The fact that electric drive did not become established in road vehicles at that time had several reasons, one of which is still relevant today: the bulky volume and the large mass of the energy storage device, the battery. However, enormous progress has been made in the meantime in the field of electronics, which is important for the control of the electric motors

M. Trzesniowski, *Powertrain*,
https://doi.org/10.1007/978-3-658-39885-9_2

and the service life of the batteries. For racing vehicles, with their requirements that differ from those of everyday vehicles,[1] the consideration of electric drives looks considerably different, especially against the background of some of the environmental criticisms of racing.

2.1 Fundamentals

If you look at the torque offered by electric motors over the speed, they are much more suitable for a road vehicle than internal combustion engines. Electric motors have an almost ideal tractive force characteristic curve above the speed and do not even require a manual gearbox. This also means an acceleration process without disturbing interruption of the tractive force (cf. Fig. 5.44). The battery can be tailored for a specific driving distance, typically known for competition vehicles. In the development of electric vehicles, even more attention is being paid to efficiency and lightweight construction than has been the case in vehicle design to date. Criteria that have always been and still are decisive for racing vehicles. An electric powertrain lends itself to racing vehicles based on these considerations alone. The lack of infernal engine noise may be a disadvantage for some, but for others it is actually another winning argument. The volume of the battery is also a considerably smaller space problem than in passenger cars. In a racing car you only need space for a maximum of two people and a trunk is not even needed. Heating of the vehicle interior is also only necessary in rally cars during the cold season. In all other racing vehicles, the waste heat of the combustion engine is not missed. On the contrary, the higher efficiency of the drivetrain enables a reduction in air resistance due to the lower cooling requirement compared to combustion engines. Heat exchangers and air intakes can be made smaller. Once again, racing could exploit the advantages of short development times coupled with the courage to come up with unusual solutions and play a pioneering role for later series developments. The current lap time record of a production sports car with electric drive (Fig. 5.125) on the Nürburgring Nordschleife, which is used by racing cars and everyday vehicles alike, is only about 30 s faster than that of sports cars with combustion engines [1].

In addition to the battery problem mentioned above, another disadvantage should not be underestimated, especially at the beginning of the era of "unfamiliar" technology, namely the serious danger posed by the high-voltage electrical system. Insulation faults can lead to electric shocks, body flow and fires.

The basic structure of the drivetrain of an electric vehicle (EV) differs only slightly from that of an internal combustion engine shown in Sect. 5.1, Fig. 2.1. A traction battery (3) stores energy on board. It therefore corresponds to the fuel tank. The storage is filled from the outside via the mains connection (1) and via a charger (2), which converts the

[1] Cf. Racing Car Technology Manual, Vol. 2 Complete Vehicle, Chap. 1

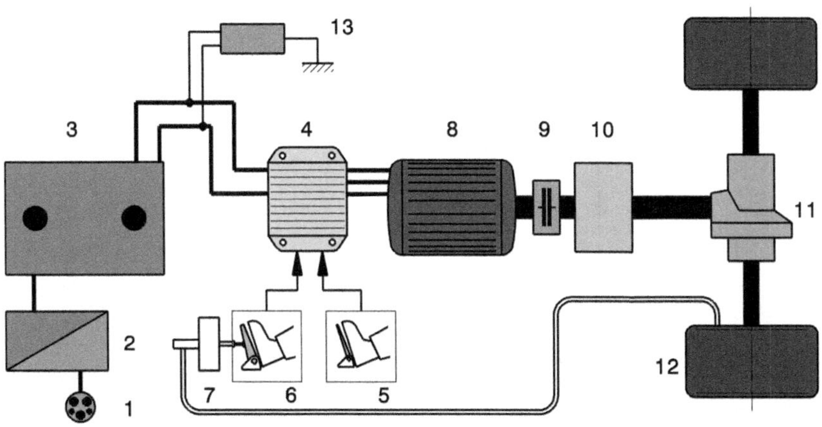

Fig. 2.1 Drive train for electric vehicles schematic. 1 Mains connection, 2 On-board charger *3* Traction battery *(HV),* 4 *motor controller,* 5 *Accelerator pedal,* 6 Brake *pedal,* 7 Mechanical *brake system,* 8 Electric motor, 9 Mechanical *safety* clutch, 10 Transmission, 11 *Final drive,* 12 Driving wheels, 13 Insulation monitoring *device*

mains current (alternating current) into a form that can be used by the battery (direct current, different voltage). The driver's wish is communicated to a control unit (4) via the accelerator pedal. In most cases, the drive torque is specified in this way, as in the case of the internal combustion engine. The control unit converts the energy from the battery into current and voltage values required for the motor. The drive control can thus be compared to the fuel metering or throttle device of internal combustion engines. The electric motor (8) delivers the torque required for propulsion. It can distribute this directly to the drive wheels (12) via a final drive (11) or the torque or speed is adjusted via a gearbox (10) and then transmitted to the wheels. Usually a single-stage gearbox with a fixed ratio is sufficient. Strictly speaking, (8) is an electric machine because it can be used as a motor and as a generator. Unlike conventional drives, electric drives can also have a signal connection between the brake pedal (6) and the motor control unit (4). During braking, part of the kinetic energy of the vehicle can thus be recovered when the motor (8) switches to regenerative operation and feeds the battery with the generated current (regenerative braking or recuperation). The usual mechanical braking system (7) is needed in any case: a regenerative braking system has certain limits (see below) and does not work to a standstill. What is not needed at all is a starting element (starting clutch) and a reverse gear. Electric motors can release torque even when the shaft is stationary and their direction of rotation can be changed by reversing the polarity of the terminals. For some types of motor (e.g. permanently excited motors), a mechanical disconnecting device (9) is required for safety reasons so that the motor torque can be quickly removed from the drive wheels in the event of a fault.

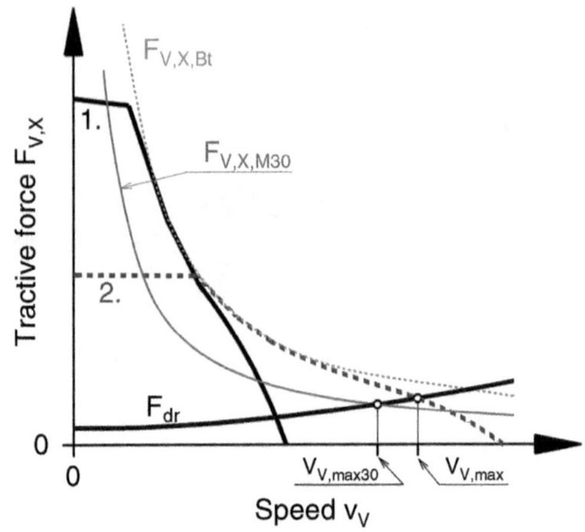

Fig. 2.2 Tractive force diagram of an electric vehicle with separately excited DC drive and 2-speed transmission [3]. $F_{V,X,Bt}$ Tractive force from maximum battery power, $F_{V,X,M30}$ Tractive force from 30-min power of the actuator, F_{dr} Driving resistance, $v_{V,\max}$ Maximum speed, $v_{V,\max30}$ maximum continuous speed over 30 min. In contrast to internal combustion engines, a distinction must be made in the case of electric motors between maximum power and continuous power over 30 min

In addition to the components shown, auxiliary systems (e.g. battery management for temperature and charge monitoring) and further safety devices (insulation monitoring, galvanic isolators, motor protection, . . .) are added.

Even if electric vehicles can manage without a gearbox, depending on the motor and the area of application, a reduction between the motor and the wheels is, however, advantageous simply because the motor can be made smaller and thus lighter. For high top speeds, manual transmissions may even be necessary. However, two gears are usually sufficient. The gear ratio i_G for two-speed transmissions is around 22 so that the vehicle can reach 250 km/h at maximum speeds of the electric motor of over 20,000 min^{-1} [2].

The operating range of the vehicle is limited by the maximum power output of the battery, the maximum power of the motor control unit or by the maximum drive power of the motor [3]. In the case of electrical machines, a distinction is made between short-term power and power that can be delivered over a certain period of time. Figure 2.2 shows an example of a design with two gears in the tractive force diagram. In addition to the tractive force resulting from the maximum power of the battery, that resulting from the continuous power of the drive over 30 min is also shown. Two gears are sufficient for this motor to reach the limit of the maximum possible tractive force from the battery power over the entire drivable range. Only for a higher final speed (intersection point F_{dr} with $F_{V,X,\text{second gear}}$) could a third gear or a different transmission ratio be introduced in second gear.

The maximum speed that can be driven over 30 minutes is lower, which is why two maximum speeds are often quoted for electric vehicles: The maximum speed that can be achieved for a short time ($v_{V,\max}$) and the speed that can be maintained for half an hour ($v_{V,\max30}$).

2.2 Drive Configurations

The wide variety of electric motors opens up an equally wide variety of conceivable drive train designs. Electric motors can drive axles or directly individual wheels. An intermediate gearbox with a constant transmission ratio makes it possible to increase the starting torque or maximum speed. Gearboxes offer the possibility of favourably influencing characteristics such as package, weight and efficiency of the electric machine. Highest driving speeds with usable starting torque can thus be represented. Figure 2.3 provides a systematic overview of drive trains with electric motors. If only individual axles are driven (first row), the electric machine can take the place of the combustion engine. Another possibility is the combination of two electric motors on one axle (tandem drive), whereby the differential can be omitted. However, the electric machine can also be accommodated directly in the wheel (wheel hub drive). The electric motor can provide the drive torque for the wheels directly or via an intermediate gearbox. Electric drives can also drive all wheels (middle row): Either conventionally as known from all-wheel drives with combustion engines (Sect. 5.6) or actually each wheel individually with wheel hub motors. This variant enables the ideal form of all-wheel drive with torque application to the wheels as required (i.e. depending on wheel load and driving condition) (torque vectoring). However, wheel hub motors have not yet been able to establish themselves. They increase the unsprung masses, are additionally stressed by temperature and vibrations and have to be accommodated in unfavourable installation space conditions.

For the sake of completeness, the diagram also includes hybrid all-wheel drives (last line), i.e. the combination of electric machines with combustion engines. The variants shown are characterized by the fact that one type of motor drives one axle at a time. The connection between the electric motor and the combustion engine is therefore only made through the road, which is why these all-wheel drives are also called through-the-road *hybrids.*

When configuring a drive train for a racing vehicle, the following aspects, among others, are taken into account:

- Overall efficiency: Mechanical (shaft to wheels) and electrical (battery via power electronics to rotor)
- Total weight
- Mass distribution: sprung/unsprung masses, mass moments of inertia
- Installation space
- Overall height: centre of gravity height and air resistance
- Mass moment of inertia of the rotating parts
- Limits of recuperation
- Number of shifting operations on a known route.

The fact that there is not just one solution is shown by the example of Formula E. Since the 2015/16 season, the teams have been free to configure the powertrain. The motor, power

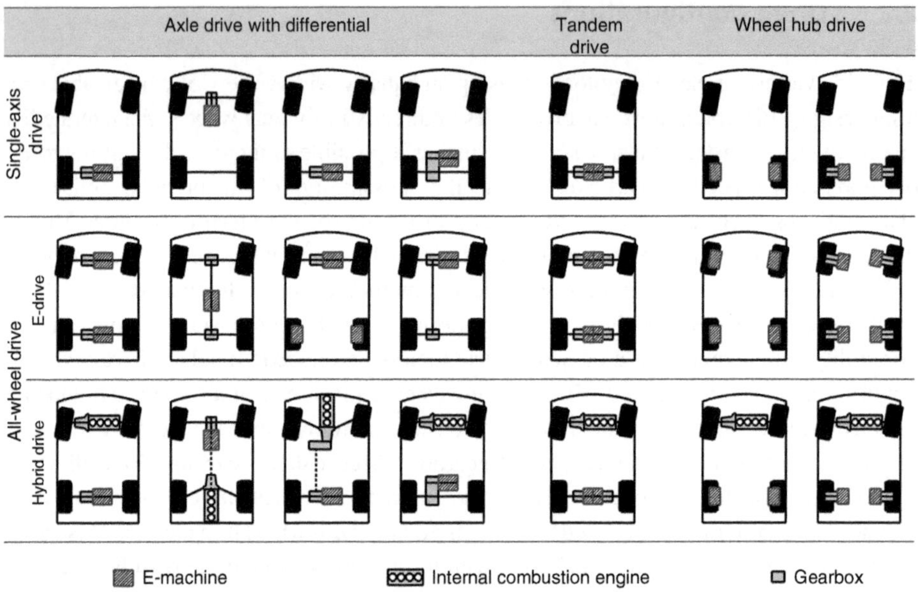

Fig. 2.3 Drive concepts for e-vehicles [2]. Different types of construction and control of electric motors allow many design variants for vehicle drives

electronics, transmission, final drive, cooling system and software are all part of the scope concerned. All participants have found different solutions for the same task [4].

A high-torque electric motor allows acceleration from a standstill to top speed with a fixed gear ratio, i.e. the manual gearbox can be dispensed with. The disadvantage of such a motor, however, is its large diameter, high mass and the strong current it draws to build up the torque. On the other hand, it is possible to recuperate during the entire braking process (without interrupting the gearshift). A lean motor, on the other hand, achieves high speeds (e.g. Formula E up to 20,000 min^{-1}) but offers only low torque, so a manual gearbox becomes necessary. The number of gears and thus the weight of the transmission depend on the motor characteristics (speed, torque curve) and the desired driving performance. The number of gearshifts in Formula E ranges from 2 to 5 gears. However, a gear change is also associated with a traction interruption and a loss of time, so that this disadvantage becomes greater as the number of gears increases.

In Formula E, the regulations create some conditions that make certain configurations of the drive appear advantageous. Torque vectoring, i.e. the targeted allocation of drive torque to individual wheels, is not allowed. Two motorss therefore only lead to more mass without being able to use the advantage of torque vectoring. The maximum motor power in the race is limited to 170 kW at 800 V nominal voltage. During a qualifying run the drive may deliver a maximum of 200 kW. The maximum speed of the vehicle is limited to 225 km/h.

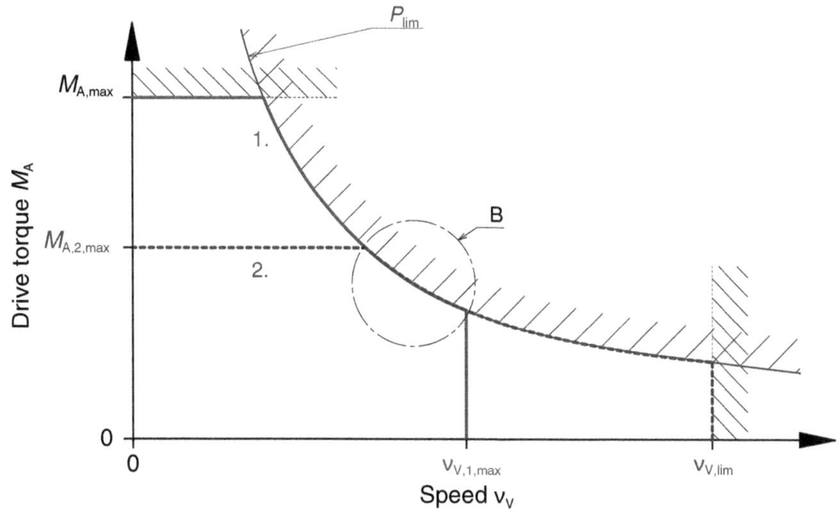

Fig. 2.4 Drive configuration in Formula E (schematic). The usable drive torque M_A at the rear axle is plotted against the road speed v_V. The regulations specify the maximum power P_{lim} and the maximum speed $v_{v,lim}$. The tyres and the axle load, together with the road surface, limit the maximum transmissible starting torque $M_{A,\max}$. Characertistics in first gear (red) and in second gear (blue dashed). In range B further gear steps can be advantageous

If we now consider possible drive configurations, the situation is as follows, Fig. 2.4. The regulations limit the maximum power P_{lim} and the maximum speed $v_{\mathrm{V,lim}}$. The tyres in combination with the road surface only permit a certain maximum drive torque $M_{A,\max}$. This marks out the usable range in a drive torque-speed diagram. A permanently excited synchronous motor permits the utilization of the specified power hyperbola, but if it is to operate with a fixed transmission ratio, it must be oversized to take advantage of the possible high starting torque. Its power would also be higher than allowed. A better solution is therefore a light, compact electric motor that works in conjunction with a manual gearbox. This also allows the motor to be operated in high speed ranges where its efficiency is higher. The picture shows the characteristics with a two-speed gearbox. With the first gear it is possible to drive up to speed $v_{V,1,\max}$. The second gear can already be engaged at lower speeds. In range B the characteristics of both gears overlap. A lower gear offers the advantage of higher motor efficiency due to the higher motor speed. Depending on the motor characteristics, other gear stages in range B may also be of interest.

2.3 Electric Motors

2.3.1 Basics

With their torque-speed characteristics, electric motors are almost ideal traction motors. They can already deliver their maximum torque at standstill and low speeds and keep the power almost constant over the speed range. As a result, in many cases the drive can be implemented without a multi-stage gearbox and a starting element (starting clutch, converter) is completely unnecessary. They also have no function-related installation position, as is the case with combustion engines due to the cooling and lubrication system.

No energy is consumed when the vehicle is stationary. In the nominal load range and at partial load, the efficiency of electric motors is very high at 89–93%. In defence of the combustion engine, however, it must be said that it first converts the energy in the fuel into mechanical energy via the intermediate stage of heat. The electric motor, on the other hand, directly receives the high-grade energy electricity.

Electric motors can be heavily overloaded for short periods of time, which can be used to advantage for acceleration processes or on extreme inclines. This short-term power is usually limited by the maximum power of the motor control (power electronics). The continuous power of the motor, in turn, is limited by the permissible motor temperature. Forced cooling of the motor can shift the power limit upwards. To describe the motor behavior, a distinction is therefore made in traction motors between the maximum power and that which can be delivered over a period of 30 minutes. The ratio between maximum and short-term power is approximately 1.5–3. In order to prevent damage to the motor, its temperature is monitored and limited by the motor control system according to thermal limit characteristics.

The two characteristic curves of a vehicle electric motor are shown in Fig. 2.5 for an inverter-fed induction (asynchronous) motor. The maximum torque is more than twice the nominal torque that can be driven over a long period of time. The powers following from

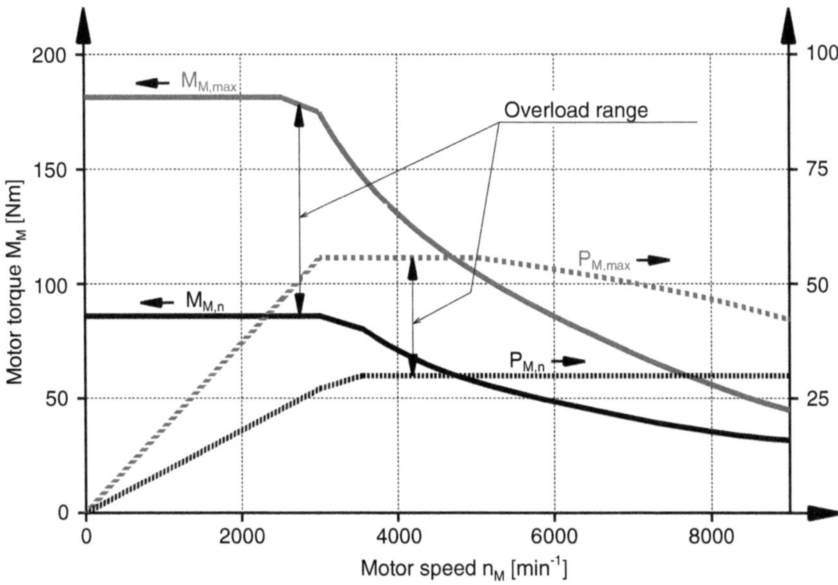

Fig. 2.5 Motor characteristics of an inverter-fed induction motor (asynchronous machine) as a traction motor [5]. $M_{M,\max}$ maximum motor torque (*peak torque*), $M_{M,n}$ Rated motor torque (*continuous torque*), $P_{M,\max}$ maximum motor power (*peak output power*), $P_{M,n}$ Nominal motor power

the speeds and the torques behave in the same way. The rated power even remains constant up to the maximum speed.

Figure 2.6 summarizes further principle-related characteristic curves of electric machines. The maximum motor torque can be set from the standstill of the rotor up to the speed n_{basic}. The power increases proportionally with the speed until the maximum power $P_{M,max}$ is reached. This power is mainly determined by the permissible winding temperature and the mechanical strength of the loaded parts. The so-called base speed follows from the relationship between power and speed to:

$$n_{M,basic} = \frac{30 \cdot P_{M,\max}}{\pi \cdot M_{M,\max}} \tag{2.1}$$

$n_{M,\text{basic}}$	Basic speed of the motor, min^{-1}
$P_{M,\max}$	Maximum motor power, W
$M_{M,\max}$	Maximum motor torque, nm

The speed increases with the stator voltage U_f and the motor torque depends on the magnetic flux Φ_f and thus on the current I_f. As soon as the maximum voltage $U_{f,\max}$ is reached, it must be kept constant and it must not increase any further despite increasing speed. This is achieved by so-called *field weakening*, e.g. by a resistor in the exciter circuit,

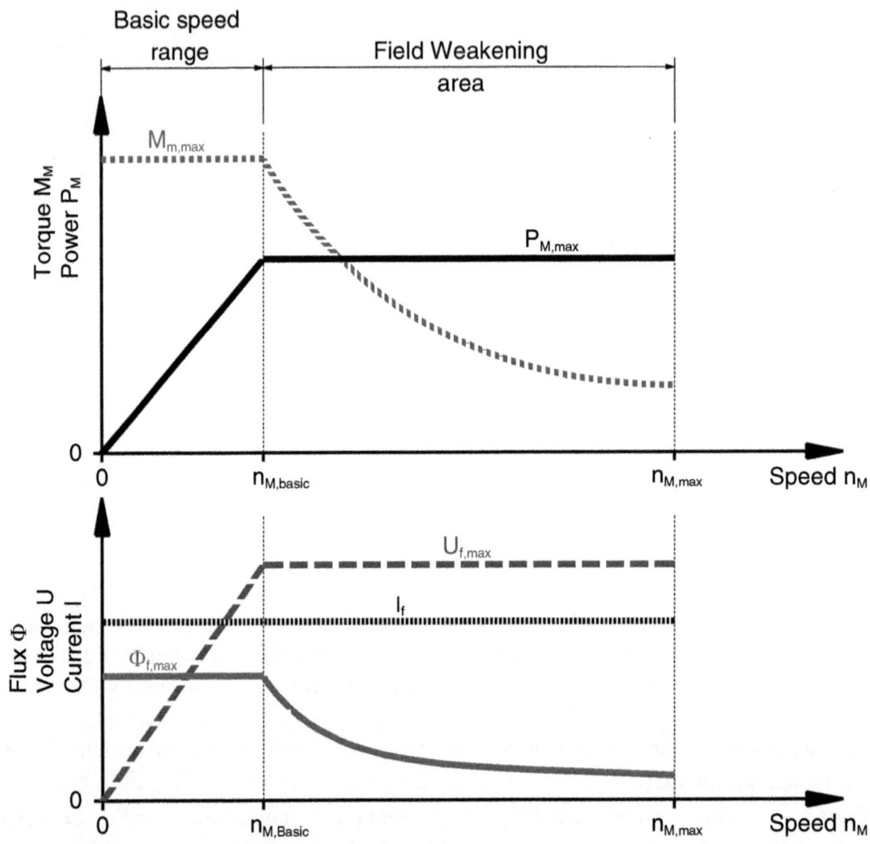

Fig. 2.6 Idealized characteristic curves of electrical machines

which reduces the magnetic flux. In the same sense as the flux, the torque also decreases. If the power is to remain constant, the torque must decrease in indirect proportion to the speed. The motor torque M_M then follows the relationship:

$$M_M = \frac{30 \cdot P_{M,n}}{\pi \cdot n_M} \tag{2.2}$$

M_M	Largest continuous motor torque in the field weakening range, N m
$P_{M,n}$	Rated motor power, W
n_M	Motor speed, min^{-1}

The base speed range is therefore the range of constant motor torque and in the field weakening range the power remains constant.

Basically, different types of electric machines can be considered as drive motors.

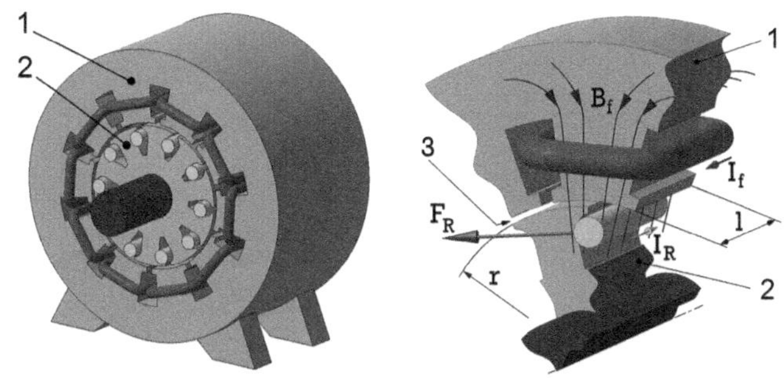

Fig. 2.7 Mode of operation of an electric motor. **a** basic structure, **b** detail for explanation. **1** *Stator,* **2** Rotor, **3** Air gap. B_f magnetic flux in the stator, I_f Current in the stand, I_R Current in rotor

Electric motors can be classified according to the type of input voltage (DC or AC machines) and further according to the type of magnetization (permanent magnets, current in stator or rotor) as well as the direction of the magnetic flux in relation to the axis of rotation of the motor (radial, axial and transverse flux).

However, the basic mode of operation is the same for all motors: The driving force is generated by the change in energy density in the air gap between a stationary part (stator, stator) and a rotating part (rotor, rotor, armature).

Figure 2.7 shows the basic structure of an electric machine (a) and the mode of operation of an electric motor (b).

Coils in the stator (1) generate a magnetic field with the flux density B_f. Current-carrying conductors in the rotor (2) also build up a magnetic field. Both magnetic fields superimpose each other in the air gap (3). Detail **b** shows a stator coil and an opposite conductor in the rotor. A force F_R acts on the conductor because it is in the magnetic field of the stator and a current I_R flows through it. The geometrical arrangement of the coils and conductor results in a torque $F_R \times r$ on the rotor (2) about its axis of rotation. By appropriately reversing the polarity of the rotor current I_R as a function of the relative position to the stator (1), an orbital rotary motion can be achieved.

The magnitude of the rotor driving force F_R is calculated from:

$$F_R = I_R \cdot B_f \cdot l \tag{2.3}$$

F_R	Circumferential force on the rotor, N
I_R	Current in rotor, A
B_f	Magnetic flux density in the stator, tesla: 1 T = 1 vs/m^2
l	Length of the machine, m

The motor torque thus follows directly:

$$M_R = F_R \cdot r \tag{2.4}$$

M_R	Rotor torque, Nm
r	Rotor radius, m

A consideration of (2.3) provides ideas on how to increase the power of an electric motor. First of all, the rotor current I_R can be increased. However, this is limited by the current carrying capacity of the windings. This can be increased by targeted heat dissipation, but this only makes sense to a certain extent, because the current heat losses increase with the square of the current, but the torque only increases linearly. With conventional cooling methods and radial flux arrangement, specific torques of about 30–40 N m/l rotor volume can be achieved in continuous operation [6]. This volume refers to the magnetic rotor volume. In short-time operation, significantly more than 100 N m/l can be delivered. The effective rotor volume can be increased. This depends, among other things, on the direction of the magnetic flux in the iron parts (Fig. 2.8). In the radial flux arrangement (a), the magnetic flux mainly runs radially between the stator and the rotor. In contrast, in the axial flux arrangement (b), the flux passes through these parts in the direction of the shaft axis. In the transverse arrangement (c), the magnetic flux passes through the iron parts both radially and axially. In axial-flow motors, the effective rotor volume is much smaller than in axial-flow machines and transverse-flow motors. Theoretically, all electrical machines could be built with these arrangements, but in practice only permanent magnet motors are used. With over 200 N m/l, these represent what is currently feasible.

Another decisive parameter for the motor torque (2.3) is the flux density B_f. This is independent of the motor type as well as the geometrical arrangement and is mainly a material property of the stator carrying the excitation windings. Industrially available laminations conduct around 1.65 T in rated operation [6]. The current I_f required for magnetization grows in proportion to the air gap. One therefore tries to keep this as small as possible.

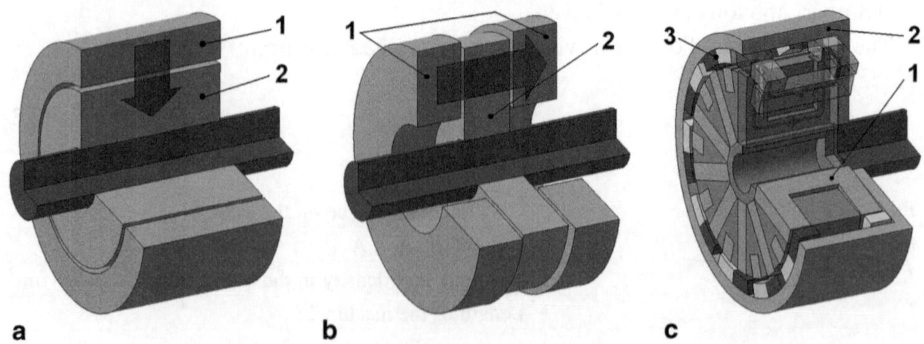

Fig. 2.8 Flux types in electrical machines (schematic). (**a**) radial *flux layout,* (**b**) axial *flux,* (**c**) transversal flux. **1** stator, **2** rotor, **3** permanent magnets. **Arrow** magnetic flux

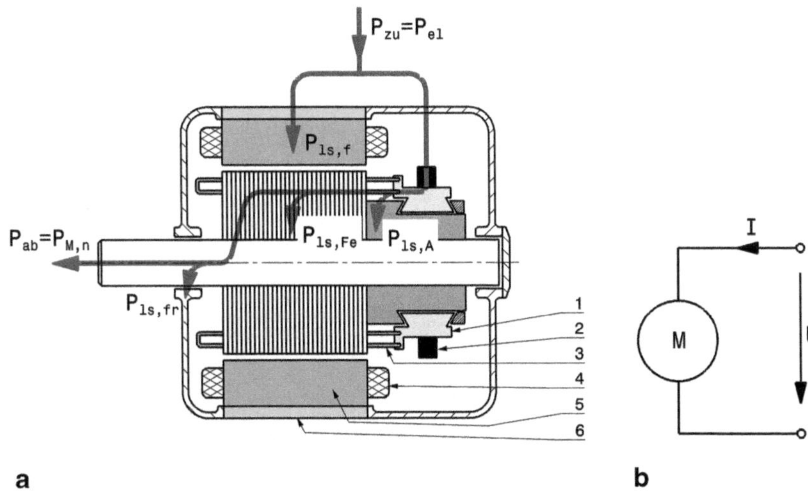

Fig. 2.9 Power flow in a direct current machine [8]. (**a**) power flow schematic, (**b**) electrical power for a motor. **1** commutator *discs,* **2** brushes, **3** rotor bars, **4** *excitation* winding, **5** excitation poles, **6** stator *yoke*

Although the motor torque increases with the length and radius of the rotor, the dimensions and mass of the motor also increase, which is why (input) gearboxes are also used in slower vehicles. By a correspondingly high gear ratio, the motor can remain small and light. This results in favourable constellations with motors that reach speeds of 10,000–20,000 min^{-1} [7].

As with internal combustion engines, another way to increase the useful power is to reduce the losses. As an example, the power flow in electrical machines can be seen in a DC motor in Fig. 2.9. The overall efficiency η_M is defined as the ratio of the power supplied to the power output:

$$\eta_M = \frac{P_{ab}}{P_{zu}} = \frac{P_{M,n}}{P_{el}} \tag{2.5}$$

η_M	Overall efficiency of the motor, –
P_{ab}	Power output, W. $P_{ab} = P_{M,n}$
P_{zu}	Power supplied, W. $P_{zu} = P_{el}$
P_{el}	Absorbed electrical power, W. $P_{el} = U\,I$

$$P_{ab} = P_{zu} - P_{ls,f} - P_{ls,A} - P_{ls,Fe} - P_{ls,fr} \tag{2.6}$$

$P_{ls,f}$	Power loss in the excitation winding, W
$P_{ls,A}$	Power loss in the armature winding, W
$P_{ls,Fe}$	Losses in iron, W
$P_{ls,fr}$	Losses due to friction, W

$$P_{ab} = P_{M,n} = \frac{\pi}{30} n_M \cdot M_{M,n}. \tag{2.7}$$

n_M	Motor speed, min^{-1}

However, friction losses occur in electrical machines only in the area of the bearings as mechanical losses and in fan wheels as flow losses, and they are subordinate in magnitude to the other losses, Fig. 2.10. The main part of the losses occurs in the windings and the materials to be magnetized (iron losses).

When the motor laminations are remagnetized, so-called hysteresis losses and disturbing eddy current formation occur. The hysteresis losses can be reduced to some extent by using complex alloys and heat treatments. A current flow is always associated with a magnetic field (this phenomenon underlies the electric motor, Fig. 2.7), but also vice versa. Thus, eddy currents form around a magnetically fluxed part. The separation of solid rotors and stators into individual sheet metal disks insulated from each other as magnetic

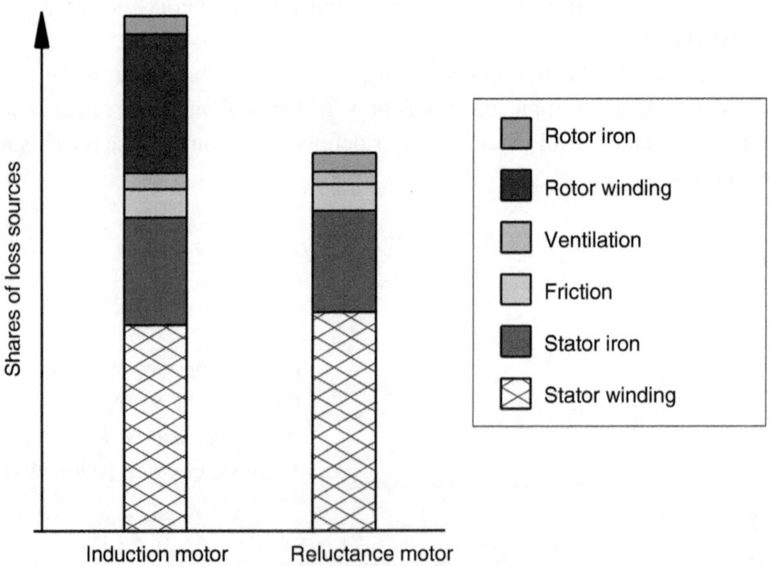

Fig. 2.10 Comparison of loss sources of electric motors [9]. A reluctance motor has considerably less losses than an induction (asynchronous) motor. This is mainly because of the missing rotor winding

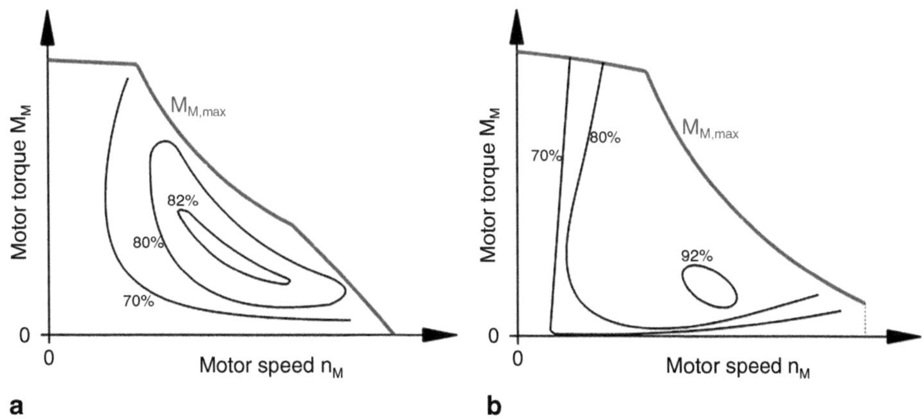

Fig. 2.11 Efficiency maps of electric motors [3]. (**a**) separately excited DC-motor, (**b**) permanently excited syncronous motor

flux paths with simultaneous suppression of the eddy currents in the transverse direction through the insulation is a measure against eddy current losses. The iron sheets exhibit high magnetic permeability with low electrical conductivity. Recent developments aim to further reduce the specific conductance of the material [6]. However, all these measures increase the cost of the materials and their processing.

A higher efficiency offers two possibilities. Either the motor heats up less with the same useful power, which benefits the bearings and the winding insulation, or it can deliver a greater useful power with the same thermal stress. At high ambient temperatures of 125 °C (e.g. in the vicinity of combustion engines), electric motors can deliver their rated power without damage. The winding temperature reaches values in the range of 200 °C due to the additional self-heating [8]. The efficiency of electric motors is similar to that of internal combustion engines in that it depends on the speed and load, but is characterised by completely different mechanisms, Fig. 2.11. Eddy currents, saturation, displacement and other electromagnetic effects primarily influence the effective power output.

The power is known to be directly proportional to the speed (cf. 2.7) and therefore a consideration of the operating frequency of the motor is helpful in this context as a further measure for increasing power. The limits of an increase in speed are set by the losses. Eddy current losses increase with the square of the frequency, hysteresis losses increase linearly with the speed. Both types of losses increase quadratically with the magnitude of the magnetic flux. The maximum operating frequency can thus be influenced primarily by the sheet thickness. Industrial sheet metal with a thickness of 0.5 mm allows about 80 Hz, while sheet metal with a thickness of 0.35 mm and a higher material quality is used in the automotive sector. This shifts the limit to 250 Hz. From 30 kg of active mass, a power of about 45 kW can be represented. Current developments are aimed at 0.2 mm sheets and thus up to 400 Hz. 30 kg motor mass is then capable of 70 kW in continuous operation [6].

2.3.2 Types *of Motors*

The following motor types can be considered or are in use for vehicle drives, Fig. 2.12:

- DC motors:
  - DC series motors
  - DC shunt-wound motors
- Three-phase motors
  - Synchronous motors
  - Induction (asynchronous) motors
- Special motors

DC *Motors*

DC motors can be powered directly from the traction battery. A magnetic flux is generated in the stator by an excitation winding (field winding). Depending on the circuit, a distinction is made between series-wound and shunt-wound motors. In a shunt motor, the same voltage is applied to the field winding as to the rotor (armature, rotor), Fig. 2.13 **a**. In the series-wound motor (**b**), the same current flows through stator and rotor windings. The rotor is also supplied with DC voltage via sliding contacts (carbon brushes). Due to the rotary motion, the direction of current in the rotor must reverse periodically so that the armature maintains its direction of rotation. The sliding contacts are therefore also called current reversers (commutators). This device is the weak point of the motor. It is subject to wear and, depending on the commutator diameter, limits the motor speed to about 7000 min^{-1} [5]. However, for low-speed vehicles, this type of motor is popularly used as a series-wound motor because of its low cost and simple motor control, although its efficiency is relatively low. DC motors can also be designed to be permanently excited (**c**). In this case, the excitation flux is generated by permanent magnets. The energy otherwise

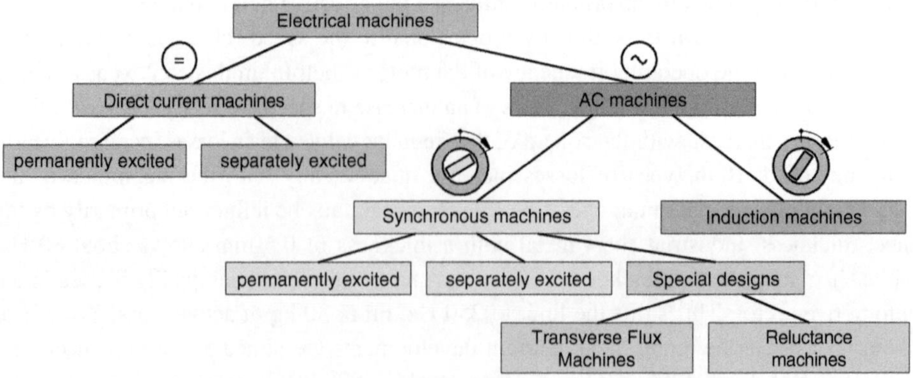

Fig. 2.12 Classification of electrical machines

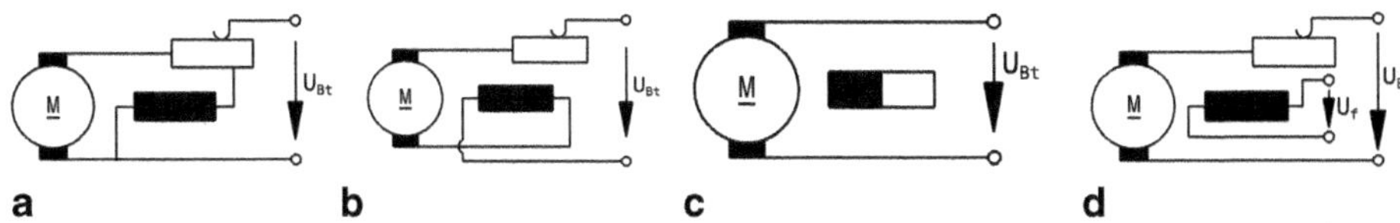

Fig. 2.13 Circuit diagrams of DC motors. (**a**) Shunt *motor,* (**b**) Series motor, (**c**) Permanently excited DC motor, (**d**) Separately excited DC motor. U_{Bt} Battery voltage, U_f Field voltage (excitation voltage)

required for excitation therefore does not have to be introduced by means of current flow, but is already stored in the motor. This increases the efficiency. A disadvantage is the higher mass of the permanent magnets (main component: *rare-earth*). These are materials that have a magnetic field in the micro range) and their price. The excitation winding can also be supplied by a different voltage source than the rotor, this is called a separately excited motor (**d**).

Three *Phase* Motors

To control the motors, the DC voltage of the traction battery is converted by converters into three-phase voltages of variable frequency and amplitude. The stator contains a three-phase winding, which generates a rotating magnetic field due to its arrangement and the voltage curve of the three-phase current. Depending on the rotor design, it rotates synchronously (with the same speed) with this rotating field or with a slip to it – i.e. asynchronously. The stator is basically the same for all three-phase motors, Fig. 2.14a. The three windings with the beginnings U, V, and W are connected to the three phases of a three-phase network L1, L2, and L3. A magnetic field (symbolized by arrow with north and south pole) circulates in the stator. The synchronous speed follows from the frequency of the alternating current and the number of windings of the stator:

$$n_s = 60\frac{f_f}{p} \tag{2.8}$$

n_s	Synchronous speed, min^{-1}
f_f	Frequency of the stator current, s^{-1}
p	Pole pair number, –. Specifies how often the three-phase winding occurs on the circumference

The 120° out-of-phase stator windings can be connected to the mains in star or delta connection, Fig. 2.15.

The rotors of synchronous machines have distinct magnetic poles. These are caused by current-carrying excitation windings (separately excited synchronous machine) or are represented by embedded permanent magnets. In the so-called full-pole rotor (Fig. 2.14a), the excitation winding is supplied with direct current via slip rings. The DC

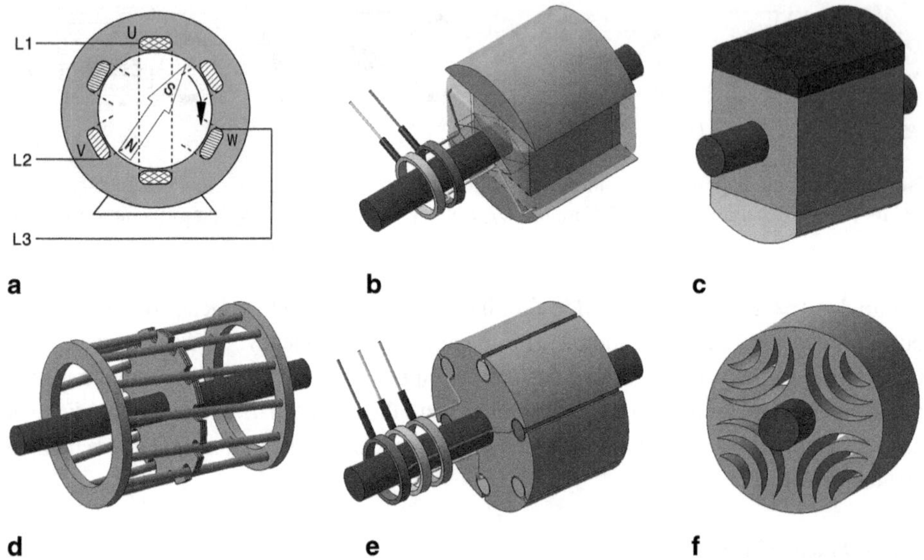

Fig. 2.14 Main parts of three-phase machines. (**a**) *Stator*, number of pole pairs $p = 1$, (**b**) *non-salient pole rotor (solid pole)*, rotor winding shown transparent. (**c**) *Permanently excited rotor*. (**d**) *Squirrel-cage rotor*, only 3 plates of the laminated core are shown. (**e**) *Slip-ring rotor*. (**f**) *Reluctance rotor*

Fig. 2.15 Circuit of stator windings. (**a**) *star configuration*, (**b**) *delta configuration*

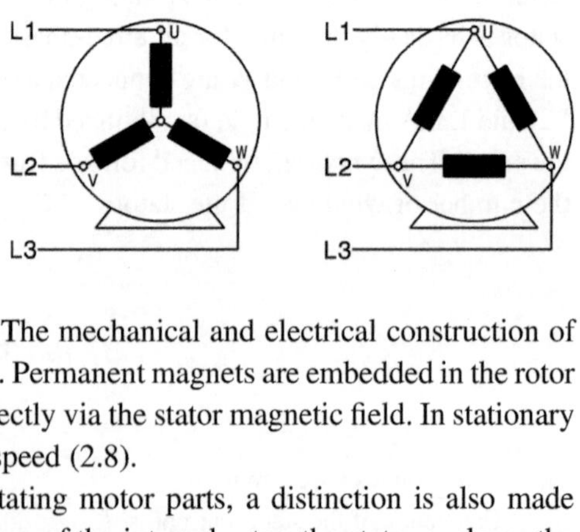

voltage is supplied by a DC/DC converter. The mechanical and electrical construction of permanently excited motors is much simpler. Permanent magnets are embedded in the rotor (Fig. 2.14b) and provide the drive torque directly via the stator magnetic field. In stationary operation, the rotors rotate at synchronous speed (2.8).

According to the arrangement of the rotating motor parts, a distinction is also made between internal and external rotor. In the case of the internal rotor, the stator encloses the rotor. The stator can be provided with large-area water cooling and the motor can therefore be subjected to greater loads. In the case of the external rotor, the heat-transferring surface is smaller, but this type of construction offers more space for the permanent magnets. Such high pole machines (eight and more pole pairs) achieve high torque and power densities at low speeds [10]. Moreover, in this arrangement, the magnets are held in position by their centrifugal forces. In the case of internal rotors, special precautions are required for this, particularly at high speeds.

Synchronous machines are characterized by high efficiencies and small construction volume. Of the motors listed above (DC, induction motors and reluctance machines), it has the lowest weight.

Induction (asynchronous) motors are designed as slip ring rotors or squirrel cage rotors. In the case of the squirrel cage rotor (Fig. 2.14d), bars are attached to the circumference approximately parallel to the axis of rotation and are short-circuited at their ends by rings. No sliding connection to stationary parts is required. During operation, the rotating field of the stator induces currents in the slower cage rotor, which in turn cause a driving magnetic field. The rotor always lags behind the magnetic field. The squirrel cage rotor never reaches synchronous speed, because in this state no more currents would be induced in the cage. This is referred to as the slip of the rotor in relation to the magnetic field. This is similar to the torque transmission of a rubber tyre, which also requires slip.[2] The greater the load on the motor, the greater the slip.

In the slip ring rotor (Fig. 2.14e), three-phase windings are also accommodated in the rotor, which are connected in star. The starts of the windings are connected to a three-phase starting resistor via slip rings and brushes. In contrast to the squirrel cage rotor, the inrush current can thus be kept low by setting the starting resistance high during run-up. During operation, the three conductors are short-circuited and the machine operates like a squirrel-cage rotor.

Squirrel cage rotors are characterised by their simple design, which permits maximum speeds of up to 15,000 min^{-1} and, apart from the bearings, has no wearing parts.

The disadvantage of all three-phase motors is the higher control effort. However, due to the rapid further development of power semiconductors, this is increasingly fading into the background.

For comparison, Fig. 2.16 shows the motor characteristics for the motor types described above.

Special Motors

Reluctance Motors

In this type of motor, the rotor requires neither windings nor permanent magnets, but operates in a three-phase stator. The rotor consists of soft magnetic material and is designed in such a way that magnetic preferred directions, i.e. different magnetic resistances (= reluctances) for different angular positions, result, Fig. 2.14f. The preferred directions and flux barriers can be clearly seen in the diagram. The stator coils are energised individually in turn and the rotor aligns itself according to the current magnetic field (switched reluctance motor or stepper motor).

The rotor can also be provided with a short-circuit winding by filling the cavities with aluminium. This type of motor shows a hybrid behaviour. It runs up asynchronously and as

[2] Cf. Racing Car Technology Manual, Vol. 4 Chassis, Sect. 1.2.3.

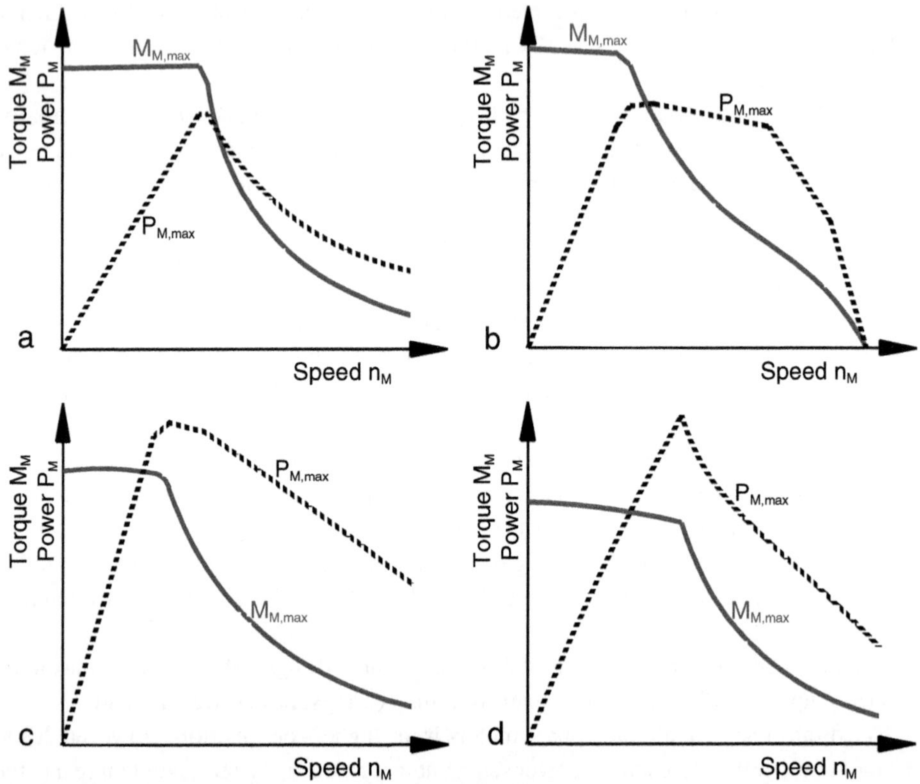

Fig. 2.16 Motor characteristics for some electric motors [3]. (**a**) DC series motor, (**b**) separately excited DC motor, (**c**) induction motor (asynchronous motor), (**d**) permanently excited synchronous motor. $M_{M,\max}$ Maximum motor torque, $P_{M,\max}$ Maximum motor power

soon as it reaches the rated speed, the reluctance torque pulls the motor into synchronous operation.

Reluctance motors can also be fed directly from a direct current source (battery, accumulator, fuel cell, . . .). Ideally, they can be controlled with the aid of a microprocessor and semiconductor switches [8].

Reluctance motors are simple in design and easy to control. They operate with high efficiency over a wide speed range (Fig. 2.10). In the passenger car sector, noise caused by the torque fluctuations can be problematic.

Transversal Flux Motor

The name of this type of motor is derived from the guidance of the magnetic flux, Fig. 2.8c. A coaxial stator winding generates flux in U-shaped stator yokes. The magnetic circuits close in the circumferential direction through permanent magnets in the rotor. This arrangement allows extremely small pole pitches without sacrificing rotor flux. This results

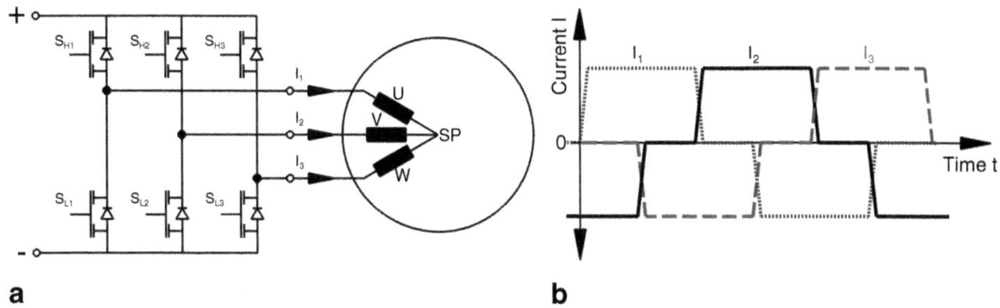

Fig. 2.17 Sketches of the BLDC motor [8]. (**a**) Circuit diagram, (**b**) Current flow in the stator coils. S_H Semiconductor switch Highside, S_L Semiconductor switch Lowside, *SP* Star *point*

in extremely high power densities, which makes the motor the first choice for wheel hub drives, i.e. direct drives without intermediate gears. In a wheel hub motor, the stator is connected to the axle and the rotor to the hub or actually to the rim.

As a disadvantage the high manufacturing costs should not be concealed, which are caused by the complicated geometry and the necessary high energy magnets.

Brushless Direct Current Motor (BLDC Motor)

BLDC motors function like synchronous machines, but are supplied with direct current. Power electronics (Fig. 2.17a) ensure that the motor is supplied with a clocked, square-wave three-phase current instead of a sinusoidal alternating current as is the case with synchronous motors (designation therefore also: electronically commutated direct current motor (Fig. 2.17b). The current is supplied to the three phases depending on the rotor position. One phase is always currentless, the other two phases are connected to the positive or negative pole of the voltage source. Of the semiconductor switches, one high-side and one low-side switch are closed. For example, in one position the current flows from the positive pole via switch S_{H1} through winding U ($I = I_1$) and back through the star point via winding V ($I = I_2 = -I_1$). The connection to the negative pole is made by the switch S_{L2}. The winding W is currentless during this time ($I_3 = 0$). The mode of operation of this motor can therefore also be explained by electronic commutation (current reversal), hence the name brushless DC motor.

The rotor is equipped with permanent magnets and follows the rotating magnetic field of the stator. This design can also be used to represent wheel hub motors.

Table 2.1 summarizes the most important characteristic values of electrical machines. For the purpose of comparison, some of the values given are not absolute values, but related values. For example, there is not an absolute torque value, but one that is related to the motor mass (specific torque). When considering the efficiency of electric machines, it is important to bear in mind that their control (power electronics) also causes losses. A realistic comparison can therefore be made for different drives if at least the overall

Table 2.1 Characteristic values of electrical machines [10]

Characteristic value	Direct current machine	Synchronous machine		Induction machine	Transverse flux machine	Reluctance machine switched
		separately excited	Permanently excited			
Maximum speed, min^{-1}	7000	> 10,000	> 10,000	> 10,000	> 10,000	> 10,000
Field weakening ratio, –	3	3–7	3	3–7		2
Specific continuous torque, nm/kg	0.7	0.6–0.75	0.95–1.72	0.6–0.8		0.8–1.1
Spec. Continuous power, kW/kg	0.15–0.25	0.15–0.25	0.3–0.95	0.2–0.55		0.2–0.62
Max. Efficiency machine, –	0.82–0.88	0.87–0.92	0.87–0.94	0.89–0.93	0.96	0.9–0.94
Max. Efficiency control, –	0.98–0.99	0.93–0.98	0.93–0.98	0.93–0.98	0.93–0.97	0.93–0.97
Max. Efficiency drive, –	0.8–0.85	0.81–0.9	0.81–0.92	0.83–0.91	0.89–0.93	0.83–0.91

efficiency of the combination of motor and control (i.e. the product of the partial efficiencies) is considered.

The power electronics to be used and the storage system must also be taken into account when dimensioning the drive. If the course of the track is known exactly or the usual characteristics of the races, these driving cycles (accelerating, braking or regenerative braking and accelerating again) are used for selection and design with the aid of a simulation. Oversized systems (especially the voluminous and heavy accumulators) represent a competitive disadvantage.

2.3.3 Choice of Motors

The following criteria can be used to select traction motors [5]:

- Compact design
- Low weight or high power density
- High efficiency
- Easy controllability over a wide speed and torque range
- Overload capacity
- Intrinsic safety
- Safety in driving operation.

For passenger cars, there are also other aspects such as low noise, low cost and low maintenance.

Table 2.2 compares these and other criteria for selecting traction motors.

Although DC motors have a high level of development and were also the first type to be used in vehicles, they are no longer important for this application. Current traction motors are fed with three-phase current because of the better efficiency and work according to the synchronous or asynchronous (induction) principle. Permanently excited motors are preferred, because with these not a part of the supplied electrical power is required for the excitation of an active machine part. The losses of the corresponding winding (field or rotor) are omitted and the efficiency increases, cf. (2.8). A disadvantage is the relatively high cost of permanent magnets. Transverse flux and reluctance motors are forward-looking types, but have not yet reached the level of development of "classic" synchronous and asynchronous (induction motor) machines. In the passenger car sector, permanently excited synchronous machines are used almost exclusively for electric drives because of their advantages [10].

The subject of the safety of electrical machines in the drive train is dealt with below.

Example

The record vehicle mentioned in the introduction (cf. also Fig. 5.125) is driven by two permanently excited synchronous motors on the rear axle (tandem arrangement). The

Table 2.2 Comparison of electric motors for vehicle propulsion, according to [5, 10]

Feature	DC motor	Synchronous motor		Induction motor	Transverse flux motor	Reluctance motor switched
		separately excited	Permanently excited			
Power density	–	0	++	+	+	+
Max. Speed	––	+	++	++	++	+
Efficiency	–	+	++	0	+	+
Volume	––	+	++	+	–	+
Weight	––	+	++	+	+	+
Cooling	––	+	++	+	+	++
Thermal overload capacity	–	+	0	++		++
Controllability (open/closed loop)	++	+	+	0	+	+
State of the art	++	–	++	++	–	0
Intrinsic safety	–	++	–	++	+	++
Motor costs	–	–	––	+	––	++
Costs of the overall system	+	0	0	++	–	+
Sound	–		++	+		–

Legend: ++ very good, + good, 0 satisfactory, – poor, –– very poor

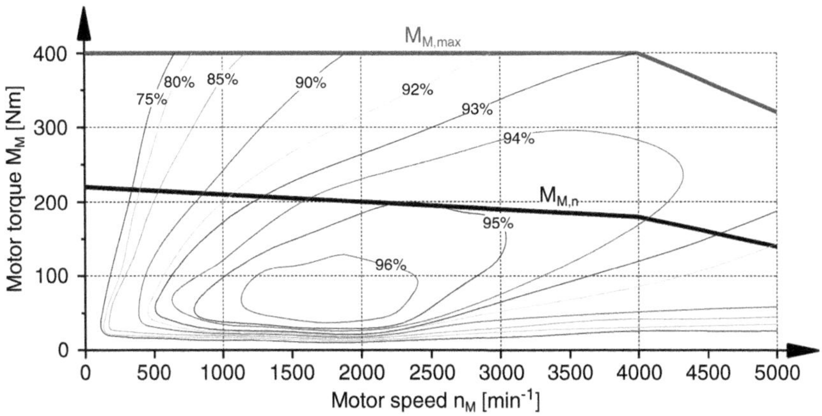

Fig. 2.18 Characteristic curves of the AFM-140-4 motor from Evo Electric [11]. In addition to the torques for 60 s ($M_{M,\max}$) and continuous operation ($M_{M,n}$), the lines of the same efficiency are entered. For 18 s the motor can even release 600 Nm. The motor is a permanent magnet synchronous motor in axial flux arrangement. These values are valid for operation with the inverter KEB 26 at a 600 V DC supply and a coolant temperature of 55 °C

motors operate according to the axial flow principle and are cooled with a water/glycol mixture (ratio 50:50). With a mass of 40 kg, one motor has a nominal output of 75 kW or 167.5 kW for 60 s and generates a maximum rated torque of 220 Nm and 600 Nm for a short time (for 18 s). These values apply to a voltage supply of 600 V DC, Fig. 2.18.

The torque and power values depend on the voltage supply and the speed. The latter is influenced by the winding configuration. The motor is shown in its external view with the connection dimensions in Fig. 2.19.

2.3.4 Cooling

Although electric motors have higher efficiencies than internal combustion engines, their unavoidable power loss must also be dissipated in the form of heat. The power loss of electric motors is between 4% and 40% of the electrical power consumed and depends mainly on [12]: Motor size (rated power), efficiency class (IE1 to IE4), stator laminations (re-magnetization), stator winding, rotor winding, bearings, seals, ventilation and operating mode (mains, inverter). This power dissipation leads to heat accumulation. The performance of the inverters as well as the motors therefore depends on the cooling during continuous operation. With simple air cooling (pure convection), the specific current load of the motor windings must not exceed 5–8 A/mm^2, otherwise the winding will be thermally destroyed. Air cooling is simple in terms of construction, but the motor takes up more space than liquid cooling because of the large cooling fins. With liquid cooling, on

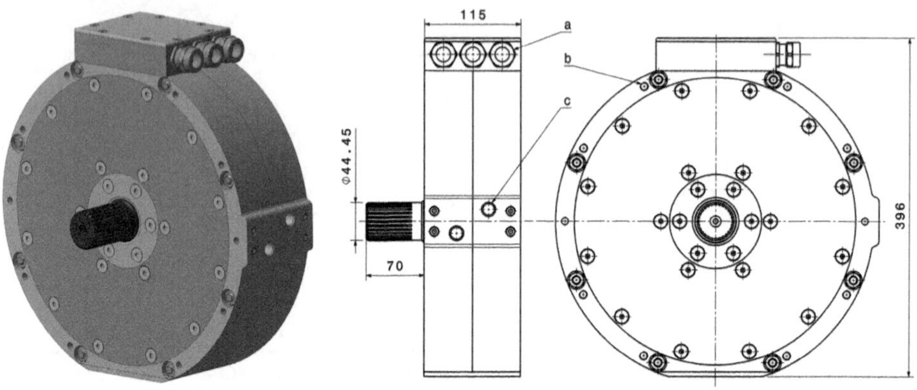

Fig. 2.19 Permanently excited axial flux motor AFM-140 [11]. Axonometric representation (left) and views (right). The motor is liquid-cooled and can be bolted axially around the circumference of the vehicle via 8 through holes (**b**). 2 additional more tightly toleranced holes are used for centering. Power is supplied via three M25 cable connections (**a**). Two threaded holes (**c**) are used for coolant supply. The cylindrical base body measures 380 mm in diameter and is 115 mm long. Its characteristic curves can be seen in Fig. 2.18. The motor weighs 40 kg

the other hand, the effect depends on how close the cooling medium comes to the winding. If a water jacket surrounds the stator, the limit of the current load is 10–15 A/mm^2. If cooling channels run between the windings, the load can be increased to 20 A/mm^2. The maximum of 30 A/mm^2 is reached when the windings are directly surrounded by a dielectric (non-conducting liquid). The outer diameter of the motor is smaller than with air cooling, but the constructional effort due to pump, heat exchanger and piping increases.

When dimensioning the cooling system, the rule of thumb is that approx. 15% of the mechanical useful power must be dissipated as heat (applies to mechanical efficiency $\eta_{mech} = 0.87$). The maximum tolerable temperature for electronics and motor is approx. 70 °C (for comparison: 120 °C for combustion engines).

The heat to be dissipated can be calculated if the operating condition of the motor is known. The required input values are taken, for example, from its characteristic diagram with entered efficiency curves:

$$\dot{Q}_M = P_{ls} = \left(\frac{1}{\eta_{mech}} - 1\right) P_M = \left(\frac{1}{\eta_{mech}} - 1\right) M_M \frac{\pi}{30} n_M \tag{2.9}$$

$\dot{Q}_M$	Heat flow to be dissipated from the motor, W
P_M	Mechanical motor power at the shaft (useful power), W
P_{ls}	Power loss, W. $P_{ls} = P_{el} - P_M$

(continued)

P_{el}	Supplied electrical power of the motor, VA = W
η_{mech}	Mechanical efficiency, –
M_M, n_M	Motor torque or motor speed, N m or min^{-1}

From the heat to be dissipated follows the required volume flow of the cooling medium, for which the medium-carrying parts must be designed:

$$\dot{V}_C = 60 \frac{\dot{Q}_M}{\Delta T_C \cdot c_{p,C}} \tag{2.10}$$

$\dot{V}_C$	Volume flow of the cooling medium, l/min
ΔT_C	Temperature difference of the cooling medium between supply and return, K. the maximum temperature increase should remain below 5 K
$c_{p,C}$	Specific heat capacity at constant pressure of the cooling medium, J/(kgK) c_p of water: 4187 J/(kgK)

If the motor is operated with a known load spectrum, the effective value of the torque can be used in (2.9) for M_M

$$M_{M,e} = \sqrt{\frac{\sum \left(M_{M,j}^2 \cdot t_j \right)}{t_t}} \tag{2.11}$$

$M_{M,e}$	RMS value of the motor torque, N m
$M_{M,j}$	Motor torque in operating range j, N m
t_j	Duration of the operating range j, s
t_t	Total time, s. $t_t = \sum t_j$

Furthermore, the current load of the motor follows from this effective value:

$$I_M = \frac{M_{M,e}}{c_M} \tag{2.12}$$

I_M	Current load of the motor, A
c_M	Torque constant of the motor, N m/A

2.4 Energy Storage

2.4.1 Introduction

In this subchapter, storage systems are considered that provide the electrical energy for driving. Their requirements and selection differ in some cases considerably from the storage systems used for energy recovery systems (cf. Sect. 3.2: Short charging time, small amounts of energy, high outputs, different energy converters). For this reason alone, only electrical and electrochemical storage systems (traction battery) are considered here, because they already provide the appropriate form of energy for electric motors. Figure 2.20 gives a systematic overview of storage systems for electrical energy.

Of the electrical storage devices, capacitors are interesting. However, they have an extremely low energy density and are therefore not suitable for traction storage. With their high power density coupled with extremely short charge/discharge times, they are tailor-made for KERS systems and are dealt with in Sect. 3.2. This leaves electrochemical storage as a traction battery.

Function

Energy storage systems have the task of storing (as much as possible) energy over a period of time and making it available again (as loss-free as possible) when required.

Requirements

The requirements for storage systems in vehicles and in particular competition vehicles result from the demand for high driving performance with adequate safety:

- High power: Especially for high acceleration and high final speed, the electric motors must be fed with large currents.
- Large energy content: If a large amount of power is to be drawn over a certain period of time, the amount of energy must also be large.
- Small mass and low volume: For vehicles, low mass means correspondingly low power input to accelerate, decelerate or cause the vehicle to deviate from the straight path (cornering, bumps). Lower power input results in lower energy input over a driven distance, i.e. the battery can be smaller. Volume also enters into the driving resistances because it can influence the air resistance depending on the arrangement in the vehicle.
- Very good dynamic behavior: If high dynamics are expected from motors, this must also be carried by the storage device.
- High reliability: The energy storage system not only provides the energy for propulsion, but also maintains all electronic monitoring systems (depending on the design of the on-board power supply). Permanently excited motors are also not intrinsically safe. A failure of the supply voltage leads to short-circuit currents in rotating motors that can exceed the rated current. A sudden failure of the battery can thus put the vehicle in an uncontrollable state.

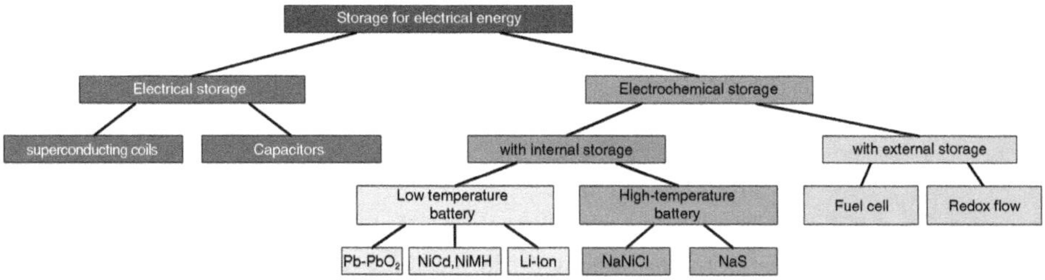

Fig. 2.20 Classification of storage technologies for electrical energy, after [13]

- Good charging characteristics (deep discharge, fast charge, efficiency, . . .): The less complicated the charging and discharging process, the simpler the environment can be designed and the less error-sensitive and safer the overall system is.
- Non-hazardous crash behavior: Vehicles can be involved in accidents. It is important that the driver or bystanders are not exposed to any additional danger. The accumulator concentrates a large amount of energy in a small space. This must not be released, or at least not abruptly. Electrochemical storage devices must not release toxic substances or react dangerously (explode, burn) when subjected to mechanical stress.

Other requirements are also of interest for passenger car applications. For example, operational reliability, functional reliability over a wide temperature range, freedom from maintenance, long service life, simple infrastructure (charging, storage, checking the state of charge, maintenance requirements and operational reliability) and acceptable costs.

Characteristic Values

The following characteristic values are used to describe energy storage devices.

Specific energy, gravimetric energy density, Wh/kg: The stored energy [Wh] is related to the required mass [kg] of the storage unit. A high value means that a vehicle can achieve a long range. A large amount of energy alone does not guarantee this. The storage unit could have a large mass and thus increase the energy requirement of the vehicle (cf. Sect. 4.1: *Rolling, climbing and acceleration resistance*) to such an extent that the range remains small. The target value is 100–200 W h/kg, depending on the distance and the vehicle.

Volumetric energy density, W h/l: Basically the same as the gravimetric energy density, except that the reference value here is the volume [l] of the storage unit. A high value means that the storage unit occupies a small volume, which is particularly advantageous for the packaging (the arrangement of the individual components) of the overall vehicle.

Specific power (gravimetric), W/kg. This is the retrievable power related to the storage mass. A high value stands for short charging or discharging times of the accumulator. This is particularly important for dynamic driving manoeuvres, e.g. acceleration or recuperation.

Power density (volumetric), W/l: Analogous to energy density.

Storage efficiency, –: Is a measure of the conversion losses during discharge.

$$\eta_{storage} = \frac{extractable\ energy}{charged\ energy}$$

$\eta_{Storage}$	Storage efficiency, –

Capacity, Ah: Is a measure of the electrical charge that the storage device can hold. The energy stored with it results from this with the nominal voltage: Energy = capacity × nominal voltage.

Calendar life, years: Due to ageing effects, storage systems suffer a loss of performance and capacity even without electrical load.

Cycle life, number of cycles: Due to side reactions during charging and other influences, memories also age during operation. The cycle life indicates how often a defined cycle (certain energy charging and discharging) can be run through until a certain ageing is reached. The calendar life and the cycle life overlap each other.

The characteristic values listed above define the extreme values of storage tanks. In addition, further characteristic values are used to describe the current state of a storage tank.

State of charge (**SOC**): State of charge of the battery as a percentage of the nominal capacity. A SOC of 100% therefore means a fully charged battery. See also Fig. 2.21.

SOC *swing*: This is the charge/discharge stroke used of the relative battery capacity in percentage. Hundred percent is the fully charged battery. In today's hybrid cars, only 20–30% SOC-swing is set to keep the battery life high, see Fig. 2.22.

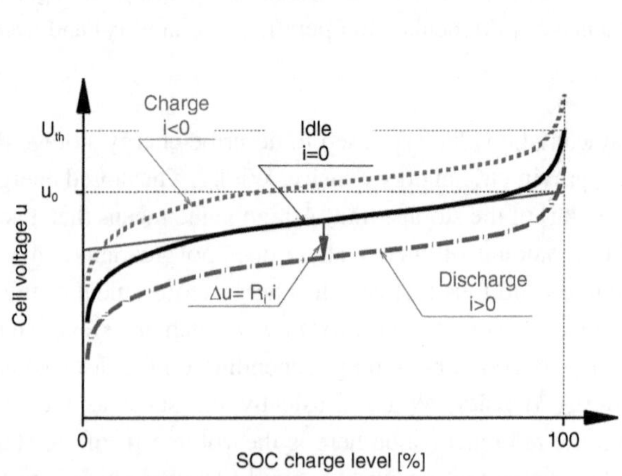

Fig. 2.21 Behaviour of battery voltage as a function of degree of charge and current, after [10]. In the unloaded state (no current flows), the cell theoretically reaches the voltage U_{th}, when the chemical energy content has been completely converted into electrical energy. The degree of charge is then 100. The open-circuit voltage u_0 can also only be determined in the unloaded state. It depends on the degree of charge. A linear characteristic can be observed over a wide range for many cell types

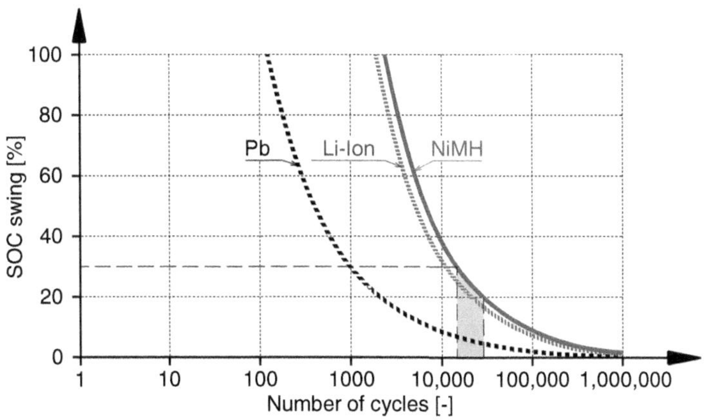

Fig. 2.22 Influence of charge/discharge swing (SOC swing) on battery life [14]. Pb lead battery. Li-Ion Lithium-ion battery. NiMH nickel metal hydride battery. *SOC swing*: This is the used charge/discharge stroke of the battery capacity in percent. In the range of current hybrid vehicles of 20–30% SOC stroke, the service life of common battery types is well above 10,000 charge/discharge operations. If one forgoes this long service life, a much larger amount of energy can be extracted from the battery

2.4.2 Forms of Electrochemical Storage (*Configuration of Accumulators*)

A basic distinction is made between primary and secondary systems. Primary systems can only be discharged once and must then be replaced. Secondary systems, on the other hand, can be used several times, and the electrochemical reaction that releases the energy (discharging) can be reversed (charging). For linguistic differentiation, primary systems were also called batteries and secondary systems accumulators. However, even in technical linguistic usage, the term battery dominates for all electrical storage devices. In this context, battery stands for the interconnection of similar devices and thus actually only describes the typical structure of high-voltage systems.

Primary Systems (*Primary Cells*)

The discharge process is irreversible, i.e. the battery is exhausted after discharge. It must be removed from the vehicle and completely replaced. Almost all types use zinc as the negative electrode. The different types result from the materials of the positive electrode and different electrolytes. Their main area of application is currently in small devices where small currents are required over long periods (clocks, data storage, . . .).

Secondary Cells

These elements convert the chemically bound energy into electrical energy and vice versa. The basic principle is based on the potential difference of two different metals – the so-called electrodes – in the electrochemical voltage series. Figure 2.23 shows the

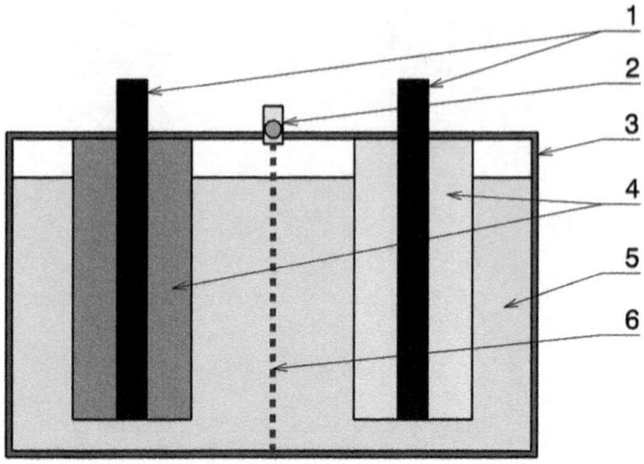

Fig. 2.23 Schematic structure of a battery cell. The main components of a galvanic cell that converts chemically bound energy into electrical energy and vice versa: 1 Pole, 2 Safety valve, 3 Housing, 4 Electrode, 5 Electrolyte, 6 Separator

schematic structure of a so-called galvanic cell. The electrodes (4) are located in a non-conducting medium (5) which reacts chemically with the metals (electrolyte). This in turn is enclosed by a container (3). The chemical reactions can produce gases, which is why a safety valve (2) may be necessary. A separator prevents electrical contact between the two electrodes. It must not allow any electrons to pass through, but must offer the lowest possible resistance for ions.

The electrolyte can be liquid, which makes the handling of the battery more difficult (leakage, gas formation, position dependence, . . .), or it is gelled or is present in solid form. Electrolytes with higher density allow an increase in mass with the same installation space. The heat capacity also increases. As a result, the sensitivity of the cell to thermal overload decreases. At low temperatures, the gel and the solid electrolyte contract and can lose contact with the electrodes.

When discharging, the electric current flows from the nobler metal (positive pole) via the consumer to the negative pole (technical current direction I). The actual, physical direction of current is described by the electrons: They move from the negative to the positive pole, Fig. 2.24. The less noble metal (anode) is dissolved (oxidized) in the electrolyte. The cations that have gone into solution migrate to the positive electrode and are deposited (reduced) there. Because ion exchange takes place in the electrolyte, a current can flow.

When charging, an external voltage source is connected to the two poles, reversing the described processes.

Figure 2.21 shows the voltage curve between the poles of a cell as a function of the degree of charge (SOC) and the current. The voltage also shows a dependence on the degree of charge of the cell in the unloaded state, i.e. when no current is flowing. During charging, energy must be supplied to the cell. The voltage is above the open circuit voltage and the current is positive. However, the cell does not store all the energy supplied. Side reactions and heating mean losses, which are recorded by the so-called internal resistance R_i of the cell or battery. A current must also flow when energy is removed. This current causes a voltage drop Δu via the internal resistance. The consequence is that the terminal

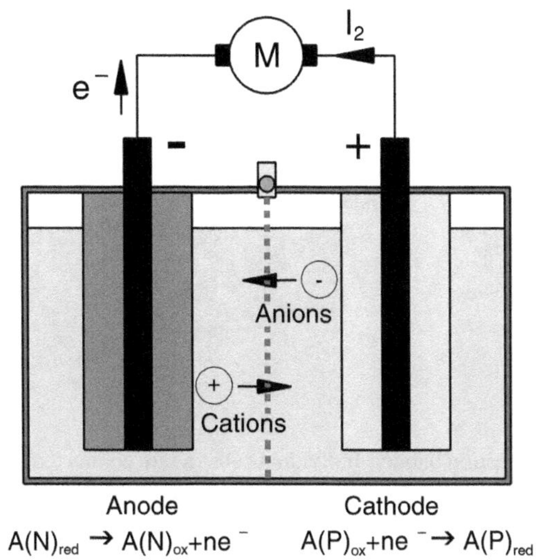

Fig. 2.24 Function of a galvanic cell. The sequences are shown for discharging, the motor is driven. The less noble metal (anode, negative pole) is dissolved in the electrolyte and deposited at the cathode. The released electrons travel via the consumer (in this case the motor) to the positive pole. I_2 Discharge current. At the anode, the active material A is oxidized and n electrons e^- are released. These n electrons are attracted by the cations of the electrolyte solution and reduce them

voltage must be below the nominal voltage and that the energy that can be extracted is less than the stored energy. A low internal resistance is therefore advantageous for the storage efficiency. In addition, a low internal resistance enables high currents, which is synonymous with high motor torques during starting and acceleration, for example.

The cell voltage follows from the potential difference of the metals involved. A lithium-ion cell, for example, has a voltage of 3.8 V. If you want to build a traction battery with 400 V from this, 105 cells must be connected in series ($105 \times 3.8 = 400$). If a large power is now to be provided, several such modules must be connected in parallel, Fig. 2.25. The voltage of 400 V remains the same, but the current increases in accordance with the number of modules. The cells are prismatic in the figure, but they can also be cylindrical. The prismatic shape has the advantage of a higher packing density. Flat prismatic shapes also facilitate heat dissipation due to their large surface-to-volume ratio [15].

When charging, all cells must reach the same voltage, 3.8 V in the above example. If some cells receive too much voltage, they will be destroyed. A balancing circuit ensures the same voltage for all cells (balancing). In addition, the battery system needs protection against overheating and spontaneous combustion. All these tasks are performed by a battery management system (BMS), Fig. 2.28. It records voltage and temperature data of the individual cells, the state of charge (SOC) and other variables and ensures that the battery does not leave a certain temperature and state of charge window. The primary purpose of this is to keep the battery's lifetime high, Fig. 2.22. The greater the battery's

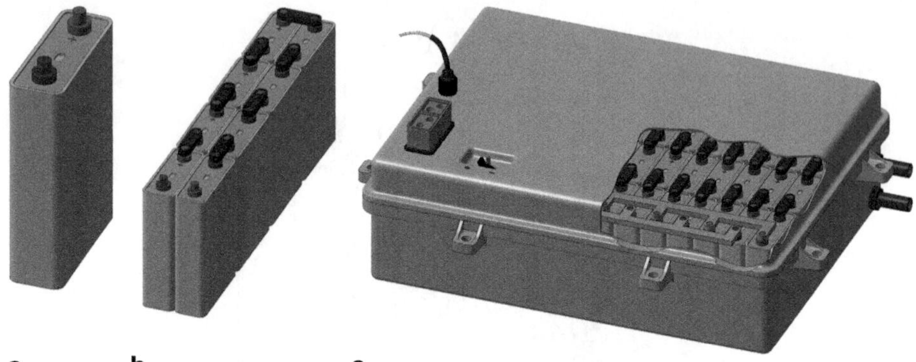

Fig. 2.25 Structure of a traction battery. Individual cells (**a**) are connected in series to form a module (**b**). The battery (**c**) consists of modules connected in parallel to achieve the desired power at the required voltage. In addition to a holder, a main switch and cable connections, heating or cooling as well as monitoring of the cell voltages is also required

state of charge/discharge (SOC) stroke, the shorter its lifetime. This can be thought of as similar to the influence of stress on the lifetime of components.[3] The storage capacity of today's batteries is only used to about 20% in series hybrid vehicles for reasons of service life [14]. If one is content with a much shorter service life and uses the battery constantly (pure electric drive), the battery can be much smaller with the same range by allowing a large SOC stroke.

2.4.3 Cooling System

Depending on the cell type, performance only suffers at very low temperatures, Fig. 2.26. For passenger cars, aspects such as service life and low-temperature behaviour essentially affect customer acceptance; for racing vehicles, these areas are subordinate. This goes so far that lithium-ion batteries are heated at low temperatures in order to maintain the desired performance and so that they are not damaged during charging. The energy for heating can be taken from the battery itself because the small amount of energy required can be delivered at low temperatures without damage. For racing cars, it is crucial that the cells have the lowest internal resistance at relatively high temperatures (around 50 °C), allowing the greatest power extraction. One will therefore strive to keep the cells within their optimum temperature window. The temperature can be controlled by air shafts or by means of a separate cooling circuit of a medium with heat exchangers similar to the internal combustion engine cooling system, Fig. 2.27. The cooling medium is advantageously electrically non-conductive (dielectric) and is pumped by a pump (1) through the internal

[3] Cf. Racing Car Technology Manual, Vol. 2 Complete Vehicle, Chap. 2, Fig. 2.26.

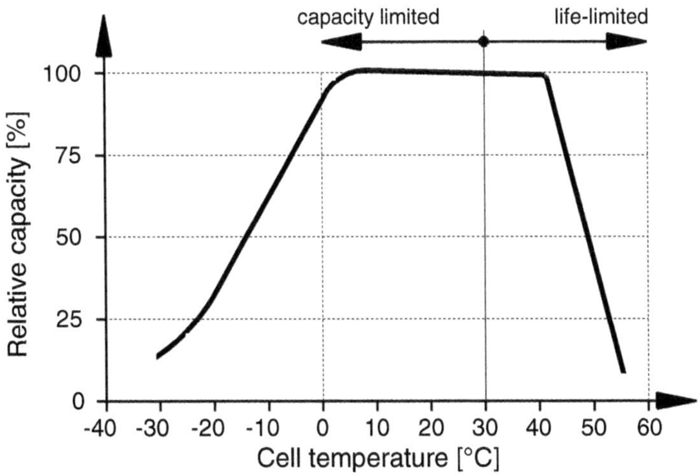

Fig. 2.26 Influence of temperature on the performance of a lithium-ion battery, after [16]. The temperature of the battery cells must be kept within a defined temperature window. Below 0 °C the capacity drops rapidly, above 45 °C the service life suffers greatly

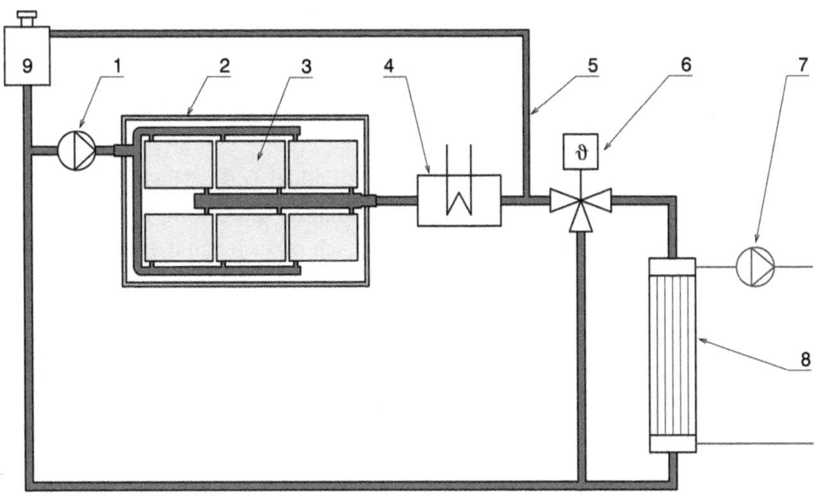

Fig. 2.27 Cooling system of a battery. 1 Pump, 2 Battery (with internal cooling system), 3 Battery module, 4 Auxiliary heater, 5 Bleeding line, 6 Thermostatic valve, 7 Air-conditioning circuit (passenger car only), 8 Heat exchanger, 9 Expansion tank *(header tank)*

battery system (2) via an auxiliary heater (4) through the heat exchanger (8). A thermostatic valve (6) ensures a constant temperature level. The heat exchanger (8) releases the heat to the ambient air or – which is the obvious choice in a passenger car – is connected to the air conditioning circuit (7).

Although a thermal management system with air is the simplest, high demands are placed on the air flowing through the battery box. It is filtered in order not to worsen the heat transfer due to deposits and dried in order to prevent creepage currents due to moisture [17].

The goal of any cooling system is to keep the temperature difference between the individual battery cells below 3–4 °C. The smaller the difference, the better for the battery's performance. The smaller the difference, the better for the performance of the battery.

For sprint competitions, convection cooling may be sufficient, i.e. no special measures are required for heat dissipation, the natural air movement is sufficient [18].

The heat loss of a battery can be assumed to be approx. 10–20% of its electrical power. The lower value applies to pouch cells, the higher to cylindrical cells. For a 300 kW battery made of pouch cells, the cooling system must therefore be designed for heat dissipation of approx. 30 kW.

2.4.4 Types of Batteries

All components of the battery are housed in an insulated container, of which crash resistance is still required specifically for use in the vehicle. Figure 2.28 gives an overview of what is included in a complete battery system for vehicles.

The combination of different electrode materials (active material) and electrolytes results in different cell and battery types, which differ not only in their voltage level but also in their properties.

Traction batteries are created by combining individual *cells*, which provide suppliers, into *modules,* which in turn are connected to the battery according to demand (voltage, power), Fig. 2.29. The cells can be connected to each other by soldering, clamp welding,

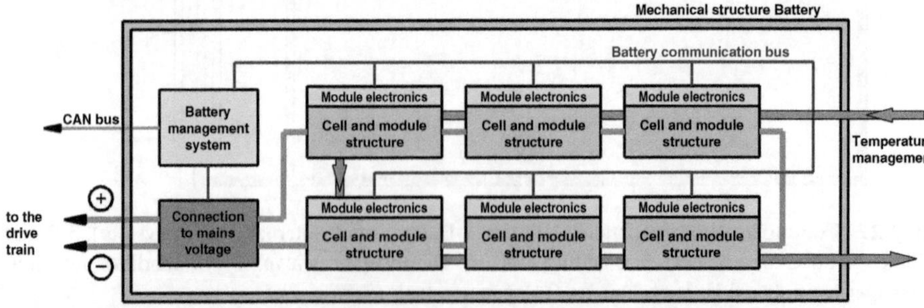

Fig. 2.28 Overall structure of a battery system, schematic [10]. Within individual modules, electronics monitor the cell condition (temperature, voltage). The battery consists of several modules whose electronic units communicate with each other and with the battery management system (BMS) via a battery bus. A temperature management system ensures that the cells and the electronic systems remain within a certain temperature window. If this is not sufficient, the BMS reduces the amount of current that can be drawn from the battery by acting on the motor control via the vehicle's CAN bus

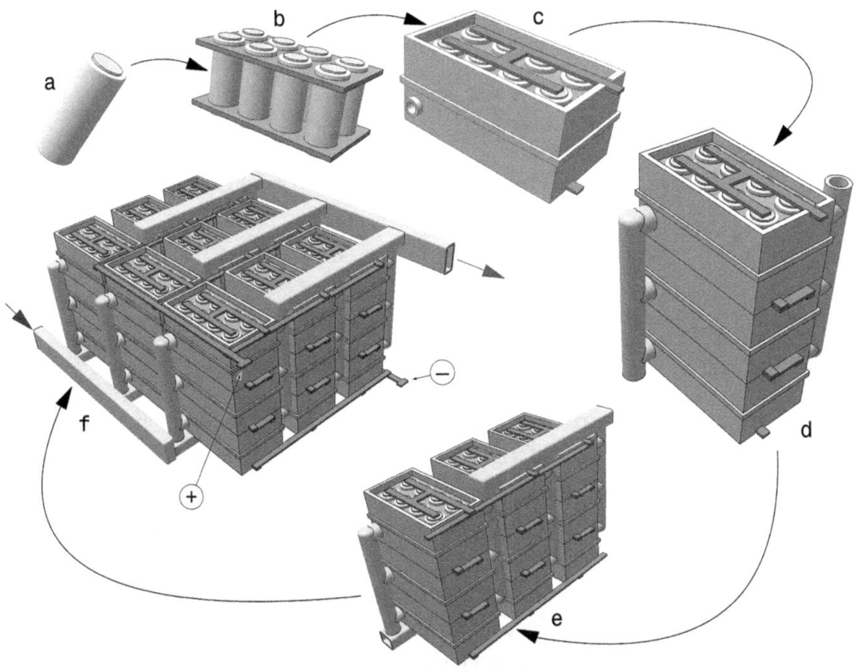

Fig. 2.29 Structure of a high-performance high-voltage battery. The production of a traction battery demonstrates the basic structure. First, several cells (**a**) are combined (**b**). In a module (**c**), cells are galvanically connected in parallel and encased so that a cooling medium can flow around the individual cells in a targeted manner. Each module has a positive and negative contact strip (*busbar*). Several modules are connected in series to form a block (*pack*) in order to increase the voltage (**d**). The resulting stacks are connected in parallel to increase the power (**e**) and combined to the battery according to the space available in the vehicle and the desired energy content (**f**). For heat dissipation, the corresponding channels for the cooling medium are tightly connected to each other. In the following steps, which are not shown here, the hardware of the battery management system is installed and the battery is enclosed

etc. The modules can also be connected to each other by means of a cable. Likewise, there are several ways to combine modules. Cooling of the cells can be by air or other suitable medium. A morphological box, Table 2.3, offers an overview of the resulting conceivable types. Not all types of accumulators are suitable for traction drives.

For high battery performance, it is crucial that the open-circuit voltages of the individual cells are the same. When assembling high-performance modules, the cells are therefore measured beforehand and only equivalent cells are installed. Equivalent in this context means a max. Voltage difference in the mV(!) range.

When designing a battery, the first thing to do is to select cells. First, the pairing of the elements (Li-ion, NiMH, NiCd, . . .) must be determined. Then the geometric shape is selected. There are cylindrical, prismatic and pouch-shaped cells. This influences the packing density and arrangement possibilities of the cells, the type of galvanic connection

Table 2.3 Morphological box: Possible battery designs

Area	Expression				
Cell elements	NiCd	NiMH	Li-Ion	Na-NiCl	NaS
Cell shape	Cylinder	Prism	Bag (*pouch*)		
Connection technology	Clamps	Soldering	Laser welding	Resistance welding	TIG welding
Cooling medium	Air	Cooling gas	Liquid	Ice	Wax
Heat dissipation	Cooling ducts	Circulation	Partial bypass	Cooling nozzles	

between them and the type of fastening of the cells in the module. In addition, these different shapes are available in different dimensions. Further criteria are energy and power density in relation to the cell mass. The availability of the cells and their cost are also decisive criteria. One will also consider the power degradation due to calendar and cyclic aging.

The individual battery modules or blocks are joined together to form the battery depending on the installation space. The external shape is therefore not completely arbitrary and will be oriented towards rectangular geometries. For installation in a vehicle, a flat battery is generally recommended, which is accommodated as low as possible, e.g. in the underbody area. This ensures a low overall centre of gravity and a position between the wheels and between the axles also represents a crash-safe area.

In the case of monoposti, they like to use the space behind the driver's seat, which is occupied by the fuel tank in the case of internal combustion engine drive.

Common battery types are considered below.

Zinc/Air Battery

It belongs to the primary elements, so it must be removed from the vehicle after use. This battery has the highest energy densities (300–400 W h/kg) at relatively low power densities. However, there are only demonstration prototypes of this type, because a special infrastructure is required for processing.

Lead-Acid Battery

Lead-acid batteries have been developed to great maturity over decades in vehicle design as starter batteries. For a starter battery, however, the focus of development is on high, short-term currents with a low cyclic continuous load. These types are therefore poorly suited as traction batteries. Lead-acid batteries are offered in various technical designs, including liquid acid and gel batteries. Common batteries have electrodes in plate construction, but a special form uses flat electrodes that are wound separately by a glass fibre fleece as a separator. This design offers higher load capacity as well as service life and thus makes them interesting for hybrid vehicles [10]. In any case, the high density of lead is a disadvantage for mobile applications, which results in a low energy density: Lead batteries with large storable energy have a large mass. Future developments are also unlikely to achieve the energy throughput required for traction batteries. This type of battery may have low acquisition costs, but the low service life cancels out this advantage in the life cycle balance.

Nickel-Cadmium Battery

NiCd accumulators have a number of interesting advantages. The low internal resistance allows large currents and therefore high performance of the traction motors. They prove to be robust against both deep discharge and overcharge. Their favourable low-temperature behaviour (−20 to −40 °C) could make them interesting for rally vehicles, because they would allow easy winter operation. However, all this is countered by one exclusionary disadvantage: The heavy metal cadmium is toxic and has already been banned by the European Community, at least for many mass applications. The NiCd battery is therefore no longer an option for future passenger car applications [10]. NiCd cells also exhibit the so-called memory effect. This is the reduction in capacity when the battery is repeatedly only partially discharged.

Nickel Metal Hydride Battery

This type of battery has a similar behaviour to the NiCd battery. The main difference is also only in the negative electrode. Instead of the toxic cadmium, a hydrogen-storing alloy (metal hydride) is used here. This type of battery is considered to be largely mature and is used today as a traction battery in buses and as a storage system for hybrid cars. The NiMH battery is characterized by high passive safety with great robustness against short-term overcharging and deep discharge. It also shows good power consumption at low temperatures. However, the cell voltage is relatively low at 1.25–1.35 V and the system weight is correspondingly high. An increase in the performance values is no longer expected, but an optimization in terms of installation space and costs.

Lithium-Ion Accumulator

Under this battery type designation, in contrast to the batteries discussed, there are a variety of active material combinations and different electrolytes. The characteristics that distinguish this type are as great as the variety. This type of cell has become the most important form in portable devices in recent years and its development is far from complete. The cell voltage is very high at about 3.6 V, but the cells tend to individualize and must be monitored and controlled individually. If a cell receives too much voltage, this leads to its electrochemical destruction. This places high demands on the battery management system of traction batteries, which consist of a large number of cells and modules depending on the overall voltage level.

The advantage is the low internal resistance, which allows large currents and thus provides high starting torque and high motor power. These batteries also show no memory effect.

The power densities of 2500 W/kg achievable today were considered impossible just a few years ago. The calendrical service life of 10 years has now also reached a level that is acceptable for vehicle applications [10].

Lithium-air technology is expected to have an energy density of 1200 W h/kg, although it is not yet ready for series production [19]. By comparison, petrol has an energy density in the range of 12,000 W h/kg.

An active battery management system is a prerequisite for maintaining the positive properties mentioned. The heat dissipated during continuous operation must be dissipated by a cooling system despite the high energy efficiency, and the cells must be heated at low temperatures. Deep discharge and overcharging must be avoided by monitoring each individual cell.

Na-NiCl Battery

(ZEBRA battery). The battery, which has become known as "Zebra" (*Zeolite Battery Research Africa Project* or *Zero Emission Battery Research Activity*), is a high-temperature battery. Its operating temperature is in the range of 270–350 °C. The open-circuit voltage of 2.6 V is higher than that of most batteries. It also impresses with its robustness and long service life. Thermal losses in the form of self-discharge have a disadvantageous effect. The high operating temperature is also a hindrance in handling. In contrast to hybrid vehicles, this battery type has already been used for electric vehicles.

NaS Battery

(Sodium-Sulfur Storage Battery) This battery is also a high-temperature battery and operates between 290 and 390 °C. The storage and power densities are similar to those of the ZEBRA battery. The values of storage and power densities are similar to the ZEBRA battery, except that the NaS battery does not exhibit self-discharge. In contrast to the ZEBRA battery, however, the electrodes are liquid and this requires good encapsulation against environmental influences so that a dangerous reaction of the sodium with water does not occur.

Fuel Cell

The fuel cell converts chemical bonding energy into electrical energy by oxidizing a fuel, such as hydrogen (H_2), with atmospheric oxygen (O_2): $2H_2 + O_2 \rightarrow H_2O$. This is basically the same as the well-known oxyhydrogen reaction, but in contrast to this the electrons involved flow via the detour of an electrical consumer (e.g. motor). This controlled combustion is aptly called "cold combustion". PEM fuel cells (polymer electrolyte membrane) operate at temperatures below 100 °C. However, the process is not reversible (irreversible) and therefore the fuel cell is not an energy store but merely an energy converter. The refuelling and storage of the fuel on board is a disadvantage, as is the poor full load efficiency. It also has a poor power-to-weight ratio and requires a large installation space, especially to achieve high efficiencies, because its part-load efficiency is good. On the other hand, it is worth noting that it emits no pollutants, contains no moving parts and operates without noise.

Because the fuel cell delivers electrical energy in stationary operation over a long period of time and with good efficiency, a combination with an electrical storage unit makes sense, which buffers the energy and ensures the transient power output for driving operation.

Redox Flow Cell

The redox flow cell basically works like the galvanic cell, except that the liquid electrolyte is stored outside the cell in separate tanks. The energy conversion takes place in a charge/discharge unit, which usually consists of a membrane with catalysts. The electrolyte is fed to this charging unit via pumps. The battery can therefore also be charged by first removing the "discharged" electrolyte from the external tank and then refilling it with electrolyte that has been charged outside. The capacity of this type of battery determines the design of the charging unit, and the energy content is dictated by the tank size. The redox flow cell has been intensively developed for stationary applications [13]. Mobile applications have been proposed [20], but no practical designs are known.

Table 2.4 compares the battery types currently used in motorsport. It should be noted that the batteries for Formula E and Electric Rallycross are used as traction batteries for purely electric drives, while the batteries for LMP1 and Formula 1 work as energy storage for recovery systems in a hybrid system.

2.4.5 Selection

In general, it is recommended to start a drive concept with the battery selection and to align any necessary compromises of the other components accordingly. The most important

Table 2.4 Comparison of motorsport batteries, according to [21]

	Series				
Property	F E (Gen. 1)	F E (Gen. 2)	eWRX	F 1 (ERS)	LMP1 (ERS)
Mass, kg	310	380	290[a]	40	–
Volume, l	≈ 310	≈ 321	260[a]	25	–
Stored energy, kW h	28	54	46.5	≈ 1.1[b]	≈ 2.2[b]
Energy delivered, MJ/lap	–	–	–	4	8
Max. Power, kW	200	250	500	120	300
Cell power density, kW/kg	1.4	2.2	2.6	12–17	10–12
Energy density cell, W h/kg	174	232	232	90 – 100	90–100

F E (Gen. 1) . . . Formula E (Generation 1), F 1 . . . Formula 1, eWRX . . . Electric Rallycross
[a]Target values for first season 2020
[b]Estimate based on published energy data. Assumes full charge and discharge per lap

considerations – if the regulations do not impose restrictions anyway – are the following, according to [22]:

1. Type of energy storage: primary or secondary system.
2. Electrochemical system: type of cell elements, pairing of electrodes.
3. Voltages: Nominal voltage, allowed minimum or maximum voltage, voltage regulation, discharge curve profile, voltage build-up latency, voltage distortion.
4. Current load and load profile: Characteristics of the load: Constant, cyclic, random dynamic.
5. Operating mode: Intermittent or continuous operation, typical curve for intermittent operation.
6. Temperature: Typical operating temperature range.
7. Service life
8. Physical quantities: Shape, dimensions, mass and connections.
9. Storage conditions: Desired state of charge during storage, typical storage time and environmental conditions (humidity, temperature, ventilation, ...).
10. Charge-discharge cycle: Typical progression of such cycles: Unregulated demand-controlled or lifetime-controlled, availability and characteristics of the charging unit, charging duration, charge/discharge losses (efficiency), ability to repeatedly absorb high charging power (for recuperation).
11. Operating conditions: Vibrations, shocks, accelerations, type and frequency of movements, environmental conditions (pressure, temperature, humidity, ...), magnetic fields.
12. Safety and reliability: permissible fluctuations in electrical parameters, failure rates, absence of leaks, leakage of (waste) gases, use of toxic or hazardous substances, behaviour under mechanical impact (penetrations, crash).
13. Maintenance and spare parts supply: Availability of cells, simplicity of battery replacement, type and availability of necessary charging stations, transport possibility, overhauling of used batteries or disposal of old batteries.
14. Costs: purchase price, operating costs (lifetime costs), use of exotic (expensive) operating materials (cooling medium, electrolyte).

First, it is necessary to estimate the order of magnitude of the required power and energy of the accumulator. The required maximum motor power $P_{M,\max}$ is dictated by the desired maximum speed $v_{V,\max}$:

$$P_{M,\max} = \frac{1}{\eta} F_{dr} \cdot v_{V,\max} \tag{2.13}$$

$P_{M,\max}$	Maximum motor power, W
η	Efficiency of the drive train, –, see section 5.1
F_{dr}	Total driving resistance, N, see Chap. 4, Eq. (4.11)
$v_{V,\max}$	Maximum speed of the vehicle, m/s

The power to be applied by the battery must be selected somewhat larger, because the power electronics of the motor control also work with a certain loss.

$$P_{Bt} = \frac{1}{\eta_C \cdot \eta_{Bt,2}} P_{M,\max} \tag{2.14}$$

P_{Bt}	Power to be delivered by the battery, W
η_C	Efficiency of motor control, –. Values see Table 2.1
$\eta_{Bt,2}$	Output efficiency of the battery, –. See Table 2.5

The current drawn by the motor at maximum power must of course be able to be borne by all current-carrying systems:

$$I_{\max} = \frac{1}{\eta_M \cdot \eta_C} \frac{P_{M,\max}}{U_{Bt}} \tag{2.15}$$

$I_{\max}$	Maximum current at rated motor power, A
η_M	Efficiency of the motor, –, cf. (2.5)
U_{Bt}	Terminal voltage of the battery, V

The energy required to be stored by the battery for a specific vehicle for a specific distance is calculated according to the physical relationship work = force times distance. The force to be overcome is the total driving resistance F_{dr}, the distance is the distance to be driven. Because the driving resistance depends, among other things, on the speed, the gradient and the acceleration and the speed changes over the distance,[4] must be added up over individual sections or, more precisely, integrated. Depending on whether the speed curve is available over the distance or the time from measurements, the following applies to the energy to be stored:

[4]See, e.g., Racing Car Technology Manual, Vol. 5 Data Analysis, Tuning and Development, Fig. 1.46.

$$W_{Bt} = \frac{1}{\eta \cdot \eta_{Bt,2} \cdot \eta_C} \int_{s_0}^{s_1} F_{dr}(s) \cdot ds = \frac{1}{\eta \cdot \eta_{Bt,2} \cdot \eta_C} \int_{t_0}^{t_1} F_{dr}(t) \cdot v_V(t) \cdot dt \qquad (2.16)$$

W_{Bt}	Energy to be stored in the battery, 1 W s = 1 J
s_0 or s_1	Start or end value of the travel distance, m
v_V	Vehicle speed, m/s
t_0 or t_1	Start or end value of the time, s

As an approximation, the fuel consumption of a similar vehicle with an internal combustion engine on the racetrack under consideration can also be used. From the calorific value H_u [kJ/kg] and the density of the fuel [kg/l] follows the energy density [kJ/l] = [kW s/l]. The equivalent amount of energy that a battery must store is then derived directly from the fuel consumption for the racedistance:

$$W_{Bt} = \eta_{M,ref} \frac{\eta_{ref}}{\eta} H_u \frac{\rho_K}{3600} C = \eta_{M,ref} \frac{\eta_{ref}}{\eta} w_K \cdot C \qquad (2.17)$$

W_{Bt}	Energy to be stored in the battery, kWh
$\eta_{M,\text{ref}}$	Effective efficiency of the internal combustion engine, –. $\eta_{M,\text{ref}} = 0.25–0.3$
η_{ref}	Efficiency of the powertrain of the reference vehicle, –
H_u	Lower calorific value of the fuel, kJ/kg
ρ_K	Density of the fuel, kg/l For gasoline $\rho_K = 0.725–0.780$ kg/l For diesel $\rho_K = 0.820–0.860$ kg/l
C	Fuel consumption, l
w_K	Energy density of the fuel, kW h/l For gasoline $w_K = 8.5–9.5$ kW h/l For diesel $w_K = 9.6–10$ kW h/l

The core of any electrochemical energy storage device is the galvanic cell. To select which electrode pairing is the most suitable, a Ragone diagram can be used, Fig. 2.30. It shows the specific values of energy and power of cells. Capacitors are also plotted for comparison. For racing vehicles, the areas of highest power densities are of interest, so that the desired driving performance can be represented. Similar is valid for car-hybrid with however less energy. In this case, nickel metal hydride battery is popularly used [10]. On the other hand, for passenger car e-drives, the areas of high energy density are targeted first because a reasonable range of the vehicle with sufficient space is an essential feature for customer acceptance. In general, the amount of energy required for a regular (i.e., non-stop) race can be well estimated from contested competitions or by simulation from the speed profile with the driving resistance and the track length. Therefore, lithium-ion high-

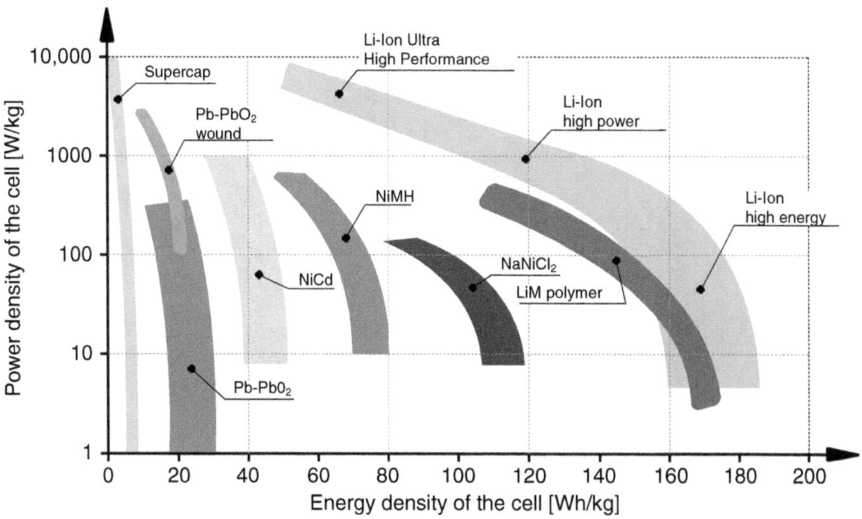

Fig. 2.30 Ragone diagram for galvanic cells [10]. Supercap Supercapacitor. Pb-PbO$_2$ Lead acid cell, NiCd nickel cadmium cell, NiMH Nickel metal hydride, LiM polymer Lithium-ion polymer cell, NaNiCl$_2$ Zebra cell

performance and ultra-high-performance batteries are the first choice, even if their energy density is not the greatest.

The selection criteria result directly from the requirements listed above. Table 2.5 provides an overview of the characteristics of known storage systems. Table 2.6 summarizes the most important characteristics in a comparative manner.

Figure 2.31 clearly illustrates the problem of storage systems with low energy density by plotting the required mass or volume for a given storage system for a driving distance of 100 km. The superiority of liquid fuels becomes clear. However, it should not be forgotten that the fuels only make their chemically bound energy available through combustion with oxygen and that the air required for this does not have to be carried in the vehicle but is taken from the environment. Similarly, the combustion products do not remain in the storage tank. The other storage systems shown in the diagram all operate with an internal storage tank, i.e. they constantly carry all the substances involved in the reactions.

In addition to the stationary characteristic values, however, the dynamic behavior is particularly decisive for vehicles. During acceleration processes, the required voltage must be built up quickly and the resulting currents must be endured, otherwise the battery (and not the drive tires or the motor) will limit the driving performance. If the battery cannot absorb (quickly enough) the power generated during subsequent recuperation, this energy is lost to subsequent accelerations and the mechanical brake must bring about the deceleration desired by the driver. In such a situation, the required seamless interaction between the braking systems poses a particular challenge for the control system.

Table 2.5 Technical data of electrical storage systems, according to [10]

Feature			Pb-Gel	NiCd	NiMH	Li-Ion	NaS	$NaNiCl_2$	Supercap
Practical. Energy density		Wh/kg	20–50	40–55	40–80	110	90–120	100–120	2–4
		Wh/l	70–100	80–110	100–270	270	100–120	160	2.5–4.5
Power density		W/kg	80–100	< 200	<200–1300	500	125–130	110–150	2000 – 4000
		W/l	160 – 200	< 360	200–700	1250	110–140	130–265	3000–5000
Voltage	Idle	V	2.1–2.5	1.3	1.3	≤ 4.2	2.1	2.59	
	Rated	V	2	1.2	1.2	3.8			2.7–3
	End of charging	V	2.7	1.55	1.45	4.2			
	End of discharge	V	1.6	0.8	0.9–1.1	2.5			
Energ. Efficiency		–	> 0.9	0.78	0.7	0.93	0.83–0.85	0.91	0.9–0.95
Self-discharge	Electrical	%/day	0.1–0.4	0.6–1	1.5 – 20	0.15	0	0	3–20
	Thermal	%/day					15–17	15–17	
Operating temperature		°C	–10 to 40	–40 to 60	–20 to 60	–20 to 60	> 300	> 300	–40 to 70
Cargo pickup		%	50 in 2 h	97 in 0.5 h	97 in 0.5 h	95 in 1 h		90 in 3.5 h	50 in 0.5–5 s
		%	100 in 5 h	100 in 1 h	100 in 1 h	100 in 4 h	100 in 3 h	100 in 5 h	100 in 1–10 s

Pb-Gel . . . lead-gel battery, NiCd . . . nickel/cadmium-B., NiMH . . . nickel/metal hydride-B., Li-Ion . . . lithium/ion-B., NaS . . . sodium/sulfur-B., $NaNiCl_2$ . . . sodium/nickel chloride-B., Supercap . . . supercapacitors

Table 2.6 Comparison of electrical storage systems, according to [10]

Feature			Pb-Gel	NiCd	NiMH	Li-Ion	NaS	$NaNiCl1_2$	Super Cap
Development status			Large-scale production	Large-scale production	Series	Initial series	Pre-series	Pre-series	Series
Service l ife	Calendrical	Years	3–5	> 4			1–2	5	> 10
	Cycles	–	700–800	2000	2000	> 2000	> 1000	> 600	> 1000, 000
Deep discharge			Limited possible	Possible		Possible	Problem	Possible	Possible
Overload			Possible	Necessary		Problem	Problem	Possible	Possible
Maintenance-free			Yes	No/yes	Yes	Yes	Yes	Yes	Yes
Comment			Mature	Temperature dependent, toxic	Fast-charging	Safety-critical	Heating/cooling system required	Heating/cooling system required	No cooling necessary

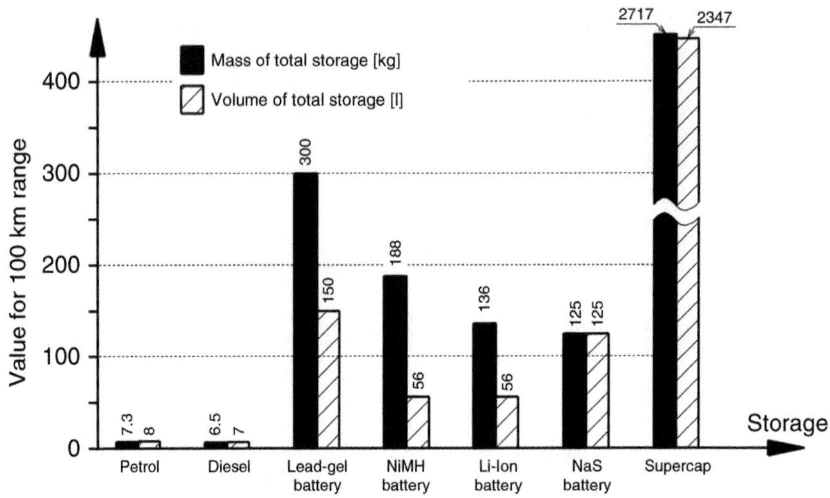

Fig. 2.31 Mass and volume of different storage systems for 100 km range [10]. Based on typical consumption values, the mass and volume of the respective entire storage system are shown. The entire storage system includes the battery trough, battery management and cooling

A dynamic stress test for traction batteries has therefore been developed for passenger cars, one cycle of which is shown in Fig. 2.32.

2.5 Charging [24]

Charging the high-voltage accumulator in the vehicle is more complex in terms of safety than it might appear at first glance. Charging involves the coupling of different systems (household three-phase current with 230 V rms and high-voltage DC) and associated protective devices. On the supply side, there are earthed three-phase networks with 230 V effective (household supply) or unearthed DC charging stations. The electric vehicle, in turn, has an isolated high-voltage DC system on the storage side. When charging, this becomes an earthed or unearthed overall system, whereby consideration must be given to the different protective measures of the systems involved. In addition, the protection regulations originate from different standardisation organisations, Fig. 2.33. In the vehicle, an insulation monitoring device reports any insulation fault that occurs. This is necessary for the vehicle's own safety alone, but also helps when coupling with the charging station. The charging station's protective device would stop charging if an insulation fault was detected. The minimum insulation resistance values specified in ISO/FDIS 6469–3 are 500 Ω/V for AC systems and 100 Ω/V for DC systems [25]. During charging, the on-board insulation monitoring system is usually switched to passive, so that the protective device of the charging station must also monitor the vehicle network.

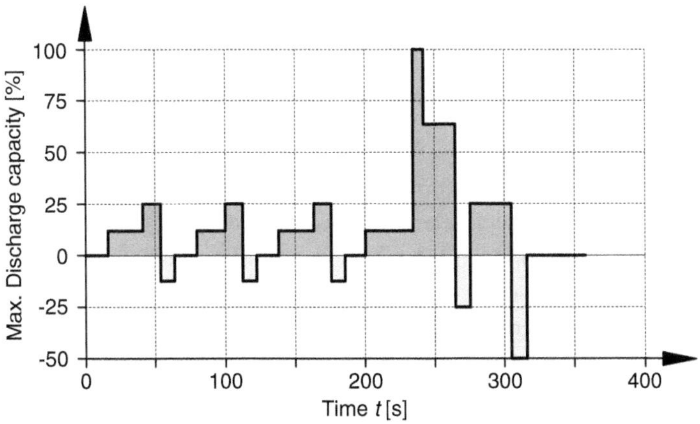

Fig. 2.32 Dynamic stress test for traction batteries of passenger cars [23]. The graph shows a cycle that simulates discharging (positive values) during acceleration and constant driving and charging (negative values) during recuperation. The discharge power in percent refers to the maximum nominal discharge power of the DUT. The cycles are repeated until the termination criterion, e.g. the minimum capacity of the battery, is reached. This allows, among other things, the range to be determined in a standardized manner. Due to the cyclic load on the battery, more realistic statements can be made about the battery behaviour than with constant load tests. The areas under the curve represent energy quantities. It can be seen that the sum of the energy extracted (light red) is significantly greater than the energy fed in (light green)

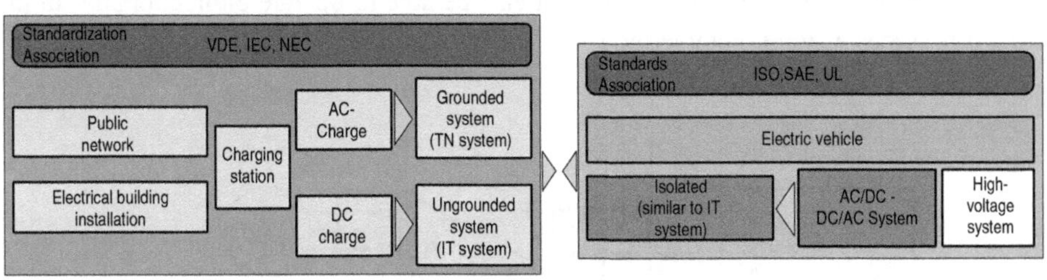

Fig. 2.33 Coupling of systems during charging. During the charging process of a vehicle, the protective devices of different systems established by different standardization organizations must be harmonized. AC (alternating current). DC (direct current)

The maximum charging power depends on the available mains voltage and the permissible mains current. With the usual 230 V three-phase mains with one phase fused to 30 A, for example, this results in a maximum charging power of 6.9 kW.

2.6 Power Electronics

2.6.1 Overview

Figure 2.34 provides an overview of the tasks of the power electronics in an electric vehicle. A high-voltage battery (1) provides the energy required for driving. A 12 V battery (5) known for the usual vehicle electrical system and the starter motor of an internal combustion engine is also on board, but its voltage and power are much too low for high driving performance. This battery supplies auxiliary consumers (lights, actuators, etc.) and control units with the required voltage. The electrical machines (3), which are used for driving and feed back braking energy in regenerative operation, generally operate as three-phase AC machines. In order for energy to flow between the high-voltage battery and the three-phase machine, a converter (2) is interposed. These power electronics convert the DC voltage into a multi-phase AC voltage (DC–AC converter). In the case of recuperation, however, the conversion also takes place in reverse: the alternating current is converted into direct current of a suitable voltage (AC-DC converter). In this case, one therefore also speaks of a bidirectional (= bi-directional~) converter.

Another power electronics mediates between two DC voltage systems which have different voltage levels. A DC/DC converter (4) reduces the high voltage of the traction battery (1) to 12 V (buck converter) and thus the energy of this battery can also be used for auxiliary consumers (6). Usually a unidirectional converter is sufficient. However, if external starting assistance from the 12 V mains is also to be possible for the high-voltage system, the DC/DC converter must also be able to operate bidirectionally. In this case, it becomes a step-up converter.

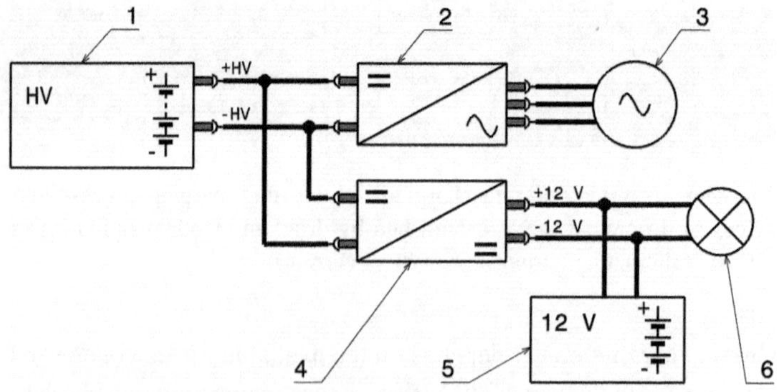

Fig. 2.34 Power electronics in an electric vehicle. 1 *High-voltage* battery, 2 Inverter, 3 Electric machine, 4 *DC-DC converter,* 5 12 V battery, 6 Additional *loads*

2.6.2 Cooling

Power converters are represented by fast switching electronic components. Basically, these are switches that conduct or block (even large) currents. In contrast to mechanical switches, however, currents flow in both switching states and become noticeable as losses in the form of heat. In addition, switching losses also occur depending on the switching frequency. The heat must be dissipated to protect the components. The simplest way to do this is to use cooling fins on the housing of the power electronics. In the case of compact dimensions and high electric motor power, liquid cooling may be necessary. The water temperature in this case is around 65 °C, with maximum values of 85 °C. Thus, when considering the efficiency of a drive train, the losses of the power electronics must also be taken into account. The power loss of converters/inverters is between 3% and 10% of the consumed electrical power and depends mainly on [12]: Device size (nominal power), power electronics, switching frequency (2–16 kHz), control method and energy recovery.

Basically, the cooling system is designed as described in Sect. 3.4. In the case of liquid cooling of electronic components, some additional notes are appropriate. With regard to the materials in the cooling system, it must be ensured that all those in direct or indirect contact are as close as possible to each other in the electrochemical voltage series, or at least that no more noble materials than the cooling plate of the electronics are present. For example, in the case of aluminium alloy cold plates, copper tubes must not be used in the cooling system. If different materials are used, corrosion protection is required for the entire cooling circuit.

Especially in the case of high humidity, another phenomenon must be taken into account. The temperature of the cooling medium must not fall below the dew point, otherwise the water from the air will condense and in the worst case lead to short circuits in the circuits. The dew point can be calculated approximately with (2.18).

$$\vartheta_\tau = c_1 \frac{\frac{c_2 \vartheta}{c_1+\vartheta} + \ln(\phi)}{\frac{c_1 c_2}{c_1+\vartheta} - \ln(\phi)} \tag{2.18}$$

ϑ_τ	Dew point temperature, °C
c_1, c_2	Constant. $c_1 = 243.12$ °C, $c_2 = 17.62$
ϑ	Ambient air temperature, °C
Φ	Relative humidity, –

2.6.3 Motor Control

The power electronics that supply the electric motor with current of the required frequency, amplitude and phase is called motor control. The main input variable is the driver's torque

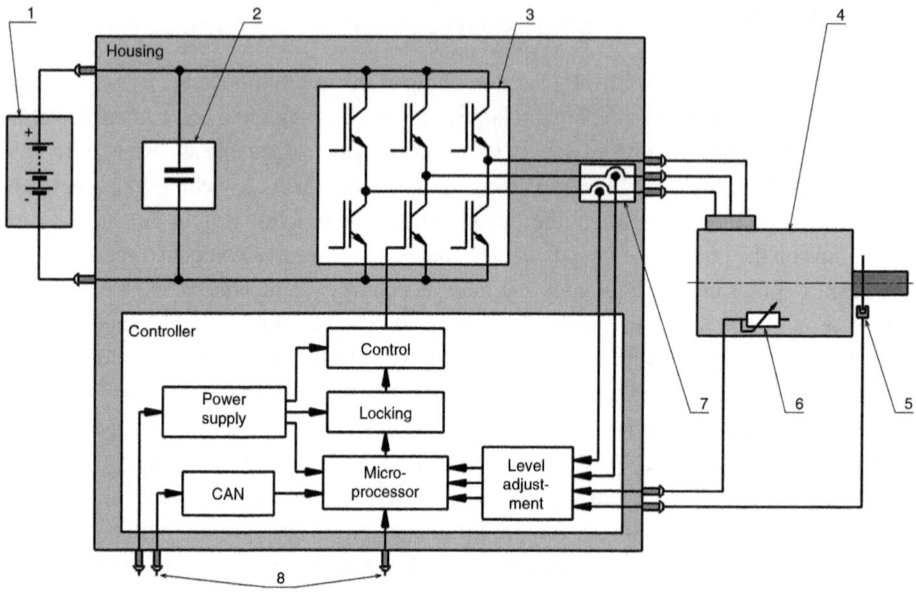

Fig. 2.35 Motor control for a three-phase motor. The motor control consists of a control computer, which switches an inverter (3) according to the driving and brake pedal and thus provides the current for the motor (4) in the required frequency, amplitude and phase position. 1 Traction battery, 2 DC link capacitor, 3 Full bridge circuit, 4 Electric machine, 5 Angle sensor, 6 Temperature sensor, 7 Current measurement, 8 External signal inputs (accelerator, brake pedal, gear lever, …)

request, which is specified via the accelerator pedal. Other variables can be the wheel slip, the lateral acceleration, but also the condition (state of charge SOC, *state of health – SOH*) *of* the traction battery. The motor control regulates the motor power output on the basis of complex algorithms. The motor control for a three-phase motor (Fig. 2.35) thus consists of the combination of an inverter (3) with an electronic control computer, which processes the decisive input signals and performs the resulting control of the inverter. External input signals (8) are analog signals, such as the accelerator pedal position, and signals from the CAN bus (ignition on, brake pedal, accelerator lever position and communication with the higher-level vehicle control unit). Internal signals are used to record the status of the electrical machine: speed (5) and temperature (6) as well as two of the three supply currents (7). For permanently excited synchronous machines, the angular position (5) of the rotor must also be known. The actual torque of the motor is calculated from this data using a machine model.

Figure 2.36 shows an example of an induction (asynchronous) motor with operating characteristics and the usable operating range. For the speed setting of the motor, the supply frequency f *of* the stator field is varied. As the frequency increases, the rotor speed increases. If the breakdown torque $M_{M,\max}$ is to be kept constant, the ratio of stator voltage U_f and frequency must be kept constant in addition to the frequency. In the field weakening

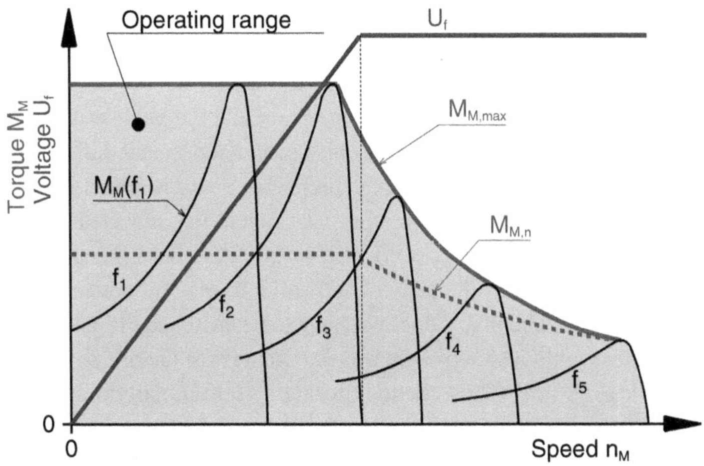

Fig. 2.36 Idealised operating characteristics of an induction motor (asynchronous machine) with variable voltage and frequency [10]. U_f Stand voltage, $M_{M,\max}$ describes the course of the so-called tilting moment, $M_{M,n}$ is the continuous torque, In addition, the moment curves for constant feed frequencies of the stator voltage f_1 to f_5 are plotted. $f_1 < f_5$. The operating range is determined by the breakdown torque and the speed limit

range (power is constant), the stator voltage U_f is kept constant and the breakdown torques decrease accordingly with increasing frequency. By controlling the motor with variable supply voltage and frequency, all points in the operating range can be approached.

The reversal of the motor's direction of rotation is also implemented purely electronically by reversing the phase sequence on the AC side. A reverse gear in the transmission is therefore not needed at all in electric vehicles. When decelerating the vehicle, the motor control unit also takes control of the electric machine running either in regenerative or currentless mode (so-called sailing). This requires the behavior of the battery (permanent damage in the event of overcharging) and the safety of permanently excited machines (dangerous short-circuit currents in the absence of terminal voltage).

2.7 Safety

In general, electric vehicles are required to offer the same level of safety as today's vehicles with combustion engines. This not only applies to reactions to mechanical impacts in a crash, but the electrical systems from the traction battery to the electrical machine must also be intrinsically safe. Intrinsically safe means that the system under consideration is safe without any additional protective device – i.e. by itself.

For "conventional" vehicle technicians in particular, electric drive systems involve a number of criteria to which little or no attention was previously paid in the case of vehicles. The voltages of traction batteries from 120 to 1000 V DC are far beyond the safety extra-low voltage, so it is necessary to monitor the insulation of these high-voltage systems with respect to the vehicle ground in order to ensure the safety of all persons who come into contact with the vehicle [26]. This function is performed by insulation monitoring devices (see also Fig. 2.1). These devices must also operate while the vehicle is in motion, where large fluctuations in the mains voltage occur due to acceleration and braking processes, and vibrations are also present. In comparison, vehicles have an unearthed IT system (strictly speaking, it is a protective separation with permanently connected consumers). The advantage here is that the first (insulation) fault in the DC network cannot yet generate a fault current. The connection of the active conductors to the equipotential bonding in the voltage source is missing. The mains does not have to be switched off in the event of a fault, which in many cases would also represent a greater danger to the driver while driving than the insulation fault itself [26]. Thus, the monitoring system initially only reports the fault, e.g. by a display in the cockpit. Of course, the fault should be located and rectified as soon as possible.

In addition to this device, all high-voltage components in the vehicle are connected to each other and to the body (if it is conductive) with solid cables to ensure equipotential bonding.

The orange high-voltage cables must also have certain mechanical properties: The conductor sheathing must be able to maintain the insulation effect in the event of deformation (e.g. accident) or at least the leakage voltage must be below the limit for electric shock. This ensures protection for the driver and helpers.

Electric vehicles require significantly fewer mechanical solutions (clutches, differential, manual transmission, . . .) than conventional vehicles. However, this is also accompanied by the need for increased software effort, because the control of the motor must be monitored and possible malfunctions must be taken into account both on the hardware side (sensors, actuators, . . .) and on the software side itself. Just imagine the situation that a seemingly small sign error in the control computer causes the drive motor to deliver the full torque in the opposite direction! Electric motors are also capable of generating braking torques that are far from the values of internal combustion engines. In the case of today's wheel hub motors, the motor cannot even be mechanically decoupled from the rim (e.g. by means of a clutch), which would make such a misconnection even worse.

In order to record and maintain safety, it is not sufficient to consider the individual components alone, but the entire system must also be analysed, because some faults only become critical in interaction with other components. In this context, the ISO 26262 standard (Road vehicles – Functional safety) is helpful. This has been developed specifically for electrical/electronic systems in motor vehicles. The standard describes a method by which the risk of a hazard can be assessed and the resulting requirements for the safety functions can be specified, Fig. 2.37.

First, the system risk is identified without taking safety measures into account by means of a hazard and risk analysis. Where necessary, a safety concept is subsequently developed that reduces the residual risk to a tolerable level. A safety case documents the effectiveness of the safety measures. The required scope of this proof follows from the safety

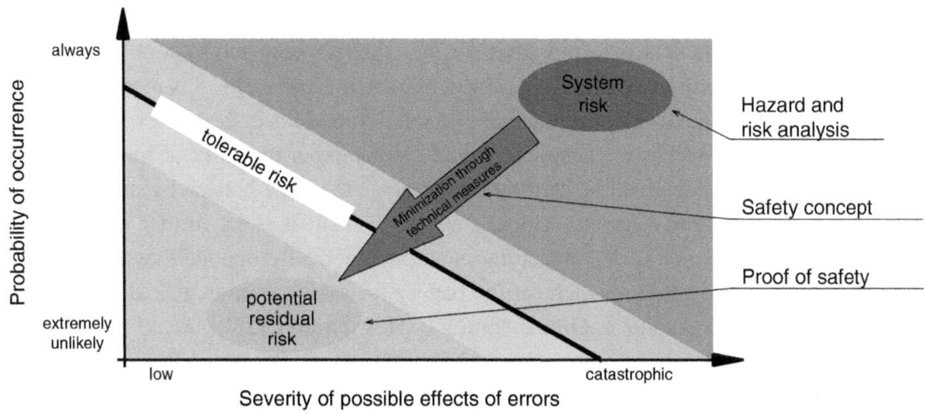

Fig. 2.37 Classification of system risk according to ISO 26262 [27]. If the system risk is high, a safety concept reduces the remaining residual risk to a tolerable level

requirement levels ASIL A to D (Automotive Safety Integrity Level). ASIL D is the highest requirement level. Non-safety-relevant requirements that lie outside of the described consideration must be dealt with by the quality assurance processes.

The risk posed by the hazard under consideration to a system under investigation is measured by the following quantities:

$$Risk = S \cdot P_E \cdot \frac{1}{P_C} \cdot P_F$$

S	Severity = extent of damage in the event of a fault
P_E	Length of stay in situations of potential risk
P_C	Security possibility
P_F	Probability of occurrence of error

The individual values are taken from tables which can be found in the mentioned standard.

A high requirement level demands a safety concept so that possible errors can be safely controlled. Which possibilities are conceivable here? The fault can be excluded according to the state of the art, e.g. by calculated mechanical design. Safety-relevant effects of faults can be excluded, e.g. by proving that the driver is not or only insignificantly affected by the fault. Or the fault is detected in good time by a protective device and the system is switched to a safe state.

In traction batteries, the central hazard emanates from the cells. Even in the event of a crash, the separator in the individual battery cells must safely insulate the electrodes from each other. Batteries can start to burn in the event of leaks or due to the ingress of foreign bodies, depending on the design and electrolyte. So they must be treated like a fuel tank in vehicle design. A firewall should separate the battery compartment from the cockpit. So there are some mechanical measures that can be taken to keep the system safety of the battery high. But in order to prove the safety in all conceivable cases, it is more purposeful to put the effort into the development of reliable safety functions in the battery management system than to map all electro-chemical action chains experimentally or mathematically [28]. The battery management system monitors the state of the battery by measuring the cell voltage and current. When a critical state is reached, it stops further extraction or supply of energy by means of contactors (electromagnetically operated switches). Even though it must be possible to switch off the battery for safety reasons, this alone does not ensure the safety of the vehicle. On the contrary! This requires [28]:

- a second independent source of electrical energy capable of supplying the vehicle's safety electronics for a specified period of time;
- that the energy stored in the inverter cannot endanger the driver;
- that switching off the motor does not cause blocking of the wheels or larger negative torques on the wheels. This makes a separation unit between wheels and motor necessary for permanently excited machines.

Other protective devices found on traction batteries or electric vehicles are: Touch-proof detachable connector, Fig. 2.38, (*manual service disconnect,* MSD) for complete galvanic disconnection of the HV battery from the vehicle electrical system, *burst discs* as a controlled expansion option in the event of overpressure in the battery trough and *douse ports* that allow the battery container to be flooded with water in the event of a fire. The battery trough must be mechanically robust and the internal parts must be sealed off in accordance with protection class IP 67.

For races at more distant competition venues, transport of the battery will be necessary. This requires additional testing to comply with international transport regulations such as UN/DOT 38.3. This regulation requires the following individual tests: altitude simulation, thermal tests, vibration test, shocks, external short circuit, impact, overcharge and forced discharge.

The processing of the signals supplied by sensors and actuators is carried out by a central control unit (central computer) which monitors the entire vehicle, Fig. 2.39. The input variables are analogue (e.g. accelerator pedal position) and digital (e.g. brake pedal, gear lever, . . .). In order to guarantee faultless control of the motor or motors (tandem drive, wheel hub motors, etc.), the central control unit has a redundant design. The current driving condition and the torque to be applied are calculated in the main computer. However, these values are not directly transmitted to the CAN bus, but a monitoring computer compares the values with results determined by simplified algorithms. The values are only passed on to the motor controller if they are plausible. If discrepancies occur, a fault handling unit becomes active and, depending on the type and duration of the fault as well as the driving condition, takes appropriate measures to maintain a safe state of the vehicle. As already mentioned, simply switching off systems is not always a viable solution.

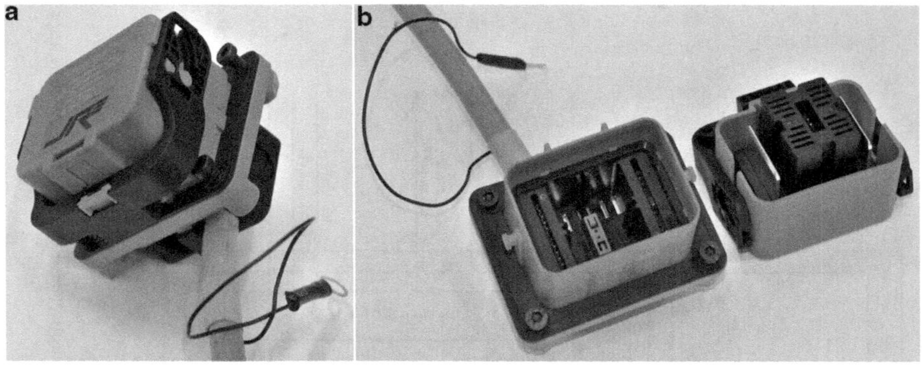

Fig. 2.38 Manual service disconnect (MSD). (**a**) Plugged state – battery connected to on-board power supply, (**b**) Fully disconnected state – battery disconnected from the mains. This connector can be pulled out by hand without any risk and thus the high-voltage battery can be completely galvanically isolated from the rest of the vehicle. The disconnection process takes place in two stages. To disconnect, the black safety clip must first be folded up, which causes the connector to move into a first detent position. Only the monitoring circuit (*high-voltage inter lock*, HVIL) is opened. To remove the plug completely, another mechanical lock (spring-loaded latch, hidden by the clip in the picture) must be released. During this time, the monitoring electronics switch off the HV circuit and the plug can be completely removed without the risk of igniting an arc. The two tongue-shaped high-voltage contacts are on the outside. In the middle between them are the much smaller HVIL contacts offset in the plugging direction, so that they are released first when the plug is pulled out. Max. Voltage DC 1500 V, max. Current DC 170 A. In the picture only one HV cable is connected. The conductor cross-section is 35 mm^2

Not only is the computer redundant, but the accelerator pedal position is also detected by at least two sensors and checked by the control unit to ensure that it is free of errors.

The behaviour of electrical machines in the event of a fault (short circuit, unintentional loss of control, . . .) is determined exclusively by the type of magnetisation [6]. Basically, there are three types: via permanent magnets, via a current in the stator and via a current in the rotor.

- Excitation via permanent magnets: When the rotor is rotating, a current is generated in the stator in the event of a short circuit. This short-circuit current easily exceeds the rated thermal current with intact cooling. As a result, such a vehicle is already exposed to a fire hazard during towing.
- Current excitation via the stator (e.g. induction motor (asynchronous machine), reluctance motor): When the inverter is deactivated, the excitation of the stator also ceases and the rotor rotates like a simple shaft.
- Current excitation via the rotor (e.g. current-excited synchronous motor): If the excitation current is deactivated, this motor cannot build up torque. As disadvantageous as

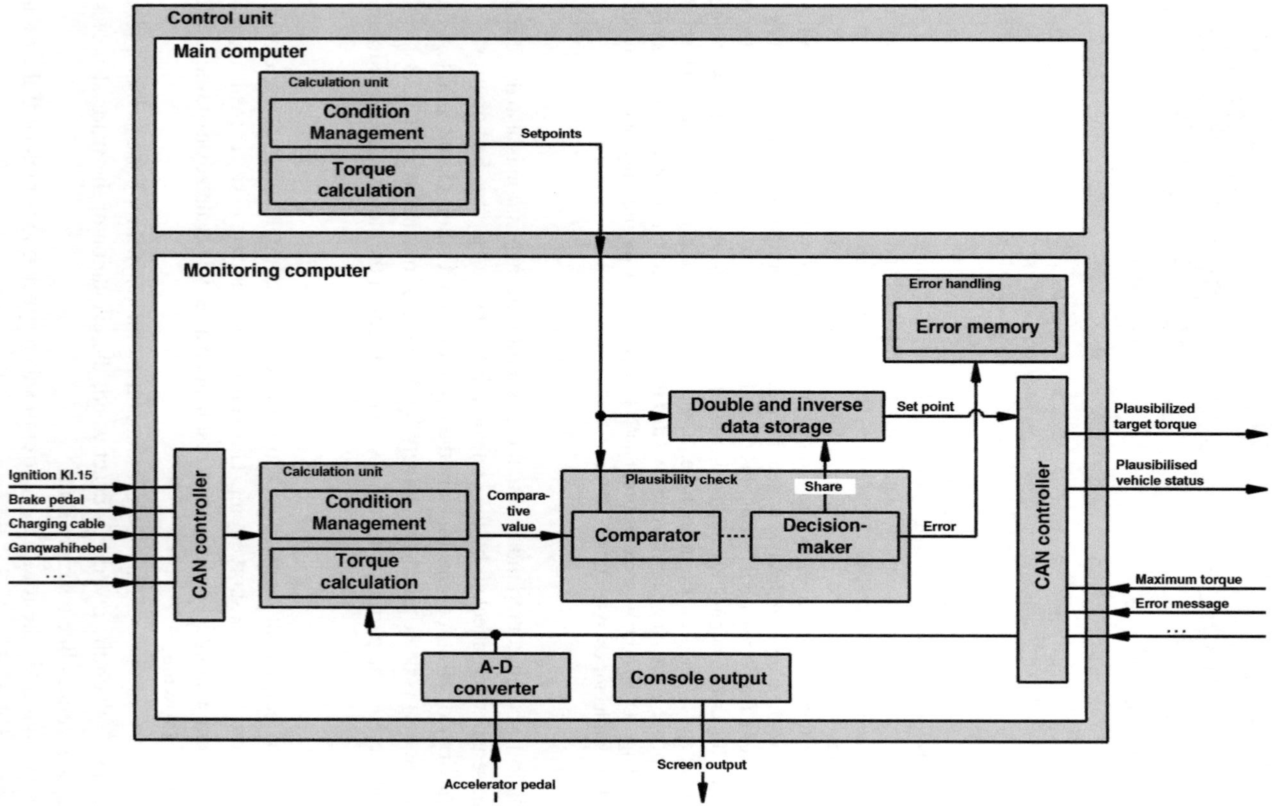

Fig. 2.39 Schematic structure of a central control unit [29]. The control unit essentially consists of the main computer and the monitoring computer. Values to the motor control are only passed on after a credibility check

current transmission via slip rings is, this design proves to be advantageous in the event of a fault.

The technical solution in the event of a fault consists in any case of an isolating contactor for rotor or stator supply lines and, in the case of permanently excited motors, an additional isolating clutch.

2.8 Special Features in Racing

The *state* of charge *of* the battery (SoC – *State of Charge*) decreases due to the output of power in the race, which in turn causes the voltage level of the battery to drop. If the power level P for propulsion is to be maintained, the current I must increase accordingly ($P = UI$), which in turn causes the losses ($= I^2 R$) due to the internal resistance R *of* the battery to increase. Due to the losses, the battery heats up and its performance decreases. This "vicious circle" of battery fatigue is shown in Fig. 2.40.

The maximum power that can be drawn from the battery is therefore a question of its temperature. Figure 2.41 shows how rapidly the maximum power drops in a current Formula E battery.

Furthermore, the amount of energy carried by the battery is limited. Therefore, the aim is to achieve the shortest possible lap times through skilful use of the stored energy and recuperation of the vehicle's kinetic energy, while still covering the full race distance. In the case of recuperation, physical conditions are added to the regulations. The state of

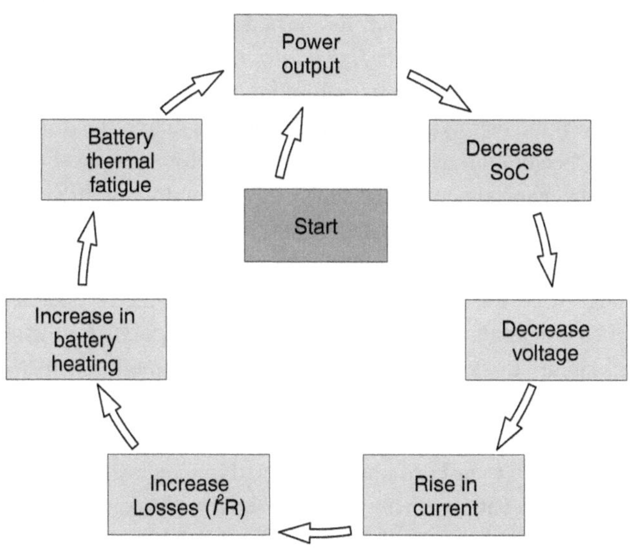

Fig. 2.40 Cycle of battery fatigue in racing operation, after [30]

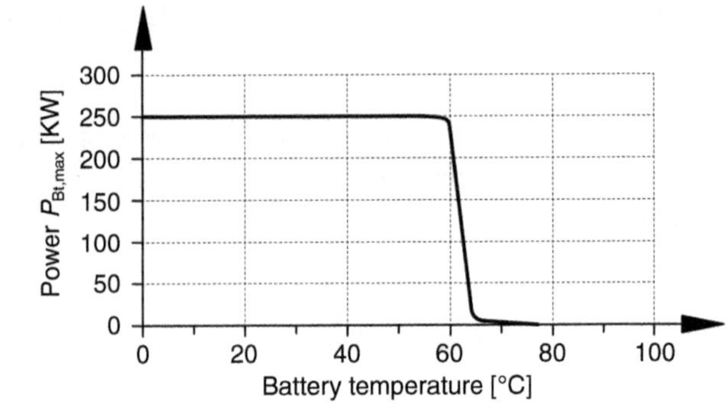

Fig. 2.41 Maximum battery power versus temperature, after [30]. $P_{Bt,max}$ Maximum power that can be drawn from the battery. From about 58 °C, the power drops rapidly from the constant value of 250 kW up to that point

charge and its temperature additionally limit the power consumption. As long as the battery is charged, it is not possible to recuperate at all. In practical terms, this means that the first effective braking with the help of the electric motor is only possible after about 1.5 laps. In addition, driving stability must be maintained, which limits the usable braking torque when the electric motor acts on the rear axle.

So how is a winning strategy for a particular race devised for Formula E? The preparation already starts in the driving simulator [31]. After a promising basic set-up has been worked out based on the previous year's data, the driver comments and the lap time achieved, the drivers simulate a qualifying lap, i.e. one that represents what is feasible without taking into account the amount of energy, tyre wear, etc. The other extreme is then determined, an energy target lap. The other extreme is then determined, an energy target lap. Here the drivers try to drive as fast a lap as possible through targeted recuperation and sailing despite a limited amount of energy (= energy target). During recuperation, part of the vehicle's kinetic energy is fed back to the battery as electrical energy when braking (regenerative braking, see also Chap. 3 *Hybrid drives*). Sailing means taking your foot off the accelerator pedal, i.e. continuing to roll without propulsion. In this driving state, no energy is drawn from the battery, but the car decelerates due to the driving resistances. In order to keep the deceleration as low as possible, the drive motor is separated from the wheels, e.g. by a mechanical separating clutch, or, depending on the electric machine, at least the induced drag torque is eliminated by disconnecting the supply connections.

The data obtained on the driving simulator (braking points, recuperation zones, sailing zones, ...) are then used for simulations. The applied and the recuperated energy in the respective sectors are determined by simulations. Different points for the start of sailing or regenerative braking are tried out and the resulting sector time is calculated, Fig. 2.42.

The results form the basis for optimization. The total lap time is optimized for different energy targets. Which energy target is used depends on the race and the race phase. Until the 2017 season, each driver in Formula E competed in a race with two vehicles, with the vehicle (and therefore the battery) being changed halfway through the race. Due to the odd total number of laps, the energy requirement for one vehicle was higher. As the race progresses, the racing line changes. If at the beginning it is still as expected from powerful

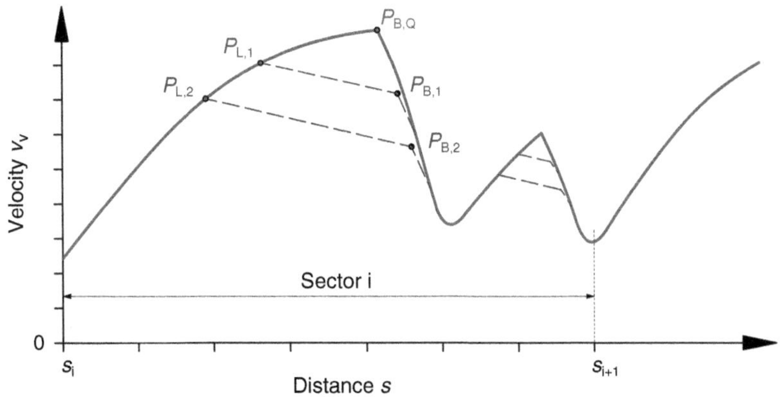

Fig. 2.42 Velocity traces formula E (schematic). The velocity curve over the route is shown. Sector i (section of the track from s_i to s_{i+1}) is examined in more detail. The red curve represents the speed during qualifying. Two variants with sailing and recuperation are shown in green and blue. The individual points are: P_L *lifting point* (start of lifting the accelerator pedal), P_B *braking point* (start of braking). The index Q stands for qualifying, 1 or 2 distinguishes the two energy target variants. In variant 2, sailing is started earlier and braking is started later. The energy expenditure in this sector is the lowest, but the running time is the largest (represented by the area under the curve – the larger, the faster)

vehicles (late turn-in), the drivers try to drive the shortest line towards the end of the competition in order to save energy.

In general, the ambient temperature influences the maximum amount of energy that the battery can absorb during recuperation. Thus, individual sectors are better suited for regenerative braking than others. If the ambient temperature is high, the temperature of the battery can become so high during charging that the vehicle can no longer complete the run. Shorter regenerative braking is more beneficial in that case. As mentioned, recuperation cannot yet be used at the beginning of the race if the battery state of charge is too high and the recovered energy cannot be absorbed. With increasing race duration, the proportion of sailing is reduced in favour of longer recuperation phases. Towards the end of the race, the temperature of the battery generally limits the amount of energy it can absorb. As a result, the lifting points (foot off the accelerator pedal) and braking points shift continuously. The driver is therefore informed of the ideal time by an acoustic signal from the race engineer via the on-board radio.

The potential lap time advantage of adaptive thermal management of the battery compared to a fixed strategy is illustrated in Fig. 2.43 for Formula E. The higher the ambient temperature, the more important a strategy becomes that adapts lap by lap to the decreasing performance of the battery. At 35 °C ambient temperature, an adaptive strategy can account for between 0.5 and 1.7 s time gain per lap.

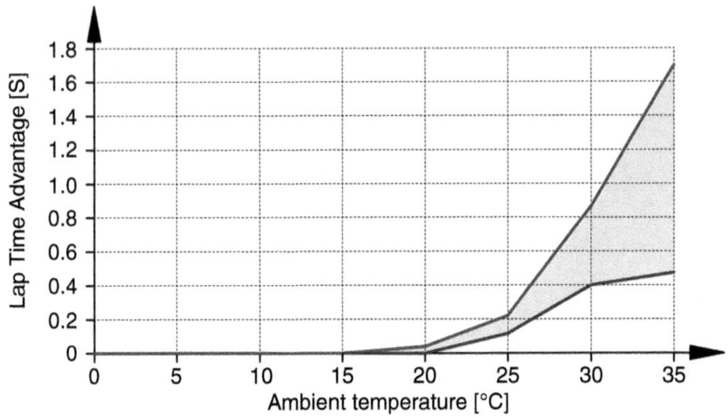

Fig. 2.43 Benefits of adaptive versus fixed battery thermal management in Formula E, after [30]. The advantage in time gain per lap increases with the ambient temperature. The upper limit is shown in red, the lower in blue

References

1. www.motorsport-guide.com. Accessed 29 Aug 2011
2. Knödel, U. et al.: Variantenvielfalt der Antriebskonzepte für Elektrofahrzeuge. ATZ **7/8**, 552–557. Vieweg, Wiesbaden (2011)
3. Bosch: Kraftfahrtechnisches Taschenbuch, interaktiv. CD Einzelplatzversion. Robert Bosch GmbH, Stuttgart (2003)
4. Opel, S., Chiocchetti, F.: Auslegung von Elektromotor und Getriebe für die Formel E. MTZ. **05**, 56–60 (2016)
5. Braess, H.-H., Seiffert, U. (eds.): Vieweg Handbuch Kraftfahrzeugtechnik, 4th edn. Vieweg, Wiesbaden (2005)
6. Mathoy, A.: Grundlagen für die Spezifikation von E-Antrieben. MTZ. **9**, 556–563 (2010)
7. Konstruktion, Zeitschrift für Produktentwicklung und Ingenieur-Werkstoffe, Heft 4, Springer-VDI, Düsseldorf. S. IW 12 bis IW 13 (2011)
8. Babiel, G.: Elektrische Antriebe in der Fahrzeugtechnik, 2nd edn. Vieweg+Teubner, Wiesbaden (2009)
9. Brosch P.F.: Der neue Energiesparer: IE4 mit Reluktanzmotor, in Konstruktion. Z Produktentwick Ingenieur-Werkst. **7/8**, 14–19. Springer-VDI, Düsseldorf (2011)
10. Hofmann, P.: Hybridfahrzeuge, 1st edn. Springer, Wien/New York (2010)
11. www.evo-electric.com. Accessed 10 Nov 2011
12. Sellschopp, K., Magyar, B., Oehler, M.: Energieeffizienz im Antriebsstrang bestmöglich ausschöpfen. Konstruktion. **07–08**, 46–52 (2018)
13. Sauer, D.: Optionen zur Speicherung elektrischer Energie in Energieversorgungssystemen mit regenerativer Stromerzeugung. RWTH Aachen, Aachen (2006)
14. Hohenberg, G.: Kann der intelligente Fahrer den Hybrid ersetzen? Hybridtechnik und Fahrereinfluss, Vortrag im Rahmen der ÖVK-Vortragsreihe (2009)

15. Fehrenbacher, C., et al.: Kühlung von Li-Ionen-Batteriemodulen, Vortrag im Rahmen der 2. Fachtagung des VDMA und der Universität Karlsruhe, Hybridantriebe für mobile Arbeitsmaschinen (2009)
16. Grebe, D.: GM's Voltec Antriebssystem – Elektrifizierung der Fahrzeuge auf hohem Niveau, in 21st International AVL Conference "Engine & Environment" (2009)
17. Neumeister, D., et al.: Integration of a lithium-ion battery into hybrid and electric vehicles. ATZautotechnology. **2**, 26–31 (2010)
18. Pickering C.: The long road to electric credibility. Racetech Mag, 66 f., Feb. (2010)
19. NN: Lithium-air chemistries to raise energiy density tenfold. Autom. Eng. **8**, 4 (2011)
20. Internet.: www.vdi.de/6390.0.html. Internetportal des VDI mit Bericht über Fachkonferenz Elektromobilität in Nürtingen vom 18.03.2010. Accessed 06 Jan 2012
21. Campling, D.: FiA Electric Rallycross – powering the future. Vortrag auf der Professional Motorsport World Expo, Köln, 08 Nov 2018
22. Linden, D., et al.: Handbook of Batteries. Part 1, Principles of Operation. Digital Engineering Library @ McGraw-Hill, New York (2004)
23. USABC electric vehicle battery test procedures manual revision 2, U.S. Advanced Battery Consortium, DOE/ID-10479, Rev. 2, Jan. 1996
24. Hofheinz, W., Sellner, H.: Elektrische Sicherheit bei der Ladung von Elektrofahrzeugen. Hanser automotive, Heft Okt. 2011, S. 76–79. Carl Hanser, München (2011)
25. ISO/FDIS 6469 Electrically propelled road vehicles – safety specification. Part 3: Protection of persons against electric shock. Mai (2011)
26. Potdevin, H.: Isolationsüberwachung in Hochvolt-Bordnetzen von Elektro- und Hybridfahrzeugen. ATZelektronik. **6**, 46–51 (2009)
27. Pfeffer, H. (ed.): Lenkungshandbuch, 1st edn. Vieweg+Teubner, Wiesbaden (2011)
28. Klausner, M. et al.: Technische Herausforderungen bei Lithium-Ionen-Traktionsbatterien und mögliche Lösungsansätze. In: Lenz (Hrsg.) Wiener Motorensymposium, VDI Reihe 12 Nr. 735. VDI, Düsseldorf (2011)
29. Heinrich P., Langer F.: Sicherheit für Elektrofahrzeuge, in Hanser automotive. Carl Hanser, München, S. 70 bis 72 (2011)
30. Catherall M.: Battery thermal management in formula E. Vortrag auf der professional motorsport world expo, Köln, 08 Nov 2018
31. Fritsche, K., Strecker, F., Huneke, M.: Simulationsbasierter Entwicklungsprozess für Rennsport und Serie. ATZ. **9**, 70–75 (2017)

3 Hybrid Drives

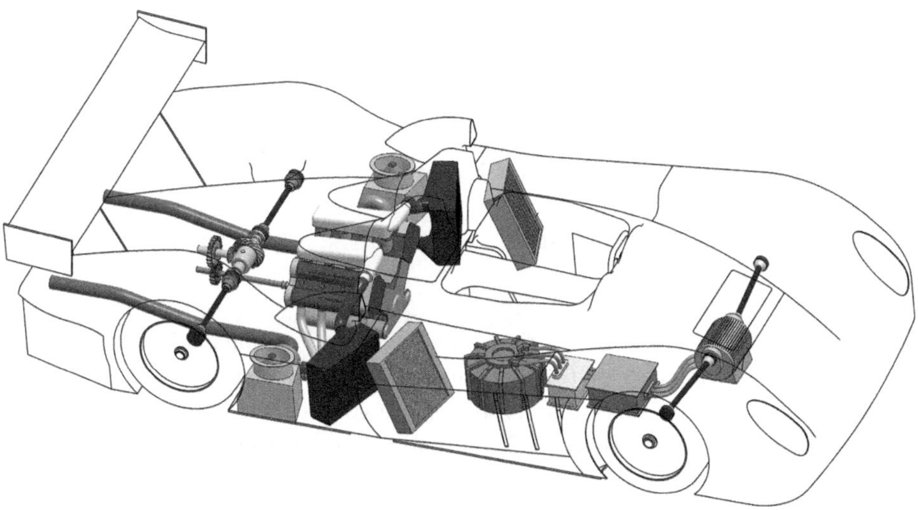

The proverbial link between classic drives with combustion engines and up-and-coming drives with electric motors is the hybrid drive, which – as the name suggests – is a mixture of both worlds.

M. Trzesniowski, *Powertrain*,
https://doi.org/10.1007/978-3-658-39885-9_3

3.1 Types of Hybrid Drives

To recover kinetic energy during braking, other energy conversion systems are also used in the vehicle in addition to the conventional combustion engine. If such a mixed arrangement of drive sources and energy storage systems is generally installed in a vehicle, it is referred to as a hybrid drive (hybrid, Latin for hermaphrodite), and the vehicle is called a hybrid vehicle or HEV (Hybrid Electric Vehicle). Thus, the combustion engine, the other motor (electric motor, hydraulic motor, . . .) or both together can be used for propulsion. The idea is to use the advantages of the individual drive sources in different driving conditions in such a way that they outweigh the obvious disadvantages of the increased technical and design effort.

If the systems involved play together skilfully, the following possibilities arise:

- Reduction of fuel consumption for a given power output, which is particularly important for long-distance competitions
- Increase in driving performance with unchanged fuel consumption.
- Acceleration support (boosting).

General advantages of hybrid systems are [1]:

- Low emissions up to locally emission-free driving
- Low-noise purely electric driving
- Increase of the functional comfort
- Ride Stabilization.

The basic possibilities of combining different drive sources are summarized in Fig. 3.1.

In the **serial hybrid,** there are two different drive sources on board, but the wheel drive is always electric. There is therefore no mechanical connection between the combustion engine and the drive wheels. This hybrid is therefore also called an “electric transmission”. The energy source for the electric motor is a battery and/or a generator driven by an internal combustion engine. For this application, the combustion engine is optimized in terms of efficiency and/or emissions to the much smaller operating range compared to driving operation (best-point instead of map operation). All components involved in the drive (combustion engine, generator, electric machine) must be designed for the maximum continuous speed. This type of hybrid therefore has the highest installed component power, which results in weight and cost disadvantages. The serial hybrid is therefore practically only found in commercial vehicles and in vehicles where the transmission of the drive power to the wheels would be mechanically too complex.

If the electric motor and the combustion engine can drive the wheels directly in parallel, this is called a **parallel hybrid**. This means that you can drive purely electrically, purely with an internal combustion engine or a combination of the two. For systems with electric motors with an output of less than 10 kW, the term “mild hybrid” has become established.

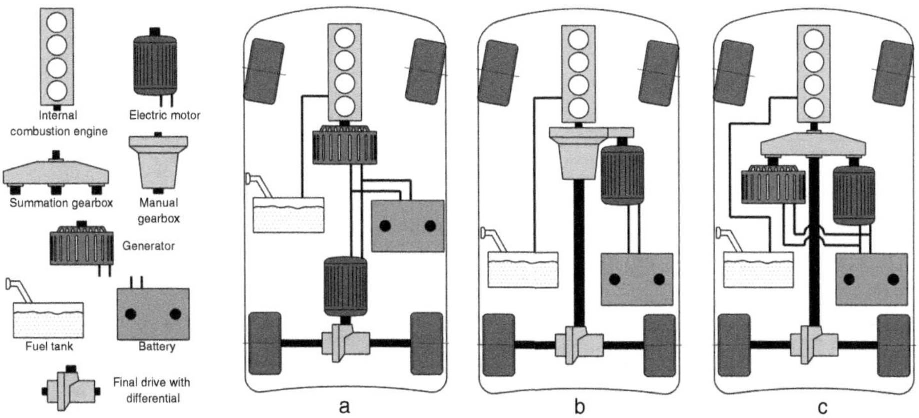

Fig. 3.1 Types of hybrid drives, according to [1]. (**a**) serial *hybrid*, (**b**) parallel hybrid, (**c**) mixed hybrid (*combined* hybrid)

With mild hybrids, purely electric driving is not possible, but start/stop of the combustion engine, energy recovery during braking and support of the combustion engine by an additional torque (boosting) are possible.

If the output of the electric machine is less than this (around 2–3 kW), it is referred to as a micro-hybrid. If the electric machine has an output of well over 15 kW, it is a full hybrid.

Mixed hybrids offer the most possibilities for combining the different drive systems. They represent a combination of parallel and serial power flow. This is why such systems are also called *power split hybrids*. With the high mechanical and electrical complexity, the main disadvantage of this combination is obvious. The control effort for a sensible interaction (*operating strategy*) of the installed systems depending on the driving condition, the predicted route, the battery condition, the thermal behaviour, etc. is also enormous.

For racing vehicles, in contrast to passenger cars, the course of the track is known exactly and exhaust emissions and battery life play a subordinate role, but simulation tools (see appendix) are still required to set up a profitable racing strategy. Here, the entire vehicle including the tires, the driver behavior and also the race track are simulated with a focus on the objectives (boosting, top speed, fuel consumption, laps between pit stops, ...) [2].

If the battery can also be recharged from the outside via the mains – i.e. through the socket – this is also referred to as a "plug-in hybrid" (socket hybrid).

In a hybrid, where purely electric driving is possible, the reverse gear in the transmission can be omitted. The reversal of the direction of rotation is represented electronically in a much more space-saving way with an electric motor with the motor control.

For this purpose, the electric machines require a disconnect clutch that decouples them from the active drive train when the limit speed is reached. Typical outputs of electric machines for hybrids are around 30 kW, which means that the specification of a desired starting torque results in a correspondingly low transmission ratio. This ratio means that the

Table 3.1 Possible functions of drive systems

Function	Hybrid				E-vehicle
	Serial	Parallel	Mixed	Plug-In	
Electric driving (pit lane, reverse)	•	•	•	•	•
Electric driving long distance	•				•
Boosting (torque support)		•	•	•	
Recuperation (energy recovery during braking)	•	•	•	•	•
Fuel consumption reduction	(•)	•	•	(•)	
Start/Stop	•	•	•	•	•
Charging on the mains				•	•

maximum speed of the electric machine is reached well before the maximum speed of the vehicle and must therefore be decoupled. With a connection ratio, i.e. a second gear, the disconnect clutch can be omitted [3].

Table 3.1 lists the most important functions of the different combinations of drive types. For comparison, the electric drive is also included in this consideration. Electric driving over longer distances would be possible with any hybrid drive, only the traction batteries and electric machines of parallel and mixed systems are designed for other functions. Strictly speaking, one can only speak of a reduction in fuel consumption (meant as energy saving) if a vehicle takes energy from outside only as fuel. A plug-in hybrid also takes in energy in the form of electricity. The vehicle consumes less fuel as a result of this alone. Whether the overall efficiency of the powertrain is higher is therefore not known.

Depending on the hybrid system, the total power available for propulsion is made up of the individual installed powers of the drive and storage components, Fig. 3.2. For this reason, the term "system power" is also used to characterise a hybrid vehicle, meaning the effective sum of the installed powers. The specification of the power of the combustion engine alone is not sufficient to represent the actual possible driving performance.

An example of the strategy for using the various systems is shown in Fig. 3.3. In this case, the operating mode selected depends on the torque required at the drive wheels and the speed of the vehicle. At low speeds and low torque requirements, the vehicle is driven purely electrically, e.g. in the pit lane. At higher drive torques, only the combustion engine drives the wheels, for example when accelerating out of slow corners. If the maximum available torque is needed, the electric motor also contributes and boosts, e.g. when accelerating on the straight and overtaking. If a negative torque is needed (braking), recuperation can be used. At high speeds, however, even a low torque leads to high braking power and thus the absorption limit of the battery and/or the electric machine is reached. If more braking is required, the remaining braking torque must be applied "as usual" by the friction brake.

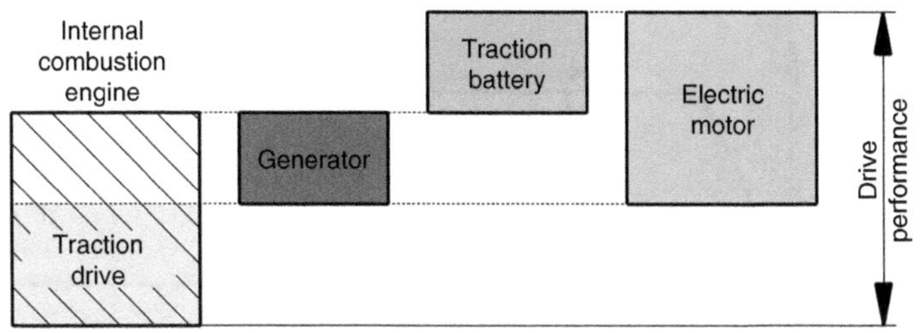

Fig. 3.2 Power addition for hybrid drives (system power). Mixed hybrid systems allow the installed power to be added together and can thus combine the vehicle's drive power from the combustion engine plus that of the electric motor. For the sake of clarity, the losses between the components are not shown in this diagram

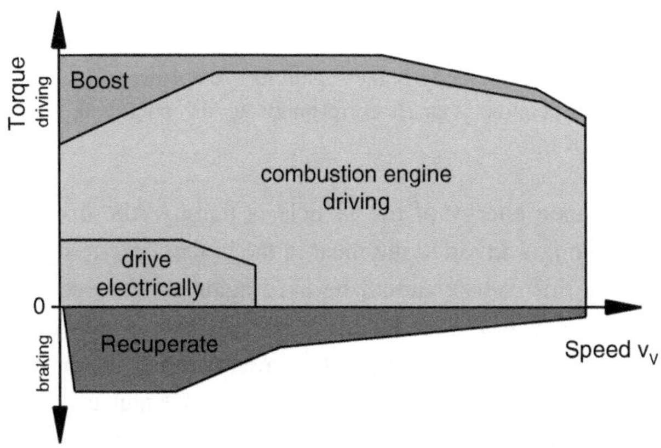

Fig. 3.3 Example of the interaction of individual operating modes depending on the travel speed and the required torque [4]

3.2 Energy Recovery: *Kinetic Energy Recovery System (KERS)*

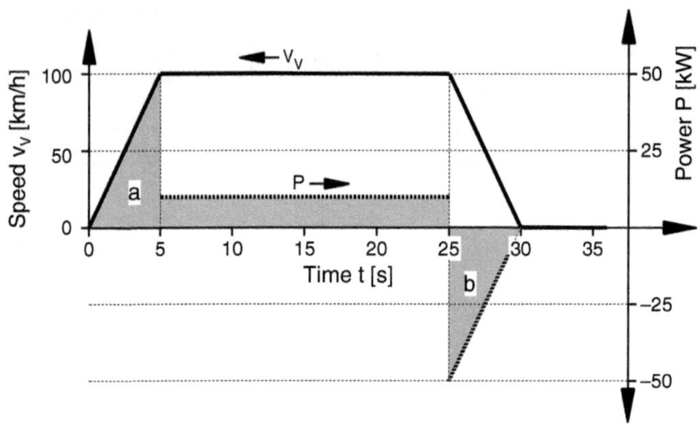

Fig. 3.4 Energy demand during driving. In the power/time diagram, the area under the curve corresponds to an energy [W s]. A vehicle accelerates from a standstill to 100 km/h in 5 s, then continues at this speed and subsequently brakes again to a standstill in 5 s. It can be seen that the energy for acceleration (triangular area *a*) corresponds to the energy dissipated during braking (triangle *b*)

When braking, the kinetic energy of the vehicle is mainly (the driving resistances also contribute to deceleration)[1] converted into heat in the brakes and dissipated to the environment. This means that this energy cannot be used again by the vehicle itself. In racing, however, the braking process is immediately followed by an acceleration phase. This is a phase in which more energy must be supplied in order to bring the car back up to a higher speed level. A similar situation can be observed with passenger cars, especially in urban areas, but braking is also necessary from time to time on the motorway. Figure 3.4 illustrates this situation.

Therefore, the idea of storing braking energy in the vehicle and feeding it back into the drive system when required (regenerative braking, recuperation) is an obvious one (especially in view of dwindling resources). Such energy recovery systems (ERS) are permitted in some racing classes, such as Formula 1 and LMP. The general structure of such a system is described in Fig. 3.5.

During braking, the kinetic energy of the vehicle coming from the wheels is directed into an energy accumulator (3). When accelerating, part of the energy from the accumulator is fed back to the wheels. This energy therefore does not have to be supplied (only) by the combustion engine. This sub-function is therefore a hybrid drive. A control unit (7) ensures that the power flow is directed in the desired direction. When braking, the corresponding energy converter (1) is switched into the drive train (clutch 8) and, if necessary, the combustion engine is disengaged (10) depending on the speed. If the stored energy is to be retrieved, the energy converter for propulsion (4) is switched into the drive train and the

[1]Cf. Racing Car Technology Manual, Vol. 4 Suspension System, Fig. 6.6.

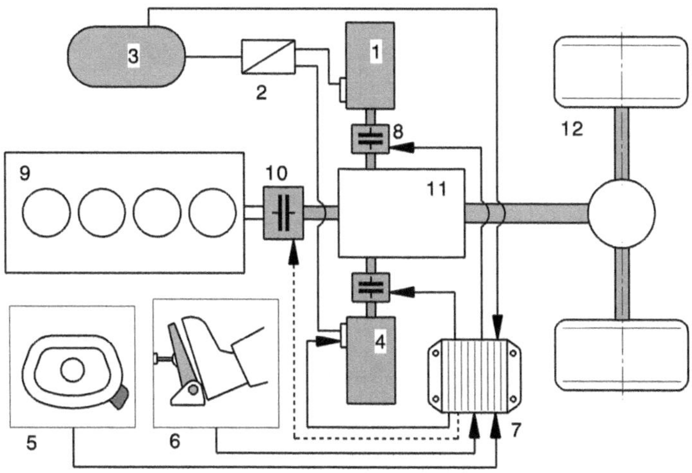

Fig. 3.5 Principle structure of a brake energy recovery system. 1 Energy converter braking, 2 Energy converter to storage, 3 Energy storage, 4 Energy converter drive, 5 Switch at steering wheel, 6 Brake or accelerator pedal, 7 Control unit, 8 Disengage unit, 9 Combustion engine, 10 Disengage unit (clutch), 11 Transmission, 12 Driving wheels

braking energy converter (1) is disengaged. The switching possibilities of the system are manifold and range from a pure amplification of the engine brake to a fully-fledged service brake. Individual elements of this system can also perform multiple tasks: An electric machine can both act as a generator to convert braking energy to electrical, and provide electrical energy as mechanical in engine operation. In Formula 1, these machines are referred to as MGU, *motor-generator units*. Depending on whether they convert energy from driving (*kinetic*) or exhaust (*heat*), they are called MGU-K or MGU-H. The energy converter (2) to the accumulator can also be omitted if both energy converters for driving and braking can be operated directly with the energy form of the accumulator. In a purely hydraulic or pneumatic system, for example, this would be the case. In an electric system, on the other hand, a generator supplies alternating voltage during braking, which must first be processed in a rectifier for the battery as direct voltage.

Depending on the application, brake energy regeneration offers the following advantages:

- Fuel saving
- Start/stop function
- Short-term increase in engine power (boosting) when overtaking or downshifting
- Protection of the service brake
- Elimination of reverse gear (shunting with the stored energy and the auxiliary drive).

However, the accompanying disadvantages should not be ignored:

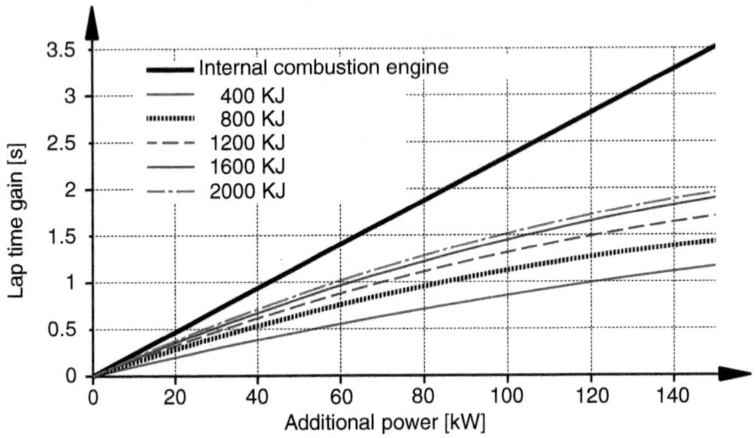

Fig. 3.6 Lap time improvement by additional power, after [2]. The simulation shows possible lap time savings as a function of the additional power and the stored energy in kJ for Formula 3 cars on the race track at Spa

- Installation space requirement of the system
- additional weight
- additional operating effort for the driver or electronic control effort in the case of an automatic control system
- Energy storage is safety critical (accident, aging)
- additional costs.

The additional power of the second drive source increases the driving performance. A simulation shows the lap time improvements that can be achieved, Fig. 3.6. With increasing additional power, a lap can be completed faster. The longer this additional power can be called up, i.e. the more energy has been stored per lap, the greater the effect. In 2009, the Formula 1 regulations allowed energy recovery (KERS) for the first time. 400 kJ per lap was allowed to be stored and the maximum additional power was limited to 60 kW. The resulting time to fully charge or empty the accumulator is 6.7 s (time = energy by power). Le Mans Prototype (LMP) cars are allowed to provide unlimited additional power, but energy is limited to 500 kJ between two braking maneuvers. LMP cars are also allowed to provide recovered energy to all wheels, whereas in Formula 1 this is only allowed on the rear axle. For an illustration of Formula 1 power units as they have been in use since 2014, see Chap. 1 *Combustion Engines,* para. 1.7.

Even though the principle of energy recovery is always the same, there are numerous different possibilities of implementation. Table 3.2 gives an overview of some possible solutions in the form of a morphological box.

By combining the individual solutions in a practical way, different systems can be created. The stored energy can basically be used for any power requirement in the vehicle:

Table 3.2 Possible solutions for the main functions in energy recovery (morphological box)

Function	Solution				
Energy converter Brake	Generator	Pneumatic pump	Hydraulic pump	Gearbox	
Energy storage	Battery	Capacitor (Supercap)	Pressure vessel	Flywheel	Spring
Energy converter drive	Electric motor	Air motor	Oil motor (hydraulic motor)	Gearbox	

To support the engine during acceleration (boosting, shift-push), to drive the ancillary units, to start the combustion engine and to start it. Of course, aligning the system to one of these possibilities changes the requirements for the individual function carriers. For starting, for example, the energy storage system must provide far less energy than for boosting. Conversely, the energy storage system must also be capable of absorbing the braking energy produced. With regard to recuperation, which is also cited as a significant advantage in hybrid passenger car applications, it should be noted that current battery systems are only capable of absorbing a small percentage of the braking power produced [5]. To get an idea of the magnitude of the storage requirement, a numerical value estimate can assist. If a 640 kg Formula 1 car brakes from 300 to 100 km/h within 3 s, the brakes must absorb about 658 kW.[2] However, the Formula 1 regulations currently allow a power feedback of only 60 kW with a maximum energy release of 400 kJ (at the rear wheels) per lap. So the (conventional) brake is still needed in this case, because the permitted energy flow (= power) is much smaller than the braking energy produced. Further on, for this example basically a storage tank with 400 kJ capacity is sufficient. Below a driving speed of 100 km/h, the stored energy cannot be returned in Formula 1 vehicles, whose driving performance is limited by the tyre traction. Figure 3.7 illustrates this concretely with a lap of the Monaco Grand Prix. During acceleration (as soon as this is no longer limited by tyre traction, the KERS helps out), the energy content of the accumulator decreases. When braking, the speed of the flywheel increases and its energy increases again accordingly.

The reason for the lap time advantage of a KERS-supported vehicle over a conventional one only becomes apparent on closer inspection. Figure 3.8 therefore presents an enlarged section of the previous image for both vehicle variants. The hybrid vehicle achieves the higher acceleration at higher speeds. As soon as propulsion is no longer limited by the grip of the tyres, the KERS system supports the combustion engine. The hybrid car also proves to have an advantage during the subsequent braking process. The energy absorption of the accumulator increases the initial braking deceleration and thus allows the driver to start braking a little later. This is also reflected in a higher final speed on the straight.

[2]Cf. Racing Car Technology Manual, Vol. 4 Suspension System, Chap. 6 *Physical Principles*.

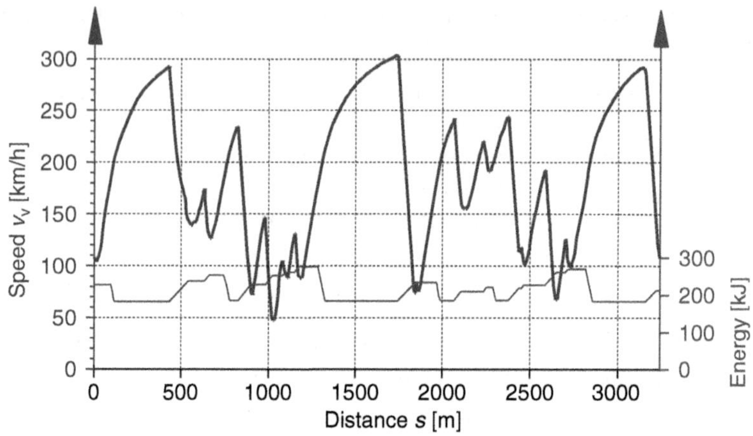

Fig. 3.7 Course of the energy of a flywheel accumulator (green curve. Flybrid system) during one lap of the Monaco Grand Prix (simulation), after [6]. In addition, the course of the vehicle speed v_V (blue) is plotted

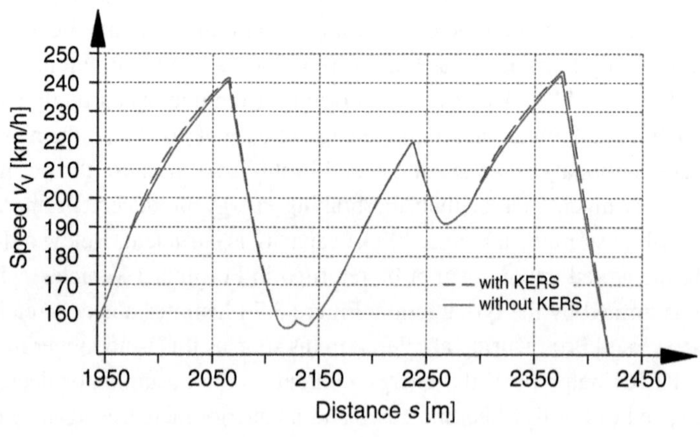

Fig. 3.8 Detail of Fig. 3.7 for a vehicle with and without KERS, after [6]

***Regenerative* Braking (Recuperation)**

The recovery of kinetic energy during braking is one of the main advantages of a hybrid drive in racing vehicles. The energy otherwise converted into heat by the friction brake would be lost for the subsequent acceleration process, Fig. 3.9. Of the energy chemically bound in the fuel, only about 27% reaches the crankshaft output of the combustion engine. The majority is lost as combustion heat and through friction. Other sources of loss on the way to the actual propulsion of the vehicle are the clutch (when shifting gears), the transmission, the drive shafts, the wheel bearings and the wheel brakes, which have a certain residual torque even when at rest. This leaves about 24.5% of the energy to overcome the resistances (rolling, air and acceleration resistance). If the vehicle continues

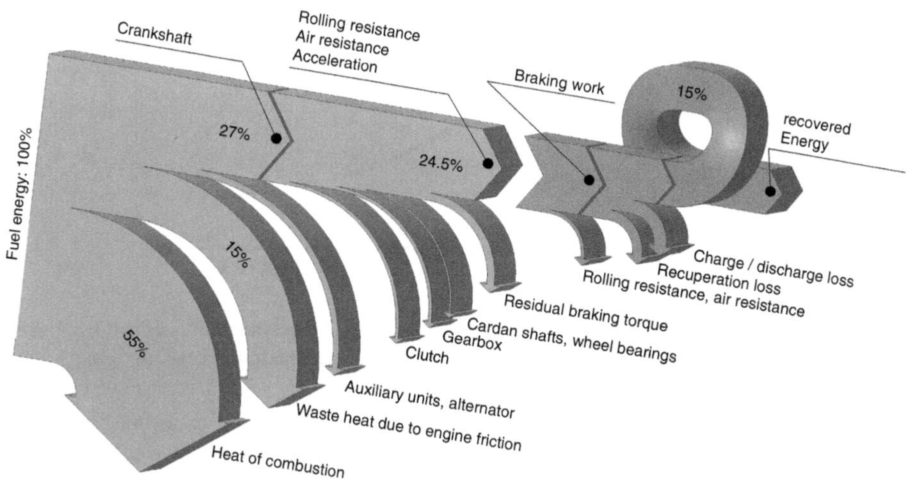

Fig. 3.9 Energy flow during recuperation. Based on the fuel energy, about 24.5% remains to overcome the driving resistance (kinetic energy). When braking, about 61% of this energy, i.e. 15% in absolute terms, can be recovered

without acceleration, only air and rolling resistance need to be balanced on level ground and less energy is required. If the excess energy is used, the car accelerates and its kinetic energy is increased accordingly. During the subsequent braking process, the braking work (this is the kinetic energy minus the driving resistance,[3] i.e. approx. 23% fuel energy) can now be used. Even during recuperation, the energy is not fed to the accumulator without loss: The engine drag (motoring) torque has an additional effect on the drive wheels without additional measures (e.g. separating clutch) and drive shafts, converters, etc. do not operate without loss. Depending on the energy converter chain, around 85% is fed to the accumulator. This in turn has a charge/discharge efficiency of around 0.77, so that ultimately 61% of the kinetic energy, i.e. 15% of the fuel energy, is available again for acceleration. During acceleration, however, a part of the efficiency chain from the drive element to the wheels must be passed through again. In the end, this results in a fuel saving of approx. 10%.

Recuperation thus makes a significant contribution to reducing fuel consumption and overall energy consumption and is therefore also used in purely electric vehicles. However, recuperation cannot completely replace the friction brake. A mechanical brake is required solely for holding the vehicle at a standstill or at the lowest speeds. Depending on the type of energy converter used for braking, further limitations arise. Figure 3.10 shows an example of the limitations of an electric machine in regenerative operation. At high speeds, which corresponds to a high vehicle speed, the maximum generator power sets the upper

[3] Cf. Racing Car Technology Manual, Vol. 4 *Suspension System*, Fig. 6.6.

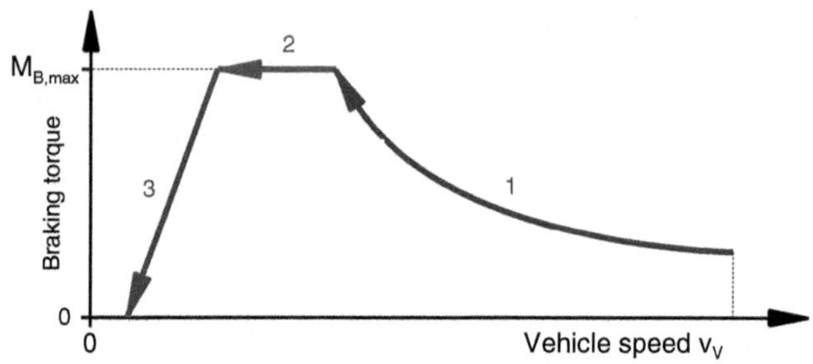

Fig. 3.10 Limits of generator operation. The operating limits of a generator are defined by: **1** *generator* power, **2 generator** current, **3 generator** efficiency. $M_{B,\max}$ Maximum braking torque

limit (1). The maximum torque of the motor or generator cannot be exceeded in any case (2). At low speeds, the braking effect decreases sharply with falling speed and finally comes to a complete standstill (3).

A further limitation arises when the energy storage is full. It cannot be charged any further and braking must be carried out with the friction brake in any case.

The maximum decelerations that can be achieved at all by an energy converter in a vehicle under consideration follow from the braking power:[4]

$$a_{X,\max} = \frac{P_{B,\max}}{m_{V,t} \cdot v_V} \tag{3.1}$$

$a_{X,\max}$	Maximum braking deceleration, m/s^2
$P_{B,\max}$	Maximum instantaneous value of the braking power, W. Here max. Power of the energy converter (generator, pump, ...).
$m_{V,t}$	Total mass of the vehicle, kg
v_V	Vehicle speed, m/s

Figure 3.11 shows results for different outputs of a generator above the vehicle speed. In addition, at lower speeds, the limitations of the electric machine explained in Fig. 3.10) are added. It can be seen that for this vehicle (total mass 1500 kg) the energy converter for regenerative braking should have at least 50 kW so that noticeable deceleration without friction braking occurs at higher speeds. At lower powers, the friction brake must be involved in any case, so that the braking distances remain short.

Due to the fact that recuperation cannot be applied for any length of time or to any extent, a strategy must be established as to which components apply the required braking torque. One possibility is to leave it up to the driver to decide whether to activate the service

[4]Cf. Racing Car Technology Manual, Vol. 4 Suspension System, Chap. 6 *Brake System*, (6.10).

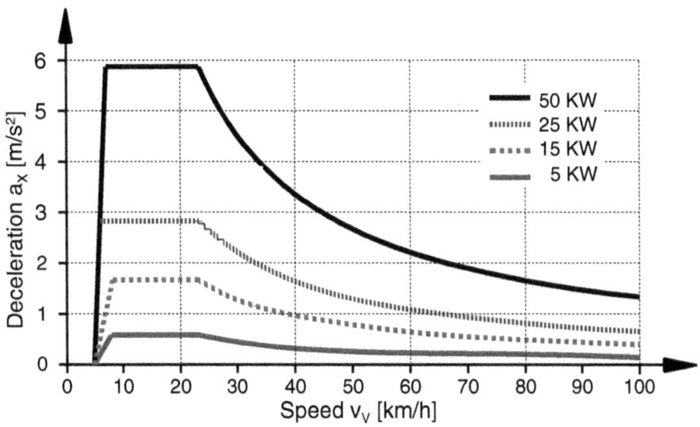

Fig. 3.11 Possible vehicle decelerations with a generator [7]. Depending on the generator power and driving speed, the maximum values of the achievable deceleration a_X change during regenerative braking. In general, the deceleration effect of the electric machine increases with decreasing speed. At standstill, no braking torque can be called up regeneratively. Total vehicle mass $m_{V,t} = 1500$ kg

brake and/or the recuperation element. This, of course, places an additional burden on the driver. If a control unit takes over the distribution of the brake components, the driver is relieved and can concentrate on other things. Some regulations require that only the brake pedal be used for all deceleration. In the case of passenger cars, this is the only solution the customer is expected to use anyway, so he has no choice at all as to which devices he uses to slow down his vehicle.

For this variant, that the driver requests a braking torque solely by depressing the brake pedal, there are two basic ways of splitting the braking torque. If all braking systems act simultaneously, this is referred to as parallel recuperation, Fig. 3.12a. The shares of the individual systems are divided according to a certain ratio. The higher the requested braking torque, the larger the share of the friction brakes. The distribution of the friction brakes on the front and rear axles is not arbitrary, but depends on the height of the centre of gravity and the deceleration.[5]

With serial recuperation, energy is fed back as long as this is possible, Fig. 3.12b. Only when the braking torque requested by the driver exceeds the generator output do the wheel brakes come into play. In this case, the brakes on the non-driven axle are activated first and only subsequently those on the driven axle when the recuperation element acts via the latter in the event of even greater braking requirements.

If the combustion engine is not disengaged, its drag torque also has a decelerating effect and thus reduces the maximum possible energy that can be fed back.

The challenge with both strategies is that the driver establishes a reproducible braking behaviour and obtains a constant deceleration when the pedal is held down. Particularly for

[5]Cf. Racing Car Technology Manual, Vol. 4 Suspension System, Chap. 6 *Brake force distribution.*

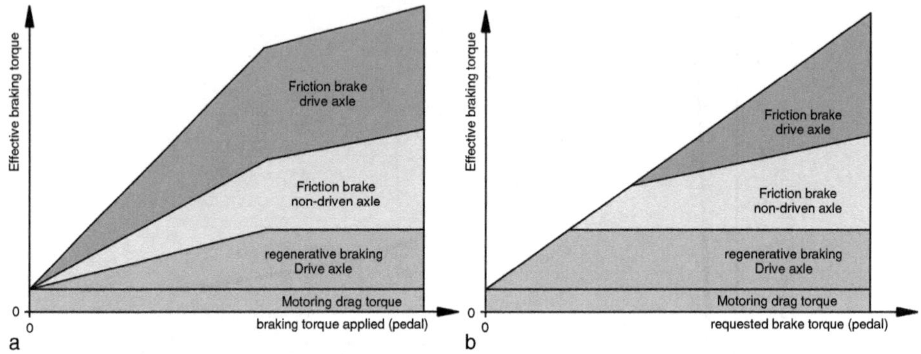

Fig. 3.12 Comparison of two regeneration strategies [4]. (**a**) parallel recuperation (*superposing regenerative braking*), (**b**) serial recuperation (*serial regenerative braking*). The driver requests a torque via the brake pedal which, depending on the strategy, is represented by the control unit as a friction brake and/or generator brake

the transition from generator to friction brake (so-called blending), precautions must also be taken with the conventional friction brake. The brake pedal must be partially or completely decoupled from the master cylinders of the hydraulic brake. The activation of the master cylinders and the generator is coordinated by a control unit depending on the pedal position (= driver's request), the driving speed, the state of charge of the energy accumulator, etc. An additional actuator must apply a counterforce to the brake pedal so that the driver feels the same sensation as with the pure friction brake.

Figure 3.13 illustrates how the interaction of generator and friction brake can look like during constant deceleration. At high speeds, the friction brake must initially take over the main part of the effective braking torque. At decreasing speeds and when the braking torque is below the maximum torque of the generator, the regenerative torque alone is sufficient for deceleration for a certain range. Only at low speeds does the friction brake become necessary again in addition (except in the case of a synchronous machine – it also provides torque at standstill). This is all the more the case the slower the vehicle becomes. Only the friction brake can be used to stop the vehicle.

Figure 3.14 shows how such a process proceeds over time, i.e. with decreasing speed. It shows the time characteristics of the braking torque on the front axle and the electrical machines, which also act on the front wheels in this vehicle (cf. Fig. 3.15). Furthermore, the braking work is shown, which on the one hand is converted into heat by the friction brake and on the other hand is supplied to the accumulator as electrical energy in regenerative operation. This part can be used again for acceleration if required.

The fact that the friction brake is partially or completely relieved in a hybrid vehicle is of course also noticeable in terms of wear. For example, brake pad wear on the front axle of the Porsche 911 GT3 R Hybrid is reduced by approximately 60% during the 24-hour race at the Nürburgring. The front brake discs also benefit from a reduction in wear of around 44% [8].

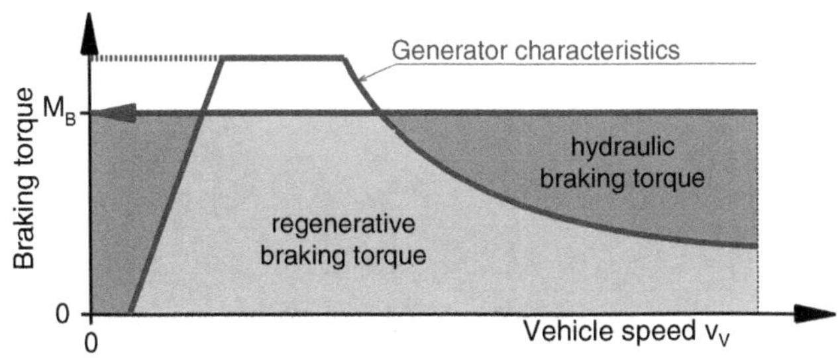

Fig. 3.13 Constant deceleration with cooperative regenerative braking system [7]. For a constant braking torque M_B the ratio between hydraulic and regenerative braking torque must be changed depending on the speed, because the generator characteristic (cf. Fig. 3.10) does not allow a constant torque. The dashed green curve applies to synchronous machines

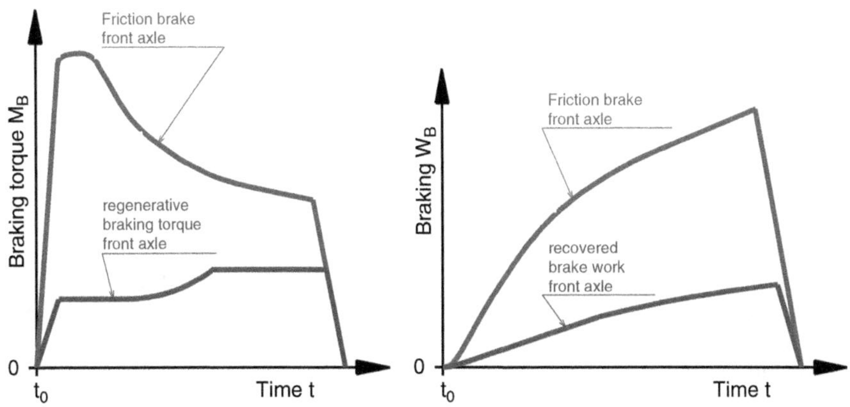

Fig. 3.14 Temporal course of the braking torques (left) and the braking work (right) at the front axle (Porsche 911 GT3 RH) [8]. At the beginning of the braking manoeuvre, the friction brake does the lion's share on the front axle. After some time, i.e. when the speed has decreased, the generators have a stronger effect and the use of the friction brake can be reduced. Generator power: 2 × 75 kW. Race weight vehicle: 1350 kg

In this vehicle, the brake pedal is partially decoupled from the hydraulic circuit of the friction brake. The influence on the friction brake required for recuperation is carried out in this case by reducing the brake pressure on the front axle, Fig. 3.16. The two electric machines also act on the front axle, which apply a proportion of the braking torque during recuperation, and the effect of the friction brake must therefore be reduced by precisely this contribution.

The driver should not feel any of this, i.e. the pedal feel must not change during regenerative braking, even if the pressure in the brake calipers is reduced. Figure 3.17 shows the systematic structure of the associated braking system. The pressure applied by the driver via the brake pedal (4) in the brake line to the front axle is reduced depending on

Fig. 3.15 System overview Porsche 911 GT3 R Hybrid [Porsche Motorsport]. 1 Portal axle with two electrical machines, 2 Power electronics for *el.* machines, 3 Power electronics for flywheel, 4 Electric flywheel storage

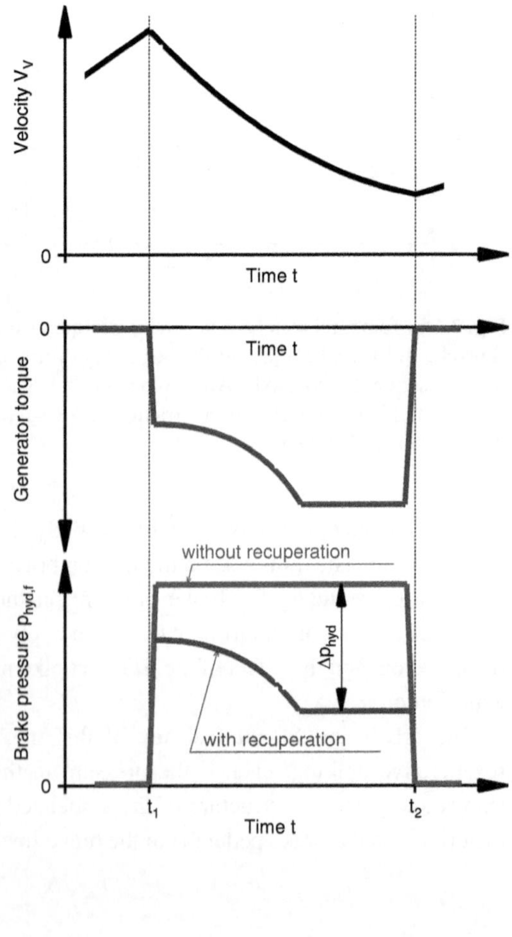

Fig. 3.16 Brake pressure reduction during recuperation (Porsche 911 GT3 RH) [8]. Δp_{hyd} Reduction of the brake pressure on the front axle during recuperation. A braking manoeuvre from the time t_1 to t_2 is shown. If recuperation is also performed with an electric machine, the brake pressure on the front axle must be reduced by the amount Δp_{hyd}, because the friction brake is relieved of the generator's contribution

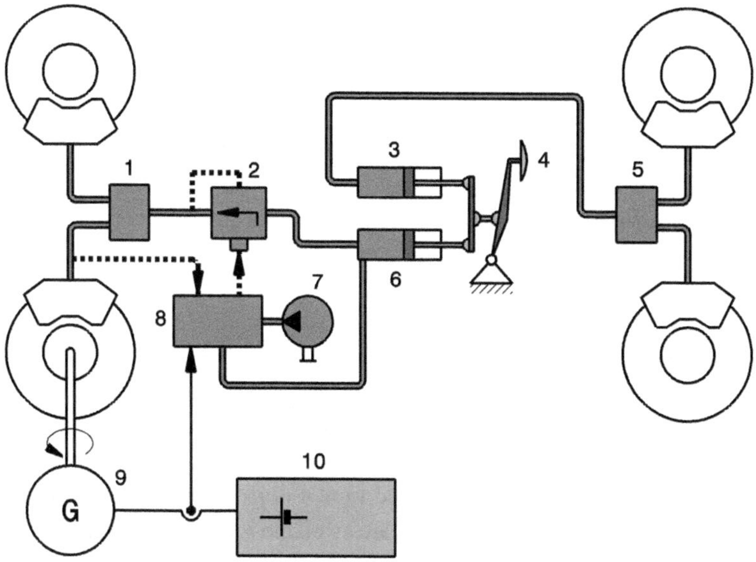

Fig. 3.17 Schematic of a braking system with brake pressure compensation, according to [8]. 1 ABS unit front axle, 2 Pressure reducing valve, 3 Master cylinder rear axle, 4 Brake pedal, 5 ABS unit rear axle, 6 Master cylinder front *axle*, 7 Hydraulic pump, 8 Valve block *(valve pack)*, 9 Electric machine, 10 *energy storage system*

the braking work performed by the electric machine (9). The pressure is reduced by a reducing valve (2) which is controlled by the hydraulic unit (8). The hydraulic unit processes the pressure in the brake line of the front axle, the pressure in the master cylinder and the braking torque of the generator. The generator torque is converted by a hydraulic pump (7) into an appropriate pressure that maintains pedal feel during recuperation. If the energy accumulator (10) becomes full during a braking manoeuvre, the friction brake alone must continue to decelerate the vehicle, the reducing valve (2) passes the full brake cylinder pressure to the front axle and the hydraulic pump compensates for the amount of brake fluid previously controlled.

In contrast to road vehicles, the use of a brake energy recovery system (KERS) in racing cars is not self-evidently advantageous. On the one hand, an additional mass of about 25–30 kg must be accommodated for Formula 1 applications. Now there are already additional masses, but the situation is different because they can be mounted in any desired position (e.g. near the front axle on the underbody). The KERS elements have a function-driven form and, depending on the design, must act on the drive train (gearbox, final drive) or on the crankshaft. This changes the vehicle's centre of gravity in height and longitudinal direction. This in turn influences the static axle load and the axle load shift during acceleration or braking. Furthermore, the accumulator must be charged during braking, which means that the service brake may only be partially applied. The correct dosage of both braking systems poses a great challenge to the driver and/or to the control system of the KERS if it is to be avoided that a KERS-less vehicle decelerates faster, i.e. brakes later. Furthermore, KERS only brakes the drive wheels, so these are usually the rear ones, which are more critical for driving stability. The additional power for overtaking at the push of a button is much easier to control, but – as Formula 1 practice shows – it can only exploit its advantage on long straights. The higher speed achieved by boosting also forces the driver of the vehicle that is the same (mass, downforce) apart from KERS to brake earlier. The benefit of KERS for racing cars therefore depends on the balance of the vehicle being as undisturbed as possible and on the control unit of the system or the skill of the driver.

3.3 Energy Storage

The energy required for driving and the energy recovered through braking must be stored so that they can be used when needed. This is the basic task of the energy storage device. An object stores energy if the work supplied to that object can be recovered in the same form [9]. In principle, a number of solutions are available for this, which can be enumerated systematically according to the type of energy, Fig. 3.18.

All of them can not be used in vehicles without restrictions. There are some additional considerations and limitations. A suitable accumulator must not only be light with a large power output, but should also be able to absorb as much energy as possible in dynamic operation, Fig. 3.19. However, the accumulator must not only absorb certain amounts of energy in the short term (approx. 0.5–3 s), it must also discharge sufficient power again (= energy per time). The discharge time follows directly from the specific values for energy and power:

$$t_2 = \frac{E_{spec}}{P_{spec}} \tag{3.2}$$

t_2	Discharge time, s
E_{spec}	Energy density, W/kg
P_{spec}	Power density, W s/kg

To select the most suitable energy storage device for a racing vehicle, the following criteria are generally considered:

- Installation space
- Cooling requirements
- Complexity (control quality in dynamic operation)
- Freedom regarding the operating strategy
- Robustness
- Handling

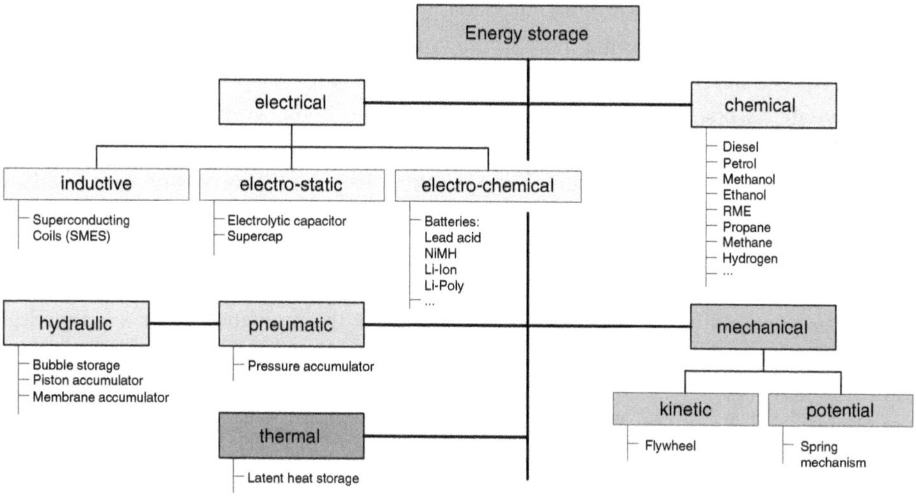

Fig. 3.18 Types of energy storage, after [4]

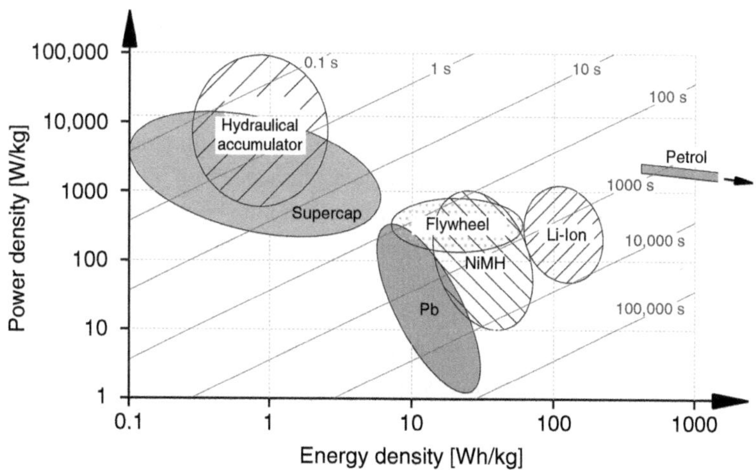

Fig. 3.19 Properties of storage systems, after [5]. The storable power and energy are related to the mass of the storage unit. Hydraulic accumulators and supercaps have the highest power densities, but can only store relatively little energy. Batteries can store much more energy, but need more time to release the energy (power = energy per time). For comparison, gasoline is also entered. Its energy density reaches far above 1000 W h/kg. In addition, the lines of equal discharge time (isochrones) are entered

- Hazard potential
- Series relevance
- Costs

The greater the installation space requirement, the more difficult it is to accommodate the storage unit in the vehicle or the more limited the place where it can be placed at all. The cooling requirement influences the aerodynamics of the car. A large heat exchanger surface area inevitably worsens the streamlined design of the body work. The less waste heat generated, the better it is for the rest of the car and neighboring parts. A reservoir is not merely charged and discharged somehow, but there is generally very little time available for this. Even at Le Mans, braking phases are hardly longer than 5 seconds. In that time, the energy that is fed back during recuperation has to be in the accumulator. Conversely, an acceleration boost is only helpful if enough power is released in a short time. In this case, the energy flow and its amount per lap are limited by the regulations. For a competitive advantage, a strategy must be devised so that the balance of expended energy (fuel, storage) and recovered energy results in a lap time gain. A storage system that is easier to control and allows greater freedom for different operating strategies (pace car phase, qualifying, rain, defect, . . .) proves to be more advantageous in this respect. In racing, large, irregular forces and accelerations act on the vehicle, so the systems on board also have to be robust enough to endure these conditions for at least a race distance. Storage systems are often swapped and teams also carry several with them as a stock at a competition. Storage systems, whose transport and handling is complicated by numerous legal requirements, require more travel time simply because of the increased expense involved. Especially in the case of cross-border transports as well as air freight, the necessary permits from individual authorities are added. And this applies in both directions, i.e. to the place of competition and back to the factory. But it is not only the handling that has to be taken into account, but also the potential danger. Accidents can happen on the track and the storage can be deformed or destroyed. The stored energy can thus be released in an uncontrolled manner or toxic substances can escape. Under no circumstances should this endanger people or the environment. For vehicle manufacturers who can afford a racing team, series relevance is also a criterion. Although the storage technology may be a purely individual solution today, it could be used in series production tomorrow – if the boundary conditions (production, recyclability, service life, safety, costs, marketing) are right. This is exactly what these teams will be looking for when choosing a storage device.

The desired properties depend on the intended use. To support the combustion engine during overtaking (overtaking push), a relatively small power (kW) but a large amount of energy (kW s = kJ) is required for a short time. Flywheels and Li-ion batteries are suitable for this purpose. The situation is different when avoiding a drop in power when shifting back (shift-push). In this case, high power at low energy is required [10]. A typical application of a supercapacitor.

3.3.1 Battery

These storage types are described in detail in Sect. 2.4. At this point, it should only be mentioned that a storage system for KERS is oriented towards partially different criteria than for electric drives. A KERS storage system must be able to absorb and release energy quickly. Its power [W] or power density [W/kg] should therefore be high and the charging/discharging time short. The power densities of batteries specially developed for KERS are nowadays around 6000 W/kg [2].

The absolute amount of energy that it should or may be able to store (regulations), on the other hand, is relatively small. For traction batteries, it is precisely the storage capacity that is an essential criterion, because the range depends directly on it.

3.3.2 Capacitors

Capacitors are gaining importance as energy storage devices for hybrid vehicles. Their power density is high (over 10,000 W/kg for supercaps) and their discharge time extremely short. This therefore represents exactly the behavior needed for brake energy storage and boost functions. Recent developments have led to an enormous increase in capacity, with a much lower internal resistance than batteries have. This allows large currents to flow, which in turn enables large electrical machine outputs or high torques. The type of construction that makes these high capacities possible is the electrolytic double-layer capacitor (*supercapacitor* or *supercap* for short). The construction schematically and technically is shown in Fig. 3.20. The electrodes are represented by a layer of aluminium (1) on which semiconducting, porous carbon (2) is deposited. The structure of the carbon is such that the effective surface area is increased many times (3000 m^2/g) [11]. A separator (5) separates the electrodes, but is permeable to charged particles (= ions, 3), which are in an electrolyte (4). This arrangement results in two capacitor layers connected in series (Fig. 3.20a, top). The resulting capacitance is not determined by the thickness of the ion conductor, but solely by the thickness of the charge layer. This is in the range of 1–2 nm for this type of construction and explains the high capacitance.

In the uncharged state, the ions are evenly distributed in the electolyte. If a voltage is applied, the positive ions migrate to the negative electrode and the negatively charged particles to the positive electrode. The energy is ultimately stored electrostatically by the polarization of the electrolyte. Thus, no chemical reaction takes place. Supercaps have a high cycle stability and a long calendar life (see Sect. 2.4).

Such supercapacitors are usually designed as cylindrical windings by rolling up the electrode foil and an insulating foil (Fig. 3.20b, c). In this way, large capacitor areas with correspondingly large capacitances are created in the smallest possible space.

Unlike batteries, charging or discharging a capacitor results in a large change in voltage. In addition, a certain threshold voltage (2.7 V) must not be exceeded for double-layer capacitors so that no chemical reaction takes place. For this reason, capacitors in vehicles are generally not connected directly to DC voltage sources, but a DC chopper keeps the supply voltage between two limits. Where the upper value is chosen so that the possible capacitor voltage is fully utilized [4]. The energy stored in the capacitor follows from:

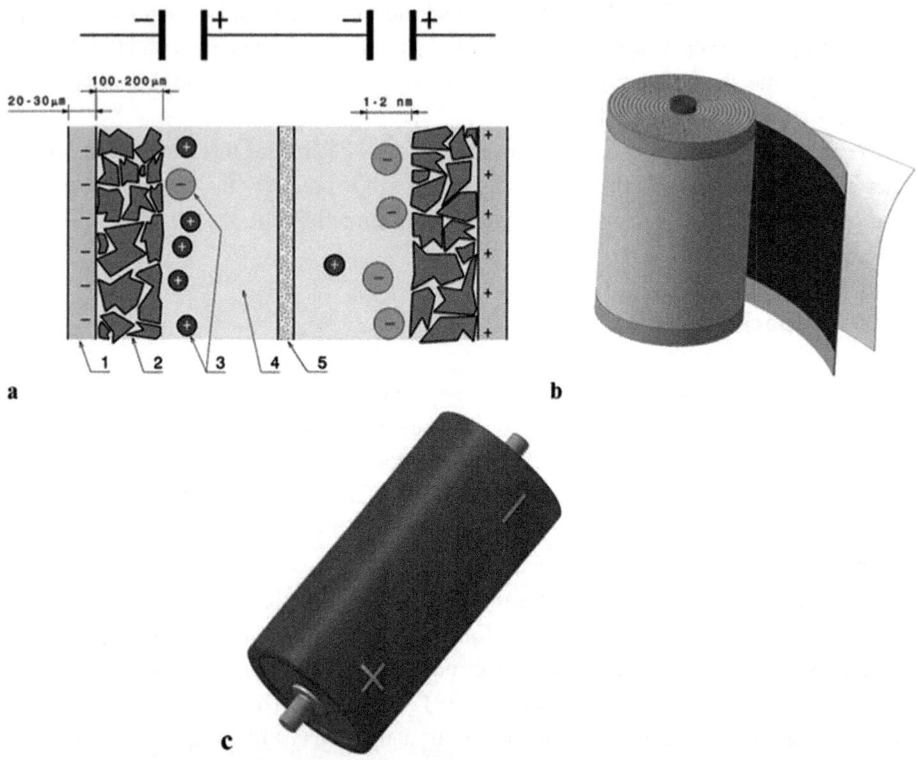

Fig. 3.20 *electrolytic double layer capacitor (EDLC).* (**a**) schematic structure: 1 *aluminum*, 2 porous carbon, 3 positive and negative *ions*, 4 *electrolyte*, 5 separator. (**b**) winding design. (**c**) technical design

$$\Delta E = \frac{1}{2} C \cdot \left(U_{max}^2 - U_{min}^2 \right) \tag{3.3}$$

ΔE	Usable energy content of a capacitor, J
C	Capacitance of the capacitor, F
U_{max}	Maximum voltage of the capacitor, V
U_{min}	Minimum voltage of the capacitor, V

3.3.3 Flywheel

When combining individual solutions from Table 3.2 to form an overall system, one of the things that must be taken into account is the compatibility of the individual systems.

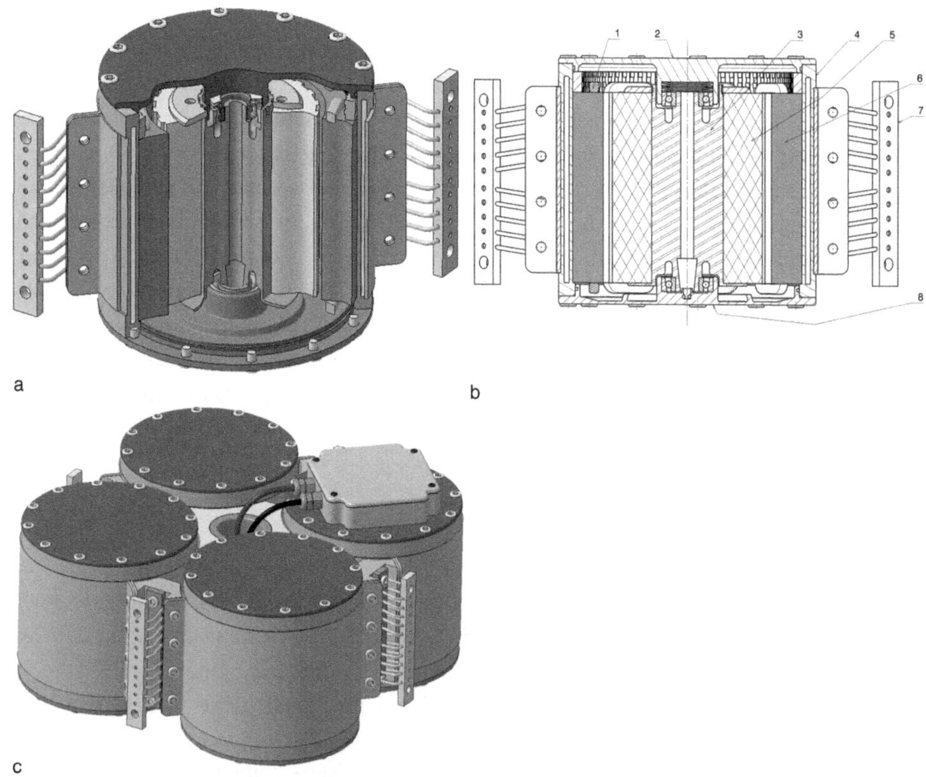

Fig. 3.21 Dynastore type energy storage system, according to [12]. The entire accumulator is made up of four such cylindrical units (**a**), arranged in a square (**c**). This cancels out gyroscopic effects and the individual flywheels can remain relatively small. Sectional view (**b**): 1 *Winding*, 2 Cover top, 3 Rotor carrier with bearings, 4 Housing with cooling jacket, 5 Rotor, 6 Stator, 7 Wire rope insulator, 8 Cover bottom with integrated coolant pump. The bearings are supplied with lubricating oil by the pumping action of the specially shaped rotor arm journal. The excess oil also serves to dissipate heat from the rotor, because in the partial vacuum the otherwise usual convection (heat transport through air) is hardly effective any more

Multiple energy conversion is clumsy, because every energy conversion involves unavoidable losses. A purely mechanical KERS follows this guideline. A flywheel (accumulator) is driven during braking via a continuously variable transmission (energy converter) (Flybrid system). Another system operates purely electrically and charges an air-cooled Li-ion battery during braking via an electric machine acting as a generator. However, this requires an inverter between the two because the generator supplies alternating current while the battery requires direct current. Energy feedback is provided by the battery-powered electric machine during engine operation, with the internal combustion engine assisted directly at the crankshaft (Mercedes-McLaren's Formula 1 system). A clever combination of both systems is realized in the so-called mechanical battery. A flywheel, which is also the rotor of an electric machine, acts as an energy storage device (Dynastore system), Fig. 3.21.

Energy is thus supplied and removed purely electrically, while storage is mechanical. In principle, flywheels have only two manipulated variables to store a large amount of energy. The kinetic energy of a rotating body is known to be $E_{kin} = J\omega^2/2$. Thus, if a large amount of energy is to be stored, the moment of inertia J and/or the angular velocity ω must be high, the latter being quadratic in the amount of energy. The speeds for executed flywheels in Formula 1 are in the range between 50,000 and 80,000 min^{-1} so that the flywheels remain small and light enough to fit in the vehicle. At these speeds, however, air friction already becomes a problem. The rotors therefore run in a partially evacuated housing. This in turn places high demands on the sealing of the shaft passage. The bearing of the rotor can also no longer be achieved by "everyday" means. In Formula 1, the hybrid bearings used are replaced after every race. For series applications in passenger cars, more durable air and magnetic bearings are being considered. The rotor itself must be made of a fiber-reinforced material so that the high energy densities can be achieved.

3.3.4 Hydraulic Storage *System*

For municipal vehicles and work machines (waste collectors, city buses, forklifts, ...) with their constant alternation of starting and braking, a hydraulic system is also offered. When braking, an axial piston unit acts as a pump to charge a bladder accumulator with hydraulic oil. For starting, the accumulator is discharged via this axial piston unit, which now pressurizes the mechanical drive train in motor operation (Rexroth system).

When selecting the energy storage systems listed in Table 3.2, the following advantages and disadvantages must be taken into account, Table 3.3.

Figure 3.22 compares important properties of energy storage systems. A flywheel with high charging power and sufficient storage capacity as well as low thermal ageing is outstanding compared to the best-known battery types. The flywheel is orders of magnitude ahead of the other energy storage systems in terms of another important criterion for energy storage systems in driving operation, namely ageing due to charging and discharging processes (cycle stability).

Table 3.3 Properties of common energy storage devices

Energy storage	Advantages	Disadvantages
Battery	High energy density, many manufacturers	Low power density, heavy, thermal ageing, short service life
Capacitor (Supercap)	High power density, maintenance-free, long service life	Low energy density, high price, large installation space requirement
Flywheel	Robust, high cycle stability, very long lifetime, high power and energy density	Behaviour in case of accident (safety), influence on driving behaviour, maintenance required, noise development during charging and discharging

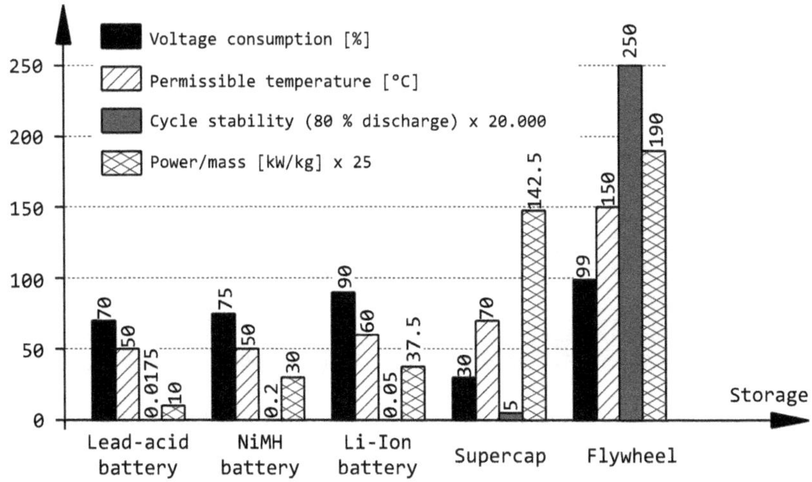

Fig. 3.22 Comparison of energy storage systems for vehicles, after [12]. For comparison purposes, some values are scaled with factors

3.4 Examples of Hybrid Drives

Toyota Supra HV-R [13, 14], Fig. 3.23

This vehicle was the first hybrid car in the world to win a race. The car, which is based on the model from the Japanese Super GT series, was victorious at the 24-hour race in Tokachi (Japan) in 2007. In addition to the 4.5-litre V8 petrol engine with 353 kW output, which is housed in the front of the vehicle as the main power unit, a 150-kW electric motor in the rear and a 10-kW electric motor in each of the front wheels also provided propulsion. This results in a system output of 441 kW for the 1080 kg car. Recuperation took place via all four wheels: the two 10 kW wheel hub motors and the rear motor were used for energy recovery during braking in generator mode. A high-performance capacitor system acted as the energy storage system, which can follow the rapid alternation of charging and discharging with high current outputs dictated by racing operations.

Porsche 911 GT3 R Hybrid [16]

The car, which is also referred to as a rolling laboratory, is based on the Porsche 911 GT3 R model year 2010 and was developed to gain experience in the design of hybrid systems in racing and to prepare high-performance hybrid drives in production vehicles. The conventional drive train in the rear of the vehicle consists of a 4-litre six-cylinder boxer engine with an output of 353 kW, which drives the rear wheels via a sequential six-speed gearbox. Two electric motors, each with an output of 75 kW, act on the front axle. This is therefore a parallel hybrid. Figure 3.15 provides a system overview. During braking, kinetic energy

Fig. 3.23 Denso Toyota Supra HV-R [15]. 4.5-liter V8 engine rated at 353 kW at 6800 min^{-1}/ 510 N m at 5600 min^{-1}. Air restrictor 2 × 29.6 mm diameter. 2 × 10-kW wheel hub motors at the front and 150-kW electric motor at the rear. Vehicle mass: 1080 kg

can be charged as electric current into the accumulator (4) via the electric machines on the front axle (1). The structure of the braking system and the recuperation strategy are explained in Figs. 3.14, 3.15, 3.16, and 3.17. The electric flywheel storage and the two generators are each controlled by power electronics (2, 3). The stored energy is fed back to the front wheels via the two electric motors during subsequent acceleration, thus increasing the power of the main drive (boosting). The car thus drives advantageously in this phase with all-wheel drive and suitable torque distribution (more at the rear than at the front). However, the axle load distribution is also balanced in the transverse direction by the location of the flywheel storage on the passenger side.

The two electric machines are not directly coupled to the front wheels, Fig. 3.24. A fixed gear stage (2) brings the rotor axes down by their centre distance (reversed portal axle), improves the centre of gravity position and allows smaller rotor diameters (6). Short drive shafts provide the connection to the wheels. Intermediate planetary gearboxes (4) support their ring gear on the housing of the axle via a multi-plate clutch (3). In the event of a fault, the hydraulically actuated clutches enable the torque flow of the permanently excited electric machines to be disconnected from the front wheels. This safety device is triggered by monitoring circuits (insulation monitors, diagnostic algorithms), but can also be activated by the driver in the cockpit.

The two electric machines are controlled by power electronics depending on rotor position, speed and load requirement. For this purpose, each machine is equipped with a position sensor (7). The power electronics and the machines are cooled by a separate low-temperature water cooling circuit. The cooling channels (8) are integrated directly into the one-piece housing of the machines. The associated heat exchanger is located centrally in the vehicle nose.

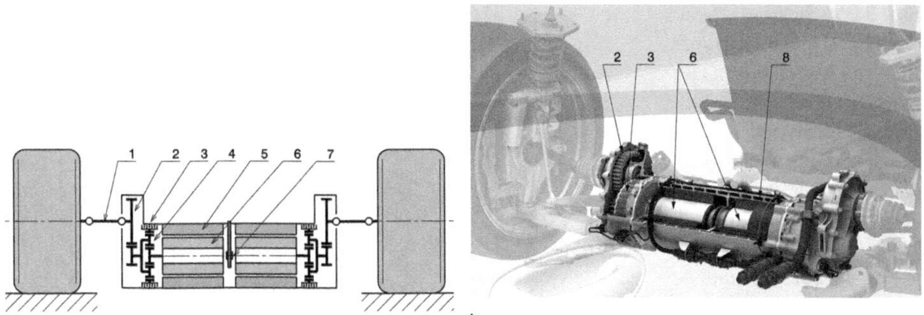

a b

Fig. 3.24 Porsche 911 GT3 R Hybrid portal axle [Porsche Motorsport]. (**a**) schematic. (**b**) axonometric view, partially cut: 1 drive shaft, 2 reduction gears, 3 multi-plate clutch, 4 planetary transmission, 5 stator, 6 rotor, 7 speed sensor (rpm sensor), 8 cooling ducts for water cooling

The two rotors (6) of the electric machines are mechanically decoupled and can thus drive the associated front wheel individually. A differential is thus not required. In addition, an individual drive torque can be assigned to the left and right wheel (torque vectoring). This is done via speed, steering angle and lateral acceleration to stabilize the driving condition. The mechanical power per machine is 75 kW at a maximum of 15,000 min^{-1} and 120 N m torque at the output shaft of the electric machine. Thus, an additional drive torque of up to 948 Nm is available at the wheel due to the reduction. In return, 68 kg of mass are added by the portal axle.

The flywheel accumulator basically works similar to the one in Fig. 3.21, except that here the rotor is outside and the stator is inside. It therefore works like an electric machine with an external rotor. The rotor is made of plastic with infiltrated magnetic particles. To enable it to withstand the high speeds of up to 40,000 min^{-1}, it is reinforced with wound carbon fibres. This means that there is no metal in the rotor, which significantly increases the efficiency compared to conventional metal rotors due to the lack of iron losses. An oil cooling circuit is sufficient for heat dissipation for the storage unit and its power electronics. The housing is evacuated during operation so that air friction at the high rotor speeds does not heat the rotor above its limit temperature. As the only wearing part, only the ceramic bearings of the rotor in the housing have to be replaced depending on the load. The charge/discharge efficiency is over 94%. With a mass of 47 kg, the system stores 736 kJ (0.204 kWh).

The main controls of the hybrid system are integrated in the steering wheel, Fig. 3.25:

- Boost paddle for calling up the additional electric drive torque
- Charge status display (SOC) of the flywheel storage in the form of an LED line under the display
- LED display for boost recommendation in manual mode. This is additionally located in the instrument cluster (LCD display).

Fig. 3.25 Steering wheel Porsche 911 GT3 R Hybrid [Porsche Motorsport]. The steering wheel also includes controls for hybrid functions: Boost rocker switch, flywheel storage charge status indicator, boost recommendation indicator, detent switch for hybrid-specific programs

- Map switch (indexed switch) for calling up hybrid-specific programs. These are racing strategies tailored to hybrid operation, such as efficiency mode, which enables the most efficient driving possible with lower fuel consumption at reduced power.

References

1. Braess, Seiffert (Hrsg.): Vieweg Handbuch Kraftfahrzeugtechnik, 4. Aufl. Vieweg, Wiesbaden (2005)
2. Schöggl, P. et al.: Hybrid im Motorsport. ATZ 2, 132–137 (2011)
3. Knödel, U. et al.: Variantenvielfalt der Antriebskonzepte für Elektrofahrzeuge. ATZ Heft **7/8**, 552–557. Vieweg, Wiesbaden (2011)
4. Hofmann, P.: Hybridfahrzeuge, 1. Aufl. Springer, Wien/New York (2010)
5. Hohenberg, G.: Kann der intelligente Fahrer den Hybrid ersetzen? Vortrag der ÖVK-Reihe (österreichischer Verein für Kraftfahrzeugtechnik), Wien (2008)
6. Cross D.: Optimization of Hybrid Energy Recovery Systems (KERS) for Different Racing Circuits. SAE Paper 2008-01-2956
7. Duval-Destin, M. et al.: The electrified chassis – impacts of an electric powertrain on vehicle dynamics control, Beitrag zum 2. internationalen Fahrwerkssymposium chassis.tech plus, München (2011)
8. Hayes, O., Armbruster, D.: Porsche 911 GT3 R Hybrid – recuperation brake system, Beitrag zum 2. internationalen Fahrwerkssymposium chassis.tech plus in München (2011)
9. Roth, K.: Konstruieren mit Konstruktionskatalogen. Band 2 Konstruktionskataloge, 3. Aufl. Springer, Berlin (2001)
10. Fiala, E.: Hybridauslegung für Personenkraftwagen, Schriftreihe des österreichischen Vereins für Kraftfahrzeugtechnik. ÖVK, Wien (2006)
11. Babiel, G.: Elektrische Antriebe in der Fahrzeugtechnik, 2. Aufl. Vieweg+Teubner, Wiesbaden (2009)
12. Sontheim, J.: Kinetic energy storage in hybrid vehicles, the mechanical battery. ATZ AutoTechnology. **5**, 24–28 (2008)
13. http://carscoop.blogspot.com/2007/11/toyota-supra-hv-r-hybrid-wins-race.html. Accessed on 25 Jan 2012

14. www.ATZonline.de. Marmorini fordert effizientere Hybridsysteme im Rennsport. Accessed on 30 Apr 2008
15. http://www.toyota.co.jp/ms/press/. Accessed on 16 Dec 2011
16. Armbruster, D., Hennings, S.: Porsche GT3 R Hybrid, Technologieträger und rollendes Labor. MTZ. **5**, 356–363 (2011)

4 Calculation of the Drive Train

A powerful engine is just as important for a racing car as a reliable drivetrain. However, it is even more important that the power of the engine is also efficiently transferred to the road so that the vehicle can demonstrate the desired high performance. To achieve this, the coordinated interaction of engine and drivetrain is crucial. The engine only works within a certain speed range – which is usually extremely narrow in the case of high-bred engines. The drive wheels, however, must provide propulsion over a wide speed range. The transmission is the assembly that mediates between the engine and the road. In order for it to do this as effectively as possible, the driving resistances to be overcome and the available engine power must be matched to each other.

M. Trzesniowski, *Powertrain*,
https://doi.org/10.1007/978-3-658-39885-9_4

4.1 Power Demand

Friction Grip, Tyre Forces F_W

The maximum force F_{rsl} that a tyre can transmit depends, in simplified terms, on the wheel load $F_{W,Z}$[1] and on the friction conditions between the tyre and the road surface.

$$F_{rsl} = \mu_W \cdot F_{W,Z} \tag{4.1}$$

F_{rsl}	Resulting tyre force, N
μ_W	Resulting static friction coefficient, –. Values see Table 4.1
$F_{W,Z}$	Wheel load, N

It can be seen in Table 4.1 that, on dry roads, worn tyres have consistently higher static friction than new tyres. Treadless tyres (slicks) are even better, achieving values of around 1.5–2.2 when warm.[2]

The maximum circumferential force $F_{W,X}$ and lateral force $F_{W,Y}$ are given by:

$$\begin{aligned} F_{W,X,\max} &= \mu_{W,X} \cdot F_{W,Z} \\ F_{W,Y,\max} &= \mu_{W,Y} \cdot F_{W,Z} \end{aligned} \tag{4.2}$$

$F_{W,X,max}$	Maximum circumferential force, N
$F_{W,Y,max}$	Maximum lateral force, N
$\mu_{W,X}$	Static friction coefficient in longitudinal direction, –
$\mu_{W,Y}$	Static friction coefficient in lateral (transverse) direction, –

If circumferential and lateral forces occur simultaneously on a wheel, the resulting force cannot exceed the maximum possible force F:$_{rsl}$

$$F_{rsl} \geq \sqrt{F_{W,X}^2 + F_{W,Y}^2} \tag{4.3}$$

F_{rsl}	Total force on the Tyre on the road, N

This fact makes the advantage of four-wheel drive obvious, Fig. 4.1. The required driving force for a vehicle is divided between two wheels with single-axle drive and four

[1] For the coordinate system used, see the appendix.

[2] For details see Racing Car Technology Manual, Vol. 4 Suspesnion System, Chap. 1 *Tyres and wheels.*

Table 4.1 Stiction coefficients μ_W of tyres on road surfaces, according to [1]

		Road condition				
Driving speed v_V km/h	Tyre condition	Dry	Wet, water level approx. 0.2 mm	Heavy rain, water level approx. 1 mm	Puddles, water height approx. 2 mm	Iced (black ice)
50	New	0.85	0.65	0.55	0.5	0.1 and smaller
	Worn[a]	1	0.5	0.4	0.25	
90	New	0.8	0.6	0.3	0.05	
	Worn[b]	0.95	0.2	0.1	0.05	
130	New	0.75	0.55	0.2	0	
	Worn[c]	0.9	0.2	0.1	0	

[a]worn to ≥1.6 mm tread height (minimum value according to § 36.2 StVZO)

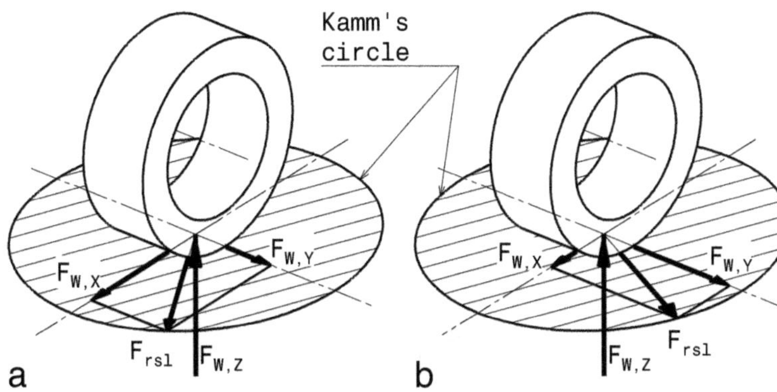

Fig. 4.1 Forces on the tyre at the adhesion limit. (**a**) *single-axle drive*, (**b**) *all-wheel* drive. The X-direction points in the direction of travel. The Kamm circle represents the maximum transmittable force F_{rsl} in all directions, i.e. the vector sum of the circumferential force $F_{W,X}$ and the lateral force $F_{W,Y}$ cannot exceed the circle. This means that the magnitude of both forces is dependent on each other

wheels with four-wheel drive. Therefore, the required circumferential force of a tire in four-wheel drive is exactly half of the circumferential force in single-axle drive. With the same total drive force, the all-wheel drive vehicle can therefore build up greater lateral forces.

The engine drives the wheels and these must overcome the driving resistances with the transmittable forces so that the vehicle can drive and accelerate. The resistances are made up of:

- Rolling resistance
- Side-slip resistance
- Air resistance
- Climbing (gradient) resistance
- Acceleration resistance.

Rolling Resistance F_R

The deformation of the tyre during rolling and the uneven distribution of pressure in the contact patch create a force which counteracts the direction of movement of the vehicle.[3] This force depends primarily on the wheel load:

$$F_R = k_R \cdot F_{W,Z} \tag{4.4}$$

F_R	Rolling resistance, N
k_R	Rolling resistance coefficient, –. Values see Table 4.2, Figs. 4.2 and 4.4

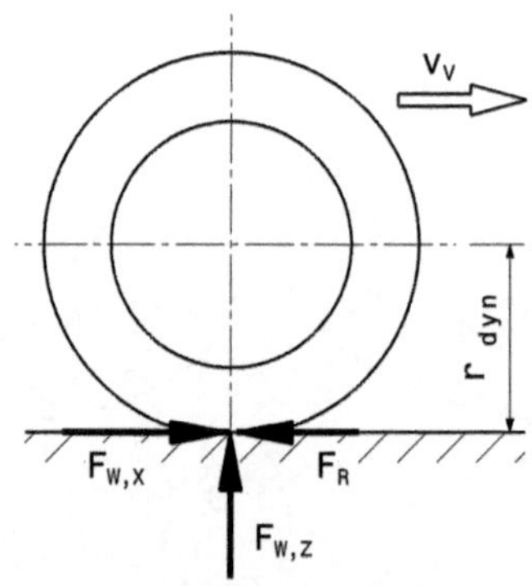

Rolling resistance increases with increasing driving speed, up to about 60 km/h it can be assumed to be constant. If the value is to be calculated more realistically for higher speeds, the following equation provides an approximation:

$$k_R = k_{R0} + k_{R1}\frac{v_V}{100} + k_{R4}\left(\frac{v_V}{100}\right)^4 \tag{4.5}$$

k_{R0}, k_{R1}, k_{R4}	Coefficients, –
v_V	Driving speed, km/h

The coefficients depend primarily on the tyre design and the tyre inflation pressure. k_{R0} is approximately equal to k_R. k_{R1} is approximately 0.0018 and increases with increasing tyre pressure. k_{R4} is approximately 0.0002 and decreases with increasing tyre pressure

[3]For more details see Racing Car Technology Manual, Vol. 4 Suspension System, Sect. 1.2.3.

Table 4.2 Guide values for rolling resistance coefficients, according to [2]

Roadway	Rolling resistance coefficient k_R
Rigid	
Asphalt	0.010
Concrete, smooth	0.011
Concrete, rough	0.014
Deformable	
Earth road, good	0.045
Earth road, bad	0.160
Loose sand	0.150–0.300

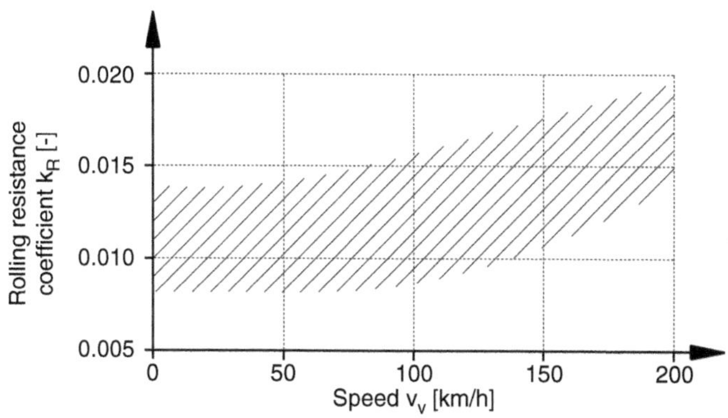

Fig. 4.2 Rolling resistance as a function of driving speed. The upper range of the value band belongs to HR, VR and WR radial tyres, the lower range corresponds to radial ECO tyres

[3]. Racing tires show a smaller increase in rolling resistance over speed than passenger car tires, wants they are designed for higher speeds. If a tyre is operated for a longer period of time in a speed range that is too high for it, its (thermal) destruction will occur.

In racing vehicles with high aerodynamic downforce, the speed-dependent air force[4] $F_{L,Z}$ also causes an increase in rolling resistance because it is added to the wheel load $F_{W,Z}$.

The rolling resistance can also be influenced within certain limits with the setup. The simplest possibility is the tyre inflation pressure. But the wheel position is also significant. Toe-in of the wheels increases the rolling resistance by about 1% per 10′ toe-in angle $\delta_{V,0}$ of a wheel [4]. Toe-in reduces the rolling resistance caused by negative camber.[5]

***Resistance Due to Tyre Slip* (Side-slip Resistance) F_α**

A rolling tyre that transmits a lateral force is deformed by the frictional forces. As a result, the direction of movement of the tyre and its centre plane no longer coincide, as is the case

[4] See Racing Car Technology Manual, Vol. 2 Complete Vehicle, Chap. 5, especially (5.5).

[5] See Racing Car Technology Manual, Vol. 4 Suspension System, Chap. 2

with straight running. The direction of movement and the centre plane include the so-called slip angle.[6] The direction of movement is then counteracted by a force component of the lateral force, the side-slip resistance (Fig. 4.3).

$$F_\alpha = k_\alpha \cdot F_{W,Z} \tag{4.6}$$

F_α	Side-slip resistance, N
k_α	Side-slip drag coefficient, –. Values: Fig. 4.3
$F_{W,Z}$	Wheel load, N

Side-slip resistance is also deliberately used for braking. On the oval tracks of the North American NASCAR and Indy Series, the speed drop in the (banked) corners is so low that many drivers drive through the corner without taking their foot off the gas. Instead of a braking maneuver, they provoke increased lean resistance with the appropriate steering angle. Indycars decelerate at approx. 3.7 m/s^2.

Drag $F_{L,X}$

The air resistance depends mainly on the dynamic pressure caused by the vehicle in the air and on the frontal area (cross sectional area):

$$F_{L,X} = \frac{1}{2}\rho_L \cdot c_W \cdot A_V \cdot v_L^2 \tag{4.7}$$

$F_{L,X}$	Air resistance (drag), N
ρ_L	Density of air, kg/m^3. $\rho_L = 1.199$ kg/m^3 at a temperature of 20 °C, an air pressure of 1.013 bar and a relative humidity of 60%
c_W	Drag coefficient, –
A_V	Cross sectional area, m^2
v_L	Incoming flow velocity, m/s. when there is no wind, $v_L = v_V$. With v_V driving speed

The incident air velocity v_L is the geometric sum of the wind velocity and the vehicle velocity v_V acting in the opposite direction to the direction of travel when there is a pure headwind on the straight line. The general case is shown in Fig. 4.4. The resulting air speed follows from the vectorial addition of negative vehicle speed (= headwind) and wind speed:

[6]For more details seeRacing Car Technology Manual, Vol. 4 Suspension System, Chap. 1

Fig. 4.3 Slip angle drag coefficient as a function of the slip angle [5]. It can be seen that from an slip angle of about 2°, the resistance is similar to that caused by rolling resistance when driving straight ahead

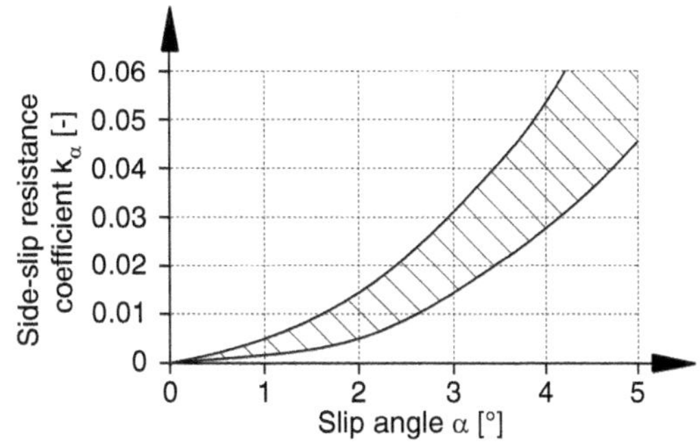

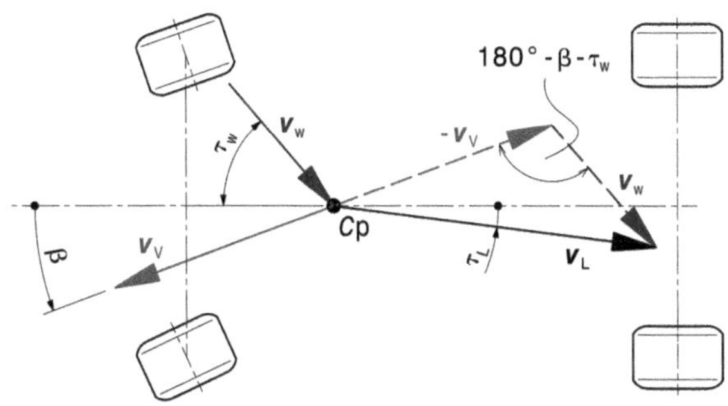

Fig. 4.4 Vector addition of the effective velocities for air resistance. The vehicle speed v_V includes the body side-slip angle β with the longitudinal axis of the vehicle. The crosswind (v_w) hits the vehicle with the angle τ_w. The resulting air velocity vector v_L is rotated against the vehicle longitudinal axis by the angle τ_L. *Cp centre of pressure*

$$v_L = (-v_V) + v_w \tag{4.8}$$

v_w	Wind speed, m/s

The amount of the resulting air velocity follows directly from the cosine theorem:

$$v_L = \sqrt{v_V^2 + v_W^2 - 2v_V v_W \cos\left(180^\circ - \beta - \tau_W\right)} = \sqrt{v_V^2 + v_W^2 + 2v_V v_W \cos(\beta + \tau_W)}$$

The magnitude of the average wind speed in Central Europe is about 4.7 m/s [3].

In (4.7), note that c_W changes with τ_L, see Fig. 4.5.

The air density for other temperatures and pressures can be determined approximately with the ideal gas equation:

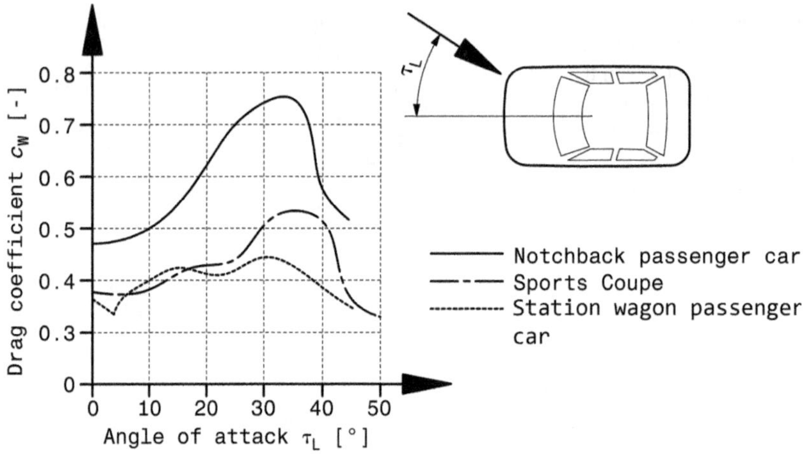

Fig. 4.5 Drag coefficient for inclined flow [5]. Depending on the shape of the vehicle, the air resistance may deviate considerably from the value for straight airflow in the case of oblique airflow

$$\rho_L = \frac{p_0}{R_L \cdot T_L}$$

p_0	Air pressure, bar
R_L	Gas constant of air, J/(kgK). $R_L \approx 287$ J/(kgK)
T_L	(absolute) temperature of the air, K

More precise values for the air density are provided by the relationship via the air humidity:

$$\rho_L = \frac{349 \cdot p_0 - 131 \cdot p_e}{T_L}$$

p_e	Partial pressure of the water vapour contained in the air, bar

$$p_e = p_{e,\max} \cdot \phi/100$$

$p_{e,\max}$	Maximum steam pressure at T_L, bar
Φ	Relative humidity, %

The drag coefficient is primarily determined by the shape of the body. Table 4.3 gives an overview of some geometric bodies and typical vehicle shapes. The direction of the airflow

Table 4.3 Drag coefficients [1]

Geometric bodies			Geometric bodies		
Body shape		c_W	Body shape		c_W
	Slice Plate	1.1	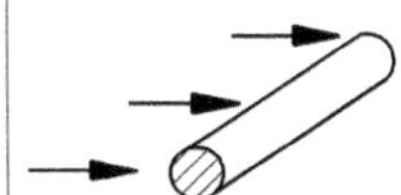	Long cylinder $Re < 200{,}000$ $Re > 450{,}000$	 1.0 0.35
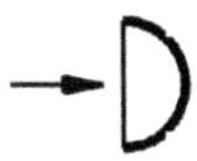	Open shell Parachute	1.4	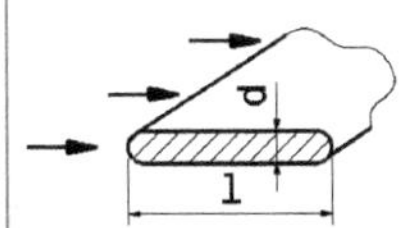	Long plate $l/d = 30$ Re 500,000 $Re \approx 200{,}000$	 0.78 0.66
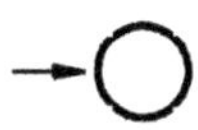	Bullet $Re^{a} <$ 200,000 $Re > 250{,}000$	 0.45 0.20	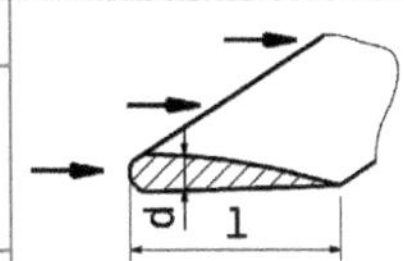	Long wing $l/d = 18$ $l/d = 8$ $re \approx 10^6$ $l/d = 5$ $l/d = 2$ $re \approx 2 \cdot 10^5$	 0.2 0.1 0.08 0.2
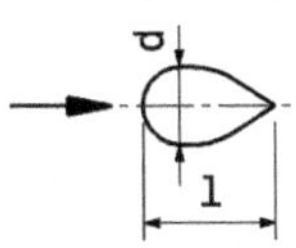	Slender body of revolution $l/d = 6$	0.05			

Vehicles		Vehicles	
Vehicle shape	c_W	Vehicle shape	c_W
Convertible top	0.5–0.7	Truck, juggernaut	0.8–1.5
Box body	0.5–0.6	Omnibus	0.6–0.7
Pontoon form	0.4–0.55	Omnibus with streamlined shape	0.3–0.4
Wedge shape: Headlights and bumpers integrated in the fuselage, wheels covered, underbody panelling, optimised cooling airflow	0.3–0.4	Motorcycle	0.6–0.7
		Racing car F1 (depending on the design of the downforce) [6]	0.6–0.75
		Racing car touring car (depending on the design of the downforce) Opel Calibra ITC 96 [7]	0.5 0.4–0.43
Favourable streamlined shape (teardrop shape)	0.15–0.2		

[a]Reynolds number (see appendix)

also has an influence. For example, crosswinds in vehicles result in an oblique inflow. The usual values of drag coefficients are related to a flow direction in the longitudinal direction of the vehicle. Figure 4.5 shows the influence of oblique airflow on drag.

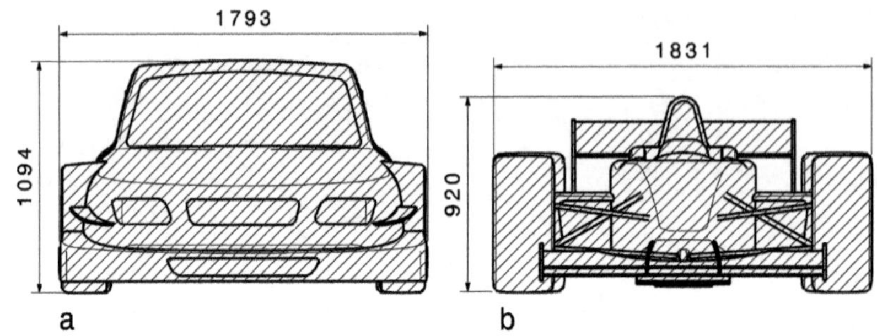

Fig. 4.6 frontal area of vehicles. (**a**) Touring car, $A = 1.59\ \text{m}^2$, (**b**) Formula car, $A = 1.24\ \text{m}^2$

The cross sectional area of a motorcycle with rider is in the range of 0.7–0.9 m^2, that of a passenger car is about 2 m^2, and that of a formula car is about 1.3 m^2, Fig. 4.6.

***Climbing* Resistance F_q**

When driving on an inclined roadway, the uphill resistance is caused by the downhill drive, i.e. by the weight component parallel to the roadway, Fig. 4.7. This counteracts the propulsion when driving uphill.

The slope of a roadway is generally expressed as a percentage (road grade):

$$q = \frac{h_Z}{s_X} \cdot 100\%$$

q	Slope of the carriageway, %
h_Z, s_X	Distances, m. on the horizontal distance s_X the height h_Z is reached by the slope q

The angle of inclination of the roadway follows directly from this:

$$\alpha = \arctan\left(\frac{h_Z}{s_X}\right) = \arctan\left(\frac{q}{100}\right)$$

α	Angle of inclination, °

The angle can be used to calculate the slope downforce from the vehicle weight:

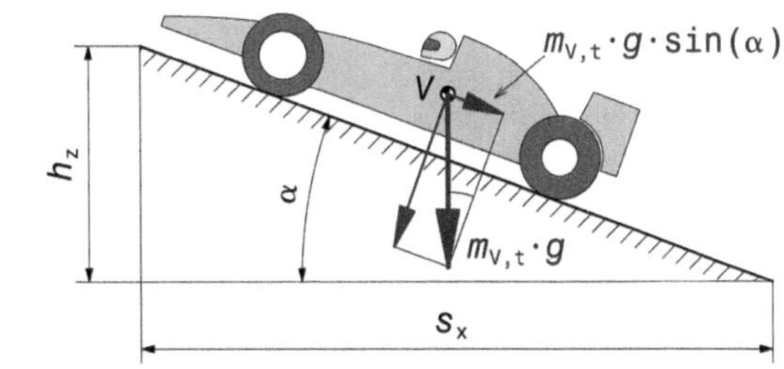

Fig. 4.7 The gradient resistance is the weight component of the vehicle that points against the direction of travel

$$F_q = m_{V,t} \cdot g \cdot \sin(\alpha) \tag{4.9}$$

F_q	Gradient resistance, N
$m_{V,t}$	Total mass of the vehicle, kg

Acceleration Resistance F_a

The resistances considered so far occur during steady-state (stationary) travel (i.e. $v_V = \text{const}$). However, if a vehicle is accelerated, the mass inertia must be overcome and a driving resistance is added, the acceleration resistance. It should be noted that it is not only the purely translationally moving mass that has to be accelerated, but also rotating masses such as the wheels, drive shafts, gearbox parts, clutch and the crank drive parts of the engine itself. However, the rotating parts do not all have the same speed, but completely different speeds depending on the axle ratio and the gear engaged. Therefore, the total kinetic energy is related to the vehicle speed and to the speed of the drive wheels and equated to the energy of a so-called reduced mass (*equivalent mass*). This results in the mass reduced (related) to the drive axle:

$$m_{red,n} = m_{V,t} + \frac{\Sigma J_{k,n} \cdot i_{k,n}^2}{r_{dyn}^2}$$

$m_{red,n}$	Mass reduced to the drive axle for gear *n*, kg
$J_{k,n}$	Moment of inertia of a rotating part *k* about its axis of rotation for gear *n*, kg m^2
$i_{\text{k,n}}$	Gear ratio (speed of part *k* to wheel speed) for gear *n*, –
r_{dyn}	Dynamic Tyre radius, m. see (4.32)

The acceleration resistance follows with this new size:

$$F_a = m_{red,n} \cdot a_X \tag{4.10}$$

F_a	Acceleration resistance, N
a_X	Vehicle acceleration in longitudinal direction, m/s^2

The reduced mass moment of inertia of a part is proportional to the mass moment of inertia about the axis of rotation of this part and the gear ratio squared. Thus, with large gear ratios, relatively small torques become significant for the vehicle. When shifting down to a gear that is too low, the acceleration resistance of the clutch and engine can cause the drive wheels to lock.

Simplified, the acceleration resistance can also be written as:

$$F_a = k_m \cdot m_{V,t} \cdot a_X$$

k_m	Torque mass allowance (surcharge) factor, –
	Values e.g. from Fig. 4.8

Total *Road Resistance (Road Load) F_{dr}* The minimum tractive force required at the drive wheels is equal to the sum of the driving resistances:

$$F_{dr} = F_R + F_\alpha + F_{L,X} + F_q + F_a \tag{4.11}$$

F_{dr}	Total driving resistance, N

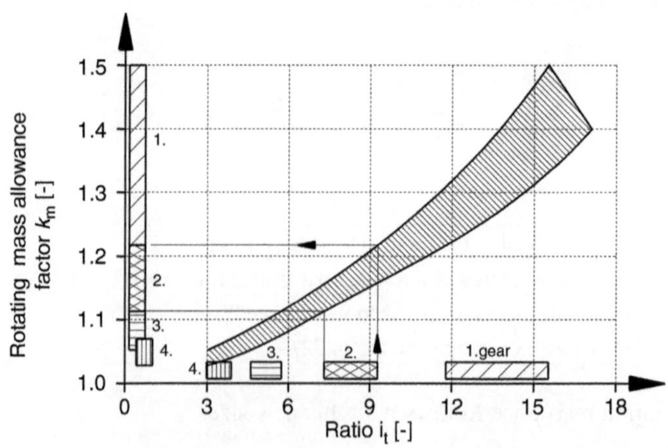

Fig. 4.8 Guide values for the torque mass surcharge factor, according to [2]. The total transmission ratio i_t follows according to Sect. 4.3. For the second gear it is additionally entered how a transmission ratio with the scatter band results in a range of values for k_m

4.2 Gear Diagram and Tractive *Effort* Diagram

Gear *Chart*

The theoretically achievable vehicle speeds as a function of tyre size and gear ratios can be clearly displayed in a gear diagram (also sawtooth diagram or speed-speed diagram). Tyre slip and tyre growth are not taken into account. The speed then increases linearly with the engine speed and is calculated from the total gear ratio (see Sect. 4.3) and the tyre size. The maximum speed for a gear *n* thus follows from the maximum engine speed:

$$v_{V,\,\max,n} = \frac{3.6\frac{\pi}{30}n_{M,\max} \cdot r_{dyn}}{i_{t,n}} \tag{4.12}$$

$vV_{,\max,n}$	Maximum speed for gear *n*, km/h
$n_{M,\max}$	Maximum speed of the engine resp. motor, $\min^{-1}$
$i_{t,n}$	Total gear ratio in gear *n*, –
r_{dyn}	Dynamic tyre radius, m. see Table 4.4

At the speed $n_M = 0$, the speed $v_V = 0$. Thus, for each gear, the speed curve can easily be represented as a straight line through the zero point, Fig. 4.9.

Traction Force Diagram

In a tractive force diagram, the tractive force available at the drive wheels and the driving resistances are plotted against the driving speed. The available tractive force depends on the gear (i.e. the overall transmission ratio) and the torque characteristic of the engine. In addition, the so-called tractive force hyperbola can be entered. This corresponds to the maximum possible tractive force at a constant power.

Torque Curve of an Internal Combustion Engine

An internal combustion engine can only be operated within a certain speed and torque range. This map is therefore limited by the extreme values of these variables. Within the map, the load of the engine is set with the accelerator pedal, the speed is then determined by the resistance that the engine must overcome. An internal combustion engine delivers its

Table 4.4 Dynamic rolling circumference and wheel radius of some tyre sizes

Dimension	Rolling circumference $C_{R,dyn}$, m	$r_{dyn,60}$, m
165/70 R13	1.730	0.275
185/60 R14	1.765	0.281
195/65 R15	1.935	0.308
205/60 R15	1.910	0.304

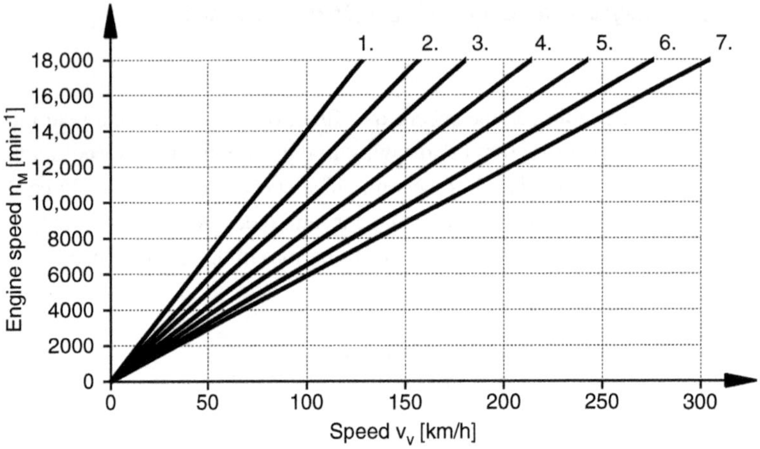

Fig. 4.9 Gearbox diagram Ferrari F1–2000, according to [8]. The maximum engine speed of the naturally aspirated V10-3 l engine is 18,000 min^{-1}. The ratios of the 7 gears are geared for the street circuit in Monaco. In the slowest corner, the engine speed in first gear drops to 6000 min^{-1}

maximum torque at full load (i.e. 100% accelerator pedal position) in a characteristic behavior over the speed. The characteristic depends, among other things, on the combustion process (gasoline, diesel), the type of air supply (naturally aspirated, supercharged), and the design (intake manifold length, valve timing, bore/stroke ratio, etc.). Basically, however, it looks as shown in Fig. 4.10. At standstill, i.e. at speed 0, the engine cannot deliver any torque; it requires a certain minimum speed to function. Due to gas dynamic effects, the full load torque first increases with the engine speed, reaches a maximum value, the rated torque, and finally drops again until the engine reaches its maximum speed. Either because it no longer receives sufficient air or because the mass-loaded valve train performs function-disturbing oscillations. The power curve is obtained by multiplying all torque values by their speeds (*P* is directly proportional to *M* and *n*). In coasting mode (accelerator pedal position 0%), the engine does not deliver any torque, but must be driven, i.e. the engine torque becomes negative. This braking torque curve increases linearly over the engine speed.

Tractive Force at the Wheels $F_{W,X,A}$

The tractive force provided by the engine at the drive wheels results from the engine torque converted by the total transmission ratio in the gear under consideration and reduced by the efficiency of the drive train:

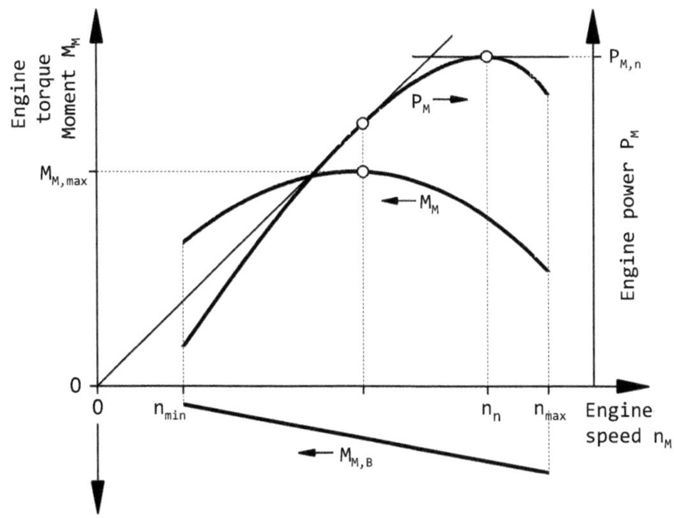

Fig. 4.10 General characteristic diagram of an internal combustion engine. n_{min} *minimum engine speed*, n_{max} *maximum engine speed*, n_n *rated speed* = speed at rated power $P_{M,n}$, M_M engine torque at full load (*WOT engine torque*), $M_{M,max}$ maximum engine torque (rated torque), $M_{M,B}$ engine braking torque, P_M *engine power*, $P_{M,n}$ *rated power* (maximum engine power)

$$F_{W,X,A,n} = \frac{3600 \cdot P_M(n_M)}{v_V}\eta = \frac{M_M(n_M) \cdot i_{t,n}}{r_{dyn}}\eta \tag{4.13}$$

$F_{W,X,A,n}$	Tractive force on the wheels in gear n, N
P_M	Engine power at speed n_M, kW
M_M	Engine torque at speed n_M, nm
v_V	Driving speed, km/h
$i_{t,n}$	Total gear ratio in gear n, –. Sect. 4.3
r_{dyn}	Dynamic Tyre radius, m. Table 4.4
η	Overall efficiency of the power train, –. See Sect. 5.1

Traction Force Diagram

The driveable range of a vehicle can be shown in a tractive force diagram, Fig. 4.11. Here, the tractive force $F_{V,X,A}$ at the drive wheels is plotted above the driving speed v_V. The limits of the drivable range (shown shaded) give the maximum engine power $P_{M,max}$, represented by the tractive force hyperbola, and the adhesion of the tires (adhesion limit given by $F_{W,X,max}$, (4.2) and aerodynamic downforce,[7] for the driven tires). The maximum speed is given

[7] See Racing Car Technology Manual, Vol. 2 Complete Vehicle, Chap. 5, (5.8).

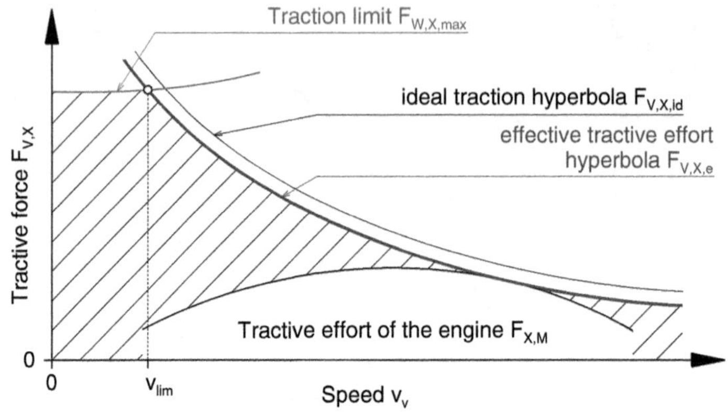

Fig. 4.11 Tractive force diagram of a vehicle without manual transmission. The ideal tractive force of a vehicle results from the maximum engine power and the driving speed. The effective tractive force follows from this with consideration of the efficiency of the drive train. The tyres cannot transmit a greater force than the adhesion limit allows. The tractive force of the engine covers only a small range of the hatched map

by the equilibrium between maximum tractive force and driving resistances (Fig. 4.12), cf. (4.23).

The ideal tractive force hyperbola results directly from the maximum power of the engine resp. motor:

$$F_{V,X,id} = \frac{3600 \cdot P_{M,\max}}{v_V} \tag{4.14}$$

$F_{V,X,id}$	Ideal tractive force, N
$P_{M,\max}$	Maximum engine power, kW
v_V	Driving speed, km/h

The effective tractive force hyperbola follows from this with consideration of the efficiency of the drive train:

$$F_{V,X,e} = F_{V,X,id} \cdot \eta \tag{4.15}$$

$F_{V,X,e}$	Effective tractive force, N
η	Efficiency of the drive train, –

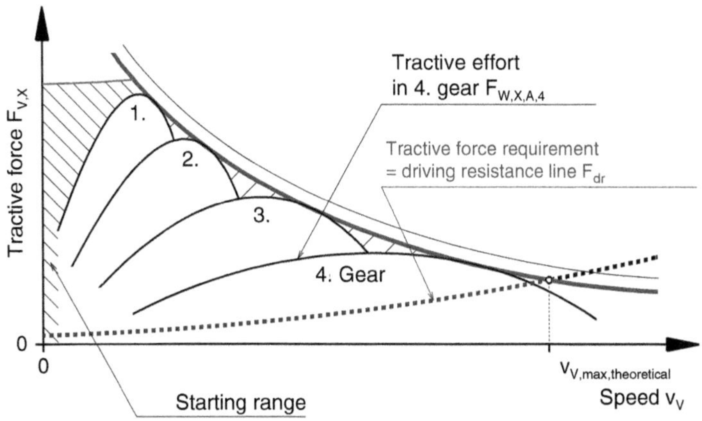

Fig. 4.12 Tractive force diagram of a vehicle with four-speed transmission. In addition to the tractive force curve of the engine scaled by the gearbox, the theoretical maximum speed $v_{V,max,theoretical}$ of the vehicle is also entered, which cannot be achieved with this ratio of the fourth gear. The entered driving resistance line applies to horizontal road surface, i.e. no gradient. ($\alpha = 0°$)

The transition from the traction-limited ((4.2) with consideration of the downforce)[8] to the power-limited part (4.14) of the drivable range, i.e. the intersection of the two curves, can be determined by equating the two relations for the tractive force:

$$\mu_{W,X} \cdot \left(m_{V,t} \cdot g + \rho_L \frac{v_L^2}{2} \cdot c_A \cdot A_V \right) \cdot \Phi_A = \frac{P_{M,\max}}{v_V} \eta \tag{4.16}$$

c_A	Downforce coefficient, –[9]
Φ_A	Axle load ratio of the drive axle, –. For an all-wheel drive vehicle, $\Phi_A = 1$

If there is no wind, i.e. if $v_L = v_V$ applies, the limiting speed $v_{\lim}$ follows from this:

$$v_{\lim} = 3.6 \cdot \left(\sqrt[3]{C + \sqrt{C^2 + D^3}} + \sqrt[3]{C - \sqrt{C^2 + D^3}} \right) \tag{4.17}$$

[8] See Racing Car Technology Manual, Vol. 2 Complete Vehicle, Chap. 5

[9] See Racing Car Technology Manual, Vol. 2 Complete Vehicle, Chap. 5

$$\text{with } C = \frac{1000 \cdot P_{M.\max} \cdot \eta}{\mu_{W,X} \cdot \Phi_A \cdot \rho_L \cdot c_A \cdot A_V} \quad \text{and } D = \frac{m_{V,t} \cdot g}{\frac{3}{2}\rho_L \cdot c_A \cdot A_V}$$

$v_{\lim}$	Limit speed at which the tractive force from traction and engine power are equal, km/h
C, D	Auxiliary variables

When a vehicle accelerates, the traction limit of the tires (in combination with the aerodynamic downforce) is decisive up to the limit speed, only then does the maximum engine power become the limiting component. In a Formula 1 car, for example, the traction of the rear tires is the limit for propulsion up to about 160 km/h. In an average passenger car, the engine already limits acceleration from about 35 km/h.

Main Function of a Gearbox

In order for the torque released by the engine to be used at the wheels over a wide range of speeds and up to the possible limits of the drivable range, an element in the driveline must adapt the actual power curve of the engine to the effective tractive force hyperbola. This element is a variable ratio transmission, i.e. a manual or automatic transmission. Figure 4.12 shows the effect of a four-speed gearbox. The tractive force curve of the engine is scaled by the transmission ratio of a particular gear. The curve is tangent to the tractive force hyperbola at the point of maximum engine power. It can be seen that the possible driving range (shown hatched) is not completely covered by the manual gearbox. On the one hand, the range from standstill to the speed corresponding to the minimum engine speed (starting range) must be made usable by a starting element (e.g. friction clutch). On the other hand, triangular areas remain between the tractive force hyperbola and the tractive force offered by the engine, which cannot be utilized. These areas become smaller as the number of gears in the transmission increases. Theoretically, therefore, a transmission with continuously variable transmission ratio changes (e.g. CVT transmission) provides the highest possible acceleration of a vehicle with a given engine.

The driving resistance lines are entered in the tractive force diagram for unaccelerated driving, i.e. $F_a = 0$ N, and after several gradients in steps.

Road Performance

In addition to the maximum speed, the tractive force diagram can also be used to determine other driving performance of a vehicle. To do this, the *excess traction* force must first be determined by comparing the existing engine traction *force* with the driving resistances to be overcome, and the possible acceleration and climbing ability then follow from this, Fig. 4.13.

The excess tractive effort that can be expended to accelerate the vehicle is the difference between the required tractive effort – the driving resistance – and the available tractive effort – the engine tractive effort:

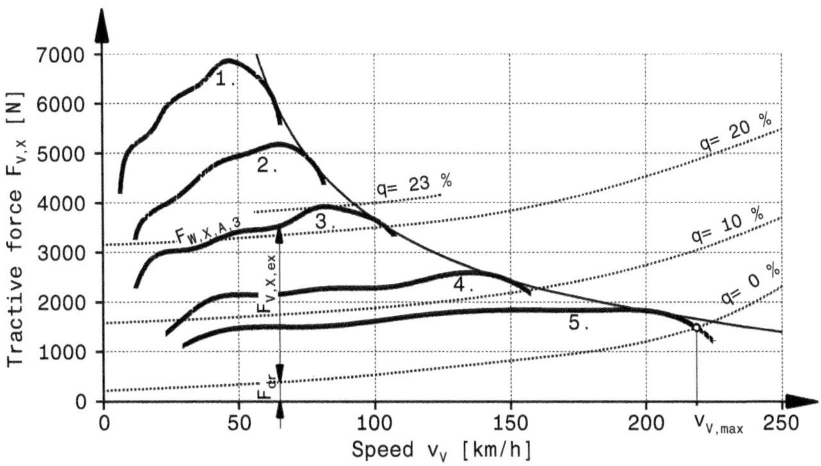

Fig. 4.13 Driving performance in the tractive effort diagram. In addition to the drive forces $F_{W,X,A}$, two driving resistance lines (road load) are drawn for the level ($q = 0\%$) and for gradients of 10 and 20%. The maximum speed on the level $v_{V,\max}$ is reached in fifth gear. For the third gear, the tractive effort surplus in the plane $F_{V,X,ex}$ at 60 km/h is plotted. The gradeability in third gear at $M_{M,\max}$ is 23%

$$F_{V,X,ex} = F_{W,X,A} - F_{dr} \tag{4.18}$$

$F_{V,X,ex}$	Tractive force surplus, N
$F_{W,X,A}$	Tractive force at the drive wheels, N
F_{dr}	Sum of driving resistances, N. see Sect. 4.1

From this generally valid equation, special cases can be derived. On the one hand, the climbing capacity of the wagon during unaccelerated travel, i.e. $F_a = 0$ N. This is achieved when the excess tractive force equals the gradient resistance:

$$F_{V,X,ex} = F_{W,X,A} - F_R - F_\alpha - F_L = F_q \tag{4.19}$$

F_q	Gradient resistance, N. see Sect. 4.1

From this follows directly the largest drivable slope:

$$\alpha = \arcsin\left(\frac{F_{V,X,ex}}{m_{V,t} \cdot g}\right) \tag{4.20}$$

α	Slope angle, °. The slope q in % is: $q = 100 \tan(\alpha)$
$m_{V,t}$	Total mass of the vehicle, kg

On the other hand, the equation for the excess tractive force yields the acceleration capacity in the plane, i.e. $F_q = 0$ N:

$$F_{V,X,ex} = F_{W,X,A} - F_R - F_\alpha - F_L = F_a \tag{4.21}$$

F_a	Acceleration resistance, N. see Sect. 4.1

From this, the possible acceleration a_X can be calculated for the speed under consideration:

$$a_X = \frac{F_{V,X,ex}}{m_{V,t} \cdot k_{m,n}} \tag{4.22}$$

a_X	Longitudinal acceleration for the speed at which the excess tractive effort $F_{V,X,ex}$ is present in gear n, m/s^2
$m_{V,t}$	Total weight of the vehicle, kg
$k_{m,n}$	Torque mass add (surcharge) factor in gear n, –. See Sect. 4.1, acceleration resistance

The maximum speed follows from (4.18) for the tractive force excess $F_{V,X,ex}$ set to zero. The driving resistances F_{dr} are then equal to the tractive force $F_{W,X,A}$ on the driving wheels. The relation between the sought velocity and the tractive force is given by (4.13). From these two relations it follows:

$$v_{V,\max} = 3.6 \cdot \left(\sqrt[3]{Y + \sqrt{Y^2 + Z^3}} + \sqrt[3]{Y - \sqrt{Y^2 + Z^3}} \right) \tag{4.23}$$

$$\text{with } Y = \frac{1000 \cdot P_{M,\max} \cdot \eta}{\rho_L \cdot c_W \cdot A_V}$$

$$\text{and } Z = \frac{F_R + F_\alpha + F_q}{\frac{3}{2} \rho_L \cdot c_W \cdot A_V}$$

$v_{V,\max}$	Maximum speed, km/h
Y, Z	Auxiliary variables
$P_{M,\max}$	Maximum engine power, kW
F_R, F_α, F_q	Driving resistances, N. see Sect. 4.1

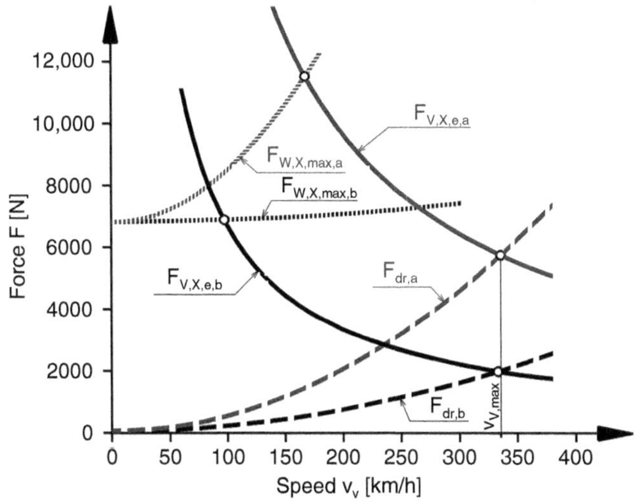

Fig. 4.14 Comparison of two vehicle designs in the tractive force diagram. Index a: Vehicle with high engine power and aerodynamic downforce. Index b: Vehicle with low aerodynamic drag and low engine power

The tractive effort diagram can also show the basic design difference between a vehicle with high aerodynamic downforce and a streamlined one, Fig. 4.14. Both rear-wheel-drive vehicles have the same total mass (600 kg) and cross-sectional area (1.24 m^2). The tyres also have the same characteristic value ($\mu_{W,X} = 2$). The engine power is matched to the driving resistance F_{dr} such that both vehicles reach the same maximum speed $v_{V,max}$. The vehicle with index b has low drag ($c_W = 0.3$) and hardly any downforce ($c_A = 0.1$). It results with a maximum engine power of 210 kW in a limiting speed of about 100 km/h. The other vehicle (index a) is designed for extremely high downforce ($c_A = 2.55$) and therefore has a correspondingly high aerodynamic drag ($c_W = 0.84$). Thanks to the high engine power (610 kW), this downforce can also be exploited. The speed limit is 170 km/h. Without downforce, this would be over 260 km/h (average with the course of $F_{W,X,max,b}$)! Up to this speed one could not use the full engine power, because the tires cannot transfer the occurring circumferential force. With a lower air resistance a much higher top speed would be possible (over 450 km/h), but this cannot be reached on usual race tracks (length of the straight lines) and is therefore not a sensible design goal. The high downforce (cf. also the adhesion limits $F_{W,X,max,a}$ and $F_{W,X,max,b}$), on the other hand, enables high acceleration in the longitudinal and lateral directions.

4.3 *Drivetrain* Overview

The conversion of the engine torque M_M to the drive torque M_A at the drive axle takes place via the drive train. With the gearbox, the total ratio can be adapted to the demand. The *total ratio* from the engine to the drive wheels follows to, Fig. 4.15:

$$i_t = i_{cl} \cdot i_G \cdot i_D \tag{4.24}$$

i_t	Total ratio, –
i_{cl}	Transmission ratio of the starting (launching) element, –
i_G	Gear ratio of the gearbox, –
i_D	Final drive ratio, –

If the starting element is a friction clutch, $i_{cl} = 1$. Hydrodynamic torque converters, which are the standard starting element in automatic transmissions, have a ratio $i_{cl} \geq 1$.

The ratios of the torques and speeds follow from the total transmission ratio i_t:

$$\frac{M_A}{M_M} = i_t \tag{4.25}$$

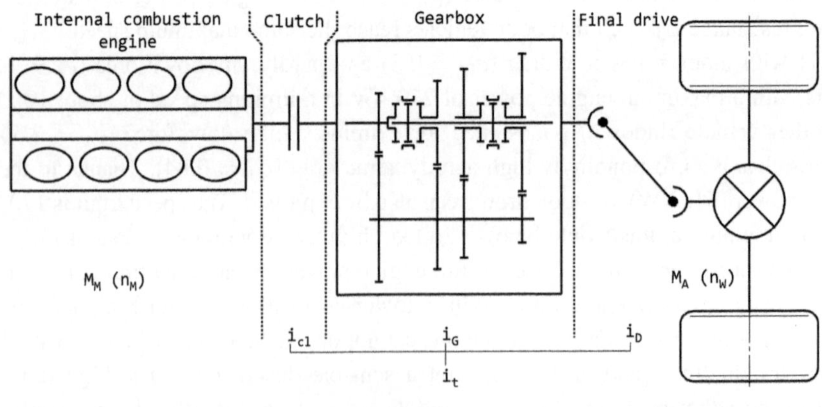

Fig. 4.15 Gear ratios in the drive train. The engine resp. motor torque M_M is translated by the drive train to the drive torque M_A

$$\frac{n_M}{n_W} = i_t \tag{4.26}$$

M_A	Drive torque at the wheels, N m
M_M	Engine resp. motor torque, N m
n_M	Engeine resp. motor speed, min^{-1}
n_W	Wheel speed, min^{-1}

4.4 Gear Ratios

Therefore, the engine power cannot be used directly for driving, but the usable speed range of the engine must be adapted by the transmission to the desired driving range of the vehicle. The gear ratios must allow for the following, depending on the vehicle, engine characteristics and intended use:

- starting on hill
- Reaching the desired maximum speed
- Achieve competitive acceleration.

For passenger cars and commercial vehicles, there is also the requirement for fuel-efficient operation.

The largest gear ratio $i_{t,\max}$ is used for starting, the smallest gear ratio $i_{t,\min}$ dictates the maximum speed. The ratio of the highest to the lowest transmission ratio is called spread, see also Fig. 4.16:

$$i_{G,t} = \frac{i_{G,\max}}{i_{G,\min}} \tag{4.27}$$

$i_{G,t}$	Gear spread, –
$i_{G,\max}$	*Highest* gear ratio, –
$i_{G,\min}$	*Lowest gear ratio*, –

Vehicles with a low specific engine power and those with engines with a narrow usable speed band need a larger gearbox spread. For comparison, Fig. 4.17 shows some reference values of different vehicles.

Selection of the Highest Gear Ratio $i_{t,\max}$

The determination of the largest gear ratio depends primarily on the power-to-weight ratio [kg/kW]. Depending on the vehicle, one of the following requirements determines the design:

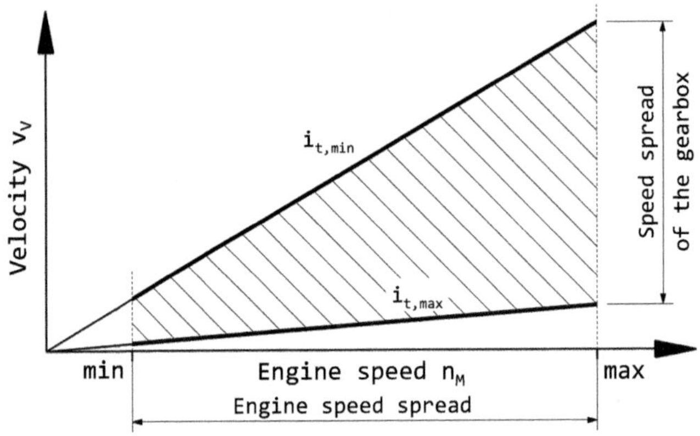

Fig. 4.16 Gearbox spreading. The drivable speed range of the engine is "spread" by the transmission to the driving range of the vehicle. The hatched area is the resulting usable range for the vehicle

Fig. 4.17 Reference values for gear spread, according to [2]

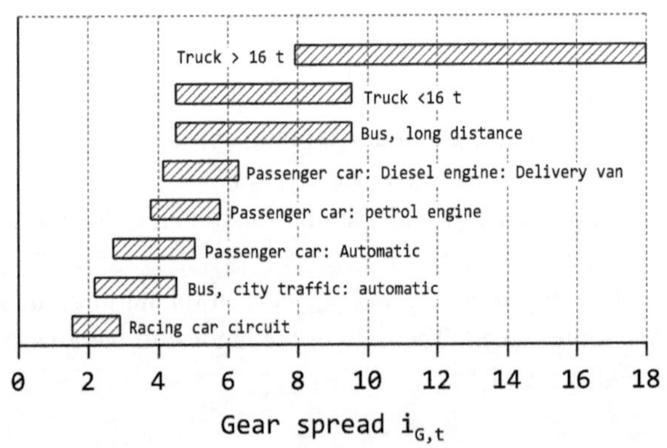

(a) Largest drivable gradient with acceleration $a_x = 0\ \text{m/s}^2$
(b) highest acceleration capacity on horizontal road surfaces

Largest Slope

If the requirement for the greatest climbing capacity is in the foreground, usually 50% for passenger cars (ramps, driveways), the greatest transmission ratio is calculated from:

$$i_{t,\max} = \frac{r_{dyn} m_{V,t} g (k_R \cos(\alpha) + \sin(\alpha))}{M_{M,\max} \cdot \eta} \tag{4.28}$$

$i_{t,\max}$	Largest total transmission ratio, –
r_{dyn}	Dynamic Tyre radius, m

(continued)

$m_{V,t}$	Total weight of the vehicle, kg
k_R	Rolling resistance coefficient, –
A	Slope angle of the roadway, °
$M_{M,\max}$	Maximum engine resp. motor torque, N m
η	Overall efficiency of the drive train, –

The dynamic tyre radius (dyn. Rolling radius) is calculated indirectly from the wheel revolutions and the distance covered in accordance with DIN 70020 at a driving speed of 60 km/h with 1.8 bar inflation pressure. Table 4.4 shows the dynamic standard tyre radius for some common tyre dimensions. The underlying dynamic tyre rolling circumference $C_{R,dyn}$ is much more decisive for the driving speed than the radius. The actual tyre radius, on the other hand, has a direct influence on the ground clearance and is thus important for aerodynamics, among other things. The dynamic rolling circumference increases with increasing driving speed. Radial tyres show a much smaller increase than cross-ply tyres due to their design (stiff steel belt). In general, the dynamic rolling circumference can be represented as a function of speed:

$$C_{R,dyn} = C_{R,dyn,60}(1 + 0.01 \cdot k_v) \tag{4.29}$$

$C_{R,dyn,60}$	Dynamic rolling circumference at 60 km/h, m
$C_{R,dyn}$	Dynamic Tyre rolling circumference above 60 km/h, m
k_v	Factor describing the increase in rolling circumference above the speed of 60 km/h, %

The factor k_v increases approximately with the square of the velocity. This relationship can be approximated with a second degree polynomial:

$$k_v = 0.00005 v_V^2 - 0.006 v_V + 0.2 \tag{4.30}$$

v_V	Vehicle speed, km/h

The values of k_v range between 0 (at $v_V = 60$ km/h) and 1.6% (at 240 km/h). They thus describe the behaviour of radial passenger car tyres. Figure 4.18 shows how the rolling circumference changes in a racing tyre. As the load increases, the rolling circumference decreases approximately linearly, while the speed has a quadratic influence. However, the increase is smaller than for a passenger car tyre. On average, the course of k_v can be approximated for this tyre:

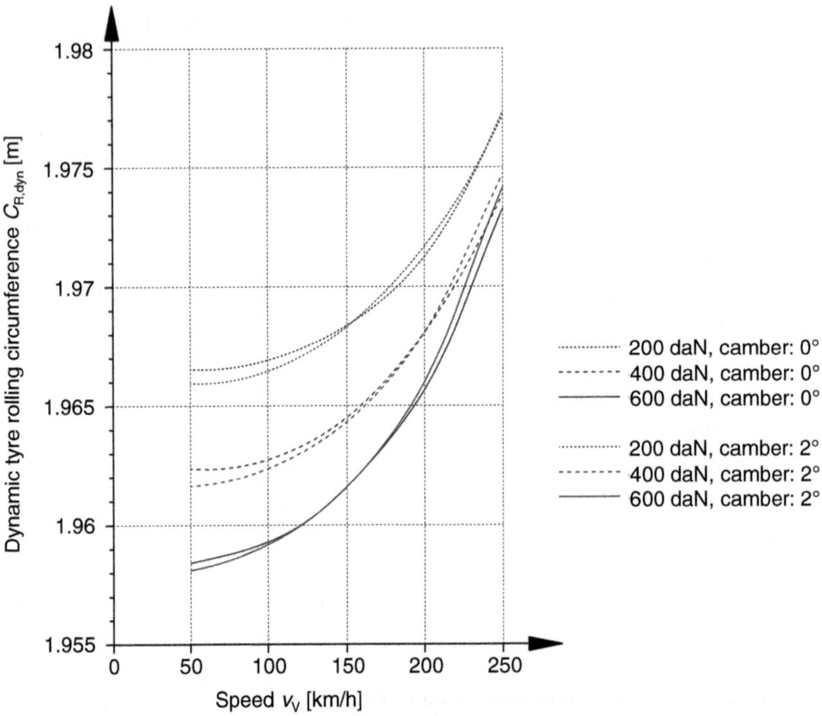

Fig. 4.18 Dynamic rolling circumference of the Formula Nissan 26/64–13 racing tyre (front axle), after [9]. The rolling circumference is plotted for 3 wheel loads and 2 camber angles versus vehicle speed. The static tyre circumference is 1.994 m. This results in an unloaded tyre radius of 0.3174 m. With a load, the radius is reduced to 0.3115 m, resulting in a dynamic rolling circumference of 1.957 m (50 km/h, 600 daN)

$$k_v = 0.000018 v_V^2 - 0.0026 v_V + 0.087 \tag{4.31}$$

The dynamic tyre radius r_{dyn} is calculated from the rolling circumference from the known relationship:

$$r_{dyn} = \frac{C_{R,dyn}}{2\pi} \tag{4.32}$$

Greatest Acceleration

The acceleration on a horizontal roadway is calculated as:

$$a_x = \frac{F_{V,X,ex}}{m_{V,t} \cdot k_{m,n}} \tag{4.33}$$

a_x	Longitudinal acceleration, m/s^2
$F_{V,X,ex}$	Excess tractive force, N
$k_{m,n}$	Torque mass add-on factor of the gear n, –

A certain excess tractive force is therefore required for a certain acceleration with a given vehicle mass. This can be taken from the tractive force diagram of the vehicle. The tractive force that can be achieved at a vehicle speed depends on the transmission ratio and the full-load engine torque curve. The calculated acceleration can also be plotted directly against the speed and a diagram is obtained as in Fig. 4.19.

For racing cars used on circuits or a fixed course, the following considerations are used to determine the highest gear ratio. If the vehicle is heavily overpowered, i.e. the adhesion limit of the tyres is the limit, a gear ratio as small as possible should be selected so that the adhesion limit of the tyres is not exceeded in the slowest turn of the circuit, Fig. 4.20. A small largest gear ratio $i_{t,\,\max}$ has the advantage that, for a given number of gears, there are smaller step changes to the smallest gear ratio and thus a better adaptation to the tractive force hyperbola is possible. The engine speed should be at the slowest point in such a way that the engine accepts the throttle without jerking when the driver accelerates again. Ideally, the speed at the exit of the corner is exactly at the speed of the highest torque. Then the driver can leave the corner with maximum acceleration.

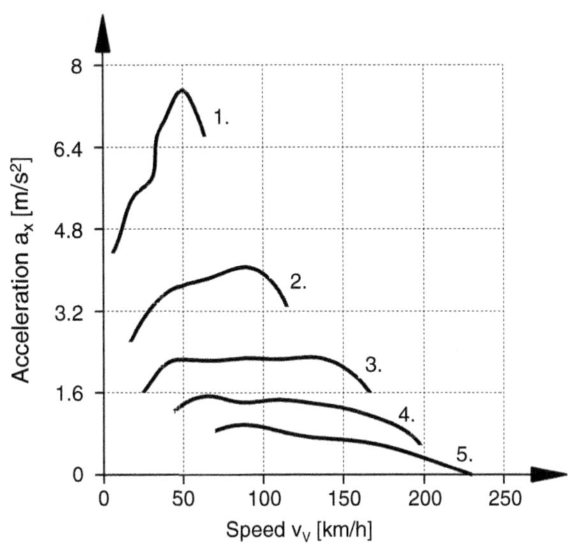

Fig. 4.19 Gear-dependent acceleration capacity. In this diagram, the influence of aerodynamic downforce on acceleration was not taken into account. As the transmission ratio becomes smaller, the tractive force surplus and thus the maximum acceleration decrease until this becomes zero when the maximum speed is reached

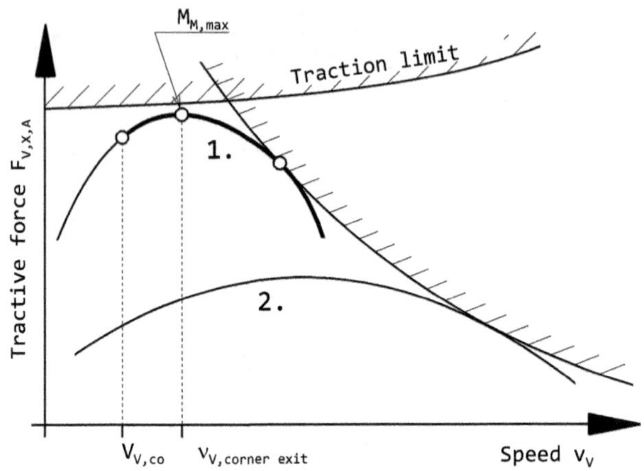

Fig. 4.20 Selection of the largest gear ratio for the narrowest turn of a circuit. The lowest vehicle speed $v_{V,co}$ occurs in the slowest corner. Here, the engine torque is before its maximum. When leaving the turn, the vehicle reaches a speed at the exit of the corner that corresponds to the engine speed at maximum torque. The tractive force is below the adhesion limit and therefore the vehicle can accelerate to the maximum

$$i_{t,\max,co} = \frac{3.6 \frac{\pi}{30} n_{M,co} \cdot r_{dyn}}{v_{V,co}} \tag{4.34}$$

$i_{t,\max,co}$	Largest gear ratio for cornering, –
$v_{V,co}$	Vehicle speed in the turn under consideration, km/h
$n_{M,co}$	Desired motor speed at speed $v_{V,co}$, min^{-1}
r_{dyn}	Dynamic Tyre radius, m

If the tyres are not the limiting link in the driveline, select the highest gear ratio so that the maximum engine torque is available at the exit of the corner.

Selection of the Smallest Transmission Ratio $i_{t,\min}$

The lowest gear ratio results directly from the desired maximum speed of the vehicle. If one neglects the tire slip and uses the maximum engine speed for reaching the theoretical maximum speed, the following follows:

$$i_{t,\min} = \frac{3.6 \frac{\pi}{30} n_{M,\max} \cdot r_{dyn}}{v_{V,\max}} \tag{4.35}$$

$i_{t,\min}$	Smallest total gear ratio, –
$n_{M,\max}$	Maximum engine resp. motor speed, $\min^{-1}$
r_{dyn}	Dynamic Tyre radius, m
$v_{V,\max}$	Maximum speed, km/h

The maximum speed of an engine depends, among other things, on the combustion process and the effect of a prescribed air restrictor. Petrol engines have a much higher final speed (up to 20,000 $\min^{-1}$) than diesel engines, which for reasons of mixture formation can only reach about 6000 $\min^{-1}$.

Other factors are also taken into account in the final determination of the smallest gear ratio. For passenger cars, fuel consumption and the service life of the highest gear are the main considerations. After all, the share of running time of the highest gear stage can be up to 80%. For racing cars, on the other hand, high excess power is preferred. For fuel consumption, the level of engine speed is decisive; a large tractive power surplus enables considerable acceleration. The following different design principles are therefore common, see also Fig. 4.21:

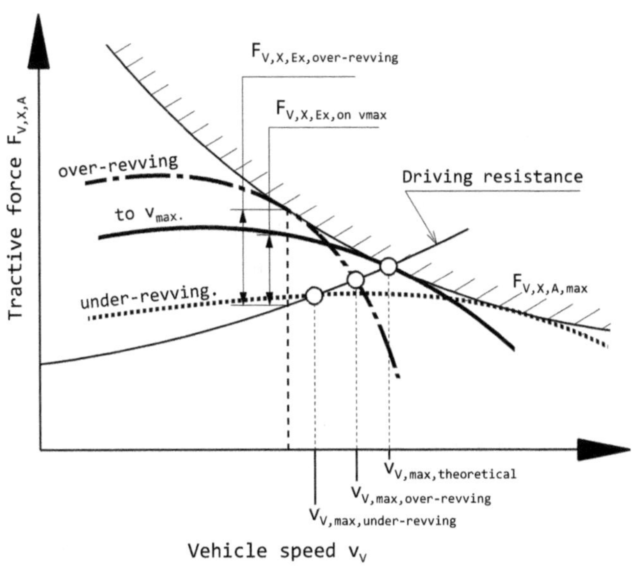

Fig. 4.21 Comparison of different design types. The highest final speed $v_{V,\max,\text{theoretical}}$ achieved is the design in which the driving resistance line is intersected by the engine tractive force curve at the point of highest engine power. If the greatest engine power is reached at a lower speed, the tractive force surplus $F_{V,X,Ex}$ is large and the final speed is lower than the theoretical maximum (overspeeding design). If the vehicle reaches the final speed before the maximum engine power, the engine speed level is lower and thus in a fuel-efficient range

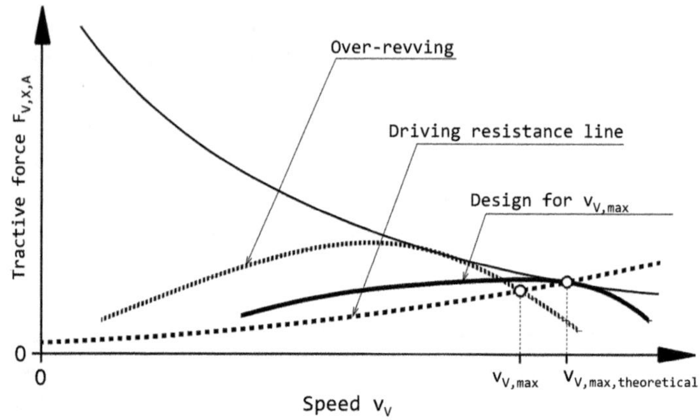

Fig. 4.22 Gearbox designs in the tractive force diagram. The theoretical maximum speed $v_{V,max,theoretical}$ is only achieved if the maximum speed is reached exactly at the engine speed with the greatest engine power. With an overspeeding design, the maximum engine power is reached at a lower speed. The maximum speed $v_{V,max}$ results from the balance of engine power and driving resistance at a higher engine speed and is lower than the theoretically possible

- Design for maximum top speed
- excessive interpretation
- underspeeding design

Design for Maximum Top Speed

If the maximum engine power is to be used to achieve the highest possible final speed, this must be achieved exactly at the engine speed at which the nominal power is applied. In the tractive effort diagram, this means that the line of driving resistance is intersected by the engine tractive effort curve at the point of maximum engine power (= tangent to the tractive effort hyperbola), Fig. 4.22.

Overspeeding (Overrevving) Layout

This design is often used for sporty vehicles. Here, the maximum speed is reached at a speed that is higher than the nominal speed. The power demanded from the engine is therefore less than the maximum power and consequently the speed achieved is below the theoretically possible speed. The advantage of this design is the large tractive power surplus $F_{V,X,Ex}$ before the maximum speed is reached. The acceleration capacity is therefore high even close to the final speed.

Compared to the design for maximum top speed, the overall gear ratio i_t is higher. This is achieved by a larger gear ratio or a larger ratio of the axle drive.

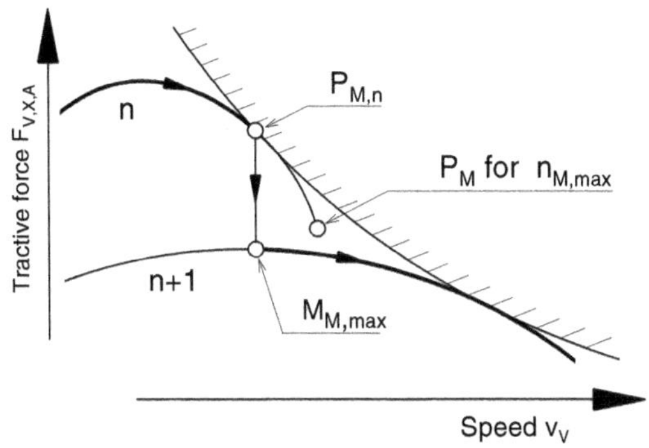

Fig. 4.23 Shifting operation in the tractive force diagram. Shifting up from gear n to $n + 1$ ideally takes place at the point of highest engine power (nominal power) $P_{M,n}$. In the next gear $n + 1$ the vehicle speed corresponds to the engine speed with the highest torque $M_{M,\max}$. Shifting down takes place exactly the opposite way from the speed of the highest torque. The maximum speed $n_{M,\max}$ must not be exceeded in gear n

Underrevving Design

If the focus is on fuel consumption, the engine speed must be reduced for high loads. The gear ratio is selected in such a way that the maximum speed is reached at an engine speed below the rated speed. This means that the vehicle cannot accelerate any further even though the engine has not yet reached its maximum power. The tractive power surplus is low, but so is the fuel consumption, because the engine is in a more fuel-efficient range of the engine map.

An underspeeding design results from a lower overall gear ratio than when designed for the highest top speed.

***Final Drive* Ratio i_D**

The final gear ratio to be realized in one stage is in the range $2 \leq i_D \leq 7$. If a larger gear ratio is needed, another gear ratio stage is provided.

Choice of *Intermediate* Gears

If the highest and lowest gear ratios are fixed, the gradations between them are determined. The gradation should be set in such a way that when shifting down when the maximum engine torque is reached, the maximum engine speed in the next lower gear is not exceeded, Fig. 4.23.

If you want to achieve the greatest acceleration, it is important that the area under the engine tractive force curve is as large as possible [10]. The shift point is then slightly above the nominal speed (speed of maximum power), Fig. 4.24. In the case of close gear graduation, the ideal shift point results from the intersection of the tractive force curves of successive gears.

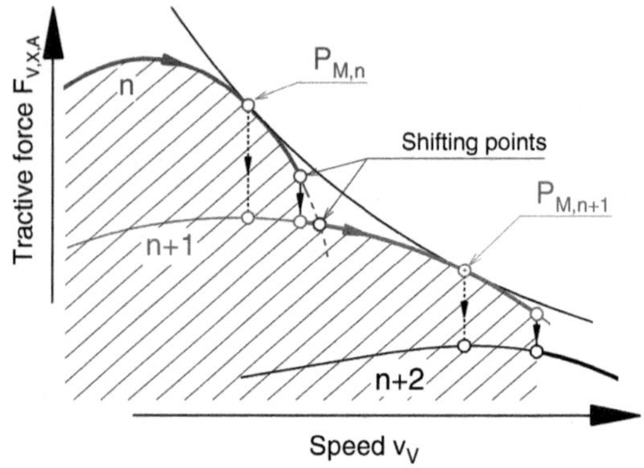

Fig. 4.24 Shifting operation for high acceleration. For high acceleration, each gear is shifted above the rated speed (solid line). For comparison, the shift point at nominal speed is entered (dotted line). In this case, the area under the tractive force curve of the engine is considerably smaller. If successive tractive force curves intersect, the point of intersection is the ideal shifting point

The steeper the torque curve of an engine and the narrower its usable speed band, the more important the gear graduation becomes. The more gears a gearbox has, the better the engine tractive force can be adapted to the tractive force hyperbola. However, the design cost and mass of a transmission increase as the number of gears increases. Small gear steps also lead to more frequent shifting, which is also associated with a certain amount of time. On the other hand, larger gear steps lead to a larger drop in speed when shifting, which places greater stress on the teeth and claws due to the correspondingly larger impact and thus reduces their service life. When shifting with traction interruption, the vehicle speed decreases during the shifting process, for more information see Sect. 5.3.1 Manual gearbox, traction interruption.

The ratio of the transmission ratios of two adjacent gears is the gradation (step jump):

$$\Phi = \frac{i_{G,n}}{i_{G,n+1}} \tag{4.36}$$

Φ	Gradation, –
$i_{G,n}, i_{G,n+1}$	Gear ratio of gear n or $n + 1$

In practice, the following methods are used to calculate the gradation:

- geometric gear gradation
- progressive gear gradation
- Tuning for a certain circuit.

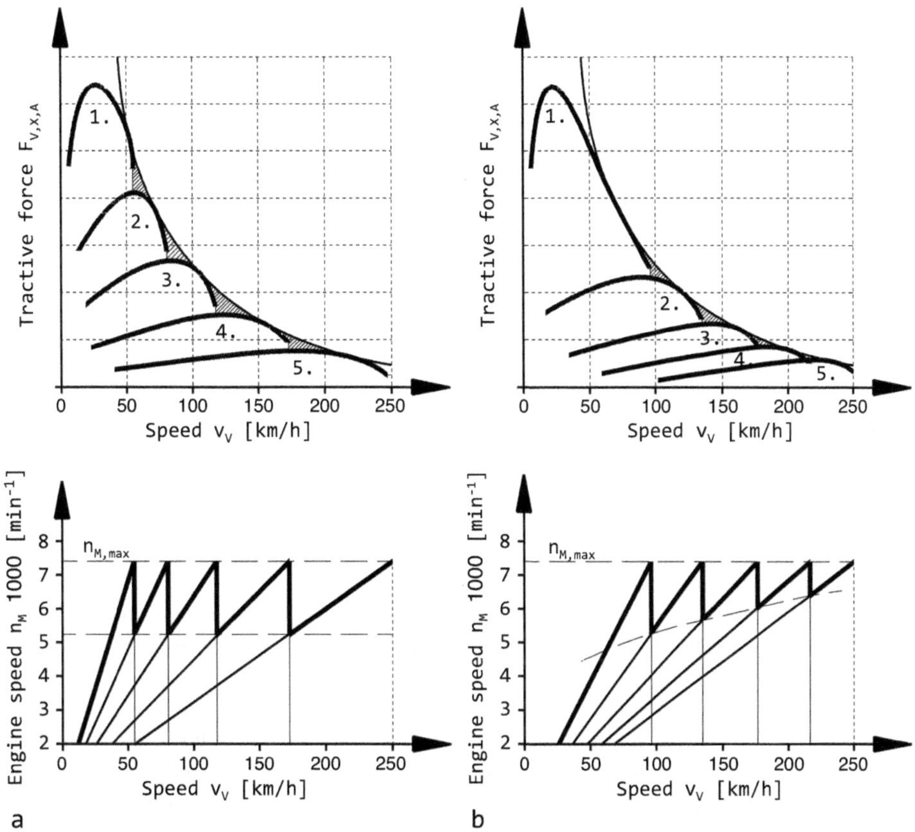

Fig. 4.25 Comparison of design types. Above tractive force diagram, below speed-speed diagram (gear diagram). (**a**) geometric design, (**b**) progressive design

Figure 4.25 provides a comparative overview of both design types. With geometric gradation, the step change is always the same. The approximation of the engine tractive force to the effective tractive force hyperbola is thus approximately the same for all gears, but with the consequence that the differences in the maximum speeds of the individual gears become greater and greater towards the top. When shifting, the engine speed always falls back to the same level.

In contrast to geometric gradation, with progressive gradation the step jump becomes smaller and smaller as the gear number increases. The speed differences at maximum engine speed thus remain more or less constant and the uncovered areas (shown hatched) in the tractive force diagram become smaller and smaller at higher gears. As a result, acceleration is better in this speed range. In the lower gears, the tractive power surplus is so large anyway that larger gaps hardly matter. When shifting, the speed drop becomes smaller and smaller with higher gears, which makes shifting easier at higher speeds.

i_G	37/14	35/18	28/18	27/21	22/20	26/27
	2.64	1.94	1.55	1.28	1.10	0.96
Δn_M	-	1909	1447	1254	1013	916
$v_{V,max}$	94.48	128.57	160.91	194.86	226.74	259.81
Gear	1.	2.	3.	4.	5.	6.

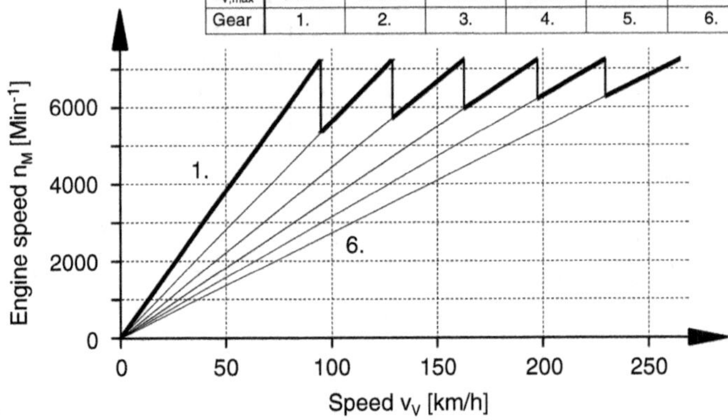

Fig. 4.26 Transmission diagram of a Formula Renault car. Shifting takes place at the maximum speed of 7200 min^{-1}. The speed drop Δn_M becomes lower with increasing gear number, it is therefore a progressive gradation. In addition, the transmission ratio i_G of each gear is given as a ratio of the number of teeth and as a rational number. The axle ratio is $i_D = 3.1$. There are two more gear ratios for this vehicle. The diagram shows the smallest gear ratios, i.e. the highest final speeds

In racing cars, the gradation is advantageously progressive. Figure 4.26 shows an example of the transmission diagram (sawtooth diagram) of a formula car. The speed curve is plotted linearly against the engine speed, i.e. no slip or tyre growth is taken into account.

The individual gear ratios are calculated from the smallest gear ratio and from the step change.

Geometric Gradation

With geometric gradation, the step change is always the same and is calculated to:

$$\Phi_{geom} = \sqrt[j-1]{i_{G,t}} \tag{4.37}$$

Φ_{geom}	Step jump with geometric gradation, –
j	Number of gears, –
$i_{G,t}$	Gear spread, –

This gives the gear ratios of the individual gears $n = 1$ to j:

$$i_{G,n} = i_{G,j} \cdot \Phi_{geom}^{(j-n)} \tag{4.38}$$

$i_{G,n}$	Ratio of the gear n, –
$i_{G,j} = i_{G,\min}$	Gear ratio of the highest gear = smallest gear ratio, –

Geometric increments are generally selected when all gears of a transmission are equivalent in driving operation.

Progressive Gradation

The gradations between the individual gears decrease steadily. So first a progression factor is chosen that determines the change in the ratio step.

k_Φ	Progression factor, –. Usual values are between 1.0 and 1.2

The first gradation step is calculated from the gear spread to:

$$\Phi_1 = \sqrt[j-1]{\frac{1}{k_\Phi^{0.5(j-1)(j-2)}} i_{G,t}} \tag{4.39}$$

Φ_1	First step, –. Usual values: 1.1–1.7
j	Number of gears, –
$i_{G,t}$	Gear spread, –

This gives the gear ratios of the individual gears $n = 1$ to j:

$$i_{G,n} = i_{G,j} \Phi_1^{(j-n)} \cdot k_\Phi^{0.5(j-n)(j-n-1)} \tag{4.40}$$

$i_{G,n}$	Ratio of the gear n, –
$i_{G,j} = i_{G,\min}$	Gear ratio of the highest gear = smallest gear ratio, –

The actual gear ratios are naturally chosen to correspond to an integer number of teeth ratio. This can result in slightly different step jumps between gears even with geometric graduation.

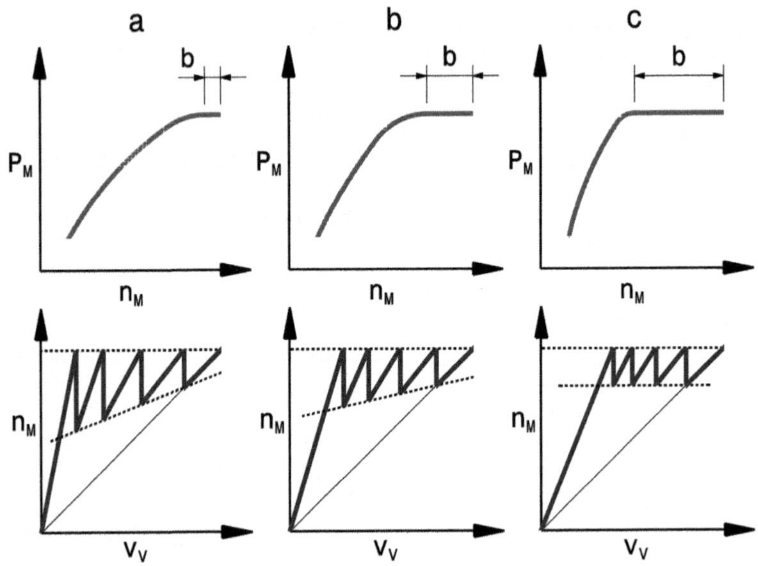

Fig. 4.27 Selection of intermediate gears based on engine characteristic curve, schematic according to [11]. The three engines differ in the width b of the speed band at maximum power. (**a**) peak characteristic curve of the engine power, (**b**) medium width of the speed band, (**c**) wide speed band. The corresponding transmission diagram is shown under each

Gearing to Engine Characteristics

The course of the maximum power P_M can also be included in the choice of intermediate gears, Fig. 4.27. Three different engine characteristics can be seen. Engine (a) shows a peaked characteristic of its maximum power and the gear is the strongly progressively stepped. Engine (c) represents the other extreme. It has a distinctly constant maximum power over the speed n_M (speed bandwidth b), such as an engine with air restrictor (surge volume limiter). The matching gearbox is linearly stepped. Engine (b) is between the two with its characteristics and its gearbox is less progressively stepped than with (a).

Circuit Graduating

On a known circuit, the speeds in the corners and at the end of long straights are known. The gear ratios are chosen so that the greatest acceleration can be achieved when leaving a turn, i.e. the speed at the exit of the corner coincides with the speed of the greatest engine torque. For long straights, the gear ratio of the corresponding gear will be chosen so that the maximum speed is reached just at the end of this section of the track. If the gear ratio were too high, the driver would have to shift up again just before the next turn and then brake.

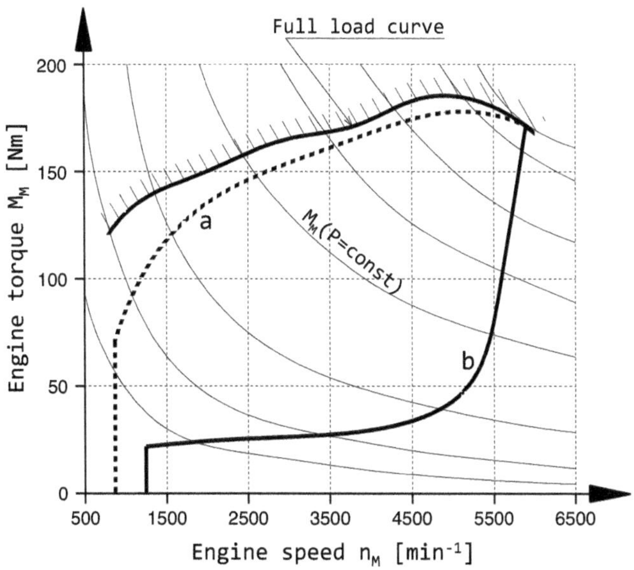

Fig. 4.28 Engine map with control characteristics of a continuously variable transmission, according to [2]. The map is limited upwards by the full load curve. This curve of maximum engine torque cannot be exceeded. With a continuously variable transmission, any speed and load points can be approached in the operating range below. Two extreme control characteristics are entered. **a** Control characteristic for minimum fuel consumption, **b** Control characteristic for high driving performance. The control for low fuel consumption is located in such a way that the consumption-favourable ranges are passed through. The tractive effort-oriented control aims for the greatest possible tractive effort surplus at every operating point

Gear Ratios and Their Changes for Continuously Variable Transmissions

As the name implies, no fixed gear steps are set in these gearboxes, but the transmission ratio can be varied as desired within a certain range. The gear spread of continuously variable transmissions, which is also called the control range in the transmission design, is $i_{G,t} = 5\text{–}6$.

Basically, the torque and speed of the engine can be set as desired with continuously variable transmissions, as long as the corresponding operating point lies within the drivable range of the engine map, Fig. 4.28. If a continuously variable transmission is to be used to achieve high driving performance, the control characteristic must run at the greatest possible distance from the full load torque curve. Then a large tractive power surplus is guaranteed.

What is gear shifting in a stepped transmission is the adjustment speed in a continuously variable transmission. The adjustment speed is the change in engine speed over time or the change in transmission ratio over time at a fixed output speed. A too low adjustment speed leads to lame driving behaviour. The adjustment should therefore take place as quickly as possible. However, there are limits to increasing the adjustment speed. The energy required for the adjustment comes partly from the kinetic energy of the vehicle. If this energy is too high, acceleration briefly becomes negative, which is noticeable as “shift jerk”.

The use of continuously variable transmissions in racing vehicles is rare. The currently most efficient design (variable pulleys with thrust link chain) has its application limit in series production at 300 N m input torque at spread 6. With power split in such a transmission, the torque capacity can be increased to 550 N m at spread 7 [12].

References

1. Bosch: Kraftfahrtechnisches Taschenbuch, 22nd edn. VDI, Düsseldorf (1995)
2. Lechner, G., Naunheimer, H.: Fahrzeuggetriebe. Springer, Berlin (1994)
3. Mitschke, M., Wallentowitz, H.: Dynamik der Kraftfahrzeuge, 4th edn. Springer, Berlin/Heidelberg/New York (2004)
4. Reimpell, J., Betzler, J.: Fahrwerktechnik: Grundlagen, 4th edn. Vogel, Würzburg (2000)
5. Wallentowitz, H.: Längsdynamik von Kraftfahrzeugen, Vorlesungsumdruck Fahrzeugtechnik I, 12th edn. Institut für Kraftfahrwesen, Aachen (2005)
6. McBeath, S.: Formel 1. Aerodynamik. Motorbuch, Stuttgart (2001)
7. Indra, F.: Grande complication, in Automobil Revue Nr. 50-1996. Automobil Revue AG, Bern
8. Wright, P.: Ferrari Formula 1. Under the Skin of the Championship-Winning F1–2000, 1st edn. David Bull Publishing, Phoenix (2003)
9. N.N.: Super Nissan V6, User Manual, Release 2.1, Dallara (2003)
10. Smith, C.: Tune to Win. Aero Publishers, Fallbrook (1978)
11. Roberts, N.: Think Fast, the Racer’s why-to Guide to Winning, 1st edn. Eigenverlag des Autors, Charleston (2010)
12. Englisch, A., et al.: Entwicklungspotenziale für stufenlose Getriebe, vol. 7–8, pp. 676–685. ATZ (2003)

Power Transmission

5

The power source – combustion engine or electric motor – provides the power needed to overcome the driving resistances. However, this power must be conducted from the engine output to the drive wheels. Sometimes, in addition to overcoming a spatial distance, it is also necessary to adapt to the engine resp. motor characteristics. The following chapter reports on how all this is done.

M. Trzesniowski, *Powertrain*,
https://doi.org/10.1007/978-3-658-39885-9_5

5.1 Function

The drive train has to fulfil the following functions. The drive torque provided by the engine must be transmitted to the wheels with the lowest possible losses. If the engine is installed longitudinally, a deflection of the torque flow by 90° is also required. If more than one axle is driven, the engine power must be distributed accordingly to several axles.

Another obvious (but still worth noting in the concept phase) feature is that the vehicle moves forward. This is because the direction of rotation of the engine is standardised and most manufacturers stick to the usual direction of rotation, namely anti-clockwise when looking at the power output side. Depending on the position of the engine and gearbox in the vehicle as well as the type of transmission (chain, gear wheel, . . .), the ring gear must be located to the left or right of the drive bevel gear when the engine is installed longitudinally, Fig. 5.1.

An overall system overview of the drive train is provided in Fig. 5.2. A starting element ensures the transition between vehicle standstill and travel when the combustion engine is running. (This element is omitted in the case of an electric motor). The torque provided by the engine is adapted in the transmission to the driving speed and thus to the driving resistance. A transfer case splits the power between the front and rear axles. In the case of a pure front- or rear-wheel drive, the transfer case is omitted. The final drive then redirects the torque flow or adjusts the speed, and the differential distributes the power to the two wheels of an axle so that they can have different speeds.

The main functions and possible principles for their representation are summarized in a morphological box in Table 5.1.

The main functions are listed in the rows and some possible solutions are listed in the columns. By combination of solution-principles to a complete working drive-train, one gets all possible designs. Naturally, not all of them are implemented in practice, for example because they are not sensible.

The drive train therefore generally comprises the following assemblies, which are important for its functional performance:

- Clutch: As a starting element and to interrupt the torque flow.
- (Manual) transmission: For adapting the engine characteristics to the requirements of the vehicle.
- Axle drive: To transmit the engine resp. motor torque adjusted by the gearbox to the wheels. Partial additional deflection of the power flow by 90°.
- Differential gear: Allows you to negotiate a tight corner with little loss of power, even with high friction coefficients of the tyres.
- Shafts: For conducting torque between said assemblies.

Apart from considerations regarding the axle load distribution and the vehicle's moments of inertia, efficiency considerations can also be useful when basically arranging the necessary drive assemblies. As always, if the driving performance is to be high, one will

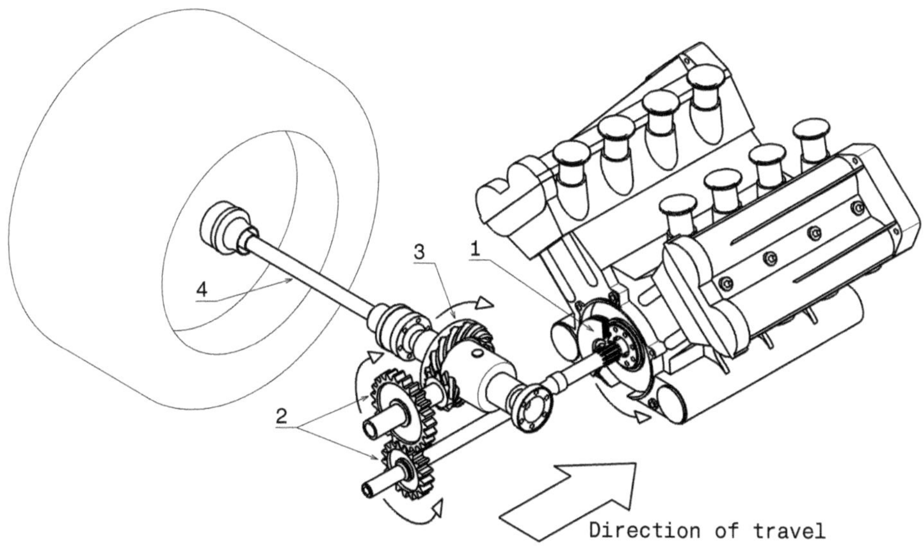

Fig. 5.1 Drive line with central engine arrangement. **1** Clutch (partially cut shown). **2** *meshing gears of mechanical gearbox.* **3** Crown wheel final drive. **4** Side shaft to left rear wheel. The arrows show the direction of rotation of the engine, shafts and gears when moving forward

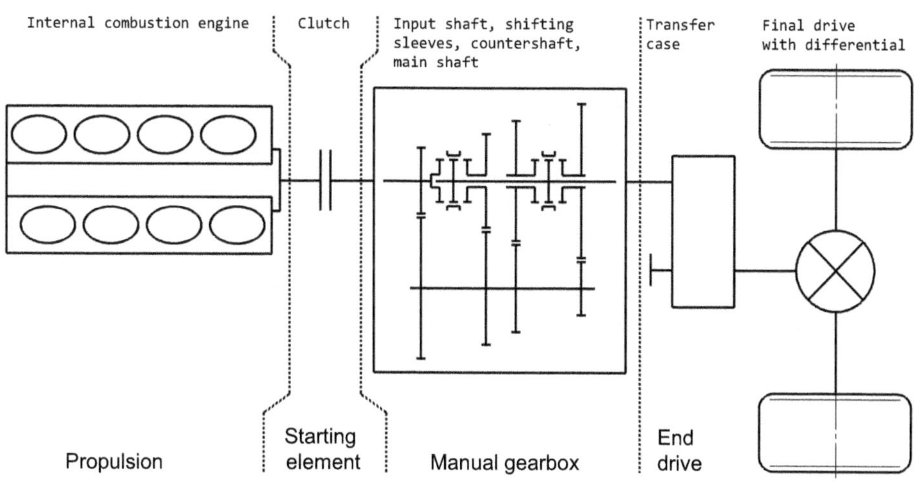

Fig. 5.2 Functional diagram of the drive train

strive to keep the losses as low as possible. Table 5.2 gives an overview of the efficiency of components in the drive train.

It can be seen that a pair of gears is the most efficient solution for converting a torque. Incidentally, the low construction effort and the large transmission ratio range also speak in favour of gears. Accordingly, vehicle transmissions based on gears are also the best choice

Table 5.1 Solution principles for the main functions of the drive train, according to [1]

Main function	Solution principle 1	2	3	4	5
Enable starting	Mechanically dry	Mechanical wet	Electro-mechanical	Hydrodynamic	Hydrostatic
Gear ratio/change speed	Spur gears	Planetary wheels	Hydrodynamic	Hydrostatic	Mechanically stepless
Shifting/power flow	Positive locking pinions	Positive shifting claw synchronized	Positive locking shifting claw unsynchronized	Multi-plate friction clutch	Friction-locked Multiple-disc brake
Operating/controlling the gearbox	Manual transmission	Manual gearbox servo assisted	Semi-automatic	Automatic hydraulic	Automatic electro-hydraulic
Axle drive (final drive)	Spur gear	Hypoid bevel gear	Spiral bevel gear	Worm wheel	Chain
Speed compensation of the two sideshafts when cornering	Bevel gear differential	Spur gear differential in planetary design	Worm gear differential	No speed compensation	
Locking possibility for differential	Open (unlocked)	Self-locking	Automatically lockable	Manually lockable	

Table 5.2 Reference values for efficiency ranges of gearboxes, [1, 2]

Gear type		Efficiency η, %
Gear pair	Spur gear	99.0–99.8
	Bevel gear	90–93
Plate link chain		95–99
Worm gear		30–98
Toothed belt		95–99
Mechanical gearbox with splash lubrication	Passenger car	92–97
	Commercial vehicle	90–97
Step automatic transmission with torque converter and lock-up clutch		90–95
Mechanical stepless transmission	Contact pressure not power demand controlled	70–80
	Contact pressure power demand controlled	80–86

if efficiency and power density are decisive criteria. As an example, Fig. 5.3 shows the distribution of the losses of a manual gearbox over the speed.

It can be seen that the largest consumers, apart from the expected gearing, are the lubrication and the bearings. It therefore pays to take a close look at the entire lubrication system when designing. Excessive oil quantities are just as bad as insufficient lubrication. Seals, as required for shaft passages, have the potential to reduce friction, especially at high speeds. Synchronizing devices are also responsible for losses and are not found at all in racing gearboxes, not least for this reason.

Different drive concepts lead to different efficiencies due to their principles, Fig. 5.4.

With the standard drive (i.e. engine at the front and drive at the rear) with coaxial gearbox, the overall efficiency in the direct gear is practically only dependent on the final drive (a). With all other gears, the balance in the gearbox looks worse, because two tooth meshes are always required in the gearbox to transmit the torque due to the countershaft design (b). Single-stage gearboxes are more favourable in this respect, as they are used in many racing cars with mid-engines (c). For all gears, only one pair of gears is ever required to change the ratio. The best design in terms of efficiency is represented by typical passenger car drives with transverse engine and front-wheel drive (d). The transmission has a single-stage design and the final drive consists of a spur gear stage, which transmits the torque with much less loss than any bevel gearing.

Figure 5.5 shows the efficiency of a drive train as a function of input torque and speed. The efficiency is poor at very low input torques. Up to about 2 N m a drag torque must be overcome, only then does the efficiency become non-zero. It increases steeply and from about 25 Nm the efficiency is above 80%. As the input torque continues to increase, the efficiency increases only slightly. For normal driving operation, it can be assumed to be approximately constant.

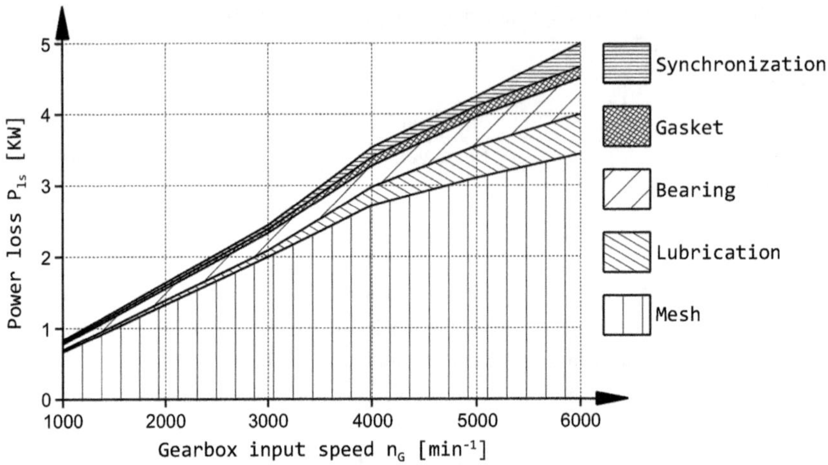

Fig. 5.3 Distribution and size of the power loss of a manual gearbox, according to [3]. The transmission is a two-stage (coaxial) 5-speed transmission with manual shift. The measurement refers to the 4th (direct) gear at maximum engine torque. The transmission oil temperature is 80 °C

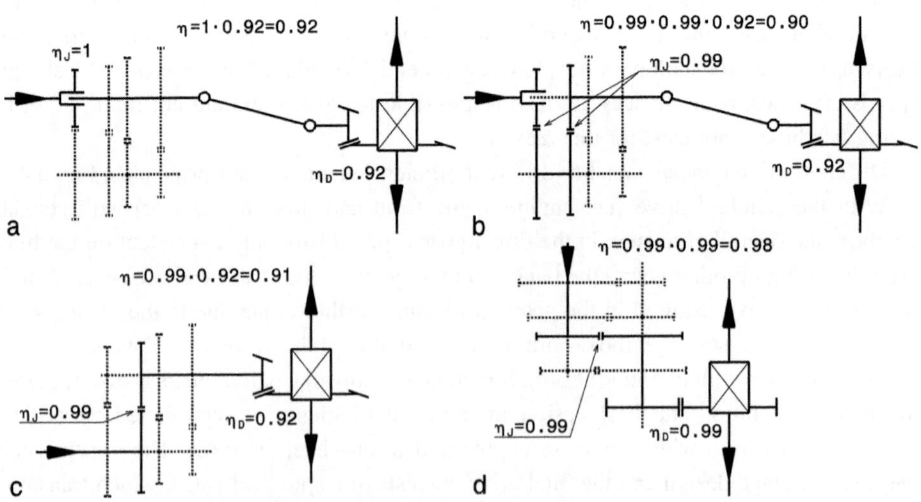

Fig. 5.4 Efficiencies of powertrains. (**a**) Standard drive, direct gear. (**b**) Standard drive in general. (**c**) Single-stage longitudinal gearbox. (**d**) Transverse gearbox for transverse engine. η Overall efficiency, η_J Efficiency of a gear pair, η_D Efficiency of the final drive

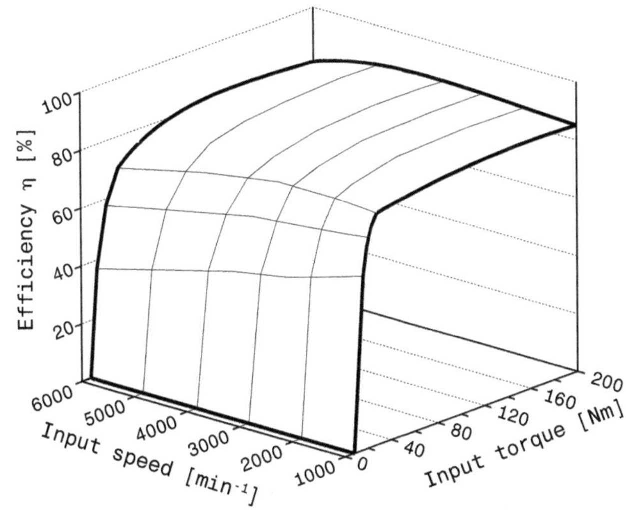

Fig. 5.5 Characteristic diagram of the overall efficiency of a drive train (engine/motor output to drive wheels), according to [3]. The map applies to the direct fourth gear of a 5-speed manual transmission. A power steering pump is taken into account as an auxiliary unit

5.2 Clutch

The clutch enables a separation of the power flow between the drive wheels and the engine. For this purpose, it is arranged between the engine output and the gearbox input.

Clutches also serve as a starting element and are subjected to the greatest stress. Heavily slipping tyres quickly lose grip, while clutches are better able to cope with slippage. However, every slipping clutch also reaches its temperature limit sooner or later, which is why the slipping of the clutch must not last too long. Electronic control units can specifically reduce the load on the clutch by means of a start-up program. This means that clutches actuated purely by the driver's foot have to be dimensioned much more strongly. There are also clutches that are practically only used as a separating element in the driveline. In this case, starting takes place by means of a bang start, i.e. the drive wheels take over the function of changing the speed. Figure 5.6 shows an idealized starting process

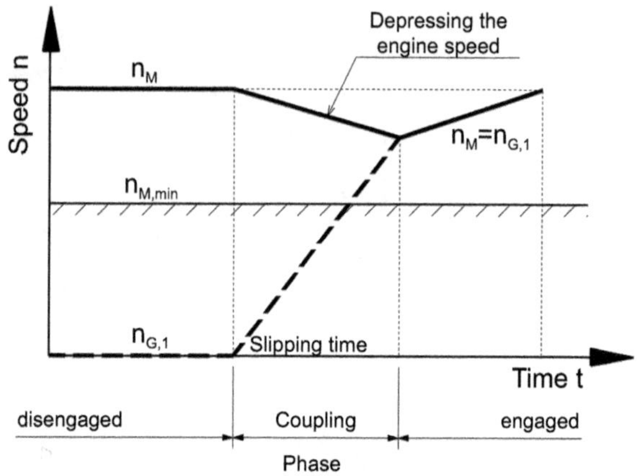

Fig. 5.6 Idealized starting procedure with a friction clutch. During starting, the speeds of the engine and wheels equalize, whereby the minimum engine speed $n_{M,min}$ must not be fallen short of at any time. n_M Engine speed, $n_{M,min}$ Minimum engine speed, $n_{G,1}$ Speed of the gearbox input shaft

with a friction clutch. The combustion engine requires a certain minimum speed $n_{M,min}$, so that it can deliver a torque. The vehicle is initially stationary and therefore the speed $n_{G,1}$ of the transmission input shaft is also zero. The clutch now has the task of equalizing these two speeds without falling below the minimum engine speed. However, this task is not performed by the clutch alone, but by the driver or a start program which regulates the engine load accordingly.

During the speed adjustment the clutch slips and thereby a part of the input power is converted into heat as power loss. The efficiency of the clutch thus results in:

$$\eta_{cl} = \frac{P_2}{P_1} = \frac{M_2 \cdot 2\pi \cdot n_{G,1}}{M_1 \cdot 2\pi \cdot n_M} \tag{5.1}$$

η_{cl}	Efficiency of the clutch, –
P_1	Input power, kW
P_2	Output power, kW
M_1	Input torque, N m
M_2	Output torque, N m
n_M	Engine speed, min-^{1}
$n_{G,1}$	Speed of the gearbox input shaft, min^{-1}

5.2.1 Types of Clutches

Of the conceivable principles for clutches, friction clutches have become established in vehicle construction. Here, the engine torque is transmitted via discs that are pressed against each other. Depending on the number of discs and the type of heat dissipation, the designs are divided into single and multi-disc clutches or dry and wet clutches. In

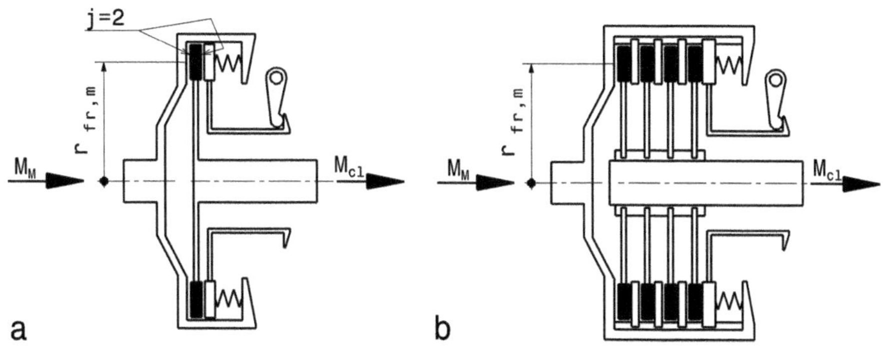

Fig. 5.7 *Shifting friction clutch.* (**a**) Single-plate clutch . (**b**) Multi-plate clutch. The clutch disc(s) are pressed against the flywheel by springs and thus transmit the engine torque. j Number of friction surfaces, $r_{fr,m}$ Average friction radius

racing, dry-running clutches with as few discs as possible are often used. Figure 5.7 shows the operating principle of friction clutches.

To disengage the clutch, the contact pressure is overcome by the disengager and the clutch discs run independently of the flywheel.

Dry Clutch Passenger Car (*Single-Plate Clutch*)

Dry clutches are highly efficient and require practically no drag torque when open. The single-plate clutch dominates in passenger cars.

Coefficient of friction between lining and friction surface:	0.3–0.45. Static and dynamic friction are practically the same. With fading, the value drops to approx. 0.2

Heat flux density: 0.3 to 3 W/mm^2 [4].

***Multi-Plate* Clutch**

Although single-disc dry clutches are predominant in passenger cars, the limit can be reached in high-performance vehicles and more – usually two – discs must be installed.

Multi-plate clutches are often used in an oil bath. This is the predominant design for motorcycles and double clutch transmissions. With the oil flowing through, the frictional heat generated can be dissipated in a targeted manner and thus the power density of the clutch can be increased.

Coefficient of static friction between lining and friction surface:	Wet: 0.1–0.2 Dry: 0.3–0.5

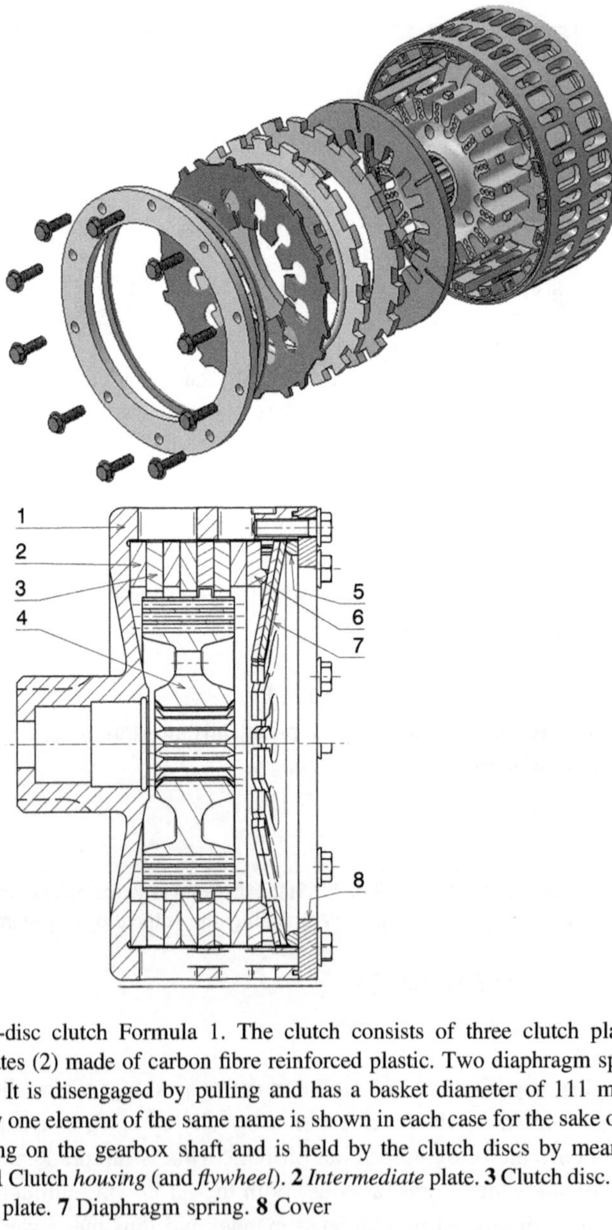

Fig. 5.8 Multi-disc clutch Formula 1. The clutch consists of three clutch plates (3) and four intermediate plates (2) made of carbon fibre reinforced plastic. Two diaphragm springs (7) provide the pre-tension. It is disengaged by pulling and has a basket diameter of 111 mm. In the shaded illustration, only one element of the same name is shown in each case for the sake of clarity. The hub is axially floating on the gearbox shaft and is held by the clutch discs by means of lugs on the circumference. **1** Clutch *housing* (and *flywheel*). **2** *Intermediate* plate. **3** Clutch disc. **4** Hub. **5** Pressure ring. **6** Pressure plate. **7** Diaphragm spring. **8** Cover

Carbon Clutch

In principle, this design is only a further development of a disc clutch, Fig. 5.8. In particular, the eponym is the material of the clutch discs and the intermediate plates. The dimensions of Formula 1 clutches have decreased over the years due to constant further development despite larger engine torques: the clutches transmit more than 600 N m, yet the clutch basket diameter decreased from 175 mm (ca. 1990) to 111 mm (2002) [5]. At the same time, the disengagement principle was changed to pulling instead of pushing. The clutch basket is made of titanium and the clutch plates, like the intermediate plates, are made of a carbon fibre-reinforced plastic. This reduced the mass of the clutch from 1800 to 1060 g (with flywheel). This combination of materials can also withstand temperatures of around 1000 °C. The service life of such clutches is in the range of 2000 km.

Antihopping Clutch (Slipper *Clutch*)

On motorcycles there is an interesting variation of multi-plate clutches, Fig. 5.9. The clutch is driven by a claw clutch with asymmetrical teeth. For this purpose the hub is divided into two parts. It consists of the actual hub (3), which receives the clutch plates, and the hub flange (4), which transmits the torque to the gear shaft and has a pressure surface for the plates. If the torque engages in the sense of forward travel, the clutch transmits the torque in the "usual" way via the friction between the plates. In overrun mode, e.g. when braking for a corner, the direction of the torque is reversed and the slanted claws between the hub and the hub flange (arrow in figure part c) release the clutch a little (this is the reason for another English name: *back torque limiter*) by the hub (3) lifting the cover (6) against the spring force (9). This prevents the full braking torque of the engine from reaching the drive wheels, which otherwise tend to lock with conventional clutches. The release travel of the cover can be adjusted via the thickness of an adjusting ring (7). The action of the anti-hopping clutch thus prevents the drive wheel from "stamping" (suspension hop). This hopping is particularly caused when downshifting at high speeds due to the disproportionately high mass inertia of the engine caused by the transmission ratio and the unloading of the rear wheel. Anti-hopping clutches are also often used in light two-track racing cars driven by a motorcycle engine. In this case, however, the effect is more to avoid engine overspeeding when downshifting and to protect the transmission input parts (primary drive and its bearings). Motorcycle engines usually have their gearbox integrated and these are designed for only one drive tyre and a lower vehicle mass. If a motorcycle engine is now used in a two-track racing vehicle, the integrated gearbox is also used. The higher vehicle mass and the two tyres cause much greater circumferential forces and impacts for which the gearbox is not designed. Its service life will therefore be shorter than in a motorcycle, but its mass and the rotating masses are lower than in a comparable separate gearbox.

Diaphragm *Spring Clutch*

A diaphragm spring can be used as a contact pressure spring for the clutch discs in cylindrical clutches. This advantageously applies its contact pressure to the entire circumference of the pressure plate and requires hardly any axial installation space. To disengage, the diaphragm spring is released by flat pressing. In comparison, with helical compression springs the force is always greater when disengaging. In addition, there is the disadvantage that the minimum contact pressure force is undercut with comparatively less wear, Fig. 5.10.

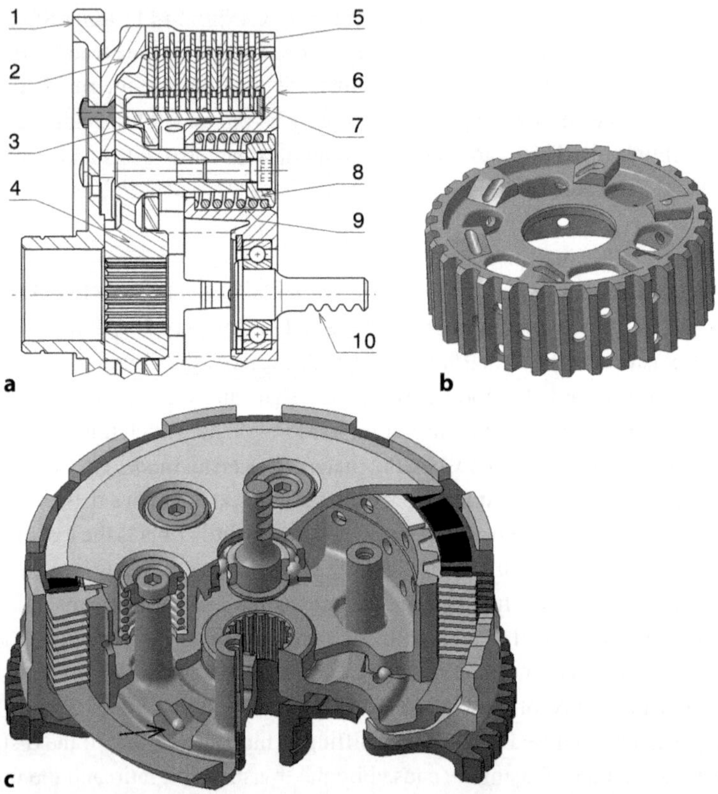

a b c

Fig. 5.9 Anti-hopping clutch. (**a**) Section. (**b**) View of the clutch hub (3) from below. The ball ramps are clearly visible. (**c**) Axonometric view of the clutch. 1 Driving gear. 2 Clutch housing. 3 Clutch hub. 4 Hub flange. 5 Disk pack (*discs assembly*). 6 Clutch cover. 7 Adjusting ring. 8 Spring collar. 9 Spring. 10 Release shaft (*actuating shaft*)

This is why diaphragm spring clutches are often used in passenger cars and also in motorsports.

In concrete terms, the arrangement of a compact diaphragm spring clutch is as follows. The clutch basket is firmly connected to the engine crankshaft via the flywheel or an adapter plate, while the clutch disc(s) is (are) positively connected (axially displaceable, but torsionally stiff) to the transmission input shaft, Fig. 5.11.

The gear shaft is centred in a locating bore of the crankshaft by a bearing. The clutch disc is mounted on a hub with splines or splined profile. It is clamped between the pressure plate and flywheel by a diaphragm spring. To disengage, the release ring is pressed against the diaphragm spring, causing the pressure plate to release the clutch disc. When the clutch disc(s) wear out, the pressure plate moves in the direction of the flywheel and the release ring must be able to compensate for the corresponding counter-movement of the diaphragm spring in the other direction, otherwise the clutch will no longer close completely.

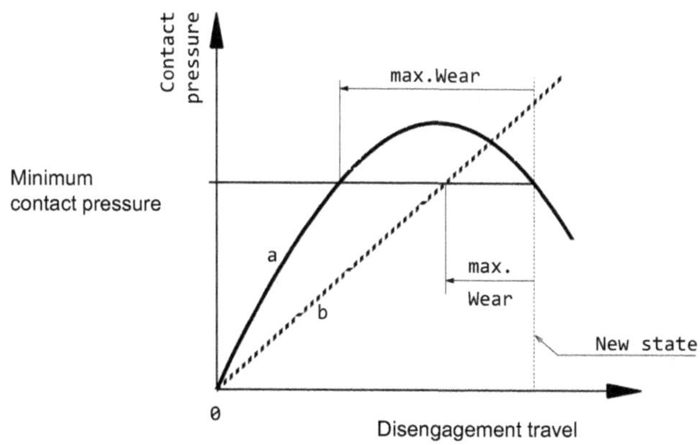

Fig. 5.10 Comparison of coil and diaphragm spring. (**a**) Diaphragm spring. (**b**) Helical spring. A minimum contact pressure is required to transmit the required minimum torque. This is reduced by wear, i.e. a reduction in the thickness of the lining and clutch plate. In the case of the linear helical spring, the minimum contact pressure is achieved with less wear

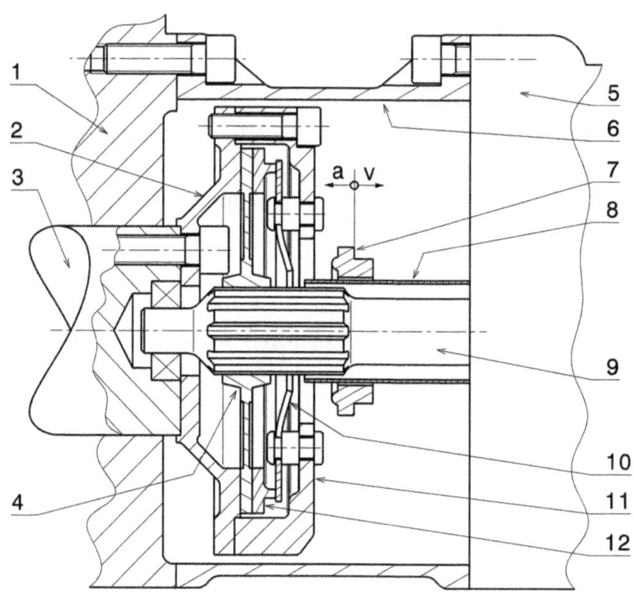

Fig. 5.11 Arrangement of a diaphragm spring clutch, diagram. **1** Crankcase. **2** Flywheel. **3** Crankshaft. **4** Clutch disc. **5** Gearbox. **6** bell housing. **7** Release ring. **8** Guide pipe. **9** primary shaft. **10** Diaphragm spring. **11** Clutch *cover* (clutch basket). **12** Pressure plate. **a** release travel, **v** wear travel. The clutch is located between the engine (1) and the gearbox (5) and switches the power flow between the crankshaft (3) and the gearbox shaft (9)

The clutch disc(s) must be able to compensate for the angular and centre misalignment between the crankshaft and the gearbox input shaft and for the flywheel wobble caused by the crankshaft bending, otherwise the hub/gearbox input shaft teeth will wear.

Torque Converter

Hydrodynamic torque converters are advantageously used in series production vehicles and in heavy commercial vehicles as a starting element in cooperation with automatic transmissions. In racing vehicles, this type of clutch is only found in vehicles for acceleration races. The importance of starting and the immense engine power let fluid couplings become the first choice for dragsters.

A torque converter is placed between the engine and transmission at the location of the flywheel. The torsionally stiff connection between the crankshaft and the converter housing is provided by a *flexplate*.

Hydrodynamic transmissions use the inertia of a fluid flow to transmit energy. The basic structure of a torque converter (Trilok converter) is shown in Fig. 5.12a. According to this, a converter consists of a fluid coupling supplemented by another fluid impeller. In a fluid coupling, the input power is converted into a mass flow of a fluid (usually ATF oil) by the blades of a pump wheel (1) and converted back into mechanical output line by an opposing turbine wheel (2). In the case of the torque converter, the oil also flows through the guide wheel (3), which diverts the oil flow and is supported for this purpose against the housing by means of a freewheel (4) before it re-enters the pump wheel. This deflection in the third wheel causes an increase in torque. The mode of operation is now to be illustrated with Fig. 5.12b. At start-up, the engine torque M_{M} directly drives the pump wheel (1) at engine speed n_M. The output shaft is still stationary (point A, $n_2 = 0$), the output power and thus the efficiency η are therefore zero. However, the output torque M_2 is more than twice the input torque M_{M}. As the vehicle accelerates, the difference in speed, i.e. the slip between the pump and turbine wheels, becomes smaller. The torque ratio M_2/M_M decreases approximately linearly. The efficiency η develops parabolically. When the point of torque equality is reached ($M_2/M_M = 1$), the guide wheel detaches from the housing and starts to rotate together with the turbine wheel (point C coupling point). The converter now works like a fluid coupling as a pure speed converter. If one wants to keep the efficiency high or bring it to 1 in the phase where the input and output speeds become equal, despite the decreasing slip, the turbine wheel must be mechanically connected to the pump shaft. This is done with the lock-up clutch (5).

Basically, there are no differences between stock converters and those for racing. However, converter housings for dragsters are bolted and not welded, so that the oil filling and the guide wheel can be changed at the venue. Both represent a significant adjustment option for the starting performance and thus the characteristics are adapted to the local conditions. While passenger car converters have an oil pressure of about 6 bar, racing versions have to withstand much higher pressures. For this reason, reinforcing plates are often welded to series housings (see Fig. 5.13).

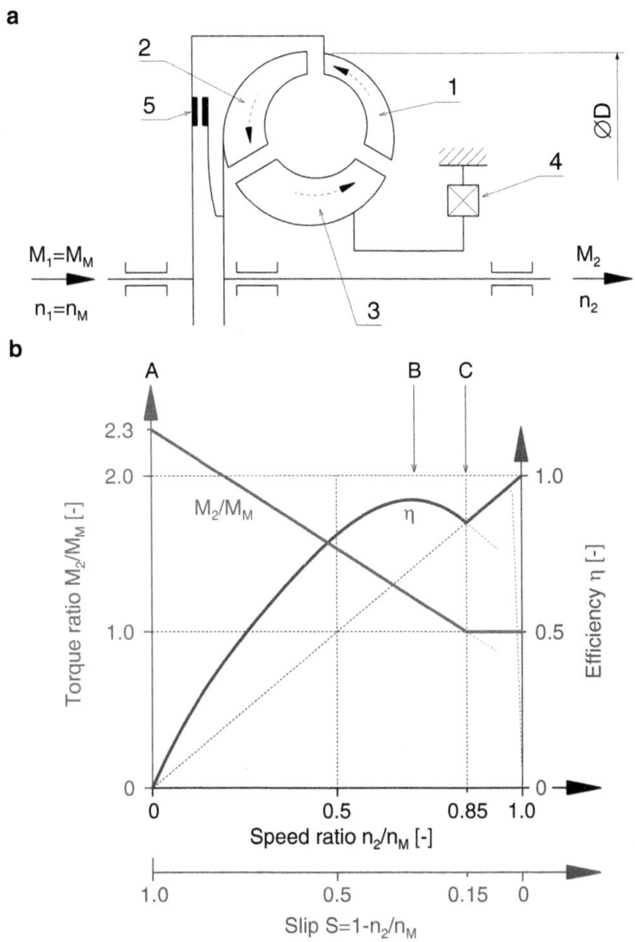

Fig. 5.12 Torque converter, a diagram, b characteristic curves. 1 impeller, 2 turbine wheel, 3 stator, 4 free wheel, 5 converter *lock-up* clutch. n_1,n_M Input or engine speed, M_1,M_M Input or engine torque, n_2 Output speed, M_2 Output torque, *D* Impeller diameter. A S*tall point* (fixed braking *point*), B *Best point* (design point), C Coupling point

Further decisive quantities result from the law for the pump torque

$$M_1 = \lambda \rho \omega_1^2 D^5 \tag{5.2}$$

M_1	Input torque, pump torque, Nm
λ	Power figure, –. A function of the speed ratio n_2/n_1 and depending on the internal structure of the converter (gaps, blades, ...)

(continued)

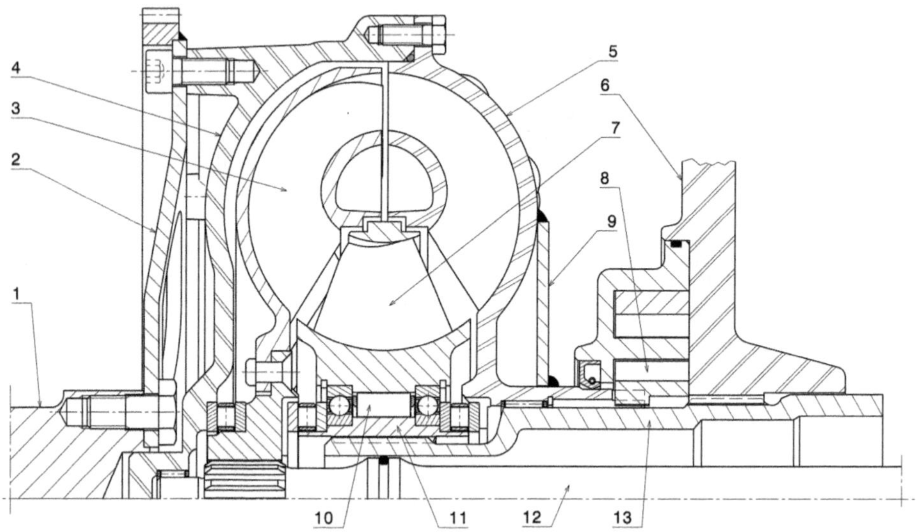

Fig. 5.13 Torque converter sectional view. 1 Crankshaft, 2 Connecting plate (*flexplate*), 3 Turbine wheel, 4 Housing, 5 Impeller, 6 Transmission housing, 7 stator, 8 oil-pump rotor, 9 Reinforcing plate, 10 Free wheel, 11 stator hub, 12 Output shaft, 13 stator axle

ρ	Density of transducer oil, kg/m^3
w_1	Circular frequency of the pump shaft, s^{-1}
D	Impeller diameter, m. see Fig. 5.12

Accordingly, the following parameters are suitable for adjusting the converter: The design and number of blades (λ), filling level and oil grade (ρ), gear ratio and nominal engine speed (ω_1) as well as the pump diameter (D). While series converters use automatic transmission fluid (ATF) oil, synthetic oils are used in racing applications. The case is filled 50–80%. For qualifying runs, the oils are approximately SAE 10 and for higher performance and thus heat input SAE 30.

Figure 5.14 shows schematically how the interaction between engine and converter can be influenced. According to the above relationship M_1 (ω_1) the pump characteristics are parabolas above the engine speed (= pump speed). If the power figure λ is not constant, but decreases with increasing speed (speed depression), this characteristic curve expands to a range. This operating range extends from start-up ($n_2/n_1 = 0$) to the clutch point ($n_2/n_1 = 0.96$). Beyond this point, the converter operates as a hydrodynamic coupling and/or is bypassed. The diagram contains operating ranges of three different designs. In the case of the hard converter, the diameter D is slightly larger (approx. 10%) and in the case of the soft converter slightly smaller than in the standard version. It can be seen that with the same converter slip – i.e. the same speed ratio n_2/n_1 – the soft converter shifts the operating range to higher engine speeds and thus to higher engine outputs. The soft converter is designed to achieve maximum engine torque at start-up. In the present example, the diameter of the

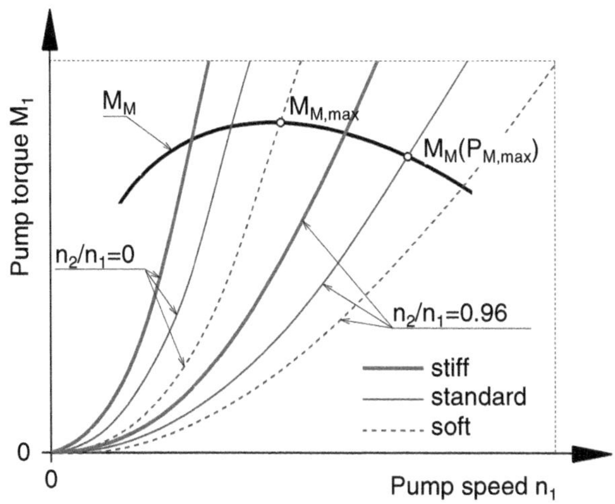

Fig. 5.14 Cooperation between engine and converter, schematic. M_M Full load torque curve of the engine, $M_{M,\max}$ Maximum engine torque, M_M ($P_{M,\max}$) engine torque at rated power. The pump torque of three converters is plotted against the pump speed. The operating ranges are spanned from starting ($n_2/n_1 = 0$) to the clutch point ($n_2/n_1 = 0.96$). The three converters (hard, standard and soft) differ in diameter D. The diameter of the hard converter is larger, that of the soft converter smaller than that of the standard version

standard converter is chosen so that its operating line at the coupling point meets the point of maximum engine power on the full load characteristic. If the power figure λ is constant, the pump torque M_1 depends only on the engine speed and the converter is characterized by the pump parabola (and no range) alone. In that case, the vehicle can be accelerated with standard converter at the power limit of the engine.

The effect of the cooperation between the engine and the converter is shown by a measurement taken during a run of a top-fuel dragster over a quarter mile (402.5 m), Fig. 5.15. The official run begins at zero second (arrow) and ends after 5.073 s. The run is then stopped at zero second (arrow). At the start, the driver rapidly depresses the throttle and the engine revs up from idle speed to about 7500 min^{-1}. The pump wheel has the same speed. The turbine wheel and thus the drive shaft are still stationary at the beginning, which causes the pressure p_{cl} of the converter oil to increase considerably within a short time. In this phase, the converter operates in the torque boost range and the speed v_V of the vehicle increases approximately linearly, so the acceleration is constant. Within 0.8 s the dragster reaches 100 km/h. The front wheels are in the air twice in this time (recognizable by the difference in speed between the vehicle and the front wheels). The contact radius of the drive wheels is not constant, but increases with their speed. This phenomenon is intentionally used as a variable gear. The effect is that the drive shaft speed and the vehicle speed are not the same in the first second, but the shaft revs up faster. As the vehicle speed increases, the slip in the converter decreases and the speed difference between the engine and the drive shaft decreases until the clutch point is reached after about 3 sec. In the

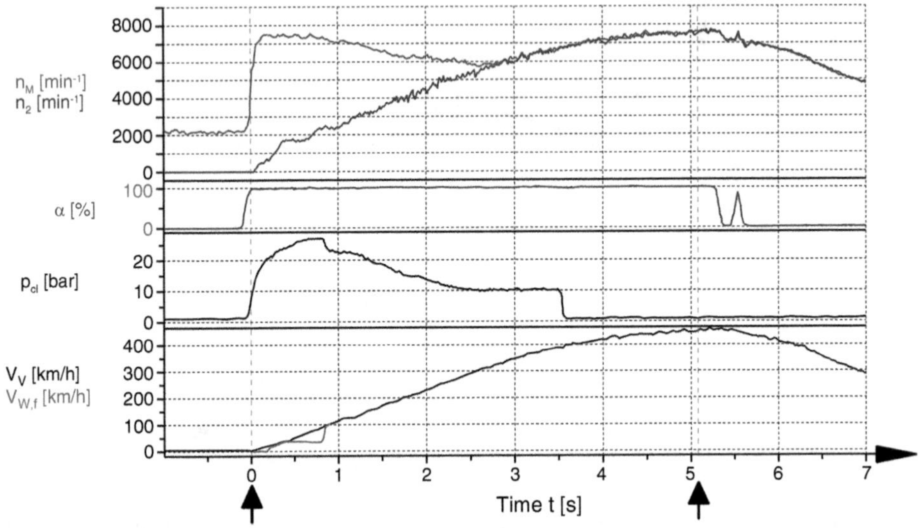

Fig. 5.15 Measurement diagram of a top-fuel dragster, after [6]. n_M or n_2 speeds of engine or drive shaft, α throttle position, p_{cl} pressure of converter oil, v_V or $v_{W,f}$ speed of the vehicle or at the circumference of the front wheels, respectively. The scored portion of the quarter-mile run takes 5.073 s. Start and end are indicated by arrows

meantime, the pressure of the converter oil has decreased to 10 bar. After about 3.5 s, a lock-up clutch is obviously activated: The oil pressure drops to the initial level and the engine as well as the drive shaft rotate at the same speed. The dragster passes the finish line at 446 km/h, but it only reaches its greatest speed afterwards because the driver is still on the gas.

The constructional design and the arrangement in the drive are explained with reference to Fig. 5.13.

The engine torque is transmitted from the crankshaft (1) to the housing (4) via the connecting disc (2), which also carries the starter ring gear. The pump wheel (5) is tightly bolted to the housing. The circulating oil drives the turbine wheel (3) and is deflected into the pump wheel by the guide wheel (7). The guide wheel is supported on its hub (11) via the freewheel (10, here sprag freewheel), which in turn is connected to the gear housing (6) via the hollow axle (13) in a torsionally rigid manner. The output torque from the turbine wheel is transmitted to the gearbox via the output shaft (12). The reinforcing plate (9) is welded onto the series housing to prevent *ballooning* under high internal pressure. The rotating housing (4 + 5) is supported in the crankshaft and on the idler shaft by a needle bearing. It also drives the oil pump (8) of the gearbox. An axonometric representation of this converter is shown in Fig. 5.16.

Housing shells are cast or milled from the solid. Impellers and blades of series converters are made of sheet metal parts and joined in slots via beads and lugs. The stator wheel is the component that determines the behaviour of the converter. Its design primarily

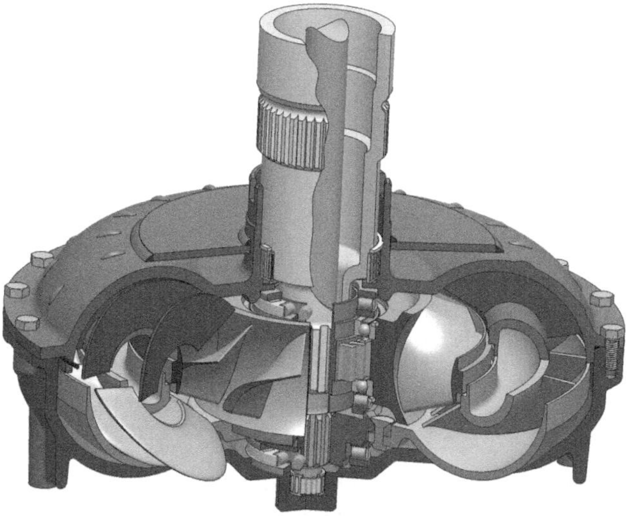

Fig. 5.16 Axonometric representation of the torque converter of Fig. 5.13. Shown without oil filling and partially cut open. The transducer rotates counterclockwise to the power output side (here: top) as seen

determines torque boost and fixed braking speed. It is milled from the solid for racing converters.

One of the most important parameters for racing is the *stall speed* of the converter. This is the pump speed that occurs when the vehicle and thus the turbine wheel are at a standstill and full throttle is applied. Thus, the stall speed corresponds exactly to the engine speed at start-up torque. If the fixed brake speed is set too high, the engine torque at start-up is too high and the tires build up too much slip. But the opposite is also undesirable: if the starting speed is too low, the dragster lacks its sprinting qualities at the start of all things. With these considerations, however, the torque overboost of the converter, which is between 2 and 4 depending on the design, must not be disregarded. After all, the tires must be able to transmit the resulting output torque. In general, racing torque converters have a smaller diameter than one would expect. However, the engines of the dragsters develop their gigantic power only at relatively high speeds, which is why the fixed brake speed must be just as high (soft design, cf. Fig. 5.14). This in turn leads to the usual diameters D between 7 and 11 inches (177.8 and 279.4 mm). A further control variable for the fixed brake speed is the number of guide vanes. This is in the order of 10–21. The more vanes are arranged, the smaller the gap through which the oil can flow. This effect leads to an increase in the fixed brake speed.

Fastening

The clutch basket is bolted to the engine crankshaft. The clutch may already contain the flywheel or this may be bolted to the crankshaft as a separate component. If a central screw

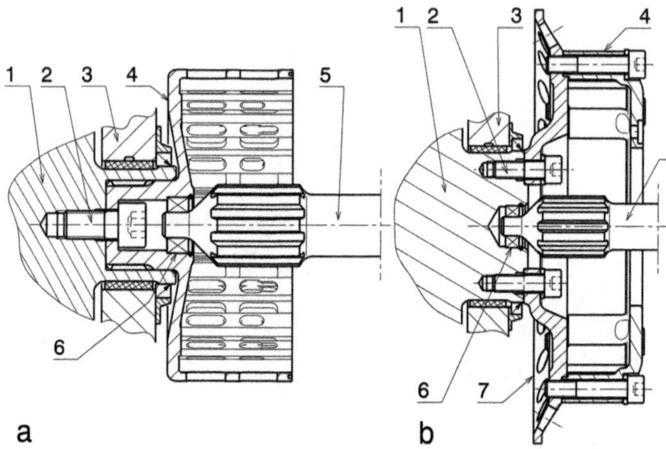

Fig. 5.17 Types of clutch attachment. **a** Central screw. **b** Peripheral screws. **1** Crankshaft. **2** Screw. **3** Crankcase. **4** Clutch basket. **5** Gear shaft. **6** Floating bearing. **7** Flywheel

is used (Fig. 5.17a), a plug-in toothing or a Hirth toothing takes over the transmission of the torque. The gearbox shaft is centred in the clutch basket (4) by means of a bearing (6). In this case, the clutch basket also represents the flywheel. If encoder splines (speed and TDC encoders) are required for engine control, the basket can also contain these.

The simpler method is a bolted connection with a sufficient number of bolts around the circumference (Fig. 5.17b). Centring is achieved by means of an appropriate collar on the crankshaft and the torque is transmitted via the friction generated by the preload force of the bolts. M8 cap screws with quality at least 10.9 and washers [5] are recommended. The gear shaft is centred in the crankshaft via a bearing.

The hub (Fig. 5.18) of the clutch or a clutch disc is seated on the gearbox shaft. A *splined or serrated-shaft profile* is used for centering (internal centering) and torque transmission. The profiles used are varied and depend on the gear shaft. DIN, SAE and factory standards can be found. All profiles allow easy axial displacement of the hub. The spline or splined shaft profile must be longer than the hub profile so that the full load-bearing width is ensured in all cases (pulley wear, tolerances).

Materials

Pressure plates, intermediate plates and the hub are made of grey cast iron, steel, aluminium or titanium. Steel is preferred for standing start. Aluminium is the lightest metal, but has a low temperature resistance. Titanium has similar properties to steel with lower density, but is extremely expensive. For carbon clutches, the intermediate plates can also be made of carbon. The clutch basket is made of metal in all versions.

In summary, Table 5.3 compares some important decision characteristics of the starting elements presented above.

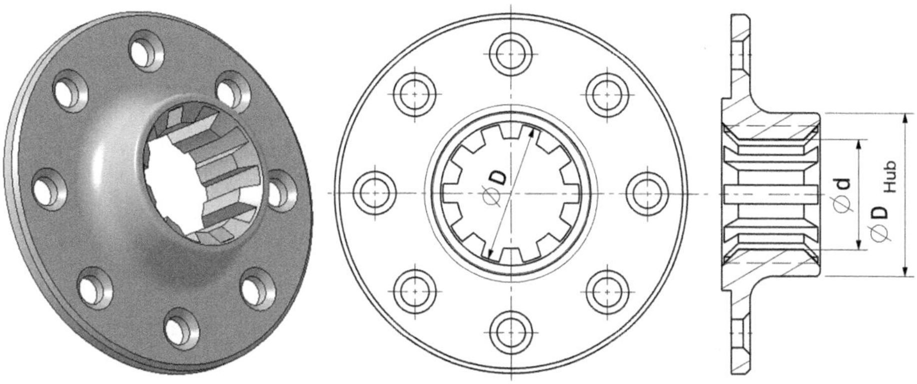

Fig. 5.18 Hub of a clutch disc. The clutch disc is riveted onto this hub. The hub has a profile with 10 teeth and straight flanks. The spline profile is characterized by the following sizes: Inner diameter *d*, Outer diameter *D*, Hub diameter $D_{Nabe.}$ Number of teeth

Table 5.3 Comparison of selection criteria for starting elements, according to [1]

		Dry clutch	Wet clutch	Hydrodyn. torque converter
Controllability	Engagement behaviour	+	++	++
	Separation behaviour	+	+	– –[a]
Thermal load capacity → continuous operation		–	+	++
Thermal load capacity → thermal shock		+	–[b]	++
(lining) wear		–	+	++
Installation space		0	++	0
Actuating energy demand		0	0	–
Crankshaft bearing load (thrust bearing)		0	++	+
Centrifugal force influences		0	++	0

[a]Separation only via additional systems
[b]Thermal shock can be avoided on the control side
Legend: ++ very good; + good; 0 satisfactory; – poor; – – very poor [1]

5.2.2 Choice of Clutch Size

The clutch must be able to transmit the maximum engine torque, which is excessive due to dynamic effects, and must be able to withstand the thermal load during frequent start-up (for formula symbols see also Fig. 5.7, for values see Table 5.4).

Table 5.4 Design parameters for clutch friction pairs [7]

	Wet run			Dry run			
Friction pairing	Sintered bronze/ steel	Sintered iron/ steel	Paper/ Steel	Steel, hardened/ Steel, hardened	Sintered bronze/ steel	Organ. coatings/ cast iron	Steel, nitrided/ Steel, nitrided
Sliding friction coefficient μ_{lo} [−]	0.05 By 0.1	0.07 By 0.1	0.1 By 0.12	0.05 By 0.08	0.15 By 0.3	0.3 By 0.4	0.3 By 0.4
Static friction coefficient μ [−]	0.12 By 0.14	0.1 By 0.14	0.08 By 0.1	0.08 By 0.12	0.2 By 0.4	0.3 By 0.5	0.4 By 0.6
Max. sliding speed [m/s]	40	20	30	20	25	40	25
Max. friction pressure [N/mm^2]	4	4	2	0.5	2	1	0.5
Max. frictional power density [W/mm^2]	1.5–2.5	0.7–1.2	1–2	0.4–0.8	1.5–2.0	3–6	1–2

The coefficient of sliding friction μ_{lo} is effective as long as the clutch is slipping. In the engaged state, static friction is effective and the static friction coefficient μ is decisive for the transmittable torque

$$M_{cl,\max} = k_{dyn} \cdot M_{M,\max} = F_{an} \cdot \mu_{cl} \cdot j \cdot r_{fr,m} \tag{5.3}$$

$$r_{fr,m} = \frac{\left(D^3 - d^3\right)}{3\left(D^2 - d^2\right)} \tag{5.4}$$

$M_{cl,\max}$	Transmittable coupling torque, N m
k_{dyn}	Dynamic advantage factor, –; k_{dyn} = approx. 2
$M_{M,\max}$	Max. Engine torque or maximum torque that can be transmitted by the drive tyres, nm. The lower of the two values is decisive
F_{an}	Contact pressure of the friction linings, N
μ_{cl}	Coefficient of friction between linings and friction surface, –
j	Number of friction surfaces, –; $j = 2$ for single-plate clutch
$r_{fr,m}$	Mean friction radius, m
d, D	Inner or outer diameter of the friction disc(s), m

The higher the contact pressure and the more discs are installed, the greater the transmittable clutch torque. However, a large contact pressure force also means a large disengagement force to separate the clutch.

Of the possible friction pairings, practically only metallic and non-metallic lamellae paired with steel lamellae are significant. In racing, friction partners made of carbon fiber-reinforced plastic are added. The metallic friction linings have a higher temperature resistance and a higher wear resistance, while the non-metallic friction linings have the advantage of lower weight, a somewhat higher coefficient of friction and, when special oils are used, a more favourable coefficient of friction curve. Steel plates with a hardness of 38 + 8 HRC and a roughness of 1–2.5 Ra are used as mating plates for metallic friction linings. For non-metallic linings ("paper linings"), unhardened steel with a surface finish of 0.2–0.5 Ra is used as the friction partner.

The coefficient of friction between the contact partners is assumed to be constant in the design calculation, but in fact it is not. It changes with the sliding speed and the lining temperature. Different friction lining materials also exhibit different behaviour. Paper-based linings show an increase in the coefficient of friction with the sliding speed, while sintered bronze linings, after an initial increase, react with a decrease in the coefficient of friction with a further increase in the sliding speed. With a further increase in the relative speed of the friction partners, the coefficients of friction of both materials remain constant at the values achieved up to that point.

The temperature rise in a clutch is due to the friction work released and the component mass that partly absorbs it. The remaining heat is dissipated via the surrounding air (dry clutch) or the oil (wet clutch). A higher clutch disc mass is more favourable in terms of temperature stability. The power loss in a slipping clutch is calculated from the relative speed and the acting torque:

$$P_{ls} = M_{cl} \Delta\omega \tag{5.5}$$

P_{ls}	Power loss, W
M_{cl}	Coupling torque, N m
$\Delta\omega$	Relative angular velocity between the friction partners, s^{-1}

The work lost over a certain period of time follows directly from this by integration over time at a constant moment:

$$W_{ls} = M_{cl} \Delta\omega \Delta t \approx M_{cl} \frac{\omega_M - \omega_{G,1}}{2} \Delta t \tag{5.6}$$

W_{ls}	Loss work, J
Δt	Coupling time (slipping time), s
ω_M	Engine angular velocity (input value), s^{-1}
$\omega_{G,1}$	Angular velocity of the gearbox input shaft (output shaft) before engagement, s^{-1}. During a starting process, $\omega_{G,1} = 0$

$$\dot{W}_{fr} = \frac{\mu_{cl} F_{an} r_{fr,m} \omega_{cl}}{A_{fr}} \tag{5.7}$$

$$q = \eta_{cl} \cdot \dot{W}_{fr} \tag{5.8}$$

$\dot{W}_{fr}$	Frictional power density, W/m^2
A_{fr}	Friction surface, m^2
ω_{cl}	Clutch angular velocity, s^{-1}
η_{cl}	Efficiency of energy conversion, –; $\eta_{cl} = 0.8$–1
q	Heat flux density, W/m^2

For the rough design, guide values for series applications with normal cooling can be used for comparison. The values are 60,000–175,000 W/m^2 for sintered bronze friction linings and 12,000–34,000 W/m^2 for organic friction linings.

When selecting a clutch, the input torque, i.e. the maximum engine torque increased by a dynamic factor, is the first criterion. If several clutchs are available for an application, further considerations are added, Fig. 5.19. As an example, a engine torque of 150 N m is considered. Three clutch diameters can be considered for this. The extreme cases have the following advantages. The smallest clutch (variant 1) saves installation space, weight and has a small mass moment of inertia. The costs of the clutch also remain small. In contrast, variant 2, the large clutch, offers other advantages. The torsional damper characteristic is more advantageous, the fading stability increases as does the lining service life. The pedal force decreases because the contact force for the same torque is smaller.

A small clutch diameter allows a lower engine installation in the vehicle. This lowers the vehicle's centre of gravity.

In the racing sector, a wide variety of friction linings are offered to cover the many different requirements, Fig. 5.20. Linings with high wear widths (see below) are used when many starts are made and long service life is required. This applies to rallying, rallycross, autocross, touring cars and endurance racing, among others. For circuit cars, there are very lightweight clutch discs that feature low mass moment of inertia and small clutch diameter. Carbon fiber clutch discs allow extremely small clutches with high temperature resistance. Another advantage results from the virtual absence of wear on the flywheel.

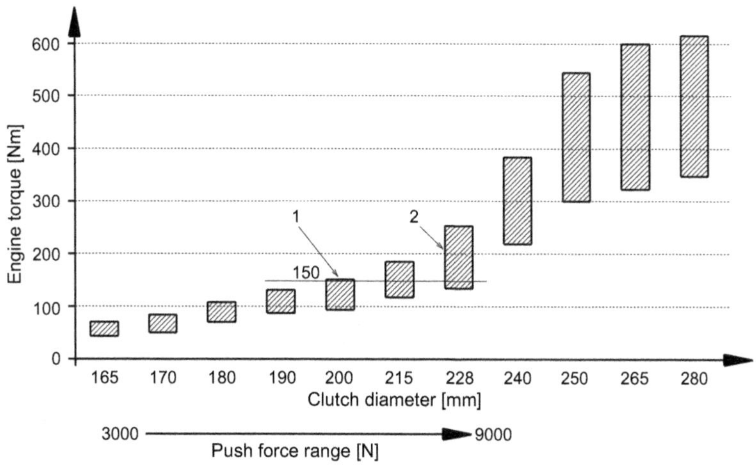

Fig. 5.19 Relationship between engine torque and clutch diameter for passenger cars [8]. Three clutchs can be considered for the 150 N m example shown. For considerations on choice 1 and 2 see text

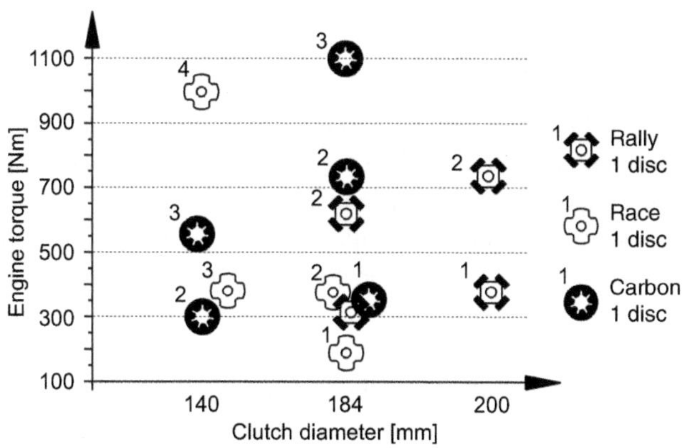

Fig. 5.20 Clutch selection for racing clutches, according to [9]. For three different versions of clutch discs (Rally, Race, Carbon) different transmittable engine torques result depending on the number of discs installed (the dynamic factor is already taken into account). The disengaging forces are in the range of 1900–4900 N

5.2.3 Clutch Actuation

The clutch is operated mechanically or hydraulically. The same fluid is used as for the hydraulic brakes. In the closed state, it is important that there is play between the release ring and the diaphragm spring. This ensures that the spring achieves the unhindered

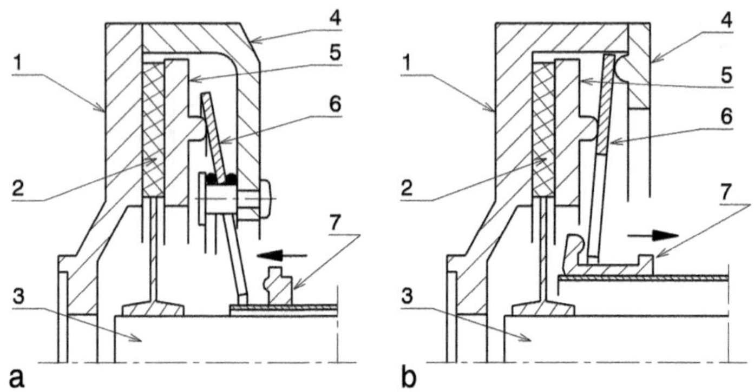

Fig. 5.21 Disengagement types of a diaphragm spring clutch. (**a**) push-type release. (**b**) pull-type release. **1** Flywheel. **2** Clutch disc. **3** Gear shaft. **4** Clutch basket. **5** Pressure plate. **6** Diaphragm spring. **7** Release button

pre-tensioning force in any case (i.e. also in case of expansion due to temperature increase or wear).

There are basically two types of disengagement. Depending on how the clutch is constructed, the diaphragm spring is pulled or pushed during disengagement, Fig. 5.21.

The pressed version has the classic design of a diaphragm spring clutch. To disengage, a simple release ring is pressed against the spring. In the case of the pulled clutch, the diaphragm spring must be pulled by the disengager in order to release the pressure plate. The advantage of the pulled clutch is the simple design of the actual clutch. The spring is not connected to the clutch basket and the clutch consists of fewer parts, so in principle it is lighter. For the same outer diameter, the diaphragm spring in this type applies a greater compressive force to the pressure plate or, for the same contact force, the spring can be designed with a smaller diameter. Current Formula 1 clutches are exclusively pulled clutches [5].

The pedal characteristic results from the diaphragm spring characteristic through the lever ratio and the ratio of the release system as well as elastic deformations, Fig. 5.22.

Due to the wear of the clutch disc and intermediate plate, the preload of the diaphragm spring changes because the entire disc pack is compressed. The spring is therefore slightly relaxed. If the wear becomes too great, the contact pressure decreases so much that the clutch can no longer transmit the desired torque and begins to slip. The wear width, Fig. 5.23, is a measure of the sensitivity of the clutch to wear. A clutch with a wide wear width can transmit the rated torque over a wider range of wear. Wear widths are in the range of 0.6–1.5 mm.

The release ring is actuated mechanically or hydraulically. A hydraulic actuation takes place via a straight-line movement and thus always presses centrally on the spring, Fig. 5.24. A mechanical lever describes a circular arc and thus inevitably deviates from the centre with large displacement movements. However, this can be prevented with an additional guide tube, Fig. 5.25. The hydraulic system works in the same way as for the

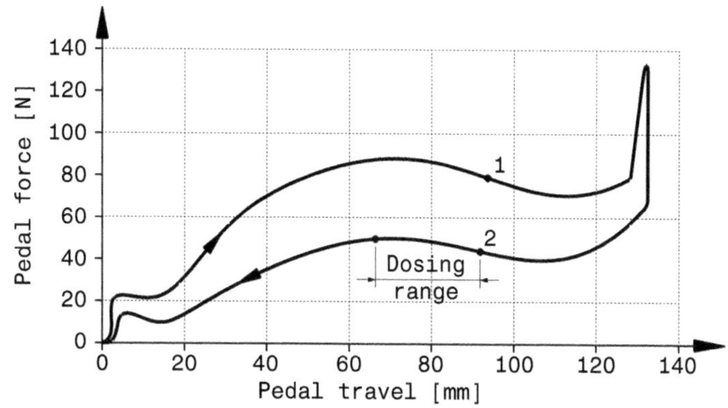

Fig. 5.22 Example of a characteristic curve of a clutch pedal [8]. The pedal characteristic was recorded with a single-disc dry clutch for passenger cars. **1** Clutch has separated. **2** Start of torque transmission. The characteristic curve is the characteristic curve of the diaphragm spring distorted by lever ratios and elastic deformations in the transmission system. A pronounced hysteresis is also formed. In the metering range, the preload force of the spring increases steadily with decreasing pedal travel

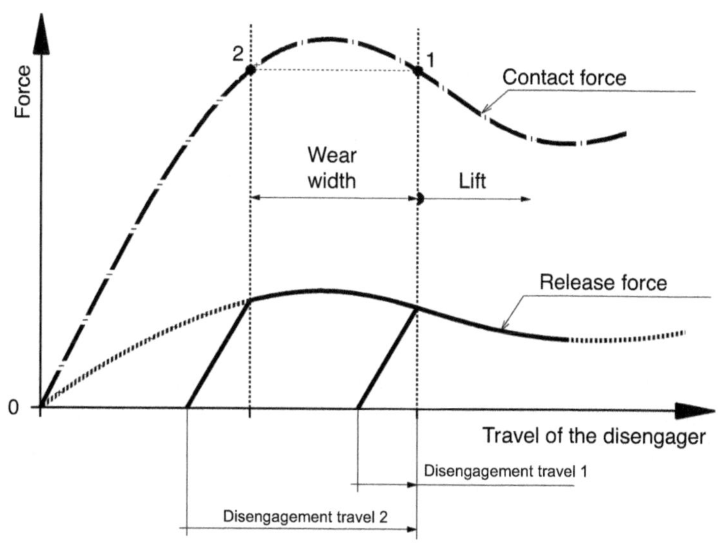

Fig. 5.23 Wear *range of* a clutch. **1** Clutch in new condition. **2** Clutch with high wear. In the new condition (1) the disengagement travel is small, the spring is preloaded above the maximum value. Due to wear, the clutch plates become thinner and the deformation of the spring decreases. As a result, the preload force of the spring initially increases and then decreases again with further wear. The wear range is reached when the preload force has dropped to the original level (2). With further wear, the preload force falls below this value and the clutch can no longer transmit the nominal torque. The disengagement travel increases with wear because the spring must be deformed until it is lifted off the pressure plate

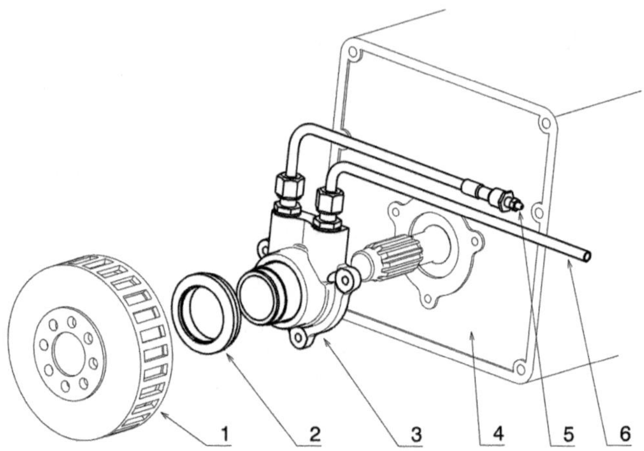

Fig. 5.24 Hydraulic disengaging device. The disengager is screwed to the face of the gearbox housing and actuates a depressed clutch. **1** Clutch. **2** Release ring with thrust bearing. **3** Release unit. **4** Gearbox housing. **5** Bleed line. **6** Pressure line from clutch pedal master cylinder or control unit when clutch is actuated. The release ring (2) is seated on the tubular piston of the release unit (3)

brakes. The clutch pedal actuates the master cylinder and its hydraulic pressure is transmitted to the slave cylinder, the releaser. The only difference to the brakes is that the displacement of the clutch spring is much greater and therefore the cylinder diameters and the lines must be designed for larger volume flows for rapid clutch engagement. The clear width of the pressure line should therefore be around 5 mm [10].

A thrust bearing is located on the release mechanism, which balances between the stationary hydraulic piston or release ring and the rotating diaphragm spring. In contrast to passenger car applications, these thrust bearings must be able to withstand significantly higher temperatures and, depending on the engine, higher speeds in racing.

The clutch is usually only used for starting or braking. The (up)shifting is done without clutch. The ignition is interrupted for this purpose in the case of manual shifting and, in the case of electronic shifting, the ignition angle is also influenced during the shifting process.

In addition to the actuated clutches described above, there are also automatic friction clutches. For acceleration races, centrifugal clutches are used in some classes (Top Fuel Dragster, Funny Cars). Here the clutch characteristics are tuned by adjustable centrifugal weights, which close the clutch when a certain speed is reached. The driver therefore "only" needs to accelerate (with the brake applied if necessary) to start the car. Whether the start succeeds as desired thus depends on the skill of the race engineer, who has set the characteristic of the clutch beforehand. Centrifugal clutches are also generally used when the race start is made standing up. A rev limiter can be set from the cockpit and activated with the start button. This allows the desired transmitted torque to be set for the start. The clutch pedal does not need to be depressed by the driver when starting and the acting torque can be controlled purely via the accelerator pedal during the slip period. When the

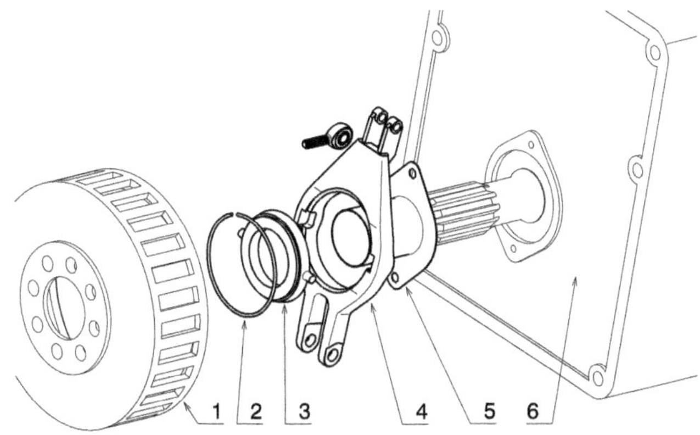

Fig. 5.25 Mechanical disengager. The release lever is located between the clutch and transmission and operates a depressed clutch. The lever has been converted from a standard part so that it can be actuated via a rod end. The release ring (3) is pressed against the clutch spring by the lever (4). To ensure that the pressure ring always remains centred despite the lever rotation, it slides on the thin-walled guide tube (5), which is screwed to the gearbox housing (6). **1** Clutch. **2** Circlip. **3** Release ring with thrust bearing. **4** Lever. **5** Guide tube. **6** Gearbox housing

accelerator pedal is in a constant position, the torque and engine speed also remain constant, thus enabling equally constant acceleration.

Figure 5.26 shows the operating diagram of such a centrifugal clutch. The (additional) clamping force due to the centrifugal masses depends on their mass, the geometry and the speed of the clutch:

$$F_{ad} = j_m \frac{\pi^2 m_{ad} r_{ad} n^2 b}{900a} 10^{-6} \tag{5.9}$$

F_{ad}	Additional clamping force due to centrifugal masses, N
j_m	Number of centrifugal masses, –. Mostly 6 to 8
m_{ad}	Centrifugal mass, g
n	Speed of the clutch (engine speed), min^{-1}
a, b, r_{ad}	Dimensions, mm. See Fig. 5.26

Such a centrifugal clutch can also be used as a purely mechanical *anti-stall clutch*. If the engine speed drops sharply, the clutch opens automatically before the engine idle speed is undershot and the engine does not stall.

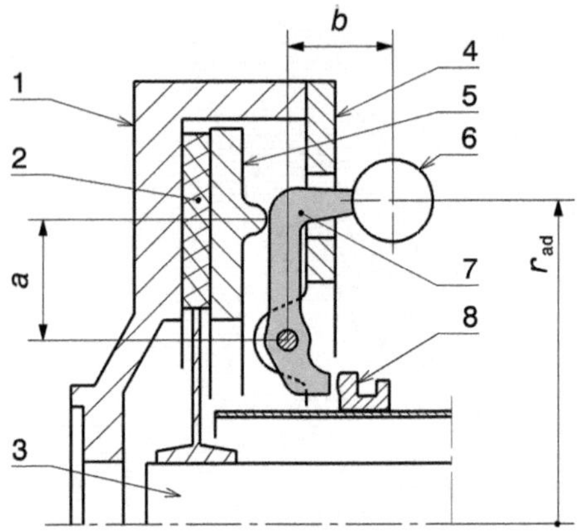

Fig. 5.26 Effect of centrifugal masses of a centrifugal clutch. 1 Flywheel, 2 Clutch disc, 3 Gear shaft, 4 Clutch cover, 5 Pressure plate, 6 Centrifugal mass m_{ad}, 7 Lever, 8 Release lever. Adjustment variables are the distance b and the mass m_{ad}. The disengaging lever 8 is not absolutely necessary. It is only used to open the clutch manually – if this is needed for shifting (depending on the gearbox type)

5.3 Gearbox

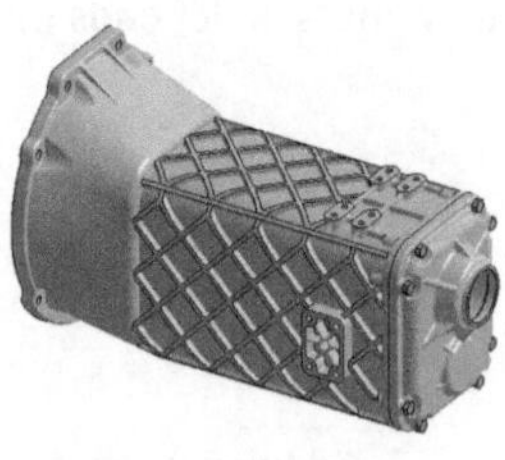

Function

The transmission allows the engine characteristics to be adapted to the vehicle and the course of the track. Reversing is also required for many vehicles.

The gearshifts should be carried out as quickly as possible and the interrupted tractive force on the drive wheels should not be restored abruptly, so that the tires are not unnecessarily stressed and the driving stability of the vehicle is not impaired. Stability is an essential criterion, especially for a racing vehicle, because it is driven at the adhesion limit of the tires. In a Formula 1 race, about 2000–2600 gear changes are necessary; during the 24 Hours of Le Mans, the drivers change gears 15,800 times.

In addition to its actual tasks, the gearbox housing also performs other functions in racing vehicles. Wheel suspension parts (wishbone connection, bearing of bell cranks and

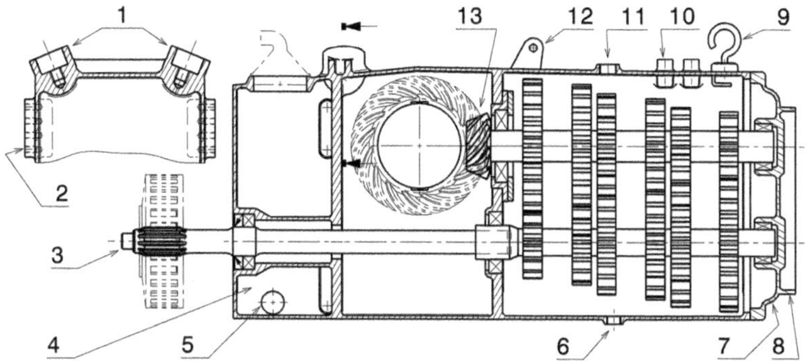

Fig. 5.27 Structure and functions of a gearbox. Designations see text

stabilizers), internal brake calipers, starter and rear wing can be accommodated by the housing. Figure 5.27 shows schematically the structure of a gearbox. The clutch sits on the gear shaft (3). The gear shaft runs through a tunnel into the actual gearbox housing. In front of it, the oil reservoir (4) for the engine is cast on. The pressure pump of the engine sucks off the oil at the opening (5). The gearbox housing accommodates the wheel sets of the individual gear ratios and the final drive (13). The rear end of the gearbox is closed with a bearing cover (7). If this cover is removed, the gear stages can be replaced. The cover also provides a receptacle (8) for the rear crash element. The oil heats up during operation and the air pressure in the housing increases. To prevent excessive stress on the seals, a vent (9) is required. This discharges in the open or in an overflow tank. The external shifting takes place via a shaft in the case of H shifting or via a lever (12) in the case of sequential shifting. In addition, there may be a switch on the outside of the housing that must be pushed or pulled to engage reverse and first gear. The housing has strong slugs (bosses) in stiff places, such as areas of bulkheads. For example, landing gear bellcranks are mounted directly on the housing (1), wishbone brackets can be offset in height (2), or spring/damper elements are bolted to corresponding slugs (10). On the upper side, such slugs can also be a base for measuring rods, which are screwed on during setup and from which the ground clearance is measured. Because the threads are used very often in this case, there are metal thread inserts in the screw pipes. The gear oil is filled via the opening (11) and drained via the threaded hole (6).

The transmission is generally flanged directly to the engine crankcase and can be arranged longitudinally or transversely in the vehicle. Longitudinal means that the transmission main shaft is parallel to the longitudinal axis of the vehicle.

A vibration-isolated suspension via elastic elements, as is standard in production vehicles, is hardly ever found in racing vehicles. Decoupling elements take up installation space, as do components that gain movement space through such a mounting. In addition, elastic elements lead to energy losses and indirectness due to additional movement space.

Layout Transmissions are either combined with an engine as a purchased part – the wheelbase can be adjusted via the length of the clutch housing – or the transmission is newly designed for a vehicle. In the latter case, important specifications for the design are the nominal engine torque, the intended use, the clutch (type and size), installation space specifications, gear ratios, the shift system (actuation type, H-shift, sequential), wheel sizes and the wheel suspension (forces can be fed in the transmission through them).

The design begins with selecting the shaft positions, designing the wheel sets and determining the wheelbase. Finally, the gearshift is designed.

Types of Transmission

Depending on the type of transmission ratio change, a distinction is made between manual transmissions and continuously variable transmissions. The type of operation results in manual transmissions, automated transmissions and automatic transmissions. A comparison of different transmission types is shown in Table 5.5. A detailed consideration of the degrees of automation of manual transmissions is provided in Table 5.6.

Figure 5.28 shows the system structure of an automated manual gearbox (ASG). The transmission and the clutch are basically the same as for manual operation, except that both are operated via actuators. They are controlled by control units that have various inputs such as speed, transmission speed, etc. for driving state detection and calculation of the shifting processes.

Figure 5.29 gives an idea of the complexity of the structural implementation of automation on a manual gearbox. The electrohydraulic unit is shown, which operates the clutch and changes gears by means of rocker actuation.

Pure automatic transmissions, which are popular in passenger cars, especially in the USA, are practically non-existent in racing cars. One manufacturer has now launched a special 8-speed automatic racing gearbox based on planetary gear sets following successful entries in the 2016 VLN Endurance Championship Nürburgring [12]. The transmission uses an internal multi-plate clutch as the starting element, and there is no torque converter with lock-up clutch – the usual solution in passenger cars. Compared to the standard transmission, from which the racing variant was derived, the spread is significantly smaller and the adapted software of the transmission control unit allows shorter shift times [13].

5.3.1 *Mechanical Gearbox*

The design of gearboxes, i.e. the selection of gears and their gradations for harmonious interaction between the engine and the wheels, is described in Chap. 4 *Calculation of the Drive Train*.

The vast majority of manual transmissions are gear transmissions. Several types can be distinguished according to the number of transmission stages, Fig. 5.30. In this case, a stage is understood as a pair of gears realising a transmission ratio.

Table 5.5 Comparison of different gear unit types, [11]

Gearbox type	Ratio	Weight	Consumption (%)[a]	Shift comfort (ATZ value[b])
Manual transmission, 5-speed	Two-shaft gearbox	Low	−10	–
Manual transmission, 6-speed	Two-shaft gearbox	Low	−12	–
Step automatic, 5-speed	Planetary gear sets	Medium	0	9
Step automatic, 6-speed	Planetary gear sets	Medium	−3	9
Continuously variable transmission	Wrap-around gear (push link belt)	High	−5	9.5
Continuously variable transmission	Wrap-around gear (chain)	High	−5	9.5
Toroidal gearbox	Friction gear	Particularly high	−7	9.5
Automated manual transmission	Two-shaft transmission with electromechanical actuation	Low	−15	6.3
Automated manual transmission	Two-shaft transmission with electrohydraulic actuation	Low	−14	6.5
Dual clutch transmission	Two-shaft transmission with electrohydraulic actuation	Medium	−8	8.7

[a]Approximate fuel consumption advantage compared to a 5-speed automatic transmission at 300 N m operation and dethrottled gasoline engine
[b]The ATZ value is a measure of the quality of the gear change. A value of 10 corresponds to an optimal (jerk-free) gear change, a value of 1 to a very uncomfortable process

Single-stage transmissions are mainly used in vehicles with transverse engines and front-wheel drive. Two-stage transmissions are suitable if the engine torque is to be transmitted coaxially, as is the case with the standard drive. One gear is the direct drive through the gearbox (in Fig. 5.30b the fifth gear, so-called direct gear). Multi-stage gearboxes can be built particularly short. Of course, they need more space in height and width.

For racing vehicles with a mid-engine arrangement, it is a good idea to integrate the axle drive into the manual gearbox. There are several possibilities to place the gearbox between the engine and the rear axle, Figs. 5.31, 5.32, 5.33, and 5.34.

Table 5.6 Degree of automation of manual transmissions [1]

Degree of automation	Starting procedure	Gearshift coupling	Changing gears
0	Foot-operated starting clutch	Foot-operated coupling	Manual operation of a gear lever
1	Foot-operated starting clutch	Automated coupling	Manual operation of a gear lever
2	Automatic start-up clutch	Automated coupling	Manual operation of a gear lever
3	Automatic start-up clutch	Gear change initiated by foot operated clutch	Manual gear selection by push button
4	Automatic start-up clutch	Automated coupling	Automated gear selection supported by engine management

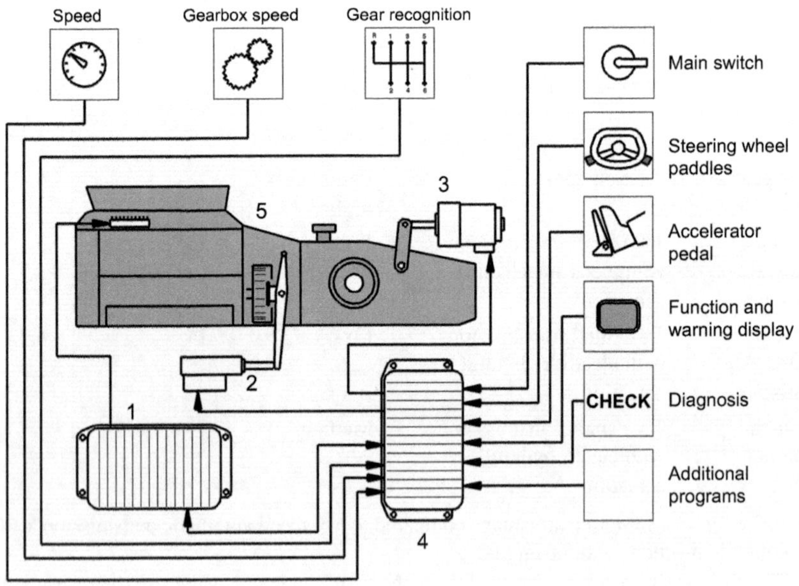

Fig. 5.28 Schematic of an automated shift gearbox (ASG). 1 Engine *control* unit. **2** Clutch actuator. **3** actuated gearshift. **4** ASG control unit (*gearbox control unit*). **5** Engine and gearbox

The transmission design influences the mass distribution. In this respect, transverse transmissions have the advantage of a short overall length and thus help to concentrate the masses around the vehicle's centre of gravity. However, greater disassembly effort when adjusting the gear ratios can become a disadvantage compared to the longitudinal arrangement.

An intermediate shaft allows the height of the gearbox to be adjusted, Fig. 5.35.

Fig. 5.29 Hydraulic control on a manual gearbox (Ferrari Formula 1). You can see the right side of the gearbox

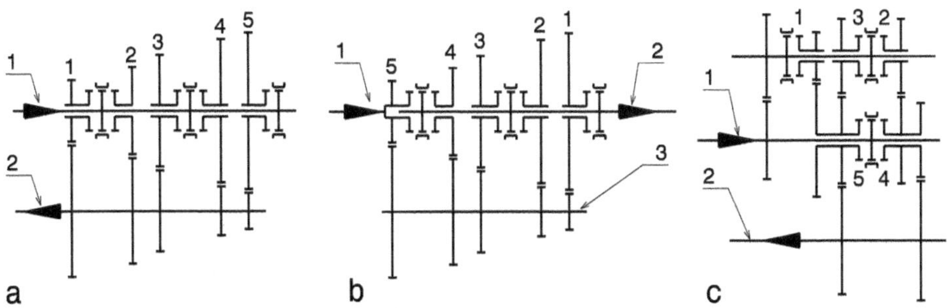

Fig. 5.30 Construction types of gearboxes. (**a**) single-stage transmission. (**b**) dual-stage transmission. (**c**) multi-stage transmission. **1** Input shaft. **2** Output shaft. **3** layshaft

Arrangement of the Individual Gear Stages

The largest gear ratio (first gear) should be close to a bearing. One will also try to keep the torsional moment in the shafts small. The input shaft must bear the maximum engine torque as the highest moment (apart from dynamic increases due to bang starts). Individual gear steps do not change this. The situation is different for the output shaft. Here the torque changes functionally with the gear. Therefore the gear of the first gear is placed closest to the final drive of the final drive. Then the highest moment must be transmitted only from the gear to the pinion or the shaft must have a larger diameter only in this area. Figures 5.36 and 5.37 show arrangements of gear stages in longitudinal and transverse gearboxes.

Shift

The gearshift system enables the change from one gear stage to the next. A distinction is made between external and internal gearshift components. All parts outside the gearbox

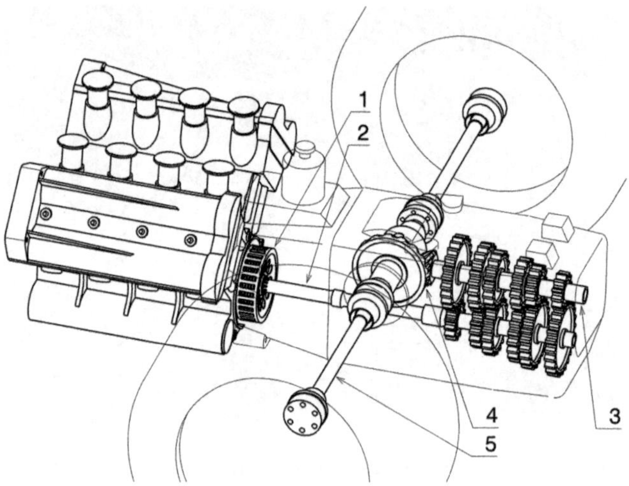

Fig. 5.31 Longitudinal arrangement of the gearbox behind the axle. **1** Clutch. **2** Input shaft. **3** Output shaft. **4** Final drive. **5** Side shaft half-shaft

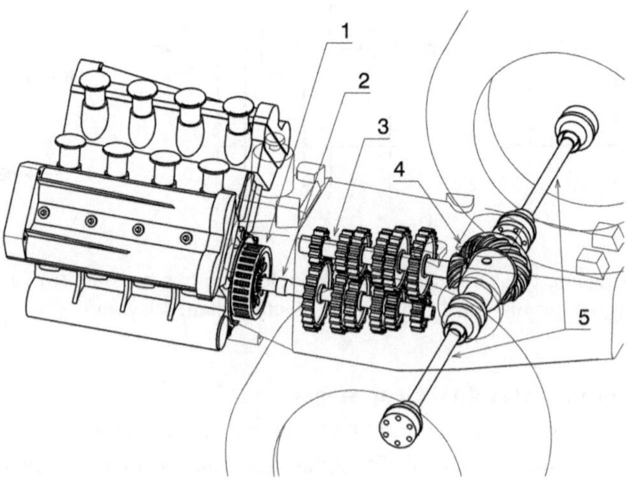

Fig. 5.32 Longitudinal arrangement of the gearbox in front of the axle. **1** Clutch *clutch*. **2** Input shaft input shaft. **3** Output shaft output shaft. **4** Final drive axle. **5** Side shaft (*half-shaft*)

belong to the outer gearshift system, i.e. gearshift levers, linkages and cables, among others. The inner gearshift enables the actual activation of a gear ratio. Usually, different gear ratios in manual transmissions are represented in such a way that a gear wheel sits firmly on the shaft (fixed gear) and the meshing counter gear (idler gear) is optionally connected to its shaft. For this purpose, a shift sleeve is located between each pair of gears, which establishes the torque connection of the left or right gear with the shaft, Fig. 5.38. In

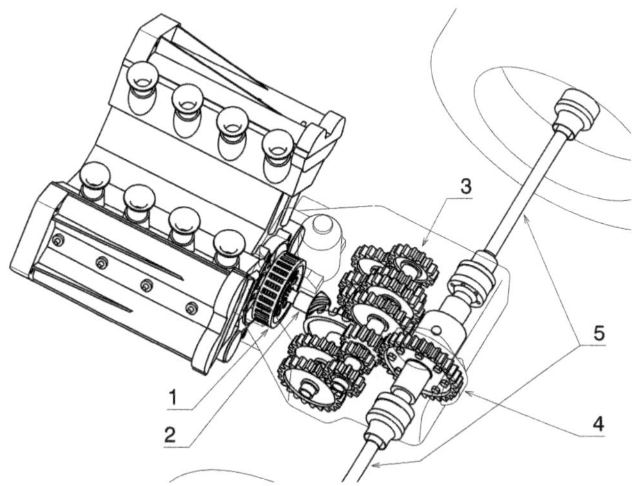

Fig. 5.33 Transverse arrangement of the gearbox. **1** Clutch. **2** Input shaft. **3** Output shaft. **4** Final drive. **5** Side shaft (*half-shaft*)

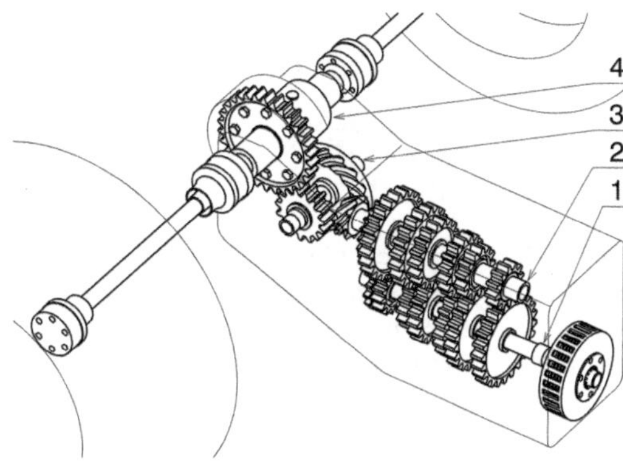

Fig. 5.34 Longitudinal gear unit with intermediate shaft. **1** Input shaft. **2** Output shaft. **3** Intermediate shaft. **4** final drive

manual transmissions with H-shifting, so-called shift aisles result. In one shift gate, first and second gears are shifted, the pairs of gears in the next gate represent third and fourth gears, and so on. A device must ensure that only one gear stage can be activated at a time.

Three shift forks are required for six gears and four in a seven-speed transmission. An internal shift is also required for reverse gear. This is done via an actuator or likewise via a shift fork, which for simplicity's sake can be operated via a cable. In any case, it must be ensured that the other shift forks are in the neutral position.

For a switchable reversal of the direction of rotation, as required for a reverse gear, there are several possibilities, some of which are listed in Fig. 5.39.

If no idler wheel is fixed by a shift sleeve, the transmission is in neutral. Because sequential transmissions offer so many options for neutral, the "planned" neutral position is

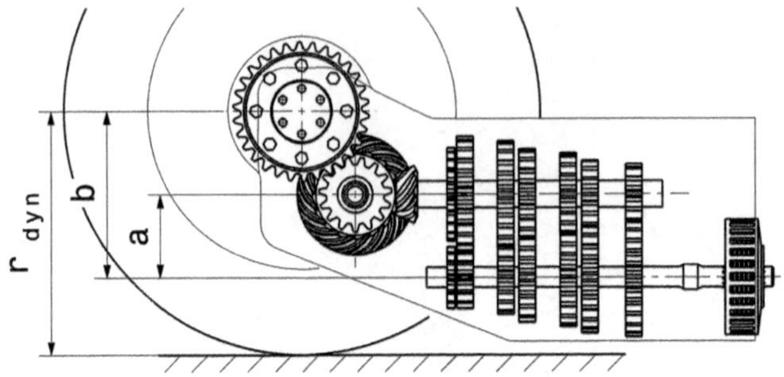

Fig. 5.35 Wheelset of a longitudinal gearbox with intermediate shaft. The intermediate shaft allows a height offset b between the transmission and the rear axle independent of the axle distance a. With a given tyre diameter r_{dyn}, the centre of gravity of the gearbox can thus be lowered and the gearbox has an aerodynamically favourable shape on the underside, i.e. the centre diffuser can be easily accommodated

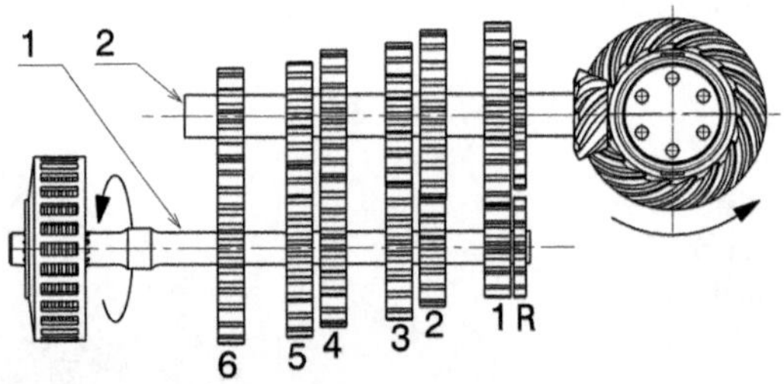

Fig. 5.36 Arrangement of gear stages in a longitudinal gearbox. **1** Input shaft. **2** Output shaft. The numbers stand for the gear steps, R for reverse gear. For reverse gear, another intermediate gear is required to reverse the direction of rotation. The idler gears are located on the output shaft. An external starter can act on the drive shaft from the rear. With the clutch closed and the engine in neutral, the starter can thus drive the engine crankshaft directly

selected via a switch in the cockpit or directly on the steering wheel. The actuator usually turns the shift drum to the neutral position between first and second gear. If the regulations require that track marshals can engage neutral from the outside in order to push the vehicle, the clutch is usually disengaged by an actuator at the push of a button.

In passenger car transmissions, the different speeds of the shaft and the idler wheel seated on it are matched by friction clutches (synchronizer rings) before the positive torque connection is fully established. Racing gearboxes are designed without synchronising devices in order to save weight and shifting time. Of course, synchronisation is also necessary with these gearboxes, but this is partly carried out by the driver by selecting the shifting speed, and partly by the shifting process itself. The necessary acceleration or deceleration of the gears and shafts is achieved by the impact of the claws. Wear and comfort are of secondary importance in a racing vehicle.

Fig. 5.37 Arrangement of the gear stages in a transverse gear unit. **1** Input shaft. **2** Output shaft. The numbers stand for the gear steps, R for reverse gear. For reverse gear, another intermediate gear is required to reverse the direction of rotation. The idler gears are located on the output shaft

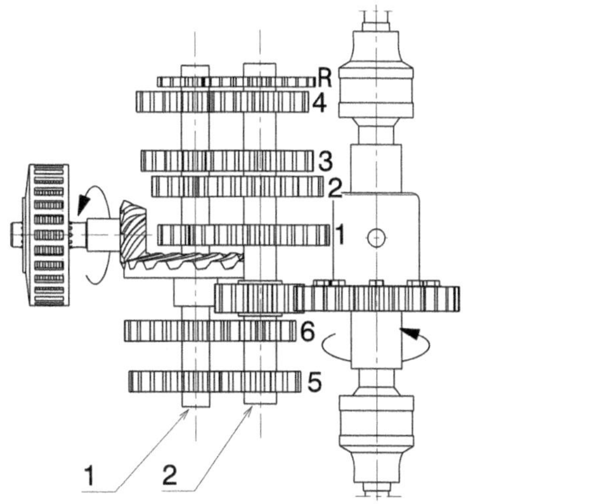

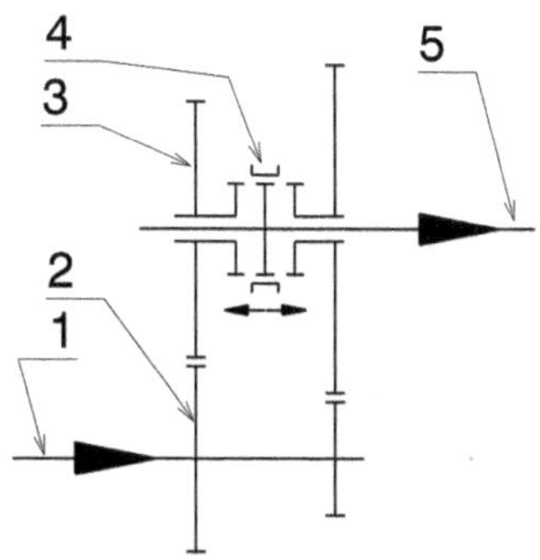

Fig. 5.38 Principle of the inner gearshift system. **1** Drive shaft (*input shaft*). **2** Fixed *gear*. **3** Idler gear. **4** dog-ring (*gearshift sleeve*). **5** Output shaft. The shift sleeve connects either the left or right idler gear to the output shaft. The gear stage consisting of fixed and connected idler gear then converts the input torque into an output torque. If the shift sleeve is in the middle position, no gear is engaged (neutral)

In a standard H-shift, the shifting motion occurs in two parts. First, the driver selects the shift gate (i.e. first gear – reverse, second gear – third gear, etc.) and then the actual shifting takes place.[1] Figure 5.40 shows the arrangement of the shift forks and their actuation by the shift finger. The selection movement is made by turning the shift shaft (1). This causes the shift finger (2) to slide into the groove of the desired shift rod (= shift gate). The actual shifting is done by pulling or pushing the shift shaft, whereby the shift fork connects the desired idler gear with the shaft. To ensure that only one shift rod can be moved at a time,

[1]See Racing Car Technology Manual, Vol. 2 Complete Vehicle, Section 4.6 *Gear linkage*.

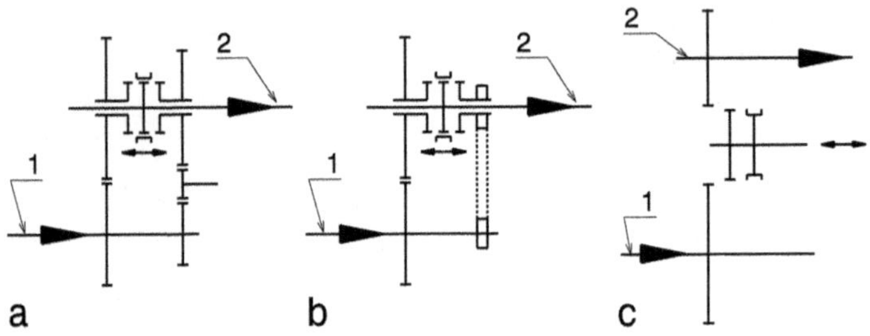

Fig. 5.39 Possibilities of a reverse gear. **1** Input shaft. **2** Output shaft. (**a**) Intermediate wheel between idler and fixed wheel. (**b**) Chain between shaft 1 and 2. (**c**) Sliding wheel between shaft 1 and 2

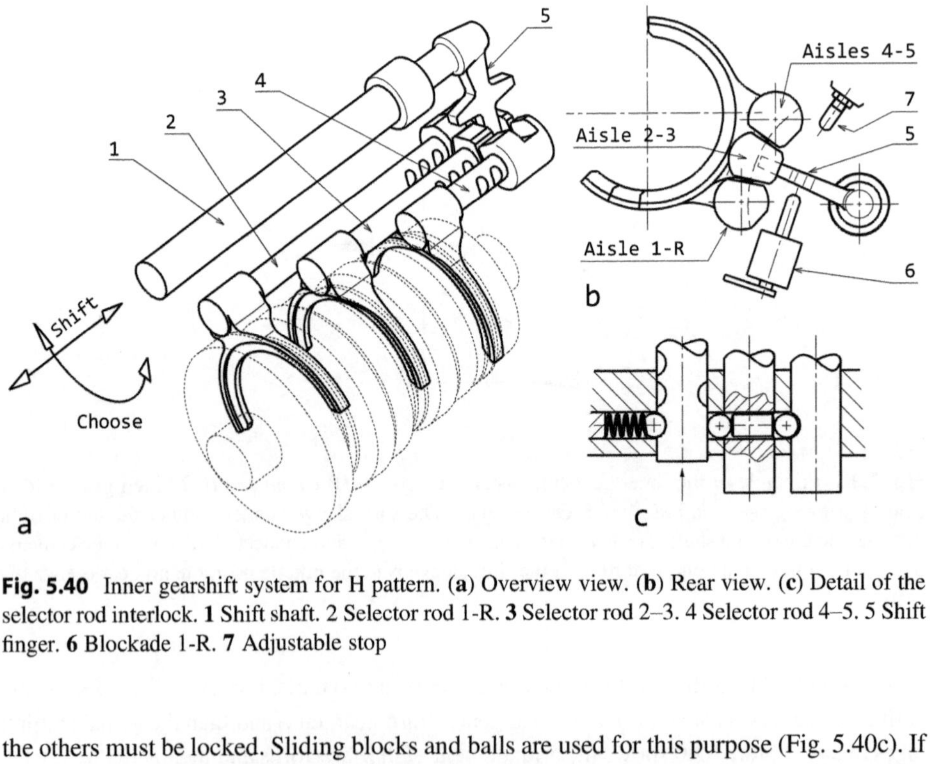

Fig. 5.40 Inner gearshift system for H pattern. (**a**) Overview view. (**b**) Rear view. (**c**) Detail of the selector rod interlock. **1** Shift shaft. 2 Selector rod 1-R. **3** Selector rod 2–3. 4 Selector rod 4–5. 5 Shift finger. **6** Blockade 1-R. **7** Adjustable stop

the others must be locked. Sliding blocks and balls are used for this purpose (Fig. 5.40c). If a shift rod is shifted from its rest position, it pushes the locking balls to the side and the other two rods can no longer be shifted. The shifted shift rod itself is locked in another groove by a spring-loaded ball. The reverse gear and the first gear are locked by a stop of the shift finger. If the driver wants to engage one of these gears, he must first unblock it via a lever in the cockpit, then shifting takes place as with the other gears using the shift lever. This blockage prevents the wrong gear from being engaged in the rush of the race.

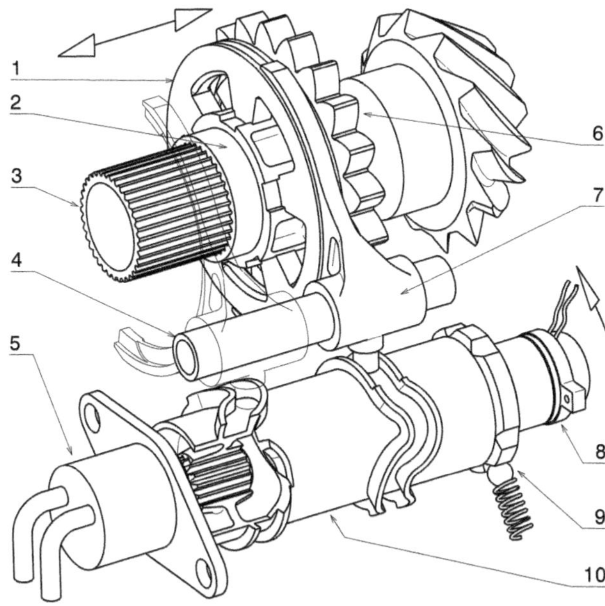

Fig. 5.41 Internal gearshift system for sequential operation (Ferrari F1 2000). **1** Gearshift sleeve (*dog-ring*). **2** Hub. **3** Output *shaft*. **4** Shaft for selector fork. **5** Rotary actuator. 6 Idler gear. **7** selector fork. **8** Position sensor (*rotary sensor*). **9** detent. **10** Selector barrel. The shift drum (10) is rotated by a hydraulic actuator (5). This causes the selector forks (7) to move along the axle (4) because they are guided by a pin in a grooved track. These paths are coordinated in such a way that in each case exactly one fork displaces the shifting sleeve (1) to the left or right and thus brings the claws of a loose wheel (6) into engagement. The shifting sleeve itself transmits the torque to its hub (2), which in turn sits on the output shaft (3) with splines. A locking device ensures that the shifting barrels remain in the desired position

With sequentially shifting gearboxes, the actual shifting is exactly the same. However, the actuation of the shifting sleeves takes place via a rotary movement. This can be done manually or electronically controlled by an actuator, Fig. 5.41.

For every two gear stages, one shift fork and one shift sleeve are required. Such a package of adjacent idler wheels with the engagening parts of the inner gear shift is shown in Fig. 5.42. The entire gear set of a shaft is thus built up by arranging a corresponding number of hubs on which the idler wheels run and the shift sleeves are seated.

Locks are also provided for the reverse gear in sequential transmissions. These are implemented electronically (through intelligent queries in the system, e.g. speed-engine speed comparison) or mechanically. Figure 5.43 shows a purely mechanical device that blocks reverse gear and also provides a parking lock. The cam wheel (1) is located at the end of the shift drum and locks its angular position and thus the engaged gear via the rocker arm (4), which is supported on the axle (7) and rests against the spring of the tensioner (6). If the reverse gear (R) is to be engaged, the locking body (5) must be pulled upwards from the cockpit via a cable so that the rocker arm can travel the required distance over the cam

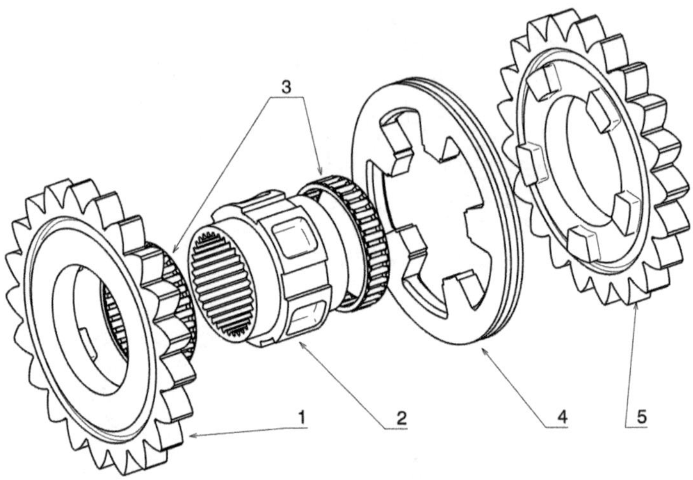

Fig. 5.42 Internal gearshift system components. These parts belong to the gearshift system of Fig. 5.41. **1** Idler for gear stage n. **2** Hub for gearshift sleeve. **3** Needle bearing. **4** Shifting sleeve. **5** Idler for gear stage n + 1. The idler gears (1, 5) are mounted on needle bearings (bearing 3) and run on the hub (2). The torque connection is ensured by the shifting sleeve (4), which is shifted to the right or left on the hub (2) and brings the claws of an idler wheel (1, 5) into engagement. The hub itself rotates with the output shaft, to which it is connected via splines

"0" unhindered. Between the first and the return gear lies the recess for the parking lock (Park). The position of the lever can be locked in this recess by moving the locking body further down than shown in the illustration. To engage the first or reverse gear, the locking body must be pulled up again. Neutral is engaged by a half shift movement starting from first or second gear.

When shifting by hand, a ratchet system is necessary so that the shift drum is rotated by exactly the required angular amount and no further. In principle, this is exactly the same as with motorcycle gearboxes, only there the gear is changed with the foot. Some kind of gear indicator is provided for the driver, because otherwise he does not know which gear is engaged. If the shifting is done by an actuator without a ratchet system, a control system is required to ensure that the gear is actually engaged in every case. In the event of wear and/or temperature changes, purely controlled systems would tend to shift incorrectly. In the case of manual shifting, the driver takes over this control by pulling or pressing the shift lever until the gear engages or he simply tries again after an incorrect shift. A position sensor is required for the control, which reports the position of the shift drum to the control unit. For safety reasons, this sensor is often designed redundantly.

Actuators enable extremely rapid and electronically controlled gear changes. Electric, electro-hydraulic and electro-pneumatic systems can be used as actuators. In terms of power density, electro-hydraulic actuators are the first choice, closely followed by electro-pneumatic ones.

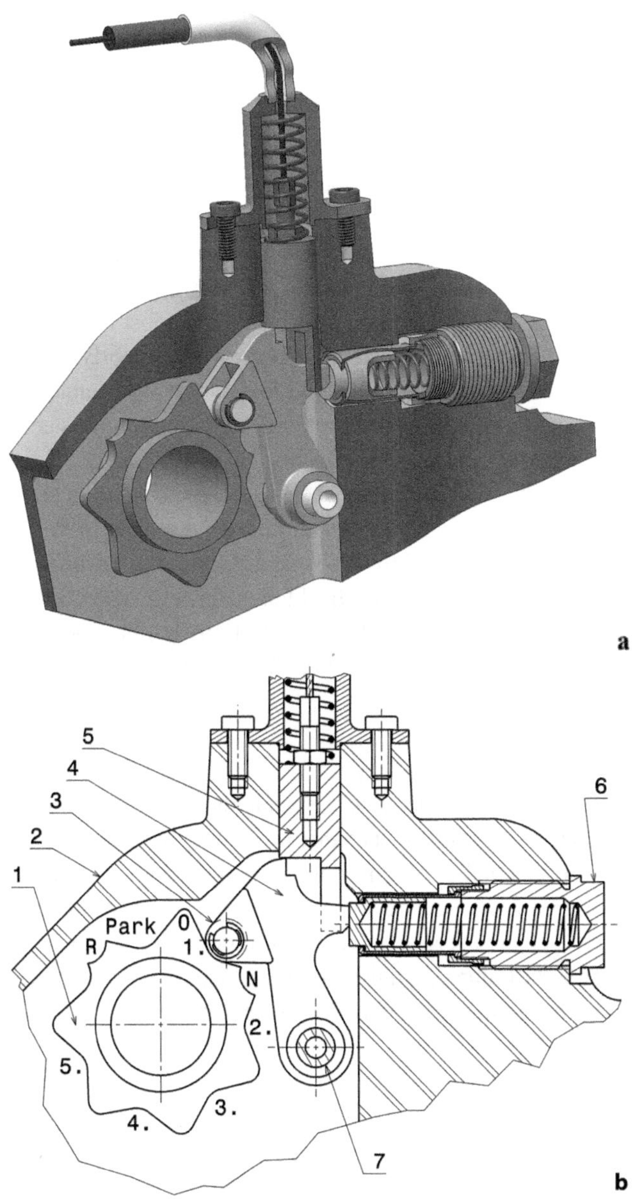

Fig. 5.43 Mechanical lock of the reverse gear (Lola system). (**a**) Axonometric view, (**b**) Sectional view. 1 Cam wheel, 2 Transmission housing, 3 Roller, 4 *cam follower*, 5 *blocker body*, 6 Screw-in tensioner, 7 Axle. The recesses on the cam wheel are designated according to the corresponding gears corresponding to this position of the roller (3)

If the necessary overall system is taken into account, there are also advantages for pneumatic actuators, which are larger due to their principle. The operating medium is the air present in the environment and leakages have less of a catastrophic effect compared to hydraulic elements.

In addition to a pump, hydraulic systems require a pressure accumulator, a cooler and a reservoir for the fluid. The system must function at engine idle (about 3000 min^{-1}) as well as at the speed limit of 19,500 min^{-1}. The hydraulic actuator is faster than the pneumatic one and for that reason alone it is preferred. In addition, synthetic oils now make it possible to use the same oil that is used as the lubricating oil for the gearbox. This eliminates the need for additional reservoirs and coolers. However, there must be a fine filter upstream of the hydraulic pump to ensure that the control valves work properly.

The internal gearshift is basically the same for manual and automated transmissions. Although there are major differences in detail, especially in the shift sleeve and gear claws (Fig. 5.42). The number of claws is eight for manual gearboxes, and only four or five for automated gearboxes. The claws are nevertheless about the same in size. The space thus gained with fewer claws makes it easier to shift through the shift sleeve. The claws also do not have an undercut on automatic shifters so that the separation of the shift sleeve from the gear can occur more quickly. The terms "facilitating shifting" and "fast" are relative. One value should illustrate how lightning-fast an electrohydraulic Formula 1 transmission works: Shifting through all gears at a standstill from first to seventh and back to first gear takes less than 0.2 s [14].

Traction Interruption

During a gear change, the torque flow between the engine and the drive wheels is interrupted. When shifting up, this means that no driving force acts on the vehicle during this time and it is decelerated by the driving resistances, Fig. 5.44. If the power flow is restored after shifting up, the tractive force – reduced by the lower transmission ratio – starts again and the vehicle accelerates again. In order that the speed drop Δv does not become too great, the shift time t_{Shift} must not be too long. Even in a passenger car, the time required for a gear change must therefore not exceed one second.

For comparison, in sporty passenger cars and sports cars with sequential automatic gearshifts supported by highly dynamic clutch and engine torque control, this results in approx. 0.15 s traction interruption [15, 16]. In racing, actuated gearshifts allow single gearshifts with 0.04 s. In Formula 1, the times for a gear change – thanks to semi-automatic transmission at the push of a button – are around 0.01–0.02 s.

This also shows the advantage of an automated gearshift. A Formula 1 car decelerates at high speed with about 1 *g* during a manual shift, which takes about 0.1 s. At the same time, its speed drop is about 3.5 km/h [14].

The sequence of an automatic shift in a touring car is shown in Fig. 5.45. When shifting up, the clutch remains closed and the throttle fully open. The ignition is briefly switched off so that the engine does not overrev during the gear change and the engine speed is even

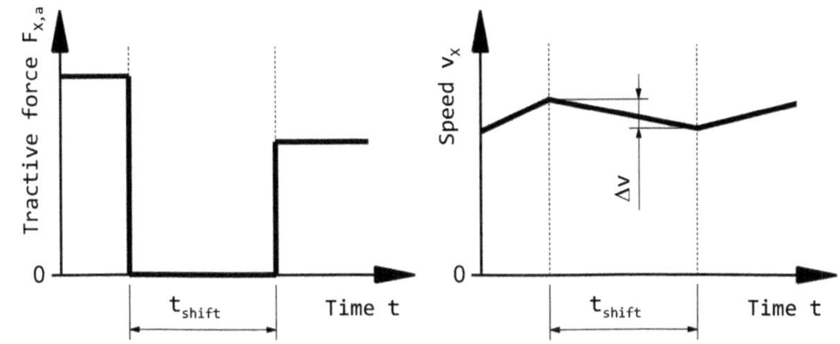

Fig. 5.44 Traction interruption during upshifting. The tractive force is completely interrupted during shifting. The vehicle speed v_X drops during the shifting time t_{Shift} by the amount Δv

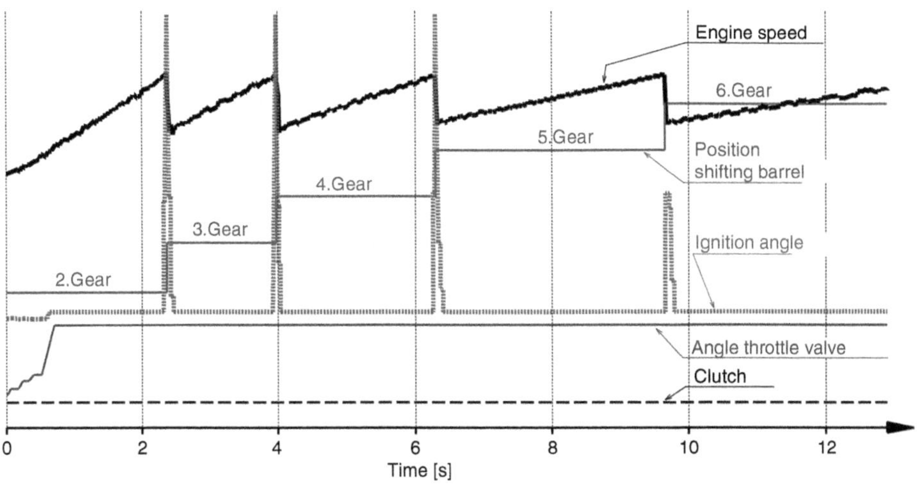

Fig. 5.45 Measurement data of a touring car when shifting up from second to sixth gear [17]. Upshifting is fully automatic when the engine reaches maximum speed. The clutch remains closed, the throttle is always fully open, only the ignition angle is reduced during shifting

reduced for easier shifting. A Formula 1 naturally aspirated V10 engine decelerated by 30,000 revolutions in one second with the throttle open, i.e. in 0.02 s of ignition interruption the engine speed dropped by 600 revolutions [14]. The remaining rotational energy is used to accelerate the vehicle when shifting through. However, in addition to the shock load on the drivetrain, this also means an abrupt reapplication of the drive torque at the wheels, which can affect the stability of the car. This is precisely what limits the shortening of the shift time. So a compromise has to be found between acceleration and driving stability. If you wanted to shorten the shifting time even further, you would have to brake the engine during upshifting, e.g. with exhaust flaps.

Fig. 5.46 Sensor box of a shifting device (KLS device applied to Formula 3000 Reynard with Judd V8). This device enables traction-free shifting without clutch. When the gearshift is actuated, the sensor box (KLS) sends a signal to the control unit and the engine ignition is interrupted until the gearshift is completed

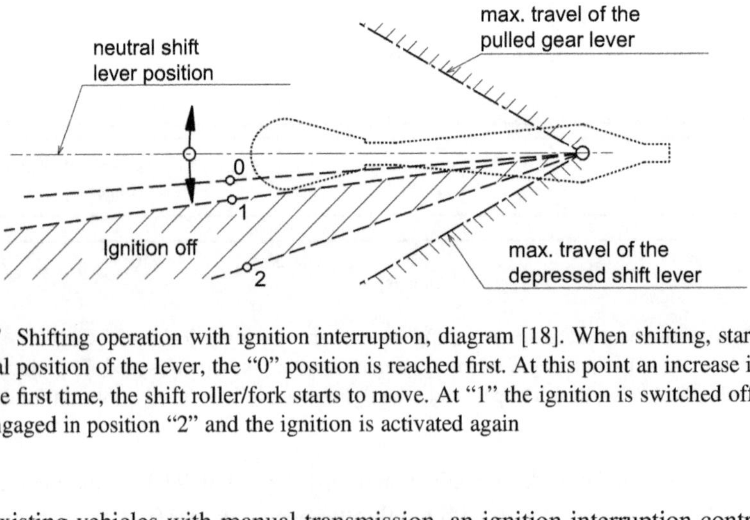

Fig. 5.47 Shifting operation with ignition interruption, diagram [18]. When shifting, starting from the neutral position of the lever, the "0" position is reached first. At this point an increase in force is felt for the first time, the shift roller/fork starts to move. At "1" the ignition is switched off until the gear is engaged in position "2" and the ignition is activated again

For existing vehicles with manual transmission, an ignition interruption controlled by the shift linkage can be retrofitted. Upshifting then takes place without a clutch and the power output of the engine is interrupted by the shifting. Figure 5.46 shows such a retrofitted control unit on a gearbox. The mode of operation is described in Fig. 5.47.

Downshifting is usually done with the throttle closed and the clutch open to free the transmission from engine braking torque. If downshifting is carried out too early, i.e. at relatively high speed, the engine may overrev when the clutch is re-engaged. This can destroy the engine or cause the drive wheels to lock up. Both must be prevented at all costs. In automated systems, the control unit therefore calculates the engine speed after the gear

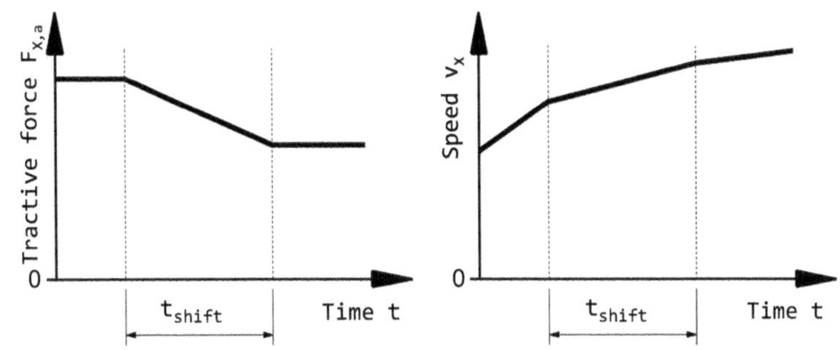

Fig. 5.48 Upshift without traction interruption *seamless upshift.* The tractive force is not interrupted during shifting, but is only reduced by the change in transmission ratio. The vehicle accelerates during the shift time t_{Shift} more

change and only allows it if this speed is below the permissible limit. To facilitate shifting through, the engine and transmission shaft can be accelerated during neutral in the transmission. To do this, the driver applies intermediate throttle in manual transmissions and in automated systems this is done by the control unit with the aid of the electric throttle control or a so-called blipper.

In the case of automated transmissions – i.e. when shifting is performed by an actuator instead of a human being – it is found that the force required must not be constant. On the contrary, at the beginning of the shifting manoeuvre, a high force is required to separate the shift sleeve from the gear wheel. For the further movement of the sleeve to the neighbouring gear, the force required is much less. The force increases again during further movement of the sleeve when the shift sleeve is in contact with the neighboring gear and the torque flow is restored [19]. Precise shifting maneuvers, as the computer can regularly perform with the help of various sensors, increase the service life of transmissions fourfold compared to manual shifting.

Even better is a gearshift that allows a change in transmission ratio without interrupting the tractive force. Such so-called powershift transmissions are, for example, conventional automatic transmissions and dual-clutch transmissions. With these, two transmission stages are engaged simultaneously when shifting, but a necessary slip is permitted by friction clutches. The previous gear stage is therefore disengaged while the next higher one is engaged. The traction is never interrupted and the vehicle continues to accelerate even during the shift, Fig. 5.48.

Continuously variable transmissions allow a continuous change in transmission ratio and thus a curve of the tractive force change without kinks. With continuously variable transmissions (e.g. CVT transmissions), the course of the tractive force and the vehicle speed thus looks like the "rounded" course of Fig. 5.48.

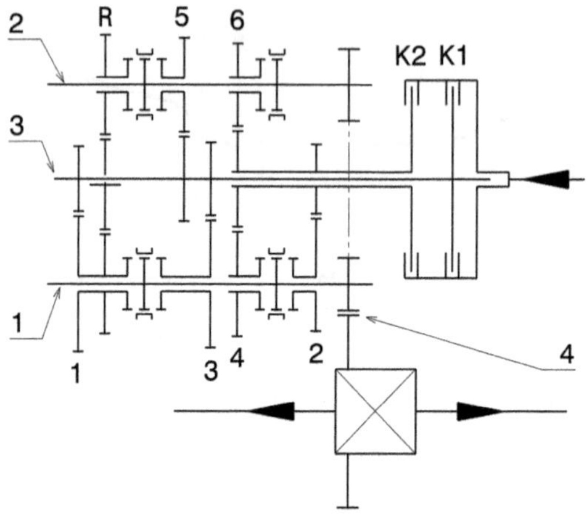

Fig. 5.49 Schematic of a double clutch transmission. **1** Output shaft 1. **2** Output shaft 2. **3** Double input shaft. **4** Axle drive. K1, K2 Clutchs. The engine torque is transmitted to the corresponding fixed wheels either via the core shaft (clutch K1 closed) or via the hollow shaft (clutch K2 closed). The axle drive (4) is acted upon by both output shafts (1, 2). The idler gears of fourth and sixth gear mesh with the same fixed gear. This saves overall length

Twin-Clutch *Gearbox* (Double-Clutch Gearbox)
A dual clutch transmission allows shifting without interrupting traction. Basically, the transmission consists of two gearboxes engaging a common final drive, Fig. 5.49. The engine torque is introduced into the two gearboxes via a double shaft (3). Two independent clutches connect or disconnect the core and hollow shafts to the engine. The gears are divided so that the core shaft serves the odd gears (1, 3, 5, R) and the hollow shaft serves the even gears (2, 4, 6). This division allows exactly consecutive gears to be engaged at the same time. An electronic control of the two clutches ensures that only one clutch is fully closed at a time, i.e. one gear is active, and that there is a traction-free transition from one gear to the next. In the latter case, one clutch is simultaneously closed while the other is opened. The clutches are actuated electrohydraulically. The internal shifting is the same as for conventional manual transmissions. Nevertheless, an additional advantage results from the parallel arrangement of the output shafts. The shifting times can be shortened considerably, because the next gear is engaged while the previous one is active. The actual shifting takes place later via the clutches, if desired.

Although shifting without interruption of traction is an advantage, shifting is also carried out with interruption. Depending on the direction of shifting (up or down) as well as the sign of the torque (pull or push operation), shifting is also advantageously carried out with interruption of the tractive force.

The reliability of the overall system stands and falls with the integration of intelligent control software and hardware. Tension conditions in the partial transmissions are a risk

Fig. 5.50 Shafts and wheel arrangement of a 6-speed dual clutch transmission. The gearbox corresponds to the one in Fig. 5.49. In the foreground is the output shaft 1, i.e. the one that drives the odd gears. The two clutches, which sit on the double shafts, are not shown

inherent in this concept and must be counteracted. A distortion condition occurs, among other things, when a gear is engaged in both partial transmissions and both clutches transmit torque. In shifting operation, this significantly increases the braking torque of the engine. Even a brief state of tension (< 100 ms) can lead to locking of the drive wheels at low friction values and thus to breakaway of the vehicle [20]. This condition must therefore be excluded under all circumstances.

The arrangement of the shafts and wheel sets of a double clutch transmission is shown in Fig. 5.50.

Transmissions of this type were already used in racing in the 1980s. As a PDK (Porsche Doppelkupplungsgetriebe), it was used in the Porsche 956 sports prototype and in the Audi Sport Quattro S1. The combination with the turbocharged engines in particular made the advantages clear. When shifting gears, the boost pressure could be maintained, which was noticeable in a reduction of the lap times. The additional weight of the transmission was thus more than compensated for [20].

The contemporary variant (direct shift gearbox – DSG) still has an immense weight disadvantage (approx. 60% heavier), but this is more than compensated for by the advantages mentioned. With a typical 25 upshifts during a lap, the lap time advantage of a Formula 1 car theoretically adds up to about 0.35 s [21].

Gears

The gears in the racing gear are spur geared. In passenger car applications, all gears are usually helical with the exception of the reverse gear. Helical gearing is used primarily because it produces less noise.

Gear wheels are designed to be time-fixed for weight reasons. The service life of the entire gearbox is in the range of a few hours. Naturally, not all wheels are equally loaded,

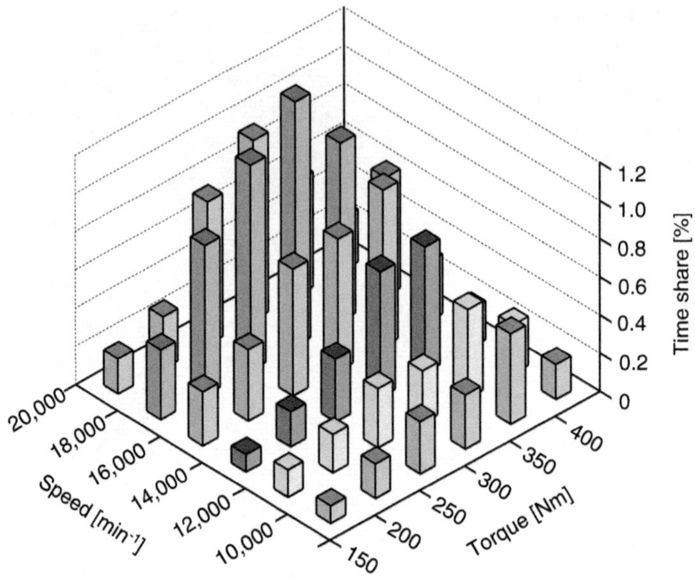

Fig. 5.51 Load spectrum for the fifth gear of a Formula 1 gearbox [22]. The frequency distribution of the load on the gear is plotted against input speed and torque. It can be seen that this gear stage is only loaded for a fraction of the gearbox service life (=100%). The values were determined with a standard 3.0-liter V10 engine at that time without a specified speed limit

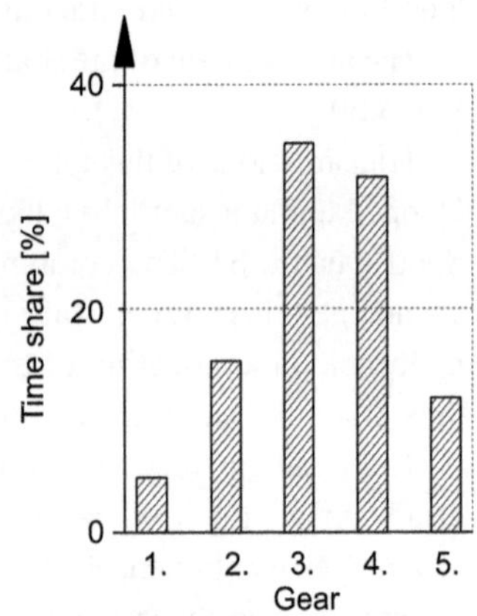

Fig. 5.52 Time fractions for the individual gears of a 5-speed passenger car transmission on a mountainous country road, after [1]. Gears 3 and 4 are used most often. First gear is used least often

but on the contrary, depending on the distance and gear, the load on individual wheels is completely different. Figure 5.51 shows an example of the load spectrum (see appendix) for the fifth gear of a 7-speed longitudinal gearbox of a Formula 1 car.

For comparison, Fig. 5.52 shows the time fractions for the individual gears of a passenger car 5-speed transmission.

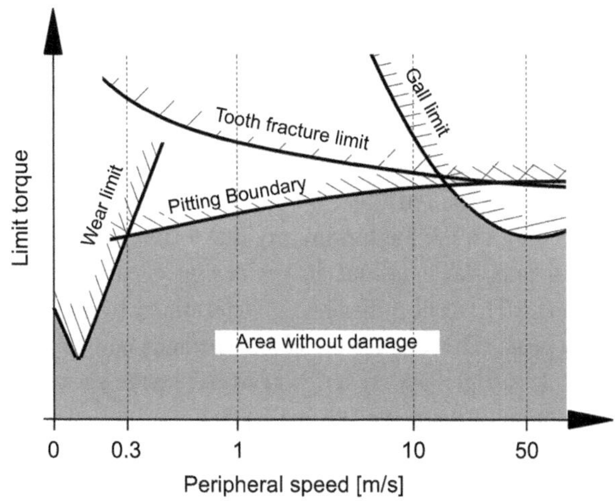

Fig. 5.53 Load capacity limits of gears, according to [1]. The limits of the load carrying capacity of gears are determined by their different causes of failure

Of the main dimensions of a gear, rolling diameter and tooth width, the width is particularly decisive because it directly influences the overall length of the gearbox. For example, if 1 mm of width is saved on the gears, a single-stage 7-speed gearbox will be 7 mm shorter. When estimating the service life of gears, the type of actuation is important. With manually shifted gearboxes, there is a risk of shifting when downshifting. If a gear is engaged that is too low, the drive wheels try to turn the engine up against its inertia. How great the forces can become due to the transmission ratio can be seen from the short-term locking of the wheels. The transmission must be able to withstand such shifting maneuvers and the gears must be dimensioned accordingly. Automated gearboxes use the logic of the control unit to prevent faulty shifting and the gears can be designed for the operating forces. The performance limit of a gear pair depends on the possible causes of failure. These can be divided into four different damage mechanisms, Fig. 5.53.

A tooth is bent by the opposing tooth. If the bending stress in the tooth root area becomes too great, the tooth fracture limit is reached and the tooth breaks off. Pitting is a fatigue phenomenon of the material. Pitting is manifested by flat breakouts on the tooth flanks, usually below the pitch circle. If the lubrication in the tooth mesh fails, the running partners seize up. Depending on the circumferential speed, this is referred to as cold or warm galling. Cold seizure (exceeding the wear limit) is purely a wear phenomenon and rarely occurs in gearboxes. At high circumferential speeds, the lubricant film can be destroyed by high loads or temperatures. This results in metallic contact between the tooth flanks and seizure.

The performance limits of gears are significantly influenced by operating conditions (type of load, peripheral speed, temperature), material selection, gear geometry, manufacturing accuracy, surface treatment and surface roughness, and lubricant.

The common calculation of gears is mainly based on empirical methods, which are standardized (e.g. DIN 3990, ISO 6336, AGMA 2001). However, non-standardized gears are preferred for racing gearboxes because this is the only way to approach the performance limit in a weight-saving manner. The tooth profiles are optimized for their behavior under load by means of FEM calculations (see appendix) and can also be realized by CNC processing machines. The end of the service life of a gear is then determined by the wear during operation. The more the actual tooth form deviates from the calculated course, the sooner the gears must be replaced. If they are not replaced, failure will occur sooner or later depending on the load, primarily due to tooth breakage.

An important variable in the design of single and two-stage gearboxes is the centre distance. The centre distance is determined by the gear with the greatest torque increase (first gear). The smaller the centre distance can be, the smaller the whole gearbox can be built. Investigations on series gearboxes have shown the following relationship for rough estimation of the centre distance [1]:

$$a = 60 + 2.08(i_{G,\max} \cdot T_1)^{0.44} \tag{5.10}$$

a	Centre distance, mm
$i_{G,\max}$	Largest gear ratio, –.
T_1	Gearbox input torque, N m

The gearbox mass can also be roughly estimated from such series tests:

$$m_G = 0.49(i_{G,\max} \cdot T_1)^{0.58} j^{0,29} \tag{5.11}$$

m_G	Gearbox mass, kg. Applies to cast iron housings
j	Number of gears, –

For an initial design, the center distance can be calculated as follows [1]:

$$a = 0.255 \cdot \sqrt[3]{\frac{T_1(u+1)^4}{u}} \tag{5.12}$$

T_1	Moment on the shaft on which the pinion (small wheel) of the first gear is located
u	Number of teeth ratio of the gear pair of the first gear, –

For a given centre distance, the pitch circle diameters of meshing gears follow from the desired transmission ratio of this pair of gears.

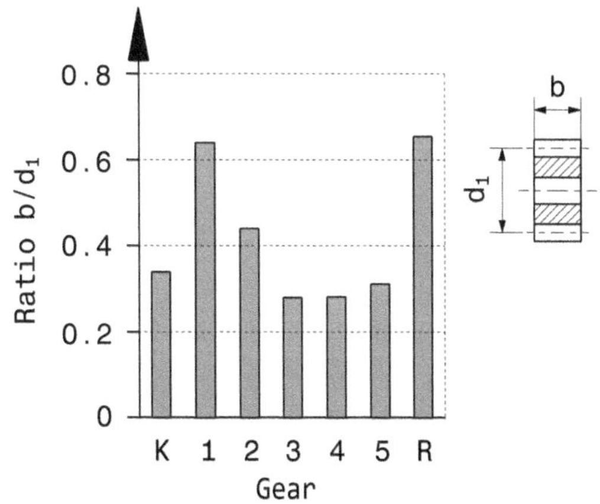

Fig. 5.54 Width ratio of designed passenger car and commercial vehicle transmissions, according to [1]. The individual gears are designated 1,2,3, . . . R (reverse gear) and K for constant ratios (i.e. gear pairs that permanently transmit torque). *b* Gear width, d_1 Pitch diameter of the pinion (small wheel)

$$i_{G,j} = \frac{n_1}{n_2} = \frac{d_2}{d_1} \tag{5.13}$$

$$d_1 = \frac{2a}{1 + i_{G,j}} \tag{5.14}$$

$i_{G,j}$	Gear ratio of gear *j*, –	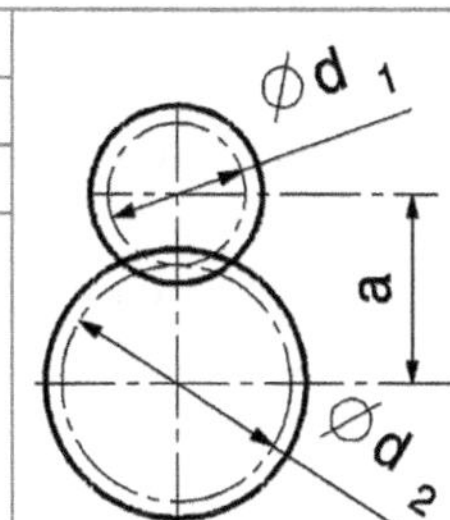
n_1, n_2	Input or output speed of the shafts	
d_1, d_2	pitch circle diameter of the gears, mm	
a	Centre distance, mm $a = 0.5(d_1 + d_2)$	

The tooth widths can be determined for an initial design from the *width/diameter* ratios b/d_1 commonly used in practice. The ratios are chosen differently in the individual gears to minimise uneven width wear, Fig. 5.54.

The tooth widths in Formula 1 are partly specified by the regulations: Min 12 mm, except first gear (15–17 mm) and reverse gear (5 mm). A gearbox must last six race weekends, with oil changes allowed. This corresponds to 3300 km at qualifying or race pace and is roughly comparable to the 24 Hours of Le Mans (4750 km).

Bearing Rolling bearings are consistently used for bearing arrangements in transmissions, both in passenger cars and in racing vehicles. One difference between the applications is the use of hybrid bearings in motorsport. In hybrid bearings, the rolling elements are made of ceramic (e.g. silicon nitride) and the rings are made of steel. Similar bearings are also used for wheel bearings.[2]

The following requirements are placed on gearbox bearings:

- Low weight: Reduces the total mass and especially the rotating masses
- Low frictional torque
- Small installation space
- Highest reliability
- Low oil consumption
- Simple assembly

A low weight and, above all, the rotating mass, whose share in the kinetic energy of the vehicle enters with the square of the transmission ratio, are important when shifting gears quickly. The transmission shaft rotates at engine speed and thus reached up to 19,000 min^{-1} with Formula 1 engines.

If the friction of the moving parts is low, little heat is released and the engine power reaches the drive wheels with less loss.

A smaller installation space for the bearings also allows a smaller and thus stiffer transmission. A slim transmission in the rear of a mid-engine car interferes less with the aerodynamics, and short transmissions leave more installation space for the engine and chassis in transverse engines.

Bearings are critical components in the gearbox. The service life of the bearings is adapted to the requirements. In Formula 1, for example, approx. 2000 km are sufficient.

A low oil requirement simplifies the design effort for lubrication. In some applications, hybrid bearings do not require forced lubrication; the existing oil mist is sufficient [23]. As a result, the oil pumps can be reduced in their delivery rate, which increases the overall efficiency of the gearbox.

The assembly of the gearbox bearings is important because the gearboxes are often disassembled and the bearings are central parts that are responsible for the position of the gears in relation to each other and the shifting elements. This function is made more difficult by the fact that the housings are made of light metal or sometimes even plastic, while the shafts are made of steel. The bearings must therefore also compensate for the inevitable differences in length that occur with changes in temperature. Figure 5.55 shows a bearing arrangement for a gear shaft or countershaft or output shaft.

The axial forces are absorbed by a double-row angular contact ball bearing (5) alone. This is clamped against the shoulder of the pinion by a shaft nut (4). A spacer ring (6) bridges the fillet between the bearing seat and shoulder. This ensures perfect contact

[2]More detailed information can be found, for example, in the Racing Car Technology Manual, Vol. 3, Sect. 2.3.4 *Wheel Bearings.*

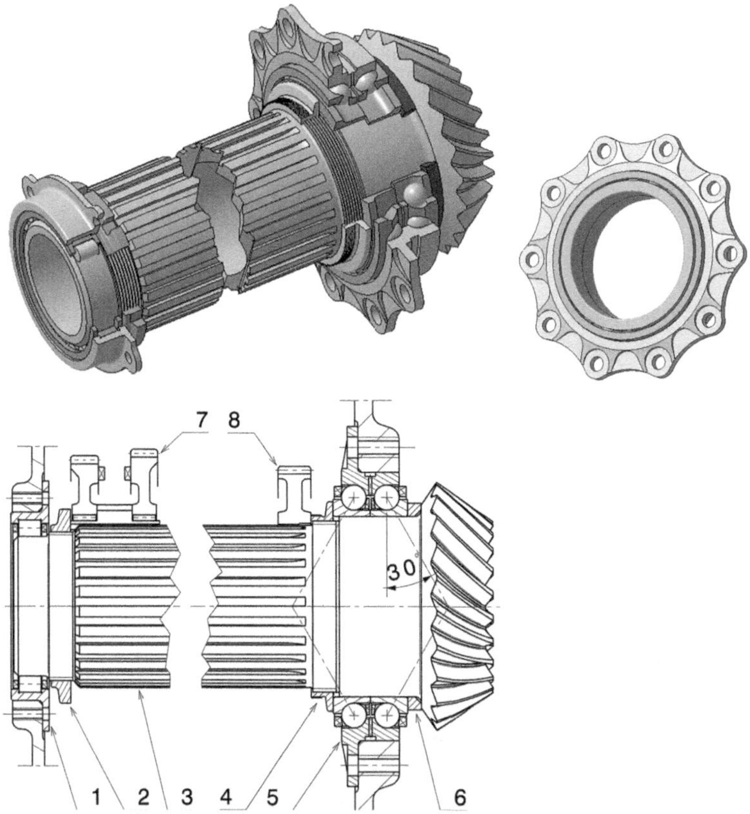

Fig. 5.55 Bearing arrangement of a shaft with hybrid bearing, according to [23]. This bearing arrangement of a layshaft (countershaft) of a Formula 1 transmission is optimized for hybrid bearings. The locating bearing is an angular contact ball bearing and the non-locating bearing is a cylindrical roller bearing whose rollers run directly on the hardened shaft. **1** Cylindrical roller bearing with flange. **2 Shaft** nut. **3** Gearbox shaft or *layshaft*. **4** Shaft nut. **5** double row angular contact ball bearing with flange . **6 spacer** ring. **7 Gear** wheel (idler). **8** gear (fixed)

with the shoulder. Due to the small distance between the rows of balls, the bearing operates almost clearance-free axially and radially over the entire operating temperature range. The outer ring of the bearing has a flange. This allows the steel ring to be screwed into the light metal housing. This improves the force flow into the bearing wall and prevents the bearing outer ring from becoming loose at high temperatures, which can be the case with pure interference fits. At the other end of the shaft is the non-locating bearing (1), which only supports the shaft radially. In this bearing, too, the outer ring is screw mounted on the housing wall via an integral flange. The bearing is particularly compact since it has no inner ring and the cylindrical rollers run directly on the shaft. The parts located on the shaft, such

Table 5.7 Permissible deflection and bending angle for shafts of gear units [1]

Shafts	Deflection	Bending angle β
Generally valid for gear wheels	$\leq 0.01\ m_n$ m_n Normal module	$\tan \beta_{zul} \leq \frac{2 \cdot d_w}{10^4 \cdot b}$ d_W pitch circle diameter b tooth width
Reference values for gear teeth	≤ 0.02 to 0.06 mm	$\tan \beta_{zul} \leq 0.005$ for cylindrical gears $\tan \beta_{zul} \leq 0.001$ for bevel gears

as the gear hubs of the idler gears and fixed gears, are axially preloaded using a shaft nut. A splined shaft profile provides the slewing ring connection.

Shafts For an initial design of a gearbox, the shaft diameters are important in addition to the centre distance. The shaft diameter of a solid shaft can be roughly determined from the comparative torque:

$$M_v = \sqrt{M_{b,\max}^2 + 0.75(\alpha_0 T_{\max})^2} \tag{5.15}$$

M_v	Comparison torque, N m
M_b	Bending moment, N m
α_0	Effort ratio, –. Depending on the time course of the load: $\alpha_0 = (\sigma_{zul}(\sigma)/\sigma_{zul}(\tau))$ Common cases include: $\alpha_0 = 1$ for σ_b alternating and τ_{ts} alternating $\alpha_0 = 0.7$ for σ_b alternating and τ_{ts} swelling
T	Torsional moment, N m

$$d_{\min} = 2.17 \cdot \sqrt[3]{\frac{1000 M_v}{\sigma_{b,zul}}} \tag{5.16}$$

$d_{\min}$	Minimum shaft diameter, mm
$\sigma_{b,\,zul}$	Permissible bending stress, N/mm^2

Gear drives react very sensitively to shaft deformations. Edge wear in the bearings and jamming of gears can be the consequences. Especially in the case of wide teeth and large rolling diameters, deviations of the shaft from the nominal position have a noticeable effect. The requirements for shaft rigidity are therefore high. Reference values for the permissible deflection and bending angle for gearbox shafts can be taken from Table 5.7.

The design is based on operational stability. Only in this way can a weight-efficient gear unit be represented. On the basis of load spectra (see appendix), the design is carried out taking into account the travel components in the various gears with regard to the desired service life.

In addition, vibrations in the drive train must be taken into account. Dynamic effects lead to considerable stress peaks and deformations (see also Sect. 5.5.1). Critical speeds are particularly interesting in this context. During rotation, unbalances generate additional inertia forces that cause vibrations. Gear shafts with several rotating gears and shifting devices have several bending critical speeds. Depending on the characteristic, two types of vibrations occur:

- Torsional vibrations: Low and high frequency vibrations primarily excited by the combustion engine.
- Bending vibrations: Higher frequency vibrations of shafts. Excited by unbalances and tooth meshing.

In order to ensure that gear shafts are as light as possible with the desired service life, the following requirements must be taken into account in their design [1]:

1. Avoid notches
2. Reduce bending moments
3. Increase critical speeds

Typical notch points on shafts are threads, transverse bores, shoulders or collars, grooves for retaining rings, grooves for retaining plates and grooves for springs as well as wedges. The following design rules help to fulfil these requirements, Fig. 5.56:

- Reduce bearing distances through compact design.
- place heavily loaded gears close to bearings so that deflections and bending moments are reduced and the bending-critical speeds are high.
- For diameter transitions, do not exceed the ratio $D/d = 1.4$. In addition, if possible, do not design the transitions with a shaft shoulder, but conical or with a rounding radius, Fig. 5.56a.
- Prefer splined shaft connections or oil press fits to a key connection, Fig. 5.56b. A slight projection of the cylindrical hub over the shaft seat reduces the notch effect.
- Defuse rectangular ring grooves by relief notches or rounding of the inner edges, Fig. 5.56c. For the design of axial restraints of hubs see paragraph 5.5.3 *Shaft-hub connections.*
- If possible, only arrange retaining rings at the shaft end, use spacer sleeves in the central shaft area for axial retaining.
- Reduction of the notch effect at shaft shoulders Fig. 5.56d: 1 Relief notch at the transition due to rounded axial single groove; 2 large rounding radius; 3 radial relief notches; 4 additional notches in the transition area.

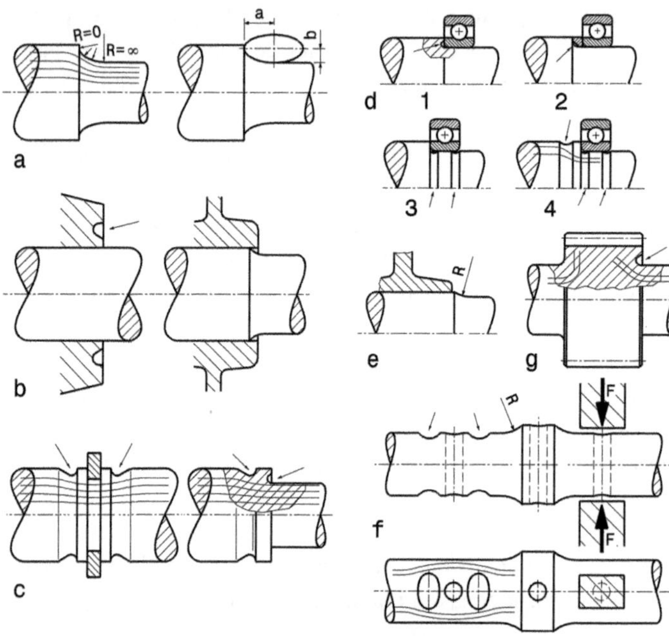

Fig. 5.56 Reduction of the notch effect on shafts. Gradual transitions and unloading notches help to make the stress curve in the shaft comparable and thus to keep the stress increase at the notches small. The lines of force are shown in some illustrations for a better understanding

- Thicken shafts with fitted hub at the hub seat, provide large transition radius and reduce hub thickness towards the edge, Fig. 5.56e.
- Transverse bores are mitigated by relief notches next to the bore mouth, by increasing the shaft diameter with large transition radii or by repressing the bore edges with a flat thrust piece, Fig. 5.56f.
- gradual redirection of force by relief notches, Fig. 5.56g.
- Shafts with an operating speed from approx. 1500 min^{-1} should be precisely balanced in order to keep centrifugal forces and the associated bending vibrations low. Balancing quality VDI 2060: Q0.4 to Q1600 (10 quality groups).
- Reduce the mass moment of inertia of the components mounted on the shaft in order to reduce deflections and increase critical speeds.

Table 5.8 gives some notch effect numbers (see appendix) for typical discontinuity points in shafts. It can be seen that round grooves cause significantly lower stress increases than rectangular grooves and this both in pure bending and torsion.

In terms of uniform shaft loading, interference fits are considerably more favourable than splined or keyed connections. However, the stress increases in notches can also be further reduced in interference fits by the design of the shaft and hub, Table 5.9. In the table, a simple interference fit is compared with a more favourably designed one. Essentially, the

Table 5.8 Notch effect numbers for shafts, [24]

Notch type	Notch shape	Notch factor K_f, –	
		At bending	At torsion
Round notch	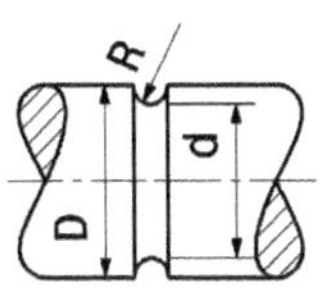	1.5 to 2	1.3 to 1.8
Circlip	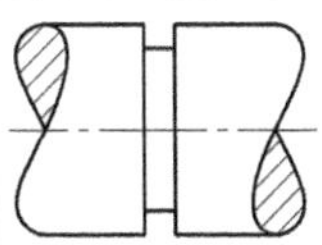	2.5 to 3.5	2.5 to 3.5
Shoulder	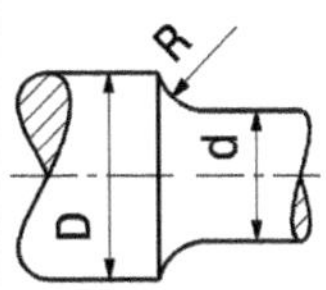	Approx. 1.5 with $R/d = 0.1$ and $d/D = 0.7$	Approx. 1.25 with $R/d = 0.1$ and $d/D = 0.7$
Cross hole	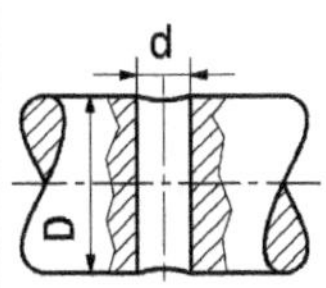	1.4 to 1.8 with $d/D = 0.14$	1.4 to 1.8 with $d/D = 0.14$
Keyway	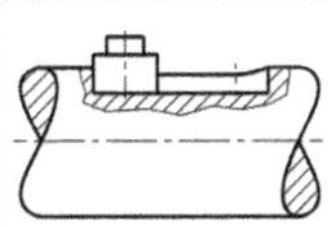	1.6 to 2	1.3
Outlet groove	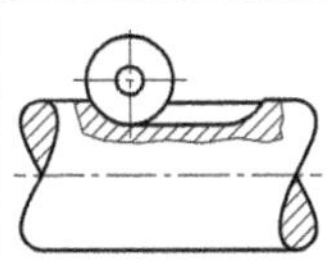	1.3 to 1.5	1.3 to 1.5
Hub seat with key or wedge	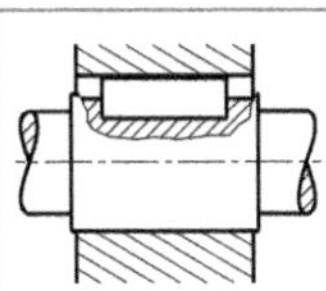	2 to 2.4	1.5 to 1.6

design of the transition between different cross-sections is decisive for the stress curve. In the case of pure bending, the comparable stress increase is always greater than in the case of pure torsion. Furthermore, it can be seen from the table that the notch sensitivity increases with increasing strength of the material (R_m), irrespective of the type of load (bending, torsion).

Shafts can advantageously be hollow bored. If the inside diameter is about half the outside diameter, the hollow shaft weighs only 75% of the solid shaft, but still has 95% of the moment of resistance to bending and torsion of a solid shaft.

Table 5.9 Notch factors for shafts with interference fit [24]

Hub form	Fit	Notch factor K_f, –					
		R_m, N/mm²	400	600	800	1000	1200
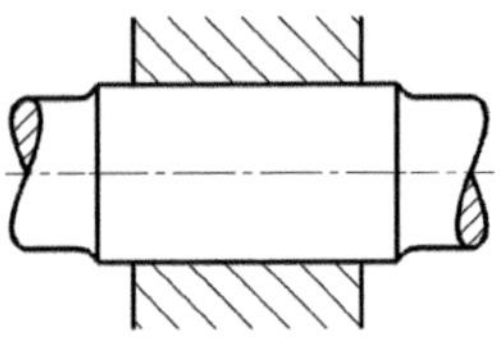	H8/u8	Bend	1.8	2.1	2.5	2.8	2.9
		Torsion	1.2	1.4	1.6	1.8	1.9
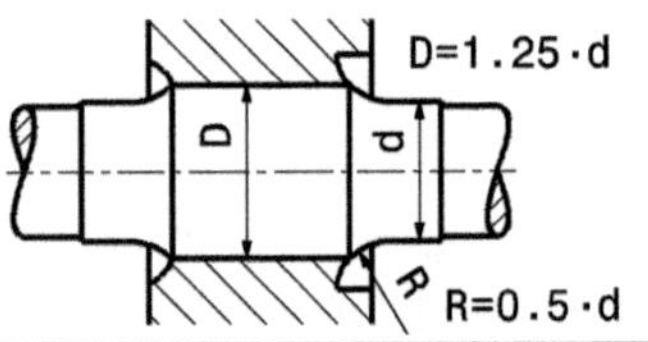	H8/u8	Bend	1.0	1.1	1.2	1.3	1.4
		Torsion	1.0	1.0	1.1	1.2	1.2

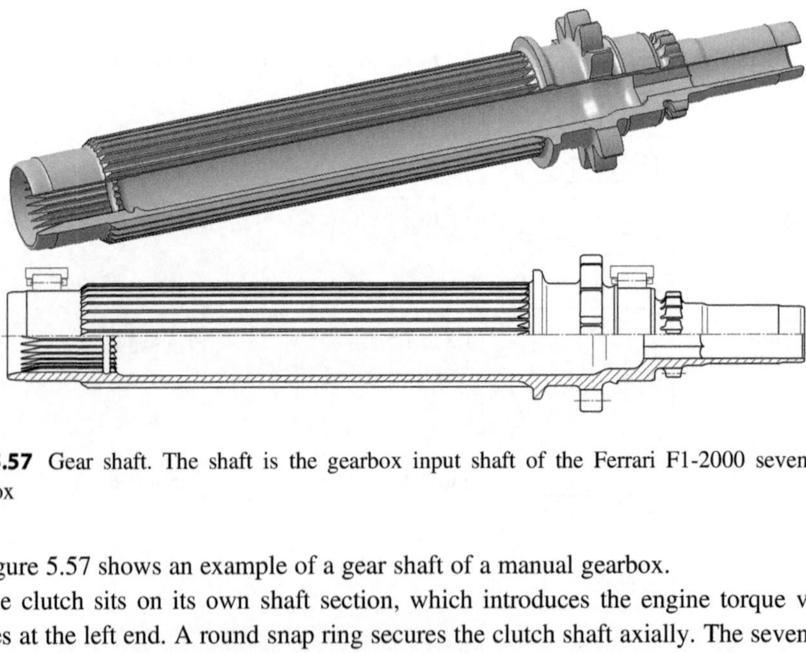

Fig. 5.57 Gear shaft. The shaft is the gearbox input shaft of the Ferrari F1-2000 seven-speed gearbox

Figure 5.57 shows an example of a gear shaft of a manual gearbox.

The clutch sits on its own shaft section, which introduces the engine torque via the splines at the left end. A round snap ring secures the clutch shaft axially. The seven fixed gears are arranged on the shaft. These transmit the torque via continuous splines, are held at a distance by sleeves and pressed against a shaft collar via a pressed-on ring. The teeth for the reverse gear are cut directly into the shaft (right end), as is the drive for the oil pump (last set of teeth on the right). At the right end, a hexagon is cut into the inside. The shaft of the external starting device is inserted into this hexagon via an opening in the gearbox housing. The shaft is supported at the marked points by two cylindrical roller bearings (thinly shown). It is sealed at the outer ends by shaft sealing rings.

Lubrication

Lubrication has two main tasks, namely the separation of two metallic running surfaces moving against each other (the actual lubrication) and the dissipation of frictional heat. Whereas an oil mist is normally sufficient for lubrication, a targeted supply and removal of larger quantities of oil is required for cooling, depending on the operating conditions and the required service life. Racing gearboxes generally feature dry sump lubrication. A suction pump conveys the oil returning in the housing via a cyclone as air separator and an oil cooler into a reservoir housed inside the gearbox. From there it flows by gravity to the lower pressure pump. The original external gear pumps have been superseded by internal gear pumps with circular gearing (gerotor, duocentric, etc.). These are self-priming, can pump air and are less sensitive to cavitation. A disadvantage of these pumps is their sensitivity to dirt. Unavoidable metal chips can cause the pump to jam. For this reason, the gaps of about 0.2 mm between the inner and outer rotors are larger than usual. The delivery pressures are in the range of 2 bar. The lubricating oil is pumped directly to the main bearing points. In addition to holes in the housing, the hollow shafts are also used for this purpose. A high-speed hybrid bearing has an oil requirement of 20–80 mm^3/h depending on the bearing size [23]. Bearings that are forced lubricated also need a defined oil drain. The diameter of oil drain holes should be at least 8 mm and is roughly derived from [25]:

$$d_{Bohrung} > 8 \cdot \sqrt{\dot{V}} \tag{5.17}$$

d_{Bore}	Return bore diameter, mm
$\dot{V}$	Volume flow of the lubricating oil, l/min

The tooth meshes and the contact points of the shift forks are specifically supplied with oil mist via spray nozzles and tubes with radial holes. In the case of the teeth, it is above all the cooling that increases the service life. The oil jet is directed at the emerging tooth flanks (injection lubrication). On the exit side, the cooling requirement is higher due to the previous friction work. The oil thrown off by the rotating parts lubricates and cools the remaining wheels and bearing points. *Baffle plates* are fitted at critical points to direct the spun-off oil to the intake point of the suction pump. Lower-lying wheel sets are also encased in baffle plates (which can, of course, be plastic) to shield them from the free oil sump. Bearings of auxiliary shafts can also be supplied by the oil conveyed by the gears, by a scraper rib feeding the collected oil to the rolling elements via a pocket in the bearing housing of the housing, Fig. 5.58.

In the development of gearbox lubrication, the focus is on oil distribution and cooling. Development goals are therefore the avoidance of splash losses, the reduction of oil foaming and a reduction of the oil volume. 1 l of oil increases the vehicle mass by about 0.85 kg. The lubricating oil volume can be reduced to less than 1.5 l for 7-speed Formula 1 transmissions with careful design. For qualifying, even less oil is poured into the gearbox.

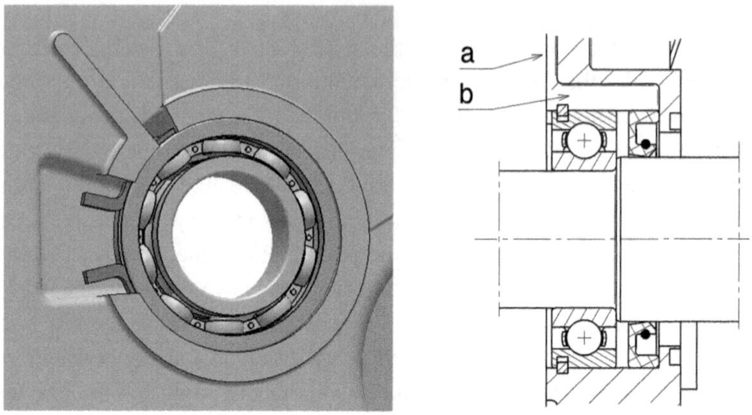

Fig. 5.58 Bearing lubrication. Left axonometric representation without shaft, right section. (**a**) rib. (**b**) pocket. The oil collected by the rib passes through the pocket to the area between the rolling bearing and the shaft seal

Venting

The shaft passages through the gearbox must be sealed so that the lubricating oil does not leak out. Such passages are the entrance of the gear shaft and the two exits of the cardan shaft heads from the differential. Generally this is done with shaft seals. These can only seal a relatively small pressure difference, and at high internal pressures the friction of the sealing lip on the shaft also increases, which increases the losses and can lead to destruction of the sealing ring due to the heat build-up. Due to the air in the gearbox housing, the pressure increases as the temperature rises during operation. To ensure that the function of the seals is not impaired, it is therefore necessary to provide ventilation of the housing. Ventilation must fulfil the following functions. When the gear unit heats up, air must be able to escape from the gear unit so that pressure equalization occurs. The escape of oil foam or mist should be avoided. During subsequent cooling of the gear unit when the vehicle is at a standstill, air from the environment must be able to flow into the gear unit so that no negative pressure can form. Moisture or contaminants should not enter the housing during this process. A vent is usually screwed into the top of the housing from the outside. On the inside there is a cast-on or riveted splash guard if the vent does not itself contain a corresponding device. By diverting the air flow and by means of throttles, the oil is separated in the vent and runs back into the housing. A filter element is interposed to protect against impurities and moisture. Figure 5.59 shows an example of a screw-in breather.

Housing

The housing accommodates all parts (Fig. 5.60) of the gearbox, ensures the position of the gears in relation to each other with the bearings and seals the system from the outside. It also provides the connection to the engine and, in many cases, the landing gear. It can also

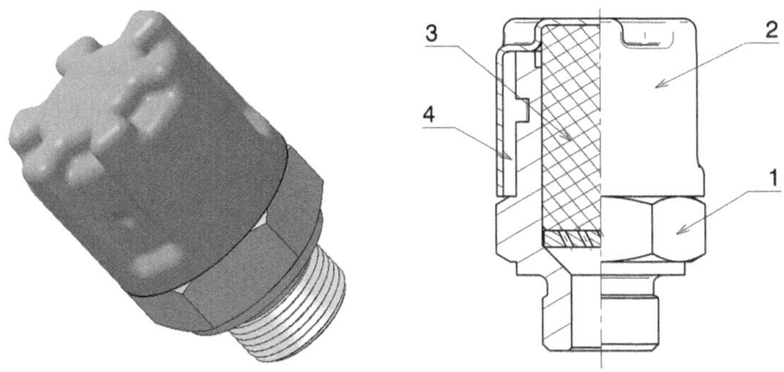

Fig. 5.59 Gearbox ventilation, according to [1]. **1** Screw-in body. **2** Splash guard. **3** Filter element. **4** Vent channel. This breather is screwed onto the top of the gearbox housing from the outside

Fig. 5.60 Parts of a manual gearbox. You can see idler and fixed gears, selector forks, idler hubs, needle bearings and the front cover that supports the input and output shafts

accommodate wing and rear impact elements. A transmission housing must therefore meet the following requirements:

- Ensuring the desired position of the shafts and gears in relation to each other
- Absorption of operating forces and torques
- Ensure heat dissipation
- allow easy assembly or disassembly and changes of gear ratios
- low weight
- Connection to the engine resp. motor or the frame.

Table 5.10 Types of gearbox housings [1]

Housing type	Advantages	Disadvantages
Trough housing 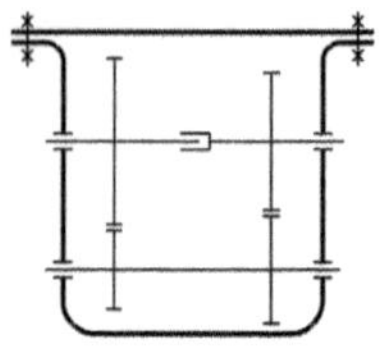	+ simple production of the bearing bores + precise manufacturing in one clamping	– unfavourable mounting conditions – no automated assembly possible
Pot housing 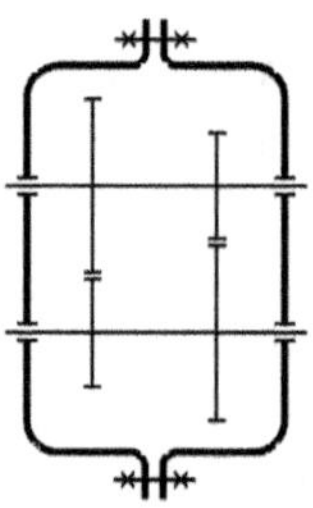	+ rigid housing + good mountable + assembly can be automated	– expensive production facility – critical holes in two setups
Box housing 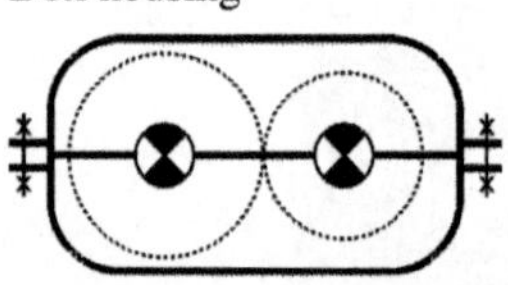	+ precise manufacturing in one clamping + very good mountable + assembly can be automated	– partial surface machining expensive – no high stiffness

In addition, the gearbox housing can accommodate the clutch housing and the lubricating oil reservoir of the engine. Due to the complexity of the design, casting and welding of individual parts are suitable for the manufacture of housings. As it must be possible to mount the shafts in any case, there are few basic types of housings, Table 5.10.

A trough housing is not split and only has a cover for closure. Split housings can be split transversely to the shaft position (transversely split) or in the direction of the shaft (longitudinally split). Transmissions of multi-track vehicles are mainly housed in transversely split trough housings. On motorcycles, longitudinally split box housings are often found, cast in one piece with the engine crankcase.

Housings are often cast because of their complex shapes. For production vehicles, light metal is die-cast, Fig. 5.61; for racing vehicles, the small quantities involved prohibit such processes, and sand casting and investment casting are used. The following recommendations for the design of cast light metal housings were determined in systematic investigations [1]:

- In principle, the wall thicknesses of castings should be as thin as possible. This avoids shrinkage cavities and the material properties are better than in thick-walled areas.

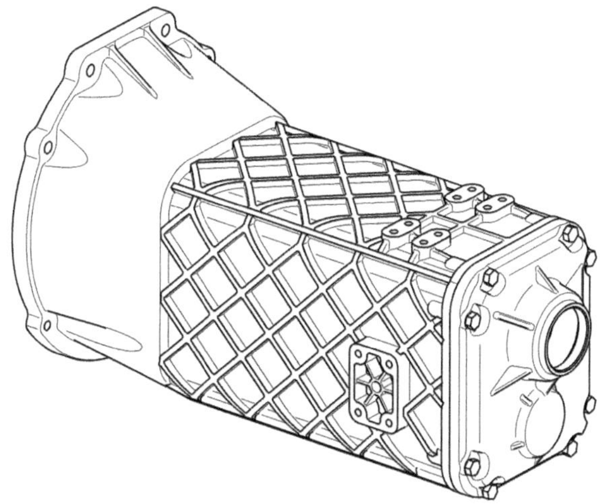

Fig. 5.61 Transmission housing of a production vehicle. The gearbox is used for a standard drive. The two-piece cup housing is die-cast and has ribs on the outside for reinforcement. The clutch bell is combined with the housing to form one part

- If reinforcements are required, ribs are provided. Housing ribs should run in the direction of the main normal stresses because cast materials are particularly sensitive to tensile stresses.
- Ribs on walls with bearings should extend in a star pattern from the bearing holes. If the wall thickness of the housing is *t*, the ribs should have the following dimensions:
  Height = 3 to 4 t
  Thickness = 1 to 2 t.
- Longitudinal walls (running parallel to the gearbox shaft) should be reinforced by wide ribs (thickness = 1 to 2 t) with a large rounding radius $R = 1.2$ t. The ribs should run at an angle of 45° to the longitudinal axis of the gearbox. The ribs should run at an angle of 45° to the longitudinal axis of the gearbox.
- A strong ribbing with rib spacing of 5 to 15 t results in a favourable acoustic transmission behaviour.

Whether the ribs are placed on the inside or outside depends on the casting process. In the case of die casting, internal ribs are costly depending on the geometry. In this case, ribs on the outside are much easier to produce. Casting processes that require a core (sand casting, investment casting, lost foam), on the other hand, allow ribs on the inside. Such a housing can be designed smooth on the outside and is then less susceptible to contamination and aerodynamically more favourable.

Materials and Manufacturing Processes

Gearbox housings are cast from aluminium, magnesium and titanium (sand casting, investment casting, rapid prototyping processes) or are welded from machined steel semi-finished products. Other (rare) designs consist of a composite of CFRP housing with cast metal and milled parts (Formula 1).

Fig. 5.62 Gearbox housing (Reynard). View from left rear. The housing is installed in a formula car and thus forms the rear end of the vehicle. You can see the position of the axle gear (far right), whose bearing is formed by the housing and a cover. This cover is dismantled at the moment. Further one recognizes the bell cranks and consoles for the wishbone connection

A disadvantage of housings made of light metals results from their different coefficients of thermal expansion: When heated, the housing expands more than the steel parts, such as the outer rings of rolling bearings. Bearings must therefore be pressed in with a particularly high overlap, which in turn increases the housing load at room temperature. In addition, the expansion of the housing also changes the centre distance, which in turn changes the tooth engagement and its efficiency. A remedy can be achieved by combining the main bearings of a bearing wall in a steel gland.

The thermal conductivity of magnesium is better than that of aluminium. The heat released by losses in the gearbox can therefore be dissipated more easily to the environment in magnesium housings, or the component temperatures are lower.

Minimum wall thicknesses (mean values): Magnesium 6 mm, aluminium 3.5 mm, steel 1.5 mm.

Gear wheels are case-hardened. Ring gears and planetary gears can also be carbonitrided.

Shafts: Case-hardened steel 16MnCr5 (DIN17210).

Quenched and tempered steel 25 CrMo4 (EN 10083-1), 34 Cr4 (EN 10083-1).

Masses: 39 kg (incl. lock, 6-speed transverse in touring cars for 450 Nm), 55 kg (Formula 1 5-speed, 1979) [26].

45 kg Formula Renault, 6-speed gearbox dry without oil.

The transmission unit is usually a supporting part of the rear of the vehicle and accommodates wishbone brackets, spring/damper strut supports and torsion stabiliser mounts, Fig. 5.62 illustrates an example.

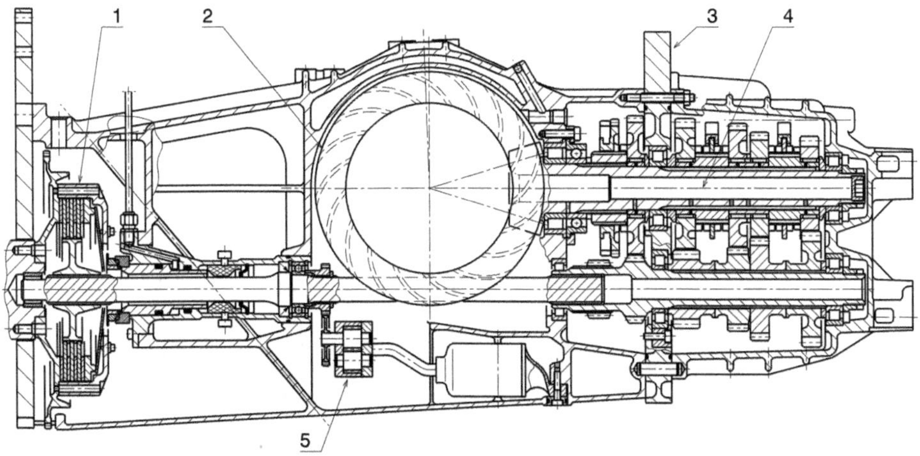

Fig. 5.63 5-speed manual gearbox of a racing car in longitudinal arrangement (Mercedes Sauber C11). **1** Clutch. **2** Crownwheel of final drive. **3** Bearing plate. **4** Output shaft. **5** Oil pump. The gearbox shafts are additionally supported between first and second gear. A bearing plate (3) is arranged between the housing halves for this purpose. The housing is a trough housing made of magnesium. A separate oil pump (5) is located under the angular gear

The structure of a racing gearbox shows the typical arrangement of a mid-engine concept (Fig. 5.63). The gearbox has been designed for a specific endurance vehicle. The engine is located in front of the gearbox. The input shaft carries the clutch (1) and passes under the final drive (2). The actual transmission gears sit behind the final drive, allowing easy access inside the vehicle, allowing for example gear ratios to be changed. Changing the transmission gears can be done in 20 minutes if access is good [26].

The next example also shows the typical arrangement of a mid-engine concept when the transmission is also used in a formula car (Fig. 5.64).

An example of a highly complex gearbox housing welded from several individual titanium parts is shown in Fig. 5.65. In addition, the gearbox is constructed from two large modules, namely the actual metal gearbox housing and a CFRP attachment which accommodates a large part of the chassis and which is bolted to the engine.

5.3.2 *Continuously* Variable Transmission (CVT)

The continuously variable transmissions in use today are almost without exception wrap-around transmissions. The transmission link is a chain which runs over tapered pulleys and transmits the power exclusively via friction. Figure 5.66 shows the central components of a wrap-around gearbox. The effective diameter of the chain can be varied by changing the distance between the drive cone pulleys. Because the chain has a fixed length, the spacing of the driven sheaves must be changed accordingly. Decisive for a useful function is the

Fig. 5.64 Gearbox housing (Hewland to Formula König). View on right side of vehicle. The housing is screwed to the cast clutch housing. The cardan shaft is dismantled, therefore the view is free on the flange of the axle gear. In the picture the actuating linkage, which is led to the rear end of the gearbox, can be seen well. If the rear cover is removed, all gear pairs can be exchanged. At the end of the transmission the rear wing is taken up over two aluminum plates, which provide also the admission for the shunting jack

control of the contact pressure of the conical pulleys. If the force is too high, the efficiency of the chain will be poor and the losses due to the contact pressure pump will increase. Conversely, too little contact pressure between the conical pulleys and the chain must be avoided at all costs, because slipping of the chain will destroy the transmission. The contact pressure must therefore be controlled depending on the performance.

The chain can be either a tension or a push link chain. The tension link chain is more efficient, while the push link chain runs more quietly and is therefore preferred for passenger cars.

The spreads of such CVT transmissions are around 5.3–6.0. If a larger spread is needed, mechanical manual transmissions are used upstream or downstream of the CVT.

The suitability of such CVT transmissions for racing is controversial. When easing off the throttle on entering a corner, the CVT briefly upshifts, giving the driver the impression that the throttle is stuck, which takes some getting used to. When accelerating out of the corner, the transmission has to "downshift" again because it has compensated for the falling engine speed in the previous shifting mode [27]. To avoid this, another transmission control must be provided for the shifting mode.

The loss of efficiency is estimated to be around 15–20%.

In the 1960s, a Variomatic was used in the F3 Tecnos. The car was successful on tight and wet courses, but showed a lack of power on the straights [27].

In [28], an improvement in acceleration compared to a 4-speed automatic is reported.

[29] calculates approximately the same acceleration from a high-revving engine with a CVT as from a standard engine with a stepped transmission.

Fig. 5.65 Transmission of a Formula 1 car (Ferrari F1-2000). You can see the left side of the gearbox. So the attachment to the engine is done with the CFRP structure on the left side of the picture, which also carries brackets for the landing gear and holds the bell cranks including torsion springs. At the rear end you can see a tripod mount which is fully supported in the gearbox housing. The cylindrical bulge that the housing has at the bottom makes room for the shift drum. At the rear end of the housing is the mount for the rear impact element including rear wing

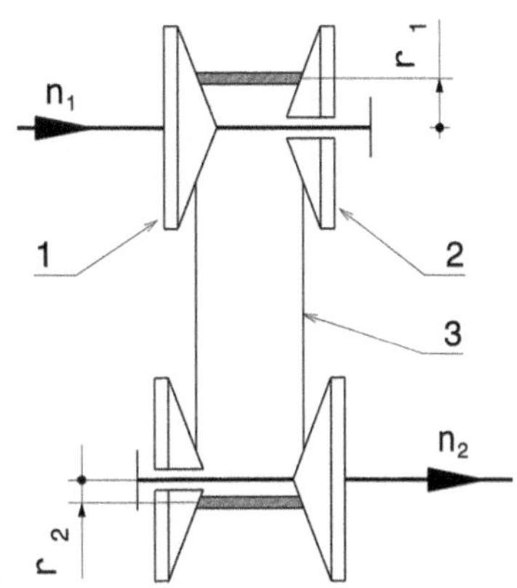

Fig. 5.66 Principle of a continuously variable wrap-around gearbox. **1** fixed conical disc. **2** movable conical pulley. **3** Sliding link chain. n_1 Input speed, n_2 Output speed, r_1 Friction radius of drive pulleys, r_2 Friction radius output pulleys. The transmission ratio i follows from the effective friction radii of the chain: $i = n_1/n_2 = r_2/r_1$

5.3.3 *Final Drive* (Axle Drive)

The axle drive has the task of diverting the torque coming from the engine, converted via the gearbox, to the drive axle. For standard drives and mid-engine arrangements, a 90° deflection is required. In the case of transverse engines, only a centre distance between the gearbox output and the drive axle must be bridged. Axle gears are usually integrated in the gearbox housing for reasons of space and strength in the case of mid-engine concepts and rear-wheel drive as well as front-wheel drive and transverse engine. Also, a common lubrication system is advantageously used. Usually, the axle drive does not drive the wheels directly, but the differential gear distributes the torque to the drive wheels. A

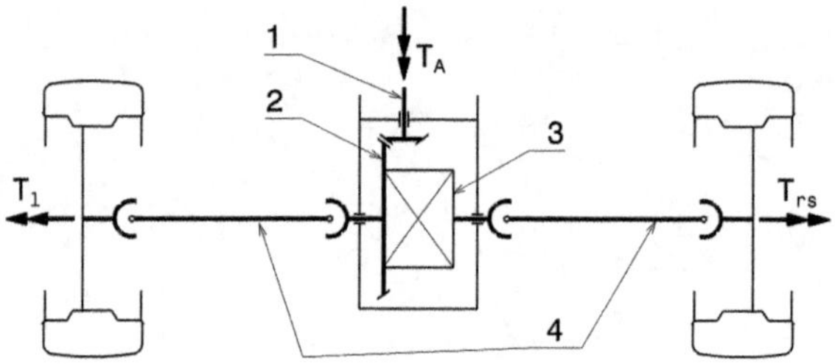

Fig. 5.67 Principle of an axle drive. **1** Drive bevel gear (*pinion*). **2** Crown wheel AE: ring. **3** Differential housing (*differential cage*). **4** Side shafts (*half shaft AE: axle shafts*). T_A Drive torque, T_l Output torque left wheel, T_{rs} Output torque right wheel

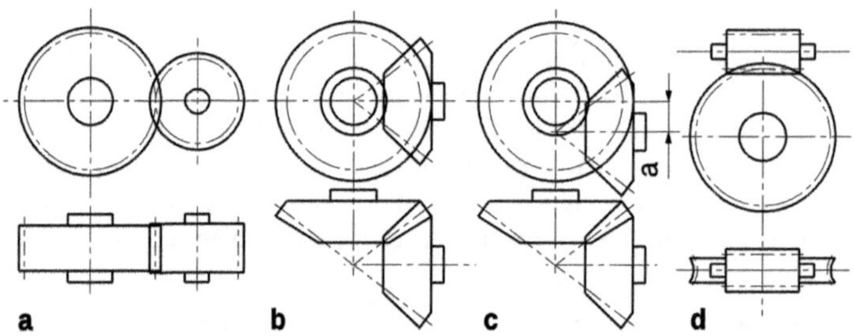

Fig. 5.68 Types of axle drives. (**a**) Spur gear axle drive. (**b**) Bevel gear axle drive with spiral bevel gears. (**c**) Bevel gear axle drive with hypoid bevel gears. a = Axle offset. (**d**) Worm gear axle drive

basic arrangement of an axle drive is shown in Fig. 5.67. The actual axle drive consists of a pinion (1) and ring gear (2), which is screwed directly onto the differential cage.

The drive torque T_A is divided between the wheels, where it is effective as drive torque on the left T_l and on the right T_{rs}. The following therefore applies: $T_A = T_l + T_{rs}$.

Figure 5.68 lists the different designs of final drives that are available depending on the engine arrangement and drive type.

A spur gear axle drive is suitable for a transverse engine and front-wheel drive. Accordingly, it is also the most common variant in passenger cars. The spur gear drive, Fig. 5.69, is characterised by a high degree of efficiency. A disadvantage due to the usual arrangement of engine and gearbox side by side transversely to the direction of travel results from the side shafts being of unequal length. The axle drive can be integrated into the gearbox housing in the case of the transverse engine.

If a deflection of the power flow by 90° is required, bevel gearboxes are used. This is the case with all drives in which the engine is installed in the longitudinal direction. The shaft drive can be integrated into the gearbox housing or designed as an independent housing. In the case of bevel gear drives, a distinction is made between those with intersecting axes (spiral bevel gears, Fig. 5.68b) and those with offset axes (hypoid bevel gears, Fig. 5.68c). Hypoid gears are often used for passenger cars. The bevel gear engages under the centre of

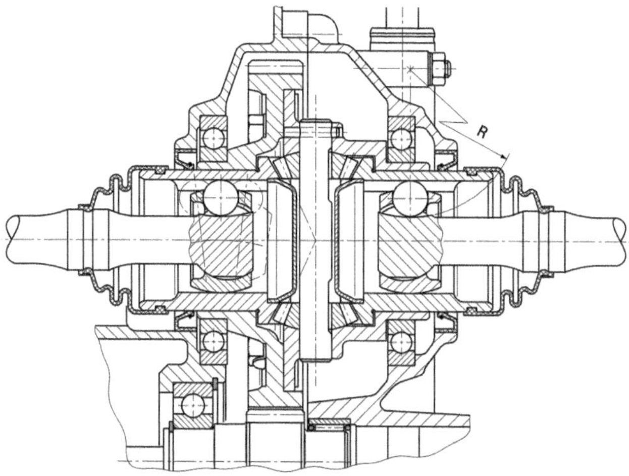

Fig. 5.69 Axle gearbox of a racing car (NSU 1100 TTS, 1967–71) [30]. The output shaft of the gearbox and the side shafts are parallel. The basket of the bevel gear differential can thus be driven via a spur gear, which offers the greatest efficiency in power transmission. The five-ball constant velocity joints (Löbro, GKN Automotive) are integrated into the axle bevel gears of the differential, which allows the sideshafts to achieve almost the greatest possible length for a given track width

the ring gear. This axis offset is of the order of 0.2× ring gear diameter. Due to the axis offset, the diameter of the drive bevel gear is larger and the ring gear can be made smaller with the same load than with the variant with cutting axes. The drive shaft is also lower and the center tunnel in the passenger compartment can be designed lower. The offset of the axes results in a sliding movement along the tooth flanks when the teeth roll. This has a noise-reducing effect, but requires the use of a special gear oil that can withstand these high sliding pressures. The large amount of friction worsens the efficiency of the power transmission.

Worm gear units allow large gear ratios in a small space. In terms of smooth running, the worm drive is superior to all other drives. Similar to the hypoid gearing, the worm also always has a sliding component in the tooth mesh, which forms an oil film between the load-bearing tooth flanks. Large axial forces occur on the worm during operation and it must therefore be supported accordingly. The worm can be arranged above or below the worm wheel and thus the drive shaft can be placed higher or lower than the wheel axis. However, the manufacture of the worm and the worm wheel is complex and expensive. There are no longer any representatives of this type of axle drive in current vehicles. In Formula 1, curved-toothed (=spiral-toothed) bevel gears are used.

Gear Ratios The desired maximum speed of the vehicle is determined by the transmission ratio of the final drive. Depending on the engine power and the design of the final drive, the ratios lie in the following ranges:

Table 5.11 Evaluation of the types of construction of final drives (axle drives) [1]

	Design			
		Bevel gear		
Criterion	Spur gear	Spiral bevel drive	Hypoid bevel gear	Worm wheel
Efficiency	++	++	++	+
Load capacity	0	+	+	++
Space requirement	+	0	+	+
Bearings	++	0	0	0
Lubrication	++	++	0	0
Lifetime	++	+	+	++
Smooth running	0	0	+	++
Production costs	++	+	0	−−

Legend: ++ very good, + good, 0 satisfactory, − poor, −− very poor

Bevel gear drives	Spur gear axle drives	Worm wheel drives
$i_D = 2.5{:}1$ to $3.5{:}1$	$i_D = 3{:}1$ to $4{:}1$	$i_D \geq 5.0$

The small ratios are used for high-performance vehicles, the larger ratios for passenger cars and all-wheel drive vehicles.

The dimensioning of the axle drives is based on the largest input torque, which is the largest output torque of the gearbox.

Finally, Table 5.11 compares some properties of individual designs.

It can be seen that with unweighted criteria, the spur gear drive performs best as an axle drive.

The service life is about 1300 km in Formula 1 [26].

Further examples of final drives are shown in the diagrams in Sect. 5.5 *Differential.*

5.4 Differential

5.4.1 Introduction

The single-axle drive is the minimum for passenger cars and commercial vehicles for reasons of driving stability and traction. For this purpose, the engine power must be distributed to a left and right driving wheel, in the simplest case by means of an undivided wheel drive shaft. When driving through a corner, however, the outer wheel travels a greater distance than the inner wheel, which in the case of a rigid drive results in tire rubbing, high wear and stress on the drive train due to tension. A gearbox is therefore required which, in contrast to a rigid through drive without a split output shaft, allows an unconstrained speed and force balance, Fig. 5.70. This gearbox must distribute the torque in the ratio 50:50% to the left and right driving wheel when driving straight ahead [1].

In principle, spur gears or bevel gears can be used for a differential, Fig. 5.71. Occasionally, worm gears are also used. However, these have a principle-based locking effect in every operating state due to the self-locking of a worm and are therefore used as self-locking differentials (e.g. Torsen differential).

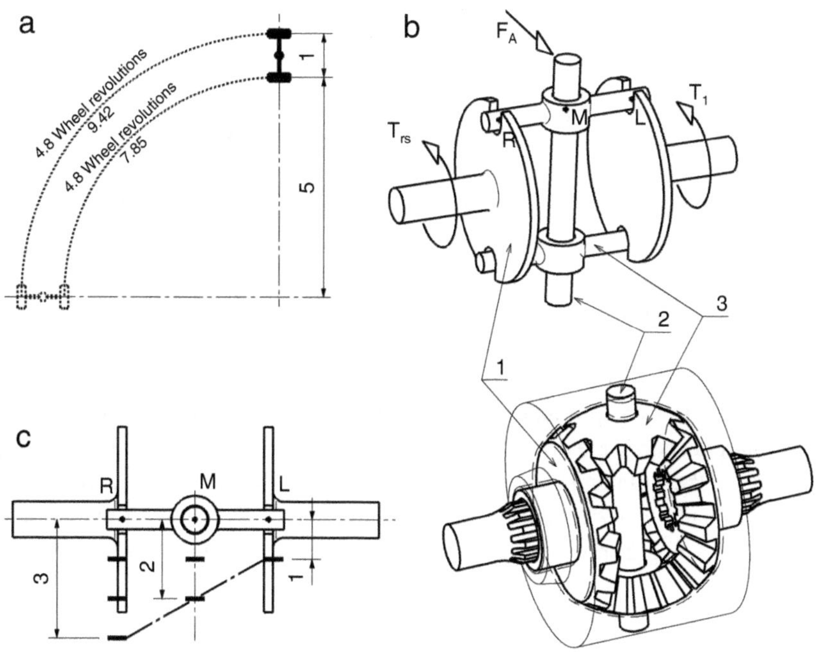

Fig. 5.70 Principle of a differential. (**a**) Different distances of two wheels during slip-free rolling through a circular arc. (**b**) Principle diagram of force distribution: The driving force F_A is distributed to the two discs via the balance beam mounted in M. The force is distributed between the two discs. If the resistance T_l of a disc is greater than T_{rs}, the beam twists and rotates the point R of the disc correspondingly further. Component equivalents for a bevel gear differential: 1 Axle bevel gear. 2 Differential pin. 3 Differential bevel gear. (**c**) View from above for the case described in (**b**). Paths and velocities of the three points always behave: $R + L = 2$ M

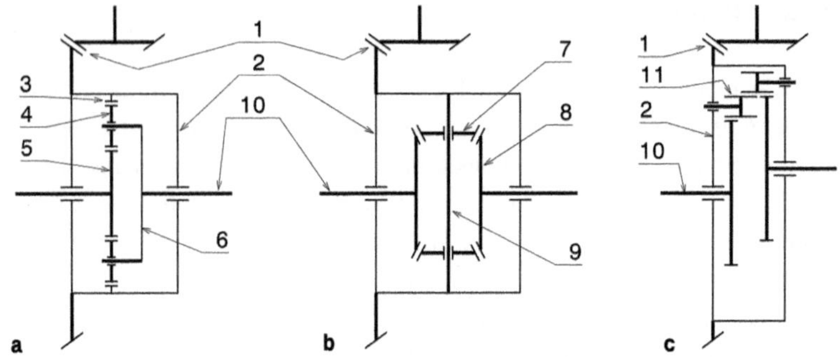

Fig. 5.71 Principle types of differentials. (**a**) Planetary-gear differential. (**b**) Bevel gear differential. (**c**) Spur-gear differential. The two sideshafts are actually coaxial and are only shown offset to illustrate the principle. 1 *Final drive* axle. 2 Differential housing (*differential cage*). 3 Ring gear (*annulus*). 4 Planet wheel. 5 Sun wheel. 6 Planet carrier. 7 planetary gear. 8 *crypto gear* (*sun wheel*). 9 Differential pin (*cross pin*). 10 Side shafts (*axle shafts*). 11 Equalising spur gears

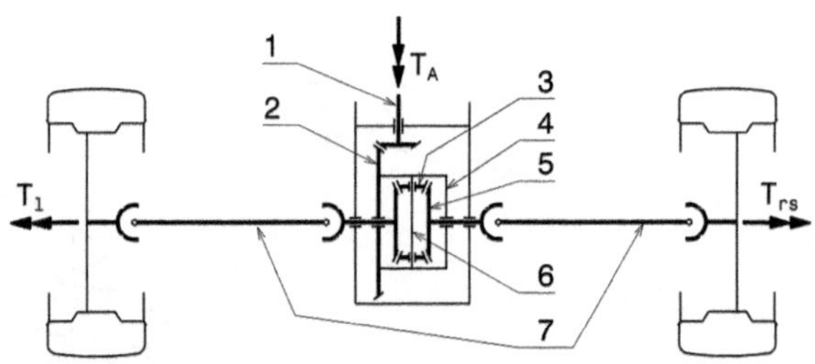

Fig. 5.72 Differential arrangement (schematic). **1** Drive shaft with pinion. **2** Ring gear. 3 Differential bevel gears. **4** Differential cage. **5** Axle bevel gears. **6** Differential pin. **7** Side shafts

The basic construction is the same for all designs. The axle drive directly drives the differential cage. This transmits the torque to an intermediate element that splits the power between the two sideshafts.

Figure 5.72 shows a split axle shaft with an intermediate bevel gear differential. The torque T_A introduced via the drive (1), for example a spiral or hypoid toothed bevel gear, is transmitted via the differential cage (4) to the balancing bevel gears (3), which act like a balance beam and always establish a torque balance $T_l = T_{rs}$ between the left and right output sides. As long as no slip occurs at the driving wheels, the following applies to the speeds:

Speed of the outer or less sticking wheel: $n_o = n + \Delta n$
Speed of the inside or more sticking wheel: $n_i = n - \Delta n$
Where n is the input speed of the ring gear and Δn is the differential speed between the output speed of the outer wheel and the input speed of the differential. When driving straight, the differential cage (4), the axle bevel gears (5), the axle shafts (7) connected to the axle bevel gears in a torsionally rigid manner, and the differential bevel gears (3) rotate inside the cage as a block. There is no relative movement between the differential pin

(6) and the differential bevel gears mounted on it. When cornering, one axle shaft must rotate faster than the opposite one; axle bevel gears and differential bevel gears roll off each other. Speed compensation between the wheels can take place [1].

Influence on the Driving Behaviour

When accelerating on a straight line, the differential has hardly any influence on the driving performance. Problems only arise when one wheel has a lower coefficient of friction than the opposite wheel. In that case, it is desirable to be able to cancel the differential function. Another problem can arise due to different tyre diameters. In this situation, a locked differential will cause the vehicle to pull to one side. When accelerating at the exit of a corner, the effect of the differential is in any case influential on the driving performance, because a wheel load shift occurs and the wheels build up different circumferential forces inside and outside the turn. The ideal differential distributes the engine's drive torque according to the wheel loads and thus supports the maximum possible acceleration. When the load changes in the turn, i.e. during the transition to overrun mode, the differential also has an influence on the behavior of the vehicle. Depending on how strongly the two halves of the axle are coupled together, it gains or loses stability. Transition behavior with variable differentials also has an influence. Abruptly applied locks, for example, generate a shock load in the drivetrain, the effect of which on vehicle behavior is difficult to assess.

A closer look at what happens in a differential is helpful in understanding vehicle dynamics. The following three equations, which result from equilibrium considerations, apply to every differential – regardless of the design [31]:

$$\frac{\omega_l - \omega_D}{\omega_{rs} - \omega_D} = \frac{\Delta\omega_l}{\Delta\omega_{rs}} = -1 \quad (5.18a)$$

$$T_D + T_l + T_{rs} = 0 \quad (5.18b)$$

$$T_D\omega_D + T_l\omega_l + T_{rs}\omega_{rs} = P_{ls} \quad (5.18c)$$

ω_l, ω_{rs}	Angular velocity of the left or right side shaft, s^{-1}
ω_D	Angular velocity of the differential cage, s^{-1}
$\Delta\omega_l$, $\Delta\omega_{rs}$	Relative angular velocity of the left or right side shaft to the cage, s^{-1}
T_l, T_{rs}	Torque of the sideshafts, N m
T_D	Torque of the differential cage (torque coming from the engine), N m
P_{ls}	Power loss in differential, W

All absolute angular velocities are always positive during forward motion. It thus follows from (5.18a) that when the sideshafts rotate relative to the cage, one rotates faster than the cage and the other slower by the same amount. If one considers for this case

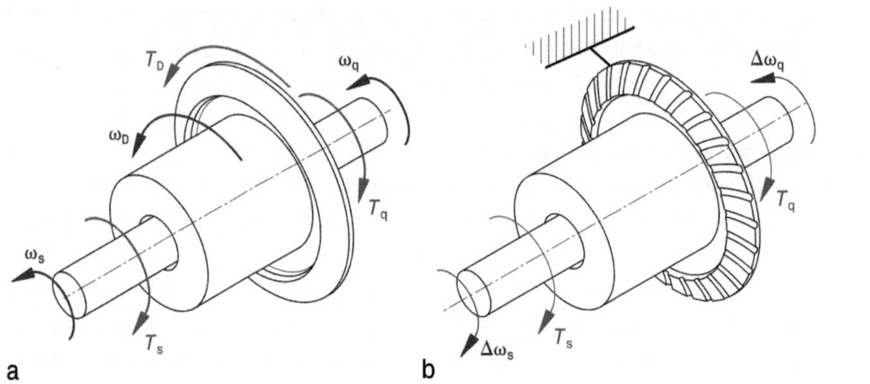

Fig. 5.73 Angular velocities and moments for a differential. Designations see text. (**a**) Absolute velocities, (**b**) Relative velocities. The relative velocities become visible by thinking of the differential cage held fixed ($\omega_D = 0$)

(cf. Fig. 5.73) the internal power balance of the differential by combining (5.18b) and (5.18c)

$$T_l(\omega_l - \omega_D) + T_{rs}(\omega_{rs} - \omega_D) = T_q(\omega_q - \omega_D) + T_s(\omega_s - \omega_D) = T_q\Delta\omega_q + T_s\Delta\omega_s = P_{ls} = P_1 - P_2 \tag{5.19}$$

Index q (*quick*)	Side shaft rotates faster than the cage
Index s (*slow*)	Side shaft rotates slower than the cage
P_1	Input power, W
P_2	Output power, W

a useful relation for the efficiency of the differential can be written down

$$\eta_D = \frac{P_2}{P_1} = \frac{P_1 - P_{ls}}{P_1} \leq 1 \tag{5.20}$$

η_D	Efficiency of the differential, –

This efficiency is the reciprocal of the TBR value (see below).

$$TBR = \frac{1}{\eta_D} \tag{5.21}$$

Depending on the effect of the torques T_s and T_q of the side shafts, this results in two different operating states. During driving, both torques are negative (the driving engine

Table 5.12 Sign depending on the operating state, see also Fig. 5.73

Operating status	T_D	$\Delta\omega_s$	$\Delta\omega_q$	T_s	T_q	P_2
Drive	+	–	+	–	–	$-T_q\ \Delta\omega_q$
Overrun	–	–	+	+	+	$-T_s\ \Delta\omega_s$

torque is positive) and in pushing operation they are positive. According to the definition of relative angular velocities (E.18a), $\Delta\omega_q$ is positive and $\Delta\omega_s$ is negative. Here, the product $T\omega$ with a negative value is the output power P_2, cf. (E.19). Table 5.12 summarizes the decisive signs and products.

The efficiencies follow directly from this to

$$\text{Drive}: \eta_D = \frac{-T_q \Delta\omega_q}{T_s \Delta\omega_s} = \frac{T_q}{T_s} \leq 1 \tag{5.22}$$

$$\text{Overrun operation}: \eta_D = \frac{-T_s \Delta\omega_s}{T_q \Delta\omega_q} = \frac{T_s}{T_q} \leq 1 \tag{5.23}$$

To understand how this affects the driving dynamics, let us consider a drive axle when cornering, Fig. 5.74. The transmissible circumferential forces $F_{W,X}$ depend on the wheel load. At low lateral acceleration, the lateral wheel load transfer is small and the inner and outer wheels are loaded vertically about the same. The outer wheel on the turn must therefore rotate faster than the inner wheel in the driving condition (Fig. 5.74a). According to (5.22), the smaller moment is applied to the faster rotating wheel. A self-locking differential counteracts the lateral force of the front wheels in this situation and thus supports the turning back of the car after the turn during acceleration. The situation is completely different when cornering quickly. A large lateral acceleration causes a strong wheel load shift (which can go so far that the wheel on the inside of the corner lifts off) and the transmittable moment of the inside wheel decreases just as strongly (Fig. 5.74b). It slips more than the outside wheel – thus rotating faster than its opposite counterpart. The outer wheel receives the greater momentum and provides corresponding propulsion. One of the main reasons why racing cars use differentials like this. A self-locking differential assists the yawing motion of the vehicle in the corner when driving. When exiting a corner, it interferes with the desired back-turning motion into the following straight.

In overrun mode the conditions are clear (Fig. 5.74c). The engine torque becomes negative and brakes the differential cage. It is true (5.23): The faster turning wheel can absorb the larger moment. This is in any case the one on the outside of the corner. Finally an extreme case, namely a completely locked differential (spool = sleeve coupling). Both wheels must necessarily have the same speed as the cage, $\omega_l = \omega_{rs} = \omega_D$. From (5.18b) and (5.18c) it follows directly $P_{ls} = 0$. Here, satisfying the moment equilibrium, (5.18b), the case can occur where the circumferential forces point in opposite directions (Fig. 5.74d). A

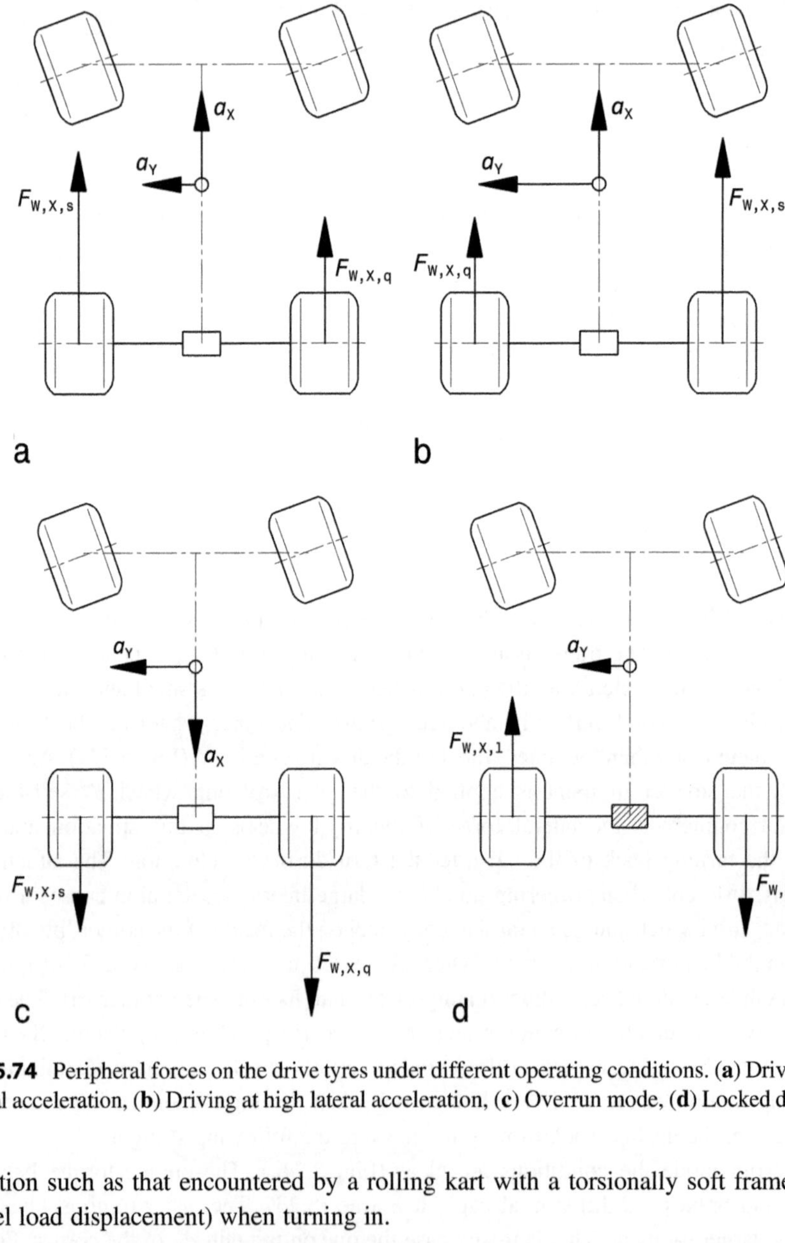

Fig. 5.74 Peripheral forces on the drive tyres under different operating conditions. (**a**) Driving at low lateral acceleration, (**b**) Driving at high lateral acceleration, (**c**) Overrun mode, (**d**) Locked differential

situation such as that encountered by a rolling kart with a torsionally soft frame (= low wheel load displacement) when turning in.

5.4.2 Types of Construction

The following differential types are found in racing vehicles:

- Open differential
- Sleeve coupling (*spool*)
- Self-locking differential with multi-plate (*clutch locker, Salisbury type*)
- Corner self-locking differential (*cam and pawl*)
- Locking differential with worm wheels (*torsen differential*)
- Torque distributing differential (*torque vectoring differential*)

Open Differential
A so-called open differential is a bevel or planetary gear differential without any locks. Accordingly, open differentials are generally used in passenger cars. The disadvantage of this design results from the characteristic that the wheel with the lower torque determines the possible total torque of the drive axle. This leads to loose spinning of the lifted inner wheel in tight, fast turns. The outer wheel then no longer contributes to propulsion. Of course, this disadvantage will not always appear in this blatant form. It depends primarily on the wheel load transfer and the power-to-weight ratio. Vehicles that increase the wheel load through aerodynamic downforce naturally have fewer problems with the described behaviour of open differentials than those without downforce aids. Strictly speaking, a distinction must also be made between corner entry and exit. When braking and turning in, a differential should be open so that the engine braking torque does not interfere with the yawing motion of the vehicle in push mode. Conversely, when accelerating out of a corner, the vehicle should be driven especially at the more heavily loaded outer wheel and remain track-stable, while the inner wheel also provides propulsion. A locked differential thus causes slight (stable) understeer at the exit of the corner.

Figure 5.75 shows a planetary differential without locks. The axle is driven by the big wheel (1) which is bolted to the ring gear (2). The torque flow is transmitted from the ring gear via three planetary gear pairs (5) to the web (3) and the sun gear (4). The number of teeth is selected so that the moment split left to right is 50–50%. At the same time, the tripod receptacle for the left-hand side of the axle is accommodated in the web. The right tripod is driven via a receptacle integrated in the sun gear. Cylinder and needle bearings are located between the individual components that perform relative movements to each other. This design allows for an extremely slim axle drive, the two tripod receptacles of the side shafts have the smallest possible distance from each other and thus from the center of the vehicle. A slim axle drive leaves room for aerodynamic underbody design and for a diffuser.

A planetary gear compensating transmission can also be locked. For this purpose, a multi-disc clutch is connected between the sun gear and the arm, which transmits a locking torque between the two components. The Formula 1 World Championship winning Ferrari F1 2000 had such a planetary gear differential with five planetary pairs with hydraulic multi-plate clutch in use [14].

Another differential design without bevel gears is shown in Fig. 5.76, a spur gear differential. The driving torque coming from the main gear 1 is transmitted via the pinions

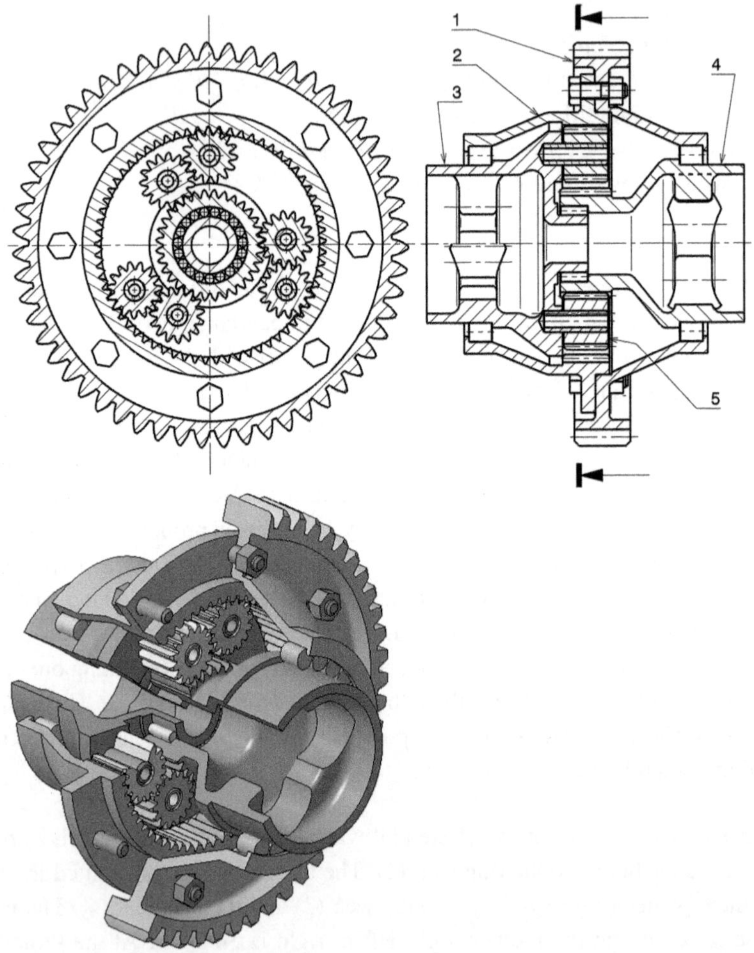

Fig. 5.75 Planetary gear differential as open differential (*epicyclic differential*). **1** Drive *gear* as spur gear. 2 Ring gear (*annulus*). **3** Arm with tripod receptacle left (*planet carrier*). 4 Sun wheel with tripod mount right. **5** Planet wheel

2 and 6 to the two receptacles (4, 5) for the side shafts. The pinions are entrained by their bearing journals (3, 7), which are located in the differential housing, and each mesh with the teeth of a side shaft receptacle and simultaneously with each other in a central overlap area. The torque is distributed equally to the left and right.

Sleeve Coupling (*Spool*)

Certain courses allow driving without a differential. The two axle shafts are then connected to each other in a torque-resistant manner using a sleeve. This is how the Mercedes sports cars drove in the 1989 24 Hours of Le Mans without a differential with rigid through-drive

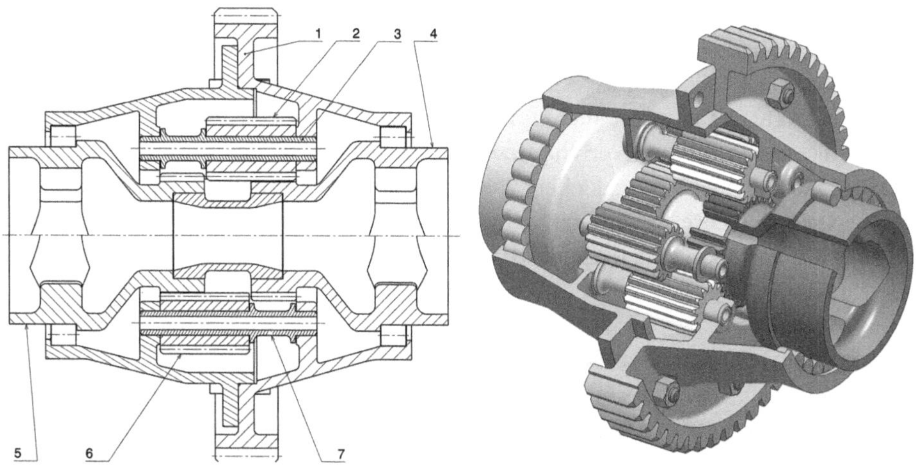

Fig. 5.76 Spur gear differential. 1 Drive gear, 2 Pinion gear for right side, 3 Bearing pin, 4 Output torque right, 5 Output torque left, 6 Pinion gear for left side, 7 Bearing pin

on both sideshafts. Thus, damage to one sideshaft or flex joint does not lead to a total failure of the system [32]. However, additional chicanes in the track layout and understeering driving behaviour led to the use of a limited slip differential in 1991 [32].

You can also drive without a differential on the oval tracks in North America. The turns have large radii and are only driven through in one sense. The vehicles therefore have different tyre diameters inside and outside (tyre stagger) and thus have a built-in taper roller effect.

Karts do not have a differential due to regulations. The drive is on a rigid axle. The necessary balance of the wheel speeds when cornering is achieved by a relatively high overall centre of gravity in relation to the track width, i.e. the driving technique aims to lift the inner wheel of the drive axle by shifting weight outwards. The front axle geometry is also designed for this circumstance. The kingpin inclination angle is up to 20° and, above all, a large caster angle provides relief for the inside cornering wheel when turning in.

The main disadvantage of a rigid drive-through becomes noticeable in corners with low lateral acceleration or on vehicles with low lateral wheel load transfer. The lateral forces of the front wheels must pivot the drive axle when turning in and out. This generates different circumferential forces on the drive wheels inside and outside the corner (cf. Figure 5.73d), which leads to an increase in rolling resistance. These additional losses increase tire wear and the corresponding power is lost for propulsion.

The other types of differential are characterised by the possibility of bridging open differentials. The most diverse measures of drive torque distribution represent the conceivable possibilities between the two extremes of open differential and sleeve coupling.

Controllable Differentials

Although the simple differential gears are necessary and sufficient for normal driving, they have serious disadvantages in special situations. If different frictional potentials occur on the left and right due to wheel load shifting or different surfaces, the wheel with the lower friction dictates the possible driving force of the vehicle. In this situation, the idea of bridging the differential, i.e. locking the compensation function, is obvious. Another consideration is to use the differential to increase the vehicle's stability by selectively directing different torques to the left and right drive wheels, thus creating a yaw moment on the vehicle.

Locking Value (Torque Bias Ratio)

Without a locking effect, an unevenly transmittable torque of the drive wheels leads to the wheel with the lower torque spinning. The vehicle can hardly accelerate in this phase. If the slipping wheel grips jerkily again, the drive train is subjected to shock loads and the vehicle can easily become unstable. Locking differentials are therefore widely used on racing cars with high engine power.

The locking value S as a design characteristic variable represents a measure for the obstruction of the balancing movement. It is defined as follows:

$$S = \frac{T_\mathrm{B}}{T_\mathrm{A}} = \frac{T_{rs} - T_l}{T_{rs} + T_l} \tag{5.24}$$

S	locking value, – or %
T_{rs}	Output torque right side shaft, N m
T_l	Output torque left side shaft, N m
T_B	Locking torque, N m
T_A	Drive torque, N m

By definition, the locking value S is between 0 and 1 (or 0 to 100%). A locking value of 0% describes a loss-free, non-locking differential gear, a value of 100% a rigid through-drive.

In front-wheel drive passenger cars, the locking values must be kept low (maximum 17%) due to undesirable feedback effects in the steering. The locking values of locking differentials are between 25 and 50% for passenger cars with rear-wheel drive and up to 75% for commercial vehicles [1]. For racing vehicles with rear-wheel drive from 45% to approx. 70% [22].

In the Anglican world, a TBR value (*torque bias ratio*) is often used to characterize the differential. This refers to the ratio $T_{rs,\ \mathrm{max}}$:$T_{\mathrm{l},\ \mathrm{min}}$. A TBR of 3:1 therefore corresponds to a value S of 0.5.

An active limited-slip differential would be ideal: when accelerating in a corner, the drive torque is shifted to the outside wheel, and when the throttle is released, the locking effect is cancelled [27].

To illustrate the locking value: With a locking differential with $S = 50\%$, a maximum of 75% of the drive torque can be transmitted to the wheel with the higher adhesion potential, while at least 25% goes to the wheel with a greater tendency to spin. The difference between these two values is $S = 50\%$, the locking value is, so to speak, the "redistribution amount" in relation to the total transmitted drive torque, i.e. $T_l + T_{rs}$. In other words: The higher the locking value, the more torque is not distributed by the differential gear, but is transmitted via the differential brake as braking or locking torque T_B. The locking value is thus also a measure of the power split that takes place between the differential and the differential brake.

The limited use of the limited slip differential as a traction aid is illustrated below using an example. A vehicle on a one-sided slippery road with $\mu_l < \mu_{rs}$ and a limited slip differential with the locking value $S = 0.3$ is considered. A maximum of $T_l = 25$ Nm can be transferred to the road with the left wheel.

From the defining Eq. (5.24) of S follows: $T_{rs} = T_l \frac{1+S}{1-S}$.

Consequently, regardless of the engine torque offered, a torque of $T_{rs} \approx 46.4$ N m can be transmitted to the right wheel. The total transmittable torque is only $T \approx 71.4$ N m. This numerical example shows the limited possibilities of the limited slip differential, because depending on the driving resistance (gradient, etc.) this may be too little torque for locomotion.

However, a differentiated view must be taken of the locking values of slip-dependent and load-dependent locking differentials: A purely load-dependent self-locking differential has a fixed, unchanging locking value. This means that regardless of the amount of drive torque, the percentage of the respective drive torque determined by the nominal locking value is always "diverted".

A purely slip-dependent limited slip differential generates a braking torque that is independent of the drive torque, depending on the speed difference that occurs. This means that higher instantaneous locking values occur with low drive torques and lower values with high drive torques. The influences of such locking differentials on the driving and traction behavior of a vehicle can therefore only be controlled by the course of the braking torque, which is exclusively dependent on the occurring differential speeds [1].

In summary, the following principles of locking can be stated:

(a) preset locking torque, e.g. claw clutch, friction disk pack with spring preload
(b) Speed-dependent locking torque, e.g. Visco clutch, friction plate pack actuated by speed difference
(c) Torque-dependent locking torque, e.g. Torsen differential
(d) Adjustable locking torque, e.g. electronically controlled limited slip differential

Limited-slip differentials show some advantages in racing cars. Accelerating out of a corner is usually faster with a limited slip differential than without. This is also seen in vehicles that do not spin the inside cornering wheel at all. Driving performance is also improved on standing starts. However, the driving behaviour can be so influenced by some designs of limited slip differential that the driver has to adapt his driving style to it [33].

Self-Locking Differential with Multi-Plate Clutch (*Limited Slip Differential*)
This differential is widely used in racing vehicles and in sporty passenger cars. The locking effect of a self-locking differential with multi-plate clutch is based on the torque-dependent internal friction generated in two multi-plate clutches symmetrically arranged in the differential cage. The self-locking effect results from a combination of load-dependence and spring loading of the multi-plate clutches. The load-dependent locking effect, Fig. 5.77, is based on the fact that the drive torque introduced into the differential cage (1) is transmitted via the differential pin (2) to two thrust rings (3) which are arranged in the differential cage (1) so as to be non-rotatable but axially displaceable. Under load, locking forces are automatically generated on the surfaces of the prism-shaped recesses (8) in the pressure rings, see detail and Fig. 5.78, which press the clutch plates together. The outer plates (5) are non-rotatably connected to the differential cage (1), and the inner plates (4) are non-rotatably connected to the axle bevel gears (6). As a result, the frictional connection between the plates provides precisely defined resistance to different rotational speeds of the axle shafts, for example when a wheel is spinning. This effect increases with increasing drive torque. Since the locking forces are proportional to the transmittable torque, the locking effect, but not the locking value, adapts to the variable engine torque and also the torque increase in the various gear steps.

The increase of the locking torque above the driving torque influences the design of the recesses in the thrust rings, Fig. 5.78. The side surfaces of the recesses work as a ramp and the differential pin as a spreading wedge. If the ramp is flat (Fig. 5.78b), a lower drive torque is sufficient to produce the same locking torque through the plates that a steeper ramp (Fig. 5.78a) achieves only at higher drive torque. In addition, with this type of locking differential, it is possible to design the ramps differently in the drive and drag torque directions. The differential then has different characteristics during acceleration and braking.

The cup springs 7 (shown in the lower half of Fig. 5.77), which can be installed to preload the multi-plate clutch, produce a constant initial locking effect independent of the transmittable torque, but which occasionally draws attention to itself by creaking noises. In a Formula Renault car, the basic locking torque of a new differential is 78 N m (with a tolerance of −14.5 to +10 N m). In use, this value drops by approx. 30%. This means that the differential can be locked even under extremely unfavourable road conditions (e.g. a wheel standing on black ice) or if a wheel lifts off. Nevertheless, the disadvantage remains that such a differential always has a slip-dependent basic locking torque. This is undesirable when parking or cornering without slip.

Figure 5.79 shows a characteristic diagram of a locking differential.

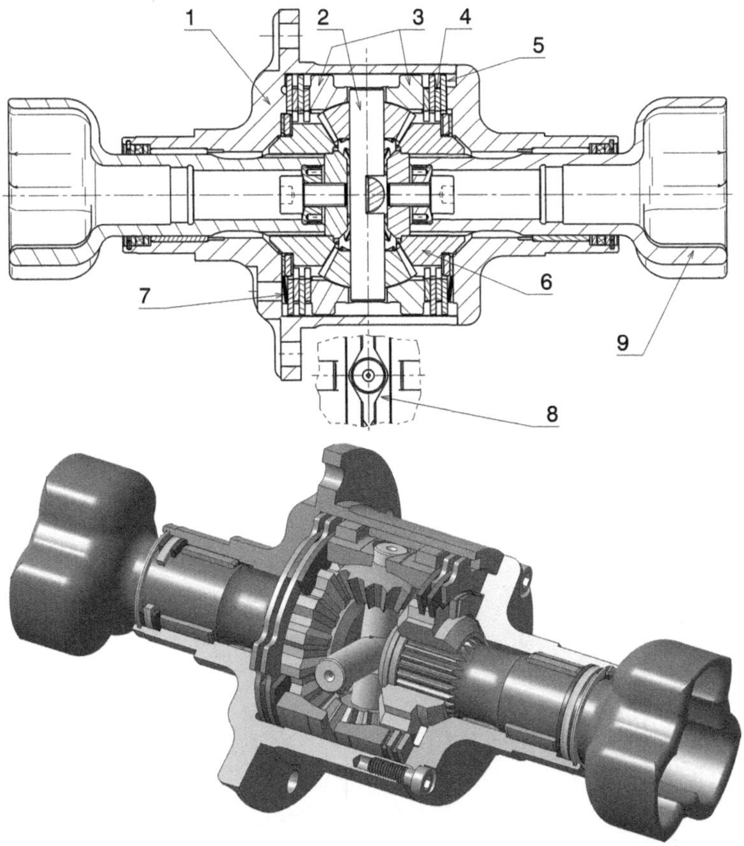

Fig. 5.77 Self-locking differential with multi-plate clutches (Drexler type) (*limited slip differential*). Upper half section: Differential without preload. Lower half section: Differential with preload via Belleville springs. **1** Differential cage. **2** Differential pin. **3** Thrust rings. **4** Inner plates. **5** Outer plates. **6** Axle bevel gears. **7** Cup springs. **8** Recess. **9** Tripod cup

The torque-dependent contact pressure can also be used as the sole means of contact pressure by the gear tooth expansion forces of the bevel gear differential. These contact forces are smaller by a factor of about 3 than those achievable with thrust rings. Furthermore, the disadvantage should be noted that the tooth engagement of the bevel gears changes negatively during the self-locking or balancing process, because the friction clutches to be pressed on must not be free of backlash.

Cam and Pawl **(Cornering Self-Locking Differential)**

This non-adjustable limited slip differential is used on many racing cars. The left and right side shafts are connected with sliding blocks via cam tracks, Fig. 5.80. The axle drive is provided by the roller cage, which transmits the circumferential force to the sliding blocks.

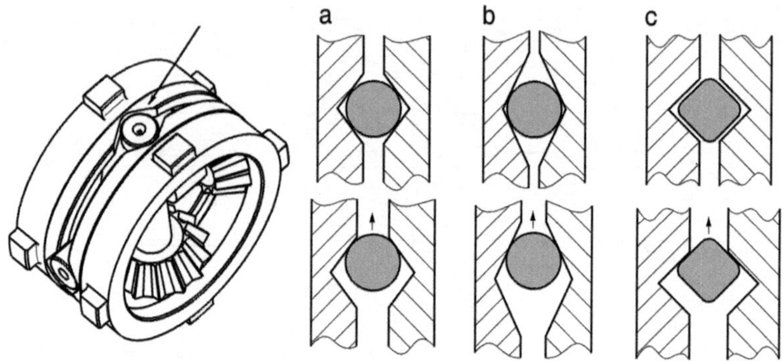

Fig. 5.78 Self-locking differential: function of the pressure rings. On the left, the thrust rings and the differential pins with the balancing bevel gears are shown. The arrow points to a recess. (**a**, **b**, **c**) represent different designs of recesses in the unloaded state (top) and under the influence of a driving torque (bottom)

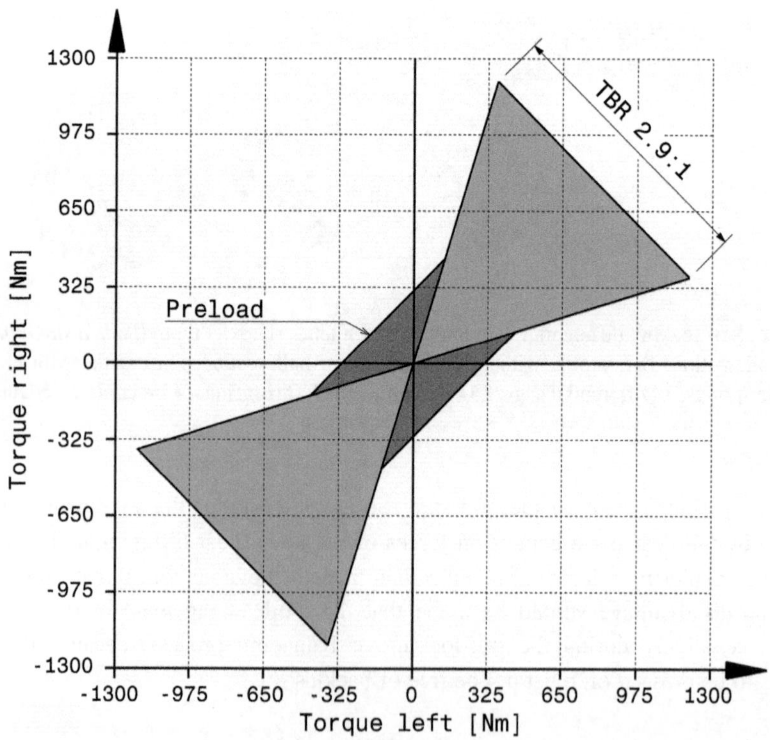

Fig. 5.79 Characteristic diagram of a limited slip differential. If an operating point consisting of left and right moment is in the grey area, the two side shafts are connected to each other. Outside the grey field, a relative rotation of the shafts to each other occurs. A preload extends the range in which only small torques occur on the wheels by a constant value

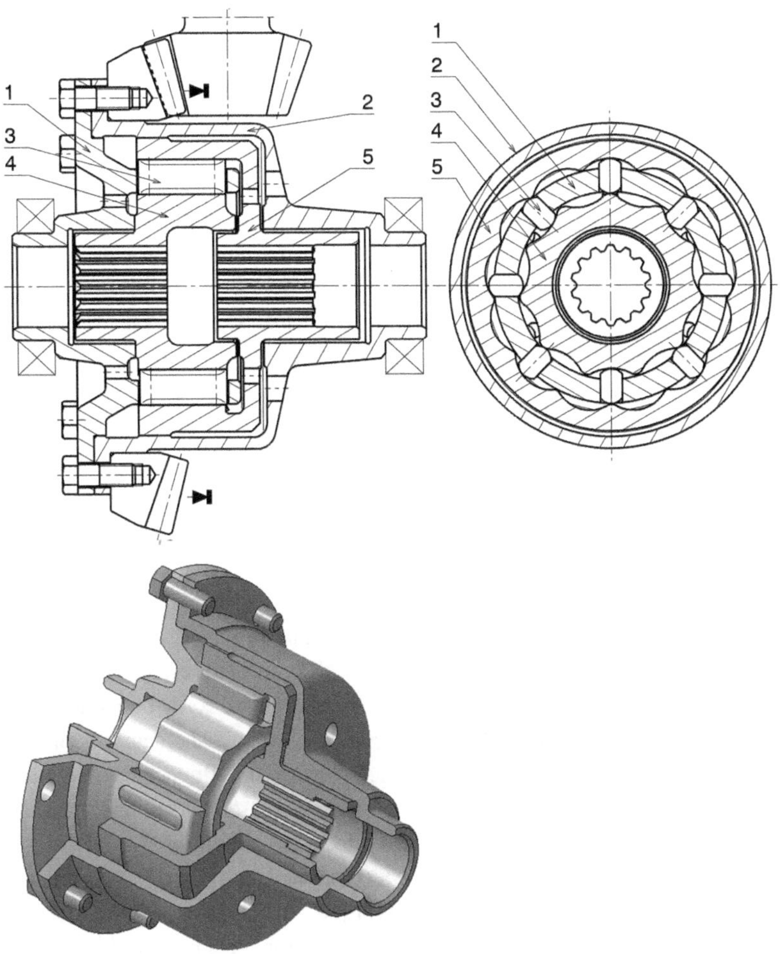

Fig. 5.80 Corner self-locking differential with radial curve paths (ZF). Some sliding blocks between the cam tracks wedge the two hubs as soon as a moment is transmitted or the difference in speed between the hubs becomes too great. **1** Roller cage. **2** Differential cage. **3** Sliding block (*pawl*). **4** Hub with inner raceway (*inner cam*). **5** Hub with outer raceway (*outer cam*)

In the traction-free state, the two drive wheels can move independently of each other. As soon as the drive torque acts on the sliding blocks, the differential locks because the number of curve apex is different for the inner and outer track. Three to four sliding blocks clamp between the ramps of the inner and outer ring and transmit the entire moment. However, it doesn't lock completely, the inner wheel of the turn can still spin depending on number of elevations and ramp angles of curve raceways. After all, the sliding blocks can follow the lagging ring in the radial direction. It is high-maintenance and shows heavy wear, especially on tyres with high adhesion. Thus, these units need to be replaced about every 600–1000 km [34]. Worn differentials operate like open differentials, but as wear progresses, the locking behavior changes, making it difficult for the driver to assess the vehicle's behavior. The locking effect starts abruptly during acceleration, which negatively affects the stability of the vehicle. Manufacturer: For example Hewland, ZF.

Self-Locking Differential with Worm Wheels (Torsen, *Quaife*)

The Torsen differential (from torque sensing) is used both as a differential gear between the wheels of an axle (lateral compensation) and as a transfer gear between the axles of an all-wheel drive (longitudinal compensation).

Drive is provided by the crown wheel of the axle drive (pos. 1 in Fig. 5.81) and thus by the differential housing (2). Six worm wheels (5) are mounted in the differential housing, which transmit the torque to the worms (3, 6) of the output shafts left (4) and right (8). The worm wheels are supported axially in the housing. The worms are guided radially by the three meshing worm wheels, which are evenly distributed around the circumference. Axially, the worms run against the housing or the adjacent worm via thrust washers (7). When driving straight ahead with the same output torque on both wheels, the entire differential rotates as a block. When cornering, there is a speed compensation, because two worm wheels of adjacent axle sides are coupled via spur gears at their ends. Although the pitch of the worms does not permit self-locking due to their size, the frictional losses resulting from the sliding movement are, as with all worm drives, so great that they can serve as a locking torque. At the same time, the circumferential force-dependent axial thrust of the two worms causes further friction between the thrust washers and the housing. The design of the thrust washers, which act like clutch plates, can influence the magnitude of the locking torque. In addition, the locking value is again increased by a deliberate mismatch of the teeth of the worms and worm wheels. The effects are shown in Fig. 5.82.

A positive input torque and a differential speed between the output shafts are required to establish the locking effect. If there is no input torque, such as in overrun mode, the locking effect is cancelled. When braking, this differential therefore acts in a similar way to an open differential and is therefore ABS-compatible.

With an open differential, the wheel with the lower coefficient of friction dictates the possible tractive force of the axle in the case of different adhesion on the left and right. Both wheels can only build up the tractive force of the "weaker" wheel, i.e. exactly twice as much in total. In the case of a Torsen differential, the locking torque between the axles is added in such a situation, so that depending on the amount of locking torque at low road friction values, the wheel on the more favourable side can transmit 3–6 times that of the neighbouring wheel, Fig. 5.82.

Torsen differentials cause the least power understeer of all self-locking differentials.

***Torque Vectoring* Differential**

In addition to hindering or locking the compensation function, there are further developments in the direction of active differentials due to the current possibilities of mechatronic systems. These can increase or maintain the stability of the vehicle by

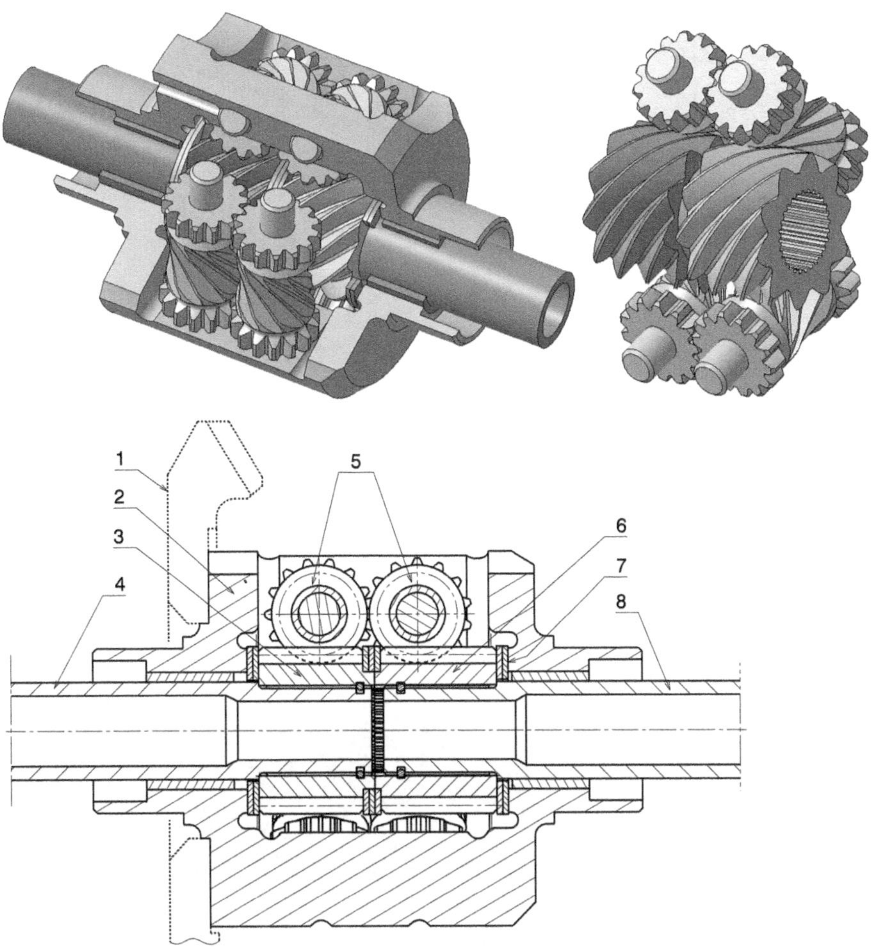

Fig. 5.81 Torsen differential. Top left: Partial cutaway view; right: wheel set, with a pair of worm wheels not shown for clarity; below: sectional view. **1** Crown wheel. **2** Differential housing. **3** Worm gear left. **4** Output shaft left (*half shaft left*). 5 Worm gears (*planetary worms*). **6** Worm gear right hand. **7** Thrust washers. **8** Output shaft right (*half shaft right hand*)

intercepting the swerving vehicle with a counter-yaw moment or by turning a vehicle accelerating out of the corner in the sense of the corner. Figure 5.83 shows an example of the function of such an active differential.

A countershaft (3) is driven by the final drive (1) via an additional spur gear stage (2). The countershaft is divided into three parts and can be connected via two friction clutches (4). In normal driving operation, both clutches (4) are open and the differential works as an open differential. If the left side shaft is to receive more torque, the left clutch is activated. The gear ratios are selected so that the countershaft makes the sideshaft rotate slightly faster

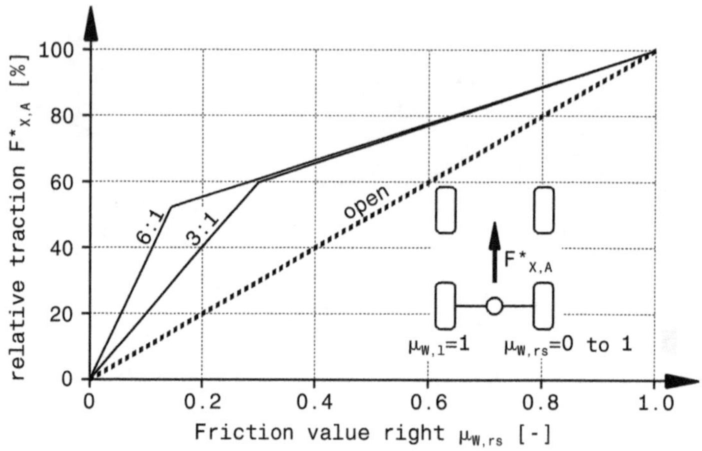

Fig. 5.82 Characteristics of Torsen differentials. In addition to two different Torsen differentials with a torque ratio (TBR) of 3:1 and 6:1, the characteristic curve of an open differential is shown for comparison. Due to the locking torque of the Torsen differential, the vehicle can build up a considerably higher tractive force, especially with low friction values on one side

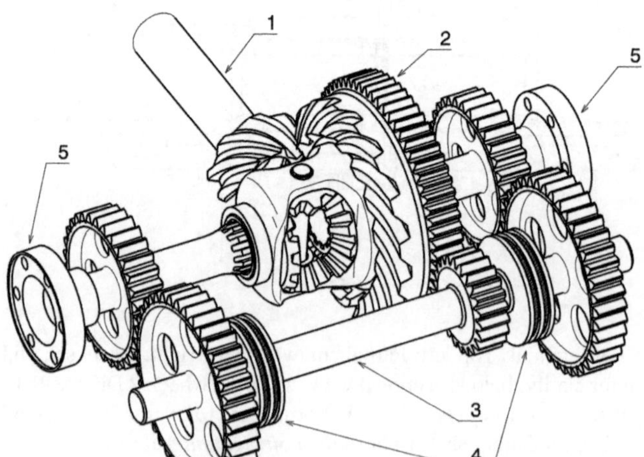

Fig. 5.83 Active differential, according to [14]. **1** input shaft final drive. **2** Spur gear, connected to ring gear. **3** layshaft (countershaft). **4** Friction *clutches*. **5** Flange *of* side shafts. The drive torque of the drive shaft (1) also reaches the layshaft (3) via spur gears (2) in addition to the usual path via the differential. From here it can be directed as required via one of the two friction clutches (4) to one of the two side shafts (5)

than the final drive (about 10% faster). So in this example, the left side shaft is accelerated while the right side shaft is decelerated by the differential gear while the driveshaft speed remains the same. In this way, increased torque can also be transmitted to the (faster

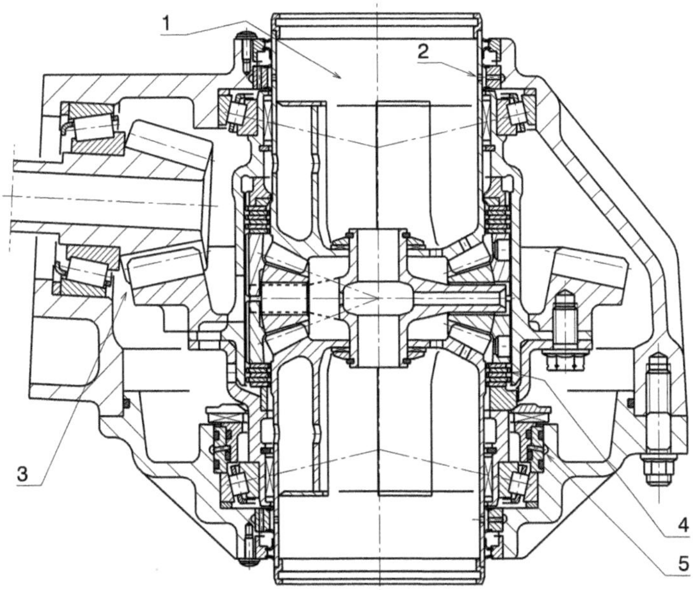

Fig. 5.84 **Final-drive cage** with integrated tripod joints (Xtrac Ltd.). **1** Tripod tulip. **2** oil feed to tripod joint. **3** Final drive. **4** differential friction plates. **5** Hydraulic piston for differential lock (*hydraulic actuator*)

turning) outer wheel on the turn. The two clutches are controlled by a control unit. The control unit detects the driving condition of the car via sensors (longitudinal and lateral acceleration, yaw movement, ...) and calculates the modulation of the clutches via a strategy. Both clutches must never be closed at the same time, because this would block the system.

Multi-disc locking differentials can also be influenced electronically via hydraulic actuators. In this case, a locking torque is set depending on the wheel differential speeds and/or the engine input torque. Figure 5.84 shows a differential whose locking effect can be influenced via hydraulic multi-plate clutches.

If two such multi-plate clutches are provided separately for each output side, the input torque can be selectively distributed to the side shafts by actuating the clutches differently. Figure 5.85 shows such a hydraulic actuator for one side of the differential. Of course, an additional electronic control effort with corresponding sensors and hydraulic components is required for the function of such a controllable differential.

However, such controllable differentials do not work wonders. On the circuit, they hardly influence the lap time or the way the car is driven. The main advantages are greater stability during acceleration and braking, and greater willingness to steer the car when turning under load [14].

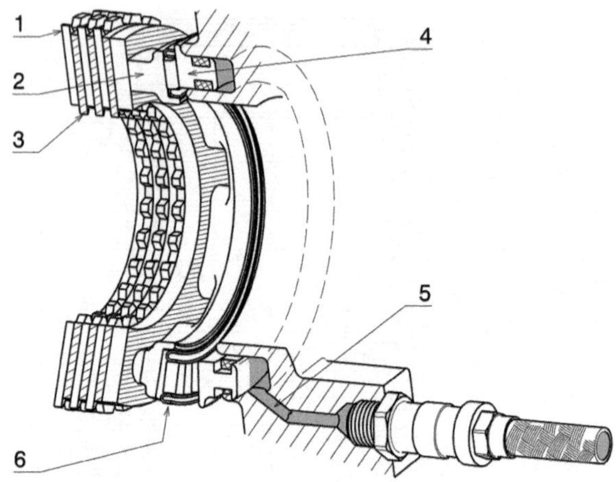

Fig. 5.85 Hydraulic actuator for a controllable differential. **1** Outer *friction plate*. 2 Thrust ring. **3** inner friction plate. **4** annular piston with sealing rings. 5 Hydraulic fluid. 6 Thrust bearing. The annular piston (4) presses the disk pack (1, 3) together due to the hydraulic pressure. As a result, the pack transmits a certain locking torque between the differential housing and the differential bevel gear or side shaft

5.5 Shaft

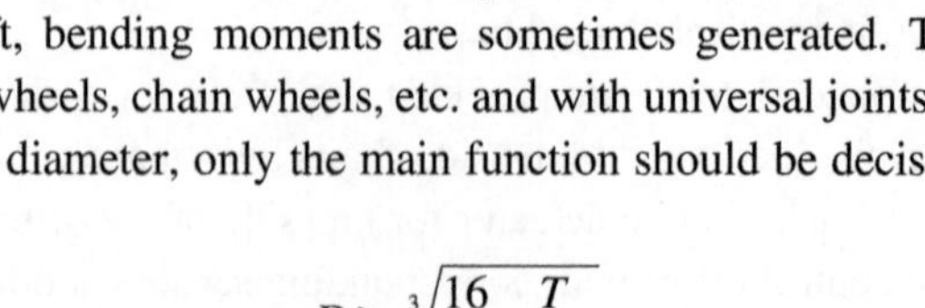

The main function of shafts is to transmit torque. Due to the actual way of introducing a torque into the shaft, bending moments are sometimes generated. This is the case, for example, with gear wheels, chain wheels, etc. and with universal joints. For a rough design of the smallest shaft diameter, only the main function should be decisive:

$$D \geq \sqrt[3]{\frac{16}{\pi}\frac{T}{\tau_{ts,zul}}} \tag{5.25}$$

$$D \geq \frac{1}{\sqrt[3]{1-(d/D)^4}}\sqrt[3]{\frac{16}{\pi}\frac{T}{t_{ts,zul}}} \tag{5.26}$$

D	Outside diameter of a shaft, mm
d	Inner diameter of a hollow shaft, mm
T	Torque, N mm Note: Due to dynamic effects, the maximum torque to be absorbed can be much higher than, for example, the max. Engine torque or the largest mathematically transmittable Tyre torque. The following applies as a realistic assumption: $T_{dyn} = 2T_{stat}$. .
$\tau_{ts,per}$	Permissible torsional stress of the shaft material, N/mm^2 $\tau_{ts,per} \approx \tau_{ts,sch/12}$ to account for not yet known bending moments and notches

Depending on the installation location in the vehicle, a distinction is made between longitudinal shafts (from the engine to the axle drive) and lateral shafts (from the axle drive to the wheel).

5.5.1 Drive Shafts (*Prop(Eller) Shafts*)

Drive shafts have the task of transmitting the torque and the rotary motion of the engine over a certain spatial distance. They are therefore required for vehicles with standard drive and for all-wheel drive vehicles (see also Sect. 5.6). Mid-engined vehicles usually have the gearbox flanged directly to the engine, thus eliminating the need for a drive shaft. There is also a mixed design where the engine is at the front and the gearbox at the rear of the vehicle. A drive shaft runs between them at engine speed (transaxle arrangement). In the case of passenger cars, it is also important that in the event of a frontal collision the propeller shaft also transmits longitudinal forces between the engine-transmission unit and the rear axle. This also requires a crash-compatible design of the drive shafts. In the event of a crash, the shaft should fail as quickly as possible so that it does not have a negative impact on the vehicle's crash behavior.

Due to the distance between the shaft ends, small (unavoidable) deformations of the frame lead to noticeable displacements of the connecting flanges. For this reason, drive shafts are designed as multi-part cardan shafts, Fig. 5.86. In addition, rigid shafts cannot be bolted without tension between mounting points that are not directly connected to each other due to the accumulation of manufacturing tolerances. This is already the case, for example, if the gearbox is bolted to the front and the axle drive to the rear of the frame. If there is a relative movement between the connection points of the shaft (e.g. if the axle drive is sprung along with the axle), length compensation of the cardan shaft is also required.

The use of a drive shaft in a touring car with four-wheel drive is shown in Fig. 5.119.

In the standard drive, the input shaft usually rotates (i.e. in the case of a coaxial layshaft gearbox) in the same direction as the engine, i.e. counterclockwise when viewed from the clutch. The direction of rotation results in the basic design of a bevel gear rear axle drive. In this case, the pinion is in the middle and the ring gear is on the left in the direction of travel.

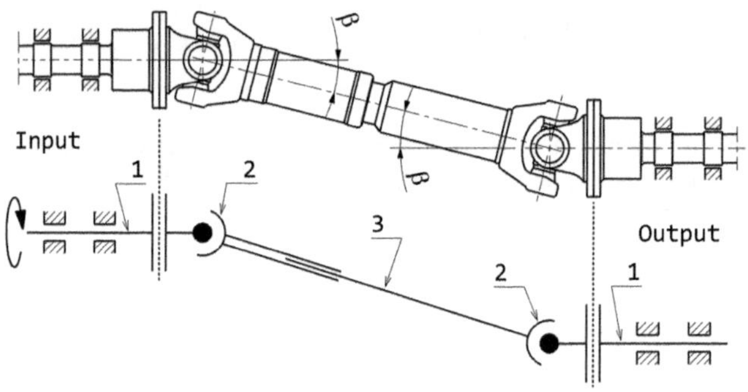

Fig. 5.86 **Cardan** shaft with universal joints and length compensation, Z arrangement. Above: View. Below: diagrammatically. **1** Connecting shaft . **2** Joint. **3** intermediate shaft. β articulation (deflection) angle

If a single-stage gearbox is used, the input shaft rotates in the opposite direction to the direction of rotation of the engine and the rear axle gearbox requires the ring gear on the right-hand side, see also Fig. 5.118.

Joints

Hook's joints, constant velocity joints (see Sect. 5.5.3) and small angle joints can be used as joints for longitudinal shafts. Hook's joints (also known as cardan joints) have a relatively simple design and operate with very low losses, but have the disadvantage that the output speed fluctuates cyclically compared to the input speed. The output shaft is therefore accelerated and decelerated again during one revolution while the speed of the input shaft remains constant. This can only be prevented by the action of two universal joints cancelling each other out. Depending on the arrangement of the two universal joints, this is referred to as a Z or W arrangement, Fig. 5.87. Even if the output shaft rotates uniformly in this way, the intermediate shaft is accelerated and decelerated in any case. For small articulation angles (<3°), however, the non-uniformity is still so small that it is tolerable. This is why this joint is still found on longitudinal passenger car shafts.

Constant velocity joints transmit the rotary motion uniformly even in the flexed state, but they are relatively heavy. If the cardan shaft only has small articulation angles, the greater mass hardly pays for itself.

Small angle joints (elastic joints) represent an interesting alternative to the established solutions, Fig. 5.88. In principle, they work like a Hardy disc: Two rigid flanges are connected with an elastic, flexible so-called flex disc. The maximum bending angle is about 1.5°. The flex-disc can be made of steel or plastic. The plastic variant (composite disc) consists of fibre-reinforced layers. The ability to bend is made possible by the elastic elongation of the glass fibres embedded in a thermosetting matrix. These joints are suitable for high speeds, have synchronous running properties and do not require additional centering. However, they should be used on a shaft together with a constant velocity

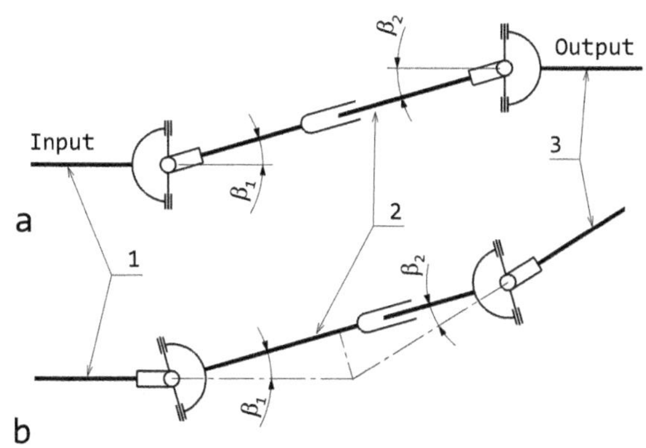

Fig. 5.87 Arrangements of two universal joints for compensation of rotational non-uniformity. (**a**) Z-arrangement (*Z-configuration*). (**b**) W-arrangement (*W-configuration*). **1** Drive shaft . **2** intermediate shaft. **3** Output shaft. To ensure that the speeds n_1 and n_3 of the input and output shafts are always the same, the angles β_1 and β_2 must be the same size

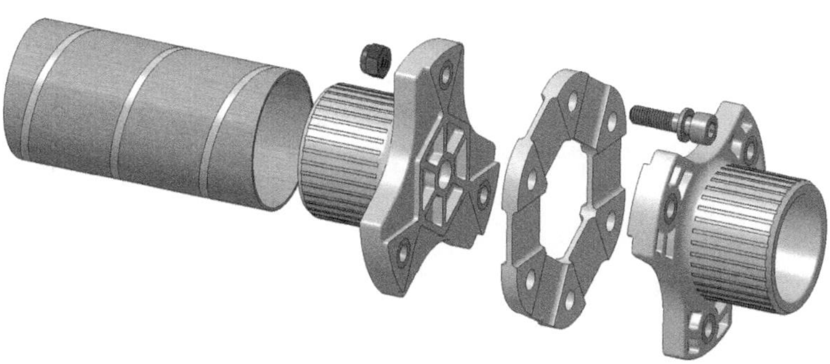

Fig. 5.88 Small-angle plastic joint. A fiber composite shaft is bonded to a plastic three-arm flange. Between the two flanges, a composite disc transmits the torque

joint. This avoids multi-axial stress conditions in the composite disc, which is especially the case during sharp start-up, where an axial force acts in addition to the torque and the articulation angle due to the lack of longitudinal compensation [35]. The three-arm flanges can also be made of plastic and accommodate the connecting bolts in metal inserts.

Typical uses of joints in longitudinal shafts are listed in Table 5.13, cf. also Fig. 5.89.

Due to dynamic processes, the maximum torque that cardan shafts have to transmit is much greater than the nominal engine resp. motor torque converted with the transmission ratio of the cardan shaft plane. As a rule of thumb, initial calculations can be based on twice the rated engine torque. As an example Fig. 5.90 shows the temporal course of the cardan shaft torque during a bang start. In this case, the depressed clutch pedal is released abruptly

Table 5.13 Joint arrangements of longitudinal shafts [36]

	Installation location		
	Gearbox side	Interim bearing	Axle gearbox side
Joint type	VL/HS	Universal joint	Flexible link
	GI/HV		

Joint designations: see Sect. 5.5.3

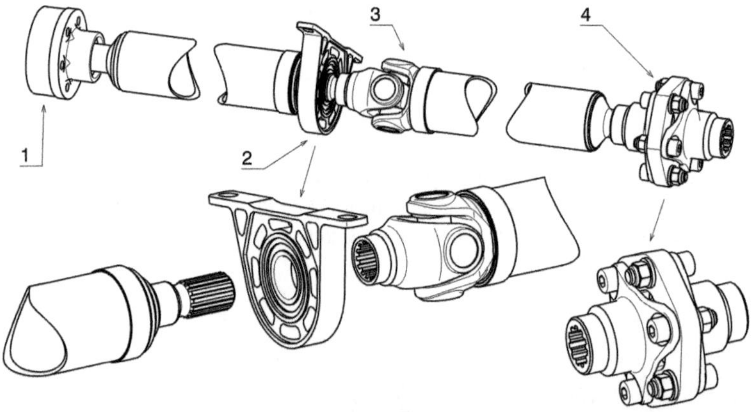

Fig. 5.89 *Two-piece* shaft with *centre* bearing. **1** Gearbox side connection (*coupling to the gearbox*). **2** Intermediate bearing (*centre bearing*). **3** Hook type joint. **4** coupling to final drive by flexible fabric

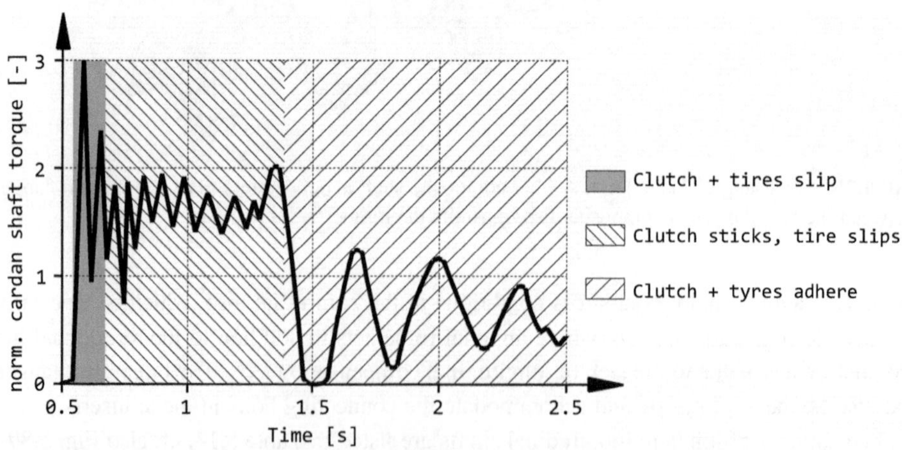

Fig. 5.90 Load increase during highly dynamic start-up, according to [37]. The torque of the cardan shaft initially reaches three times the nominal engine torque and, at the transition from sliding to sticking of the tyres, twice the nominal engine torque

Table 5.14 Bending-critical speeds of smooth steel shafts [38]

Support	$n_{crit,1}$, min-^{1}	$n_{crit,2}$, min-^{1}	$n_{crit,3}$, min-^{1}	$n_{crit,4}$, min-^{1}
Freely supported ("spherically" supported) shaft, e.g. self-aligning bearing	$122.5 \cdot 10^6 \cdot d/l^2$	$4n_{crit,1}$	$9n_{crit,1}$	$16n_{crit,1}$
Clamped at both ends, e.g. very rigid bearings	$277.7 \cdot 10^6 \cdot d/l^2$	$2.8n_{crit,1}$	$5.49n_{crit,1}$	$8.9n_{crit,1}$
One end clamped, one end free ("flying" shaft)	$43.6 \cdot 10^6 \cdot d/l^2$	$6.276n_{crit,1}$	$17.55n_{crit,1}$	$34.41n_{crit,1}$
	D	Shaft diameter, mm		
	L	Bearing spacing or shaft length, mm		

at high engine speed in first gear. In the diagram the cardan shaft torque is related to the engine torque and converted to this position.

Immediately after the abrupt clutch engagement, the torque of the cardan shaft reaches the maximum value with three times the nominal engine torque (i.e. actually three times the engine torque times the gear ratio of the first gear). It can be seen from the time curve that an oscillation is formed. In the first phase, the engine and vehicle masses are still decoupled because the clutch and tires are slipping. The frequency of the torque oscillation is therefore still relatively high at 13 Hz. As soon as the clutch starts to stick, the engine mass is part of the oscillating system and the frequency drops to about 10 Hz. Finally, the tires reach their adhesion limit and thus the vehicle mass also comes into play. The frequency drops further to about 3 Hz. Jerk oscillations are formed. At the transition to the phase of tire adhesion, the cardan shaft torque reaches a second notable overshoot, namely to about twice the rated engine torque.

Whirling Speed (Critical Bending Speed)

An important criterion for cardan shafts is the bending-critical speed. These are the speeds at which the vibrating system shaft plus additional masses starts to resonate. These critical speeds depend on the stiffness (deflection) and the mass. The stiffness is influenced by the bearing distance, the shaft cross-section and by the shaft material (Young's modulus). For a system with n partial masses, there are n critical speeds, whereby the smallest is usually of interest (fundamental frequency). In the case of smooth continuous shafts, the system can be imagined as consisting of an infinite number of partial masses and therefore there are also an infinite number of critical speeds. Deflections caused by tooth forces or other forces acting radially on the shaft have no influence on the critical speed because they do not cause any "centrifugal forces". The mounting position of the shaft is also of no significance. Reference values for the first four critical speeds of smooth steel shafts are shown in Table 5.14.

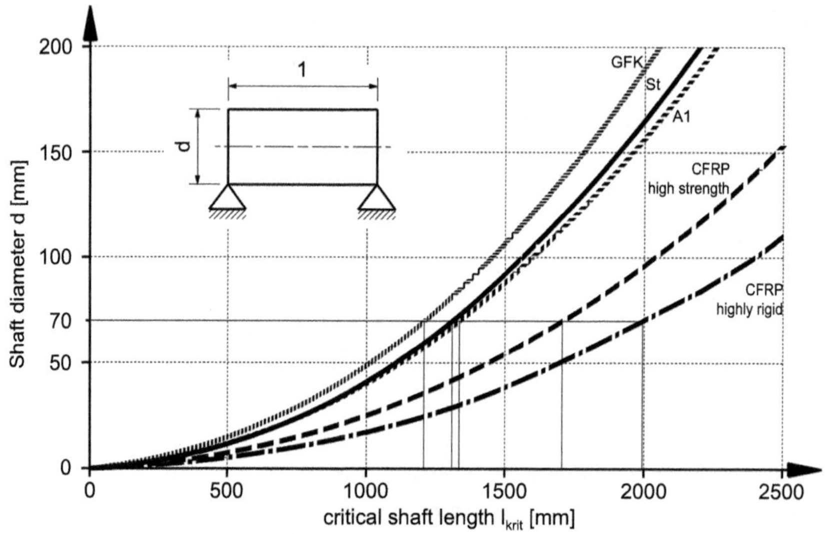

Fig. 5.91 Material influence on the critical length of cardan shafts, according to [37]. The diagram is based on a critical bending speed of 7000 min^{-1}. The reading example shows the differences for a shaft diameter of 70 mm. A shaft made of GRP may be a maximum of 1200 mm long, one made of highly rigid CFRP can be almost 2000 mm long before it starts resonating at 7000 min^{-1}

Figure 5.91 shows how the material of a cardan shaft affects the critical bending speed. The bearing distance is changed in such a way that the critical speed becomes 7000 min^{-1}. For a given shaft diameter, this results in a critical length from which the shaft starts to resonate at 7000 min^{-1}. The shaft under consideration would therefore have to be made shorter or with a larger diameter in the case of use. If neither of the two measures is possible, e.g. because there is not enough space, the only solution is to use one or more intermediate bearings. The picture shows that CFRP shafts can still be designed in one piece even with larger vehicle wheelbases, whereas steel or GRP shafts may already require an intermediate bearing.

The fact that steel and aluminum shafts exhibit approximately the same vibration behavior results from the similar modulus of elasticity/density ratio of the two materials.

When designing the shaft, it must be ensured that the maximum speed has a "safety distance" from the first critical bending speed.

Torsional Critical Speed

Just as bending vibrations are possible, an elastic shaft can also perform torsional vibrations. For a shaft with two masses at its ends, a torsional critical speed of:

$$n_{crit,ts} \approx 72.3\sqrt{\frac{T}{\beta}\left(\frac{1}{J_1}+\frac{1}{J_2}\right)} \tag{5.27}$$

$n_{crit,ts}$	Torsional critical speed, $\min^{-1}$
T	Torque, N m
β	Angle of twist of the shaft at moment T, °
J_1, J_2	Mass moment of inertia of the masses at the shaft end, kgm^2

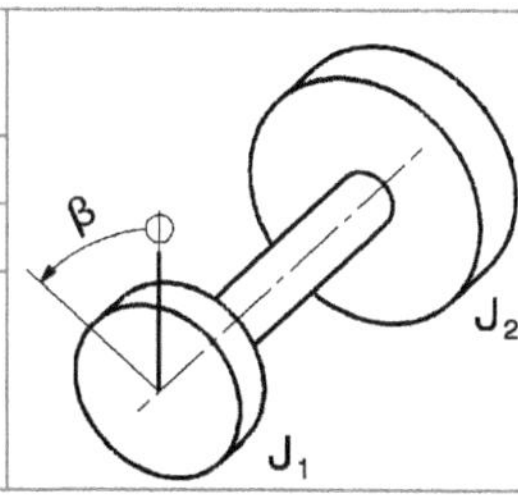

A one-piece longitudinal shaft has weight and cost advantages over a two-piece design. Nevertheless, the two-piece variant with intermediate bearing is used in the majority of passenger cars. The reason for this is the critical bending speed and acoustic requirements which, with the given wheelbases, can only be met with steel shafts with intermediate bearings. Figure 5.89 shows such a longitudinal shaft with intermediate bearing. The gearbox connection (1) is usually formed by sliding joints in constant velocity or tripod design. At the intermediate bearing (2) the shaft is split for mounting reasons. A universal joint (3) is used in connection with this bearing. The connection to the rear axle drive is established via an elastic joint (4).

To ensure that shafts fulfil their function with as little mass as possible, they are constructed in a differential design. Each shaft section can be made of the most suitable material. The connecting pieces, i.e. joint forks, three-arm flanges or splined shaft journals, are made of steel or aluminium alloys. The actual shafts are manufactured in tubular form from steel, aluminium and titanium alloys or fibre composites. In the case of the latter, the winding process results in a wide range of bending and torsional properties, which are influenced by the winding angle (fibre laydown angle), the mixing ratio of different fibres and the matrix material. Besides the directional stiffnesses, axial thermal expansion can be designed to zero [39]. In addition, shafts made of fibre composites have the advantage that they can bridge larger distances without resonance compared to steel shafts with the same shaft diameters (Fig. 5.91), i.e. they can be designed in one piece. In addition, there is the advantage of lower density compared to metallic materials. With CFRP hollow shafts, a weight saving of approx. 60% can be achieved. Another advantage of this type of construction is that balancing is usually not necessary because the unbalance of finished shafts remains tolerably small due to the winding process on ground cores.

Great importance is attached to the connection technology for such shafts in differential design. The following methods are used for the connection between hollow shafts and connecting pieces:

- Adhesive joints
- Clamp connections
- Press connections
- Friction welded joints.

The choice of joining method depends not only on manufacturing aspects but also on the materials to be joined.

Adhesive Joint

Connecting parts made of plastic, steel or aluminum can be connected to fiber composite shafts by bonding. Different methods exist. A method from the company GKN is presented as an example [35], Fig. 5.92a. The joint at the connecting part (1) for the tubular shaft is formed by two centring collars (3) with a conical annular gap between them. The shaft (4) is pushed over the locating point and then the adhesive is introduced under pressure through the bore (2). The adhesive initially spreads in the axial direction until it abuts the opposite centering collar, where the annular gap is wider than at the entry point. From then on, the flow path is in the circumferential direction. When the annular gap is completely filled with adhesive, it emerges again at the opposite hole (5). This process ensures an air-free connection. The adhesive used is a one-component epoxy resin. The connection is cured in the oven.

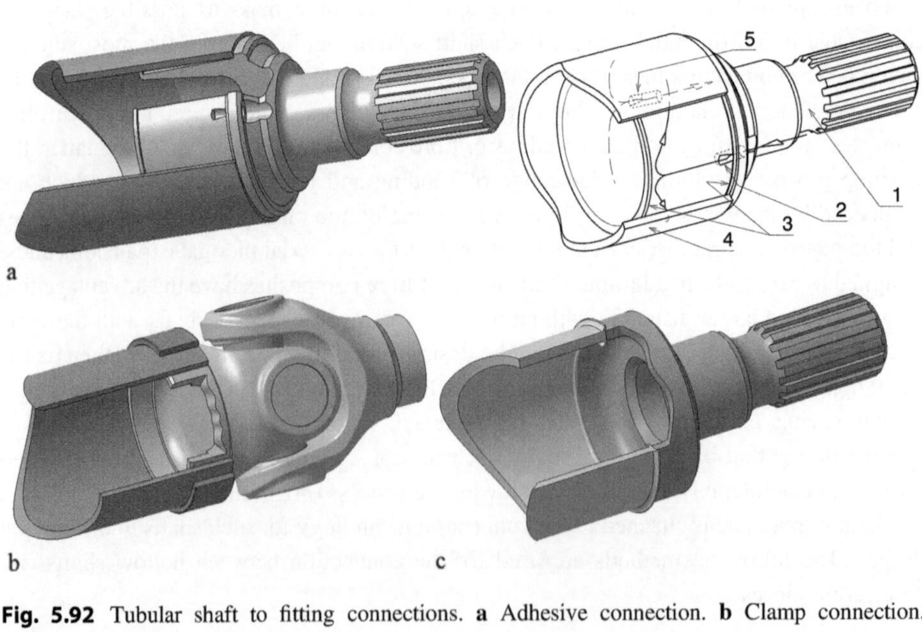

Fig. 5.92 Tubular shaft to fitting connections. **a** Adhesive connection. **b** Clamp connection. **c** Friction welded connection

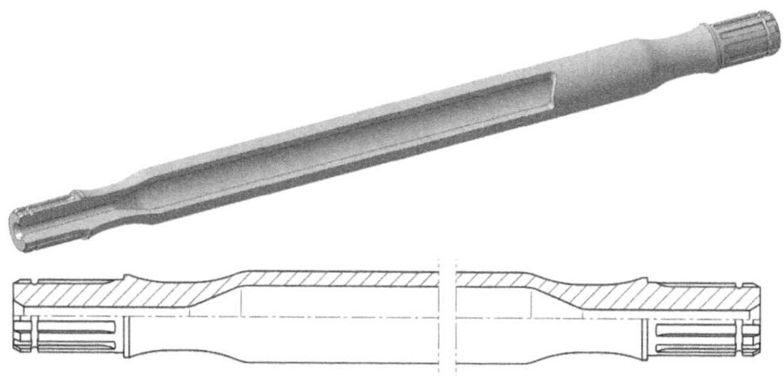

Fig. 5.93 Lightweight monobloc shaft, after [40]. The wall thicknesses are optimized for the load to save weight. The shaft is manufactured by forming processes

Clamp Connection

This technique was developed for the connection of steel and aluminum connecting parts with fiber composite pipe. However, this connection is also possible for aluminium shafts on steel journals. The locating point of the connecting piece is inserted into the shaft with a transition fit. In addition, a steel support ring is pushed over the outside of the shaft. Then the support ring and thus the shaft is hydraulically pressed onto the connecting piece (transverse press fit), Fig. 5.92b.

Press-Fit Connection

This method was developed for the same application as the clamp connection. The main difference to clamping is also only in the joining process. During joining, the end piece of the shaft is pressed into the shaft with a defined interference fit (longitudinal press fit). A steel support ring on the shaft restricts the expansion of the shaft and ensures the required preload to transmit the desired torque.

Friction Welded Joint

With the aid of this process, not only aluminium but also steel end pieces can be joined to aluminium tubes in a process-safe manner, Fig. 5.92c. During the actual welding process, the rotating fitting is pressed against the shaft until a certain amount of energy has been introduced into the contact zone by friction. After stopping the rotating movement, the contact pressure is increased. Only the aluminium tube is plastically deformed in the process. The metallurgical changes in the weld area (hardness and strength reduction) also primarily affect the aluminium shaft, but can be kept low by targeted process control. The static and dynamic torsional strength of this connection is higher than that of the tube.

In contrast to welding, there are no material changes with the press and clamp connections. However, this is disadvantaged by the steel support ring and the required overlap of shaft and connecting piece, which increases the mass of the connection.

In addition to shafts in differential design, there are also one-piece lightweight shafts (monobloc shafts), see Fig. 5.93.

Fig. 5.94 Side shaft on a formula car (Formula Renault 2000). View from the rear with the rear wheel removed. The shaft shown leads from the axle drive output to the left wheel carrier (left in the picture). A tripod joint is used on the gearbox side and a constant velocity ball joint on the wheel side. At the lower edge of the picture you can see the diffusor of the underbody

5.5.2 Sideshafts, *Half Shaft* (*AE: Axle Shafts*)

Side shafts transmit the drive torque from the axle drive to the wheels, Fig. 5.94. If the axle drive is located in the middle of the vehicle, both shafts can be the same length. This is advantageous because one spare part covers both sides of the vehicle and because the torsional stiffness and, in particular, the torsional behaviour are the same.

The torsional stiffness has an effect on the load change impact. Winding up the drive train, e.g. due to a gearshift manoeuvre, results in the so-called bonanza effect. If load alternation and bonanza effect occur, a thick sideshaft is selected. In the case of torsional vibrations caused by the drive train, a relatively thin sideshaft is selected. In the case of a long shaft, a large shaft diameter or an absorber is selected. An absorber has the disadvantage that it represents an additional mass, which is why it is hardly ever used for shafts in passenger cars.

In the case of shafts of different lengths, such as those frequently found in transverse engines, different joint angles occur and there are generally effects on the self-steering behaviour of the vehicle. Further consequences are reaction forces in the steering. There is also a change in the secondary torque. This can be remedied by an intermediate shaft that bridges the offset of the axle drive to the center of the vehicle and is mounted on the engine. In this way, the two propeller shafts can be designed with the same length.

Acoustically, there is a hum with stiff shafts. With soft shafts, the wind-up is negative.

The stiffnesses must be individually adjusted for each vehicle type. Parameters that can be changed are: Shaft diameter and shaft design (hollow shaft, monobloc, etc.). Absolute

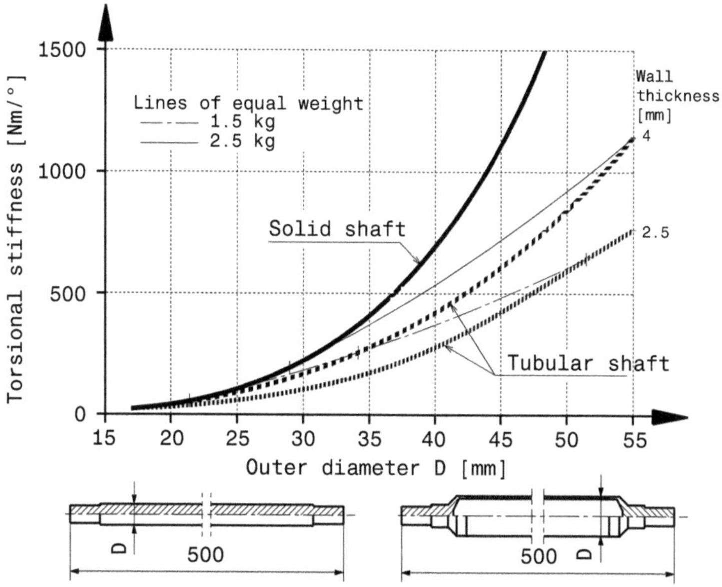

Fig. 5.95 Influence of the shaft design on the mass, after [40]. A solid shaft with an outside diameter of approx. 37 mm has a torsional stiffness of 500 Nm/° at a mass of over 2.5 kg. The same torsional stiffness is achieved by a hollow shaft with an outside diameter of approx. 47.5 mm and a wall thickness of 2.5 mm at a mass of less than 1.5 kg

values, from when stiffness differences are noticeable, therefore do not exist. The stiffness of the sideshafts is also more of a comfort issue than a driving dynamics problem.

A hollow shaft has a much lower mass than a solid shaft. By choosing a correspondingly large outer diameter, the torsional stiffness of a hollow shaft can be matched to that of the solid shaft without reaching the weight of the solid shaft, Fig. 5.95. Besides, the bending stiffness of the hollow shaft is considerably greater than that of the solid shaft due to the larger diameter.

Short shafts are usually designed as solid shafts and long shafts as hollow shafts. The connecting stubs on hollow shafts are welded on or directly forged on. Forged parts are reserved for series parts for cost reasons. The so-called monobloc shafts represent a further development of the welded tubular shafts. The weld seam and thus a zone of material change is omitted. The monobloc shaft can also be hardened in the total length. For these reasons, much lighter weight shafts can be represented than with the conventional methods. For cost reasons, different ways of manufacturing such shafts have been developed. This starts with a simple tube that is formed by drawing and hammering processes. The wall thickness of the shaft is largely constant. For high strength and lightweight requirements, shafts are formed to the desired cross-sectional shape by a combination of hydraulic expansion, drawing and hammering. The largest accumulations of material thus only occur in the spline area of the end journals. The rest of the tube contour can be dictated

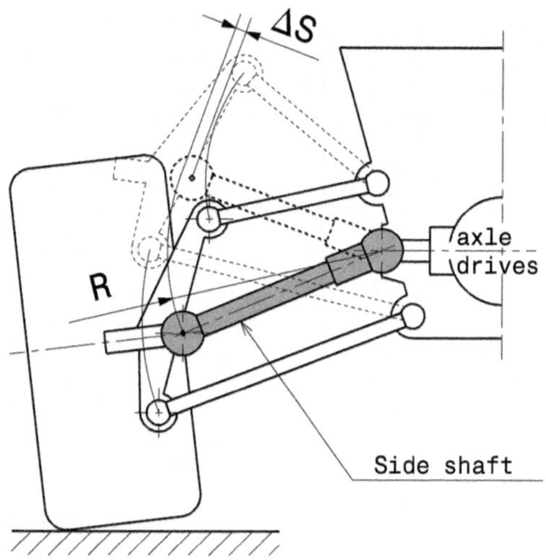

Fig. 5.96 Length change of a side shaft during springing. When springing, the wishbones and the cardan shaft move on different circular paths. This changes the distance between the wheel carrier and the output of the axle drive by the amount Δs. The side shaft must therefore be movable in length, otherwise it blocks the spring movement

not only by the stress profile but also by the space available in the vehicle, Fig. 5.93. The minimum wall thickness should not be chosen significantly below 2.5 mm because of the risk of corrosion and buckling.

Length Change

(*Length Variation*) In the case of independent wheel suspensions, the distance between the output of the final drive and the shaft flange at the wheel carrier changes when the suspension is sprung, Fig. 5.96. Side shafts must therefore ensure some form of length compensation (*axial plunge accommodation*). In older vehicles, a two-piece shaft with an axially displaceable spline was used. However, when a drive torque was transmitted, friction caused by the circumferential forces obstructed the length compensation, Fig. 5.97.[3] Elastic couplings made of elastomer materials compensate for axial and angular deviations, but they are limited in their torque capacity and require a relatively large amount of space in the diameter. The present solution is a flexure joint as a floating bearing, which operates with very low friction if properly designed and lubricated.

Side shafts that drive front wheels also require length compensation because steering also results in a change in the distance between the flanges of the shaft that are fixed to the wheel and those that are fixed to the car.

Side shafts rotating in the free airflow generate air turbulence and disturb the inflow of the rear wing. In addition, they generate lift due to friction effects on their surface. In the case of single-seaters with open (free-standing) wheels, an attempt is therefore made to

[3] See e.g. Racing Car Technology, Vol. 4 Chassis, Chap. 4, Fig. 4.75.

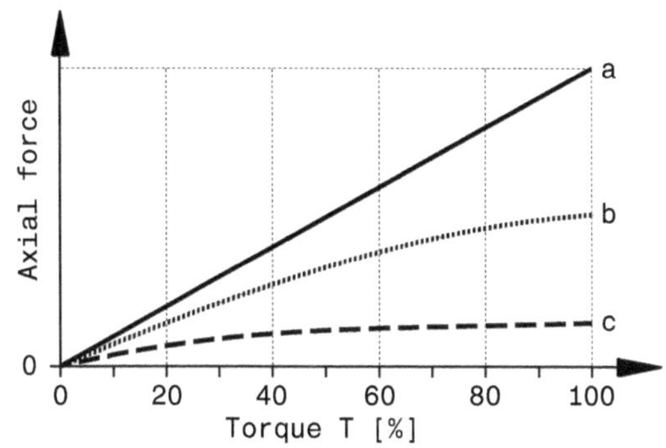

Fig. 5.97 Resistance to length change as a function of torque and surface condition of length compensators [30]. (**a**) Wedge profile steel on steel. (**b**) Polyamide-coated (Rilsan) splined shaft connection. (**c**) Length compensation with balls. In a splined shaft profile, the frictional force that counteracts length compensation increases linearly with the transmitted torque. Friction can be significantly reduced by coatings or by rolling elements between the shaft parts that are inserted into each other

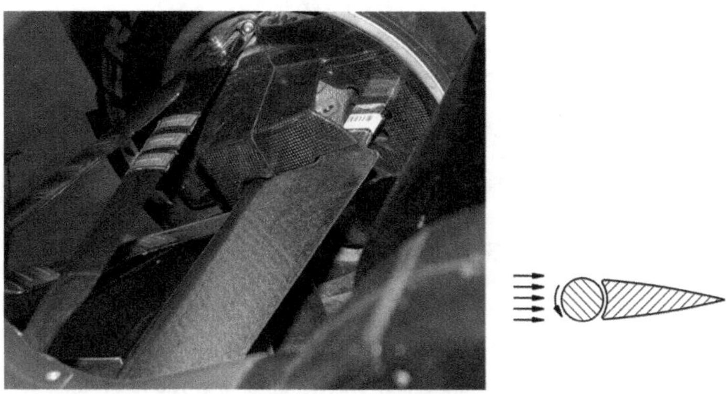

Fig. 5.98 Shielded drive shaft on a Formula 1 car (Ferrari). The right rear wheel can be seen. Behind the side shaft, the tie rod is designed in such a way that a flow-favourable drop profile is created in the cross-section of the shaft and tie rod

keep the shaft diameter small (wound CFRP shafts are at a great disadvantage in this respect) and to shield the shaft (Fig. 5.98) or to clad it (Fig. 5.99).

5.5.3 Materials

Case hardening steel C45E (was Ck45 DIN 17210) with hardness 52 + 6 HRC with hardening depth Rht 450 HV = 2.5 + 2.0.

41SiNiCrMoV7-6 (DIN) heat treated.

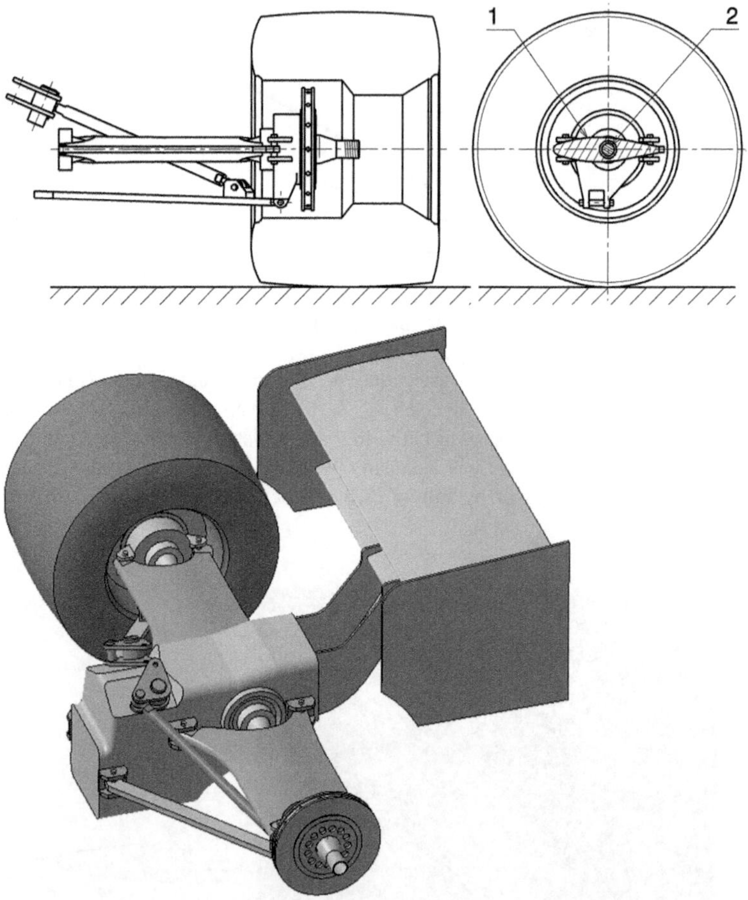

Fig. 5.99 Clad drive shaft on a Formula 1 car. This solution was found on the McLaren MP4/15 (2000) and the Williams FW16 (1994), among others. The upper control arm is made of CFRP and designed to completely enclose the side shaft. To do this, it has to be mounted relatively low, at the height of the wheel centre. The control arm has an aerofoil profile. **1** upper wishbone. **2** side shaft

45NiCrMo16 (1.2767).

Spring steel with 0.2% V; 0.5% C; 1% Cr; through hardened to 1500 N/mm^2 and locally nitrided in the spline area [41].

Titan Ti 10 2 3.

5.5.4 Shaft Joints (*Universal Joints*)

Homokinetic joints with uniform angular velocities between input and output are used in the wheel drive. The length compensation of the shaft, which is required when the wheel is sprung, is performed by a sliding joint, which is arranged on the gearbox side for weight reasons (sprung masses). The lighter fixed joint, which is one of the unsprung masses, is located on the wheel side. Even if the unbent position is the ideal one for low losses, the shafts should still run with a small deviation from the stretched position during operation. This will ensure that the transmission elements (balls, rollers) move in their tracks and do not strike each other.

The load on the joints is composed of the moment to be transmitted, the articulation angle and the speed. All three influencing variables increase the load in the event of an increase. If several loads occur simultaneously, their product provides a comparative value for the total load. Thus, the product of shaft speed and articulation angle should not exceed a maximum value depending on the design. The higher the shaft speed, the smaller the articulation angle must be. In general, the joints are grease-lubricated and convection is sufficient for cooling. The operating temperature should not exceed about 80 °C, short-term peaks should not exceed 120 °C. However, if the joints are not directly exposed to an air flow, but on the contrary are shielded by the underbody and bodywork, oil lubrication with cooling may be necessary. For this purpose, the joint mounting must be integrated into the transmission housing, Fig. 5.84. Such solutions can be found in Formula 1 cars as well as in Dakar vehicles, where their large suspension travels result in extreme articulation angles and thus high heat generation.

Different types of constant velocity ball and tripod joints are manufactured in series.

The transmission efficiency of joints is not constant but depends on the articulation angle and the design, Fig. 5.100. The efficiency of tripod joints increases with increasing torque and decreases with increasing speed. Ball joints, on the other hand, show decreasing efficiency with both increasing torque and increasing speed [42]. In addition, the losses can be reduced by friction-reducing high-temperature greases (up to 150 °C). These greases are definitely required if the joints are heated by the exhaust system. But even without a heat

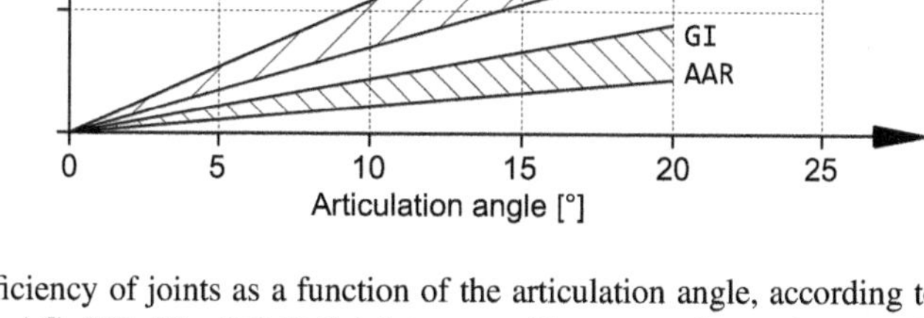

Fig. 5.100 Efficiency of joints as a function of the articulation angle, according to [36]. GI, AAR Tripod species. AC, UF, VL, DO Ball joint types. The torque losses increase linearly with the articulation angle. Tripods are more favourable than ball joints in terms of transmission efficiency

source in the vicinity, the temperatures of the joints rise to about 80 °C during operation due to the frictional work.

Regardless of whether grease or oil lubricated, the joints need a seal to the outside. This is done with an elastic bellows. When arranging the bellows, it should be noted that its diameter increases at high shaft speeds and contact with the pressure rod or the hot exhaust pipe can occur if the distance to these components was selected too small. If space is a problem, stiffening rings around the circumference of the bellows can help.

The targeted service life in racing is between 5000 and 15,000 km for sports cars and GT3 vehicles.

Constant *Velocity* Joints (*CV Joint*)

Ball constant velocity joints consist of a hub and a bell-shaped outer part. The hub and outer part are connected to each other via balls – usually six. A cage holds the balls in the intended position. For the classic case of a locating/non-locating bearing arrangement of a cardan shaft, constant velocity joints exist as locating bearings, which do not permit axial displacement, and sliding joints, which permit a change in the distance between the bearings when the wheel is sprung.

Fixed *Joints*:

AC:	Fixed ball joint on the wheel side (*angular contact*). The cross-section of the path for the joint balls is elliptical. The balls therefore make point contact with the track. Possible articulation angles up to 47°
RF:	Fixed ball joint on the wheel side with a toroidal path for the joint balls. The balls have line contact in their path. Possible articulation angles up to 47°
UF:	Fixed ball joint without *undercut* (*undercut free*). The ball track is extended compared to the AC version. This and other measures allow articulation angles of up to 50° and more

With articulation angles of almost 50° for the fixed joint, there are no disadvantages for front-wheel drive concepts in terms of steering.

Plunging Joint (Displacement Joints):

DO:	Double offset sliding ball joint. The ball tracks run parallel to the axis of the outer part. This design allows large displacement ranges
VL:	Sliding joint Löbro. The ball tracks run diagonally (crossing) to the axis of the outer part. The joints can be made short and light. Max. Max. Articulation angle: 22°, max. Displacement between 16 and 30 mm. VL joints have an axial stop and can therefore be used simultaneously at both ends of a rear axle shaft

A constant velocity universal joint allows an angle between the shafts and at the same time allows axial displacement. Maximum articulation angles of sliding joints are 15°, preferably 25°. In extreme cases, 30° are also realized. The largest displacement travels are between 45 and 55 mm.

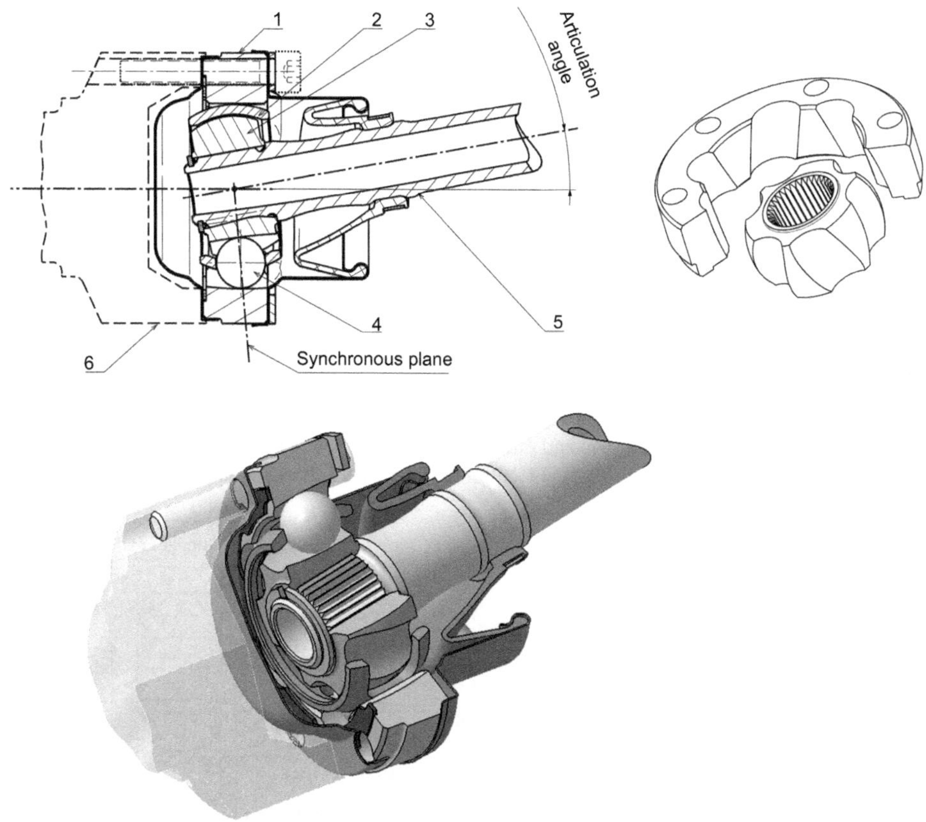

Fig. 5.101 Constant velocity joint in disc design (type VL), also: high speed joint. The ball raceways of hub and flange are crosswise inclined, thus the axial displacement is limited and this design can be used at both ends of a shaft at the same time. The max. Articulation angle is 12.5°, the max. Speed 8000 min^{-1}. **1** flange with outer race. **2** Cage for balls. **3** Hub with inner race. **4** Ball. **5** Drive shaft. **6** Flange for hub or differential output shaft

Figure 5.101 shows the structure of a constant velocity joint. Balls (4) transmit the rotary motion from the hub (3) to the flange (1). The balls (4) are held by a cage (2) and move on the so-called synchronisation plane. This is normal to the bisector of the input and output axes. This means that there are no fluctuations in the speed on the output side. The joint flange (1) is bolted to the wheel hub or differential output.

Figure 5.102 shows a constant velocity joint on the rear axle of a formula car in the installed state. Figure 5.69 shows the use of constant velocity joints on an axle drive, which are integrated into the differential bevel gears.

A complete sideshaft for front wheel drive is shown in Fig. 5.103. It has a fixed and a displacement constant velocity joint. The fixed joint is installed on the wheel side, the displacement joint is located on the gearbox side. The maximum articulation angles are 22°

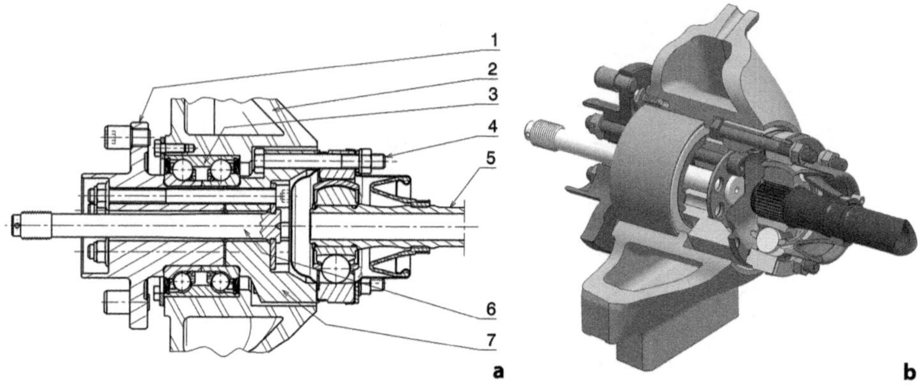

Fig. 5.102 Connection of a constant velocity joint (Renault 2000 formula). (**a**) section, (**b**) axonometric representation. *1* Wheel flange and outer hub part, 2 Wheel carrier (upright), 3 Wheel bearing, *4* Hub screw connection, *5* Drive shaft, *6* Central bolt for wheel screw connection, *7* Inner hub part

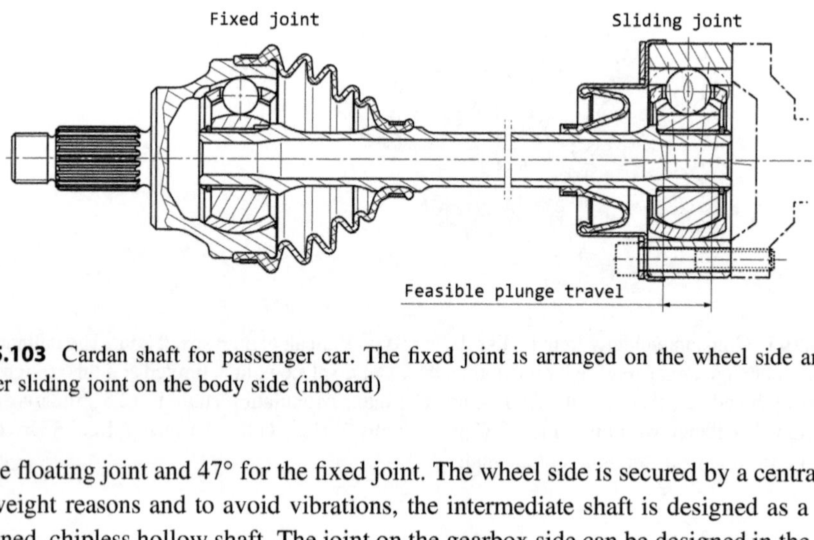

Fig. 5.103 Cardan shaft for passenger car. The fixed joint is arranged on the wheel side and the heavier sliding joint on the body side (inboard)

for the floating joint and 47° for the fixed joint. The wheel side is secured by a central nut. For weight reasons and to avoid vibrations, the intermediate shaft is designed as a case-hardened, chipless hollow shaft. The joint on the gearbox side can be designed in the same way as the fixed joint, i.e. with a closed cup and forged-on pin. This design saves weight. The joint is then inserted directly into the differential and axially fixed by means of a shaft retaining ring. For more details see section Connection Types of drive shafts.

Materials

Hub and joint flange: Induction hardenable steel C35G (was Cf53 according to DIN 17212) with hardening depth Rht 550 HV = 1.1 + 1.0.

C45, C55, 20MnCr5, 42CrMo4 (designations according to DIN EN 10027-1) with surface hardnesses around 58 HRC or 700 HV10. Strength values of the base materials are in the range of 800–1000 N/mm^2.

Fig. 5.104 Structure of a tripod, type GI. The side shaft is connected to the tripod star via splines. This carries three rollers with needle bearings and a ball profile, which run in corresponding cylindrical tracks of the flange. The flange is bolted to the wheel hub or differential output

Cage: Case hardening steel 20NiCrMo2 (DIN 17210) with case hardening depth Eht = 0.6 + 0.4.

Ball: Rolling bearing balls class II/III (DIN5401) made of rolling bearing steel 100 Cr6 (DIN 17350), hardness HRC 63 ± 3.

Tripod Joints

Tripod joints consist of a hub, on which there are generally three journals (name!) with spherical rollers, and a tubular counterpart, which has three cylindrical tracks for the rollers.

Displacement Joints:

AAR:	**Angular adjusted rollers:** The journals on which the rollers run are themselves spherical and thus allow a certain angle between the journal and the roller. The rollers thus always run with the roller axis normal to the track axis. As a result, the displacement force is lower than that of the GI type. Max. Max. articulation angle: 26°, max. Displacement: up to 55 mm
GI:	**Glaenzer interior tripod**. "Classic" design in which the rollers are mounted on a hub with three journals and move with the shaft when flexed. Max. Max. Bending angle 23°, displacement up to 55 mm

Fixed joint:

GE:	**Glaenzer exterior tripod**. The tripod journals are mounted on the outside, i.e. in the tulip and not on the shaft. Allows larger articulation angles than type GI and can thus be installed with steered front wheels. Max. Max. Articulation angle: 45°

The basic structure of a tripod joint is illustrated in Fig. 5.104.

The articulated shaft is centered by two of the three tripod rollers. The centre of the tripod hub does not coincide with the imaginary centre of the shaft. The shaft thus performs

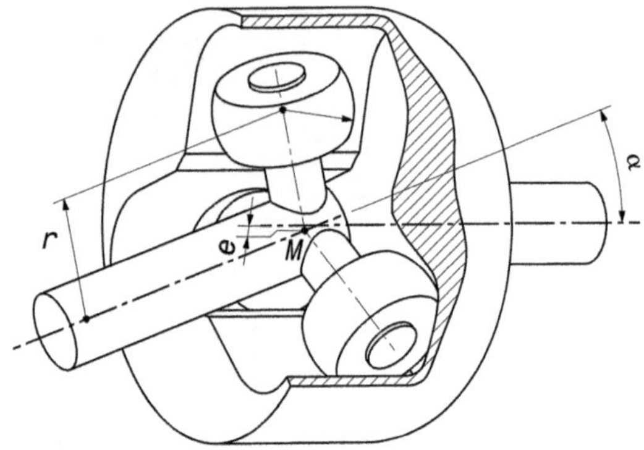

Fig. 5.105 Sketch of tripod movement. Designations see text

an eccentric movement, Fig. 5.105. The centre *M* of the tripod star rotates in the same direction with three times the shaft speed and the eccentricity *e:*

$$e = \frac{r}{2}\left(\frac{1}{\cos(\alpha)} - 1\right)$$

e	Eccentricity of the centre point, mm
r	Pitch radius of the tripods, mm
α	Articulation angle of shaft ends, °

The two tripod stars of a shaft should therefore be offset from each other by half an angular division (i.e. e.g. 60°, Fig. 5.106) for reasons of oscillation.

If a sideshaft is installed with two sliding joint tripods, it requires a device which restricts its axial position because there is no fixed joint. Of the known possibilities, Fig. 5.107 shows a simple one. The locking element can be arranged on the gearbox or wheel side.

Alternatively, in rare designs, springs hold the floating shaft in axial position. The springs are located on the end faces of the shaft and press it into the desired position, see Fig. 5.108. In the design, inertial forces of the shaft and joint reaction forces must be taken into account so that the shaft also maintains the desired position during travel.

The tripod tulip can be connected to the differential output or the wheel flange via a flange (Fig. 5.104) or it can be integrated directly into the differential output, Figs. 5.84 and 5.109. This design saves weight and installation space and allows oil lubrication of the tripod joints.

The tripod tulips in Fig. 5.84 directly support the teeth of the differential bevel gears on their front sides. This makes the differential as narrow as possible. The locking effect of the differential gear can be influenced via hydraulic multi-disc clutches.

Oil lubrication instead of permanent grease filling has the advantage that oil can be fed specifically in the circuit to the lubrication points. This reduces losses due to the lower

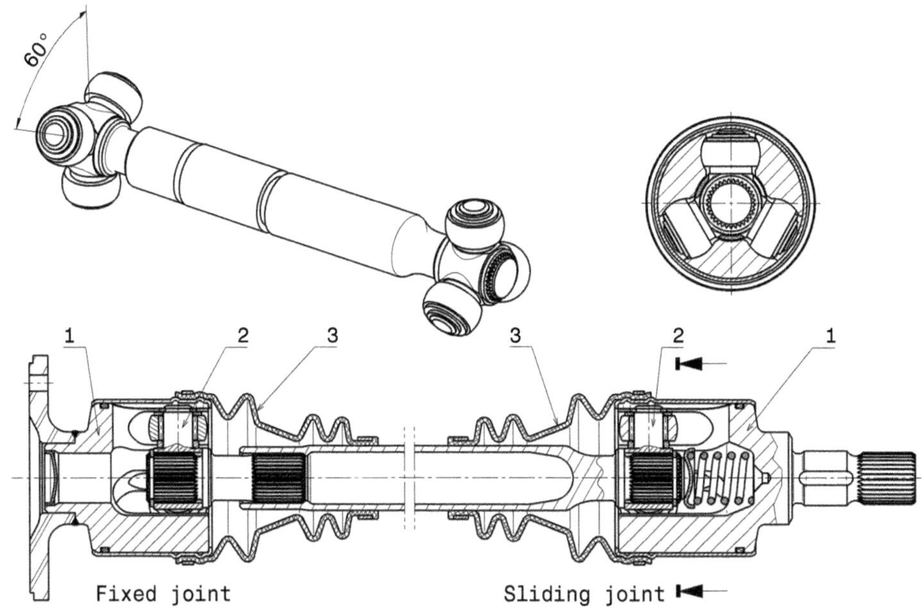

Fig. 5.106 **Cardan** shaft with tripod joints. **1** Tulip, **2** Tripod, **3** Boot (bellows)

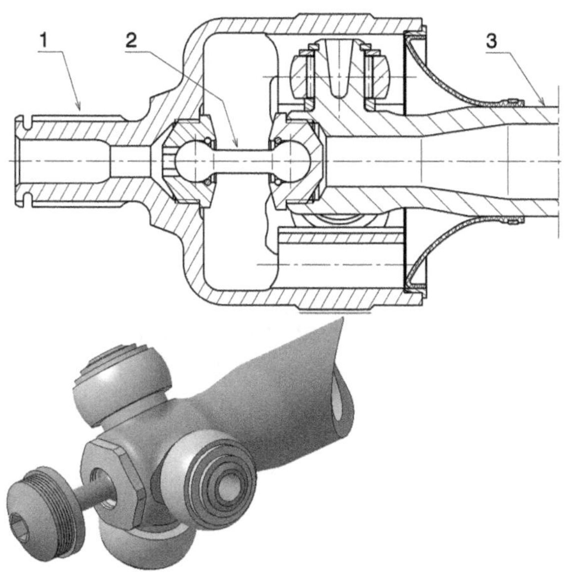

Fig. 5.107 Axial securing of a tripod (Pankl system). **1** Tripod tulip. **2** Retaining element. **3** Side shaft (*half shaft*). The element consists of a dumbbell-shaped connecting piece held in sockets on both sides. The element is first screwed into the shaft and then screwed into it from the outside through the hollow axle journal of the tulip using a hexagonal pin spanner

viscosity of the lubricating oil and at the same time enables heat to be dissipated via the oil. The latter is particularly important for transmissions that are shielded from the cooling air flow by the diffuser and underbody.

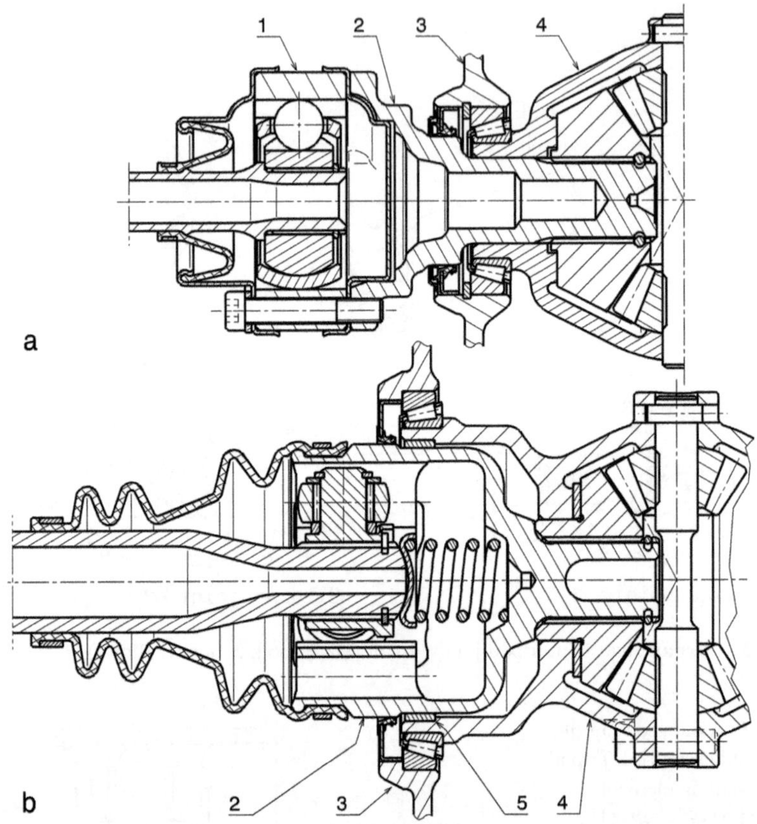

Fig. 5.108 Shaft connection to the gearbox. (**a**) Flange connection (*flange mounting*). (**b**) Integral connection. **1** Flange. **2** stub axle resp. tripod housing. **3** Final drive housing. **4** Differential housing. **5** Bronze slide ring

Materials

Hub with trunnion (tripod): Case-hardened steel, journal hardened and ground.

Flange (multi-piece version) and tulip (one-piece): Carbon steel, induction hardened: C45, C60. Tempering steel: 42CrMo4, 42CrMoS4 (DIN EN 10083).

41SiNiCrMoV7-6 (DIN) heat treated. The sheets are broached or milled.

Rollers: Case hardened steel.

30CrNiMoMn24 (EN).

Choice of Joints

The choice of the different joints depends on the place of use (steered front wheels, non-steered rear wheels), the required displacement (fixed joint, sliding joint, off-road wheel suspension) and other criteria (vibrations, . . .). Table 5.15 shows typical places of application of joints for orientation.

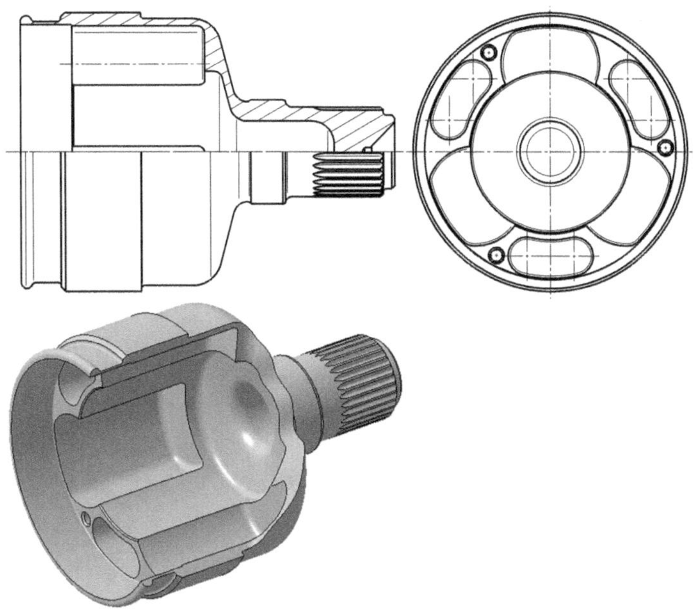

Fig. 5.109 Tripod tulip in integral design. The component is inserted directly into the axle bevel gear of the differential and accommodates the tripod star. Sealing to the outside is provided by a shaft seal and a bellows

Table 5.15 Joint selection for sideshafts, according to [36]

		Installation location	
Rear-wheel drive:		Wheel side (fixed joint)	(Axle) gearbox side (sliding joint)
Joint type		VL monobloc	VL disc joint
Front wheel drive:			
Joint type	Ball joint	AC, UF	VL disc joint, VL monobloc, DO
	Tripod joint	GE	GI, AAR, Triplan

Shaft-Hub Connection (*Inner Race Shaft Connection*)

The hubs of constant velocity joints and of assembled tripod stars are generally connected to the side shafts via splines. In the case of ball constant velocity joints, separation of the shaft and hub is necessary in any case for assembly reasons. During assembly, the hub must be bent over so that all balls can be inserted. The splines have splined profiles, serrated profiles and involute flank profiles, Fig. 5.110. Serrated shaft profiles (b and c) can transmit larger and more shocking moments than splined shafts (a) due to the higher "number of teeth" for the same shaft diameter. Compared to the splined profile, the shaft and hub are also weakened less, which means that the connection can be made shorter and the mean diameter smaller.

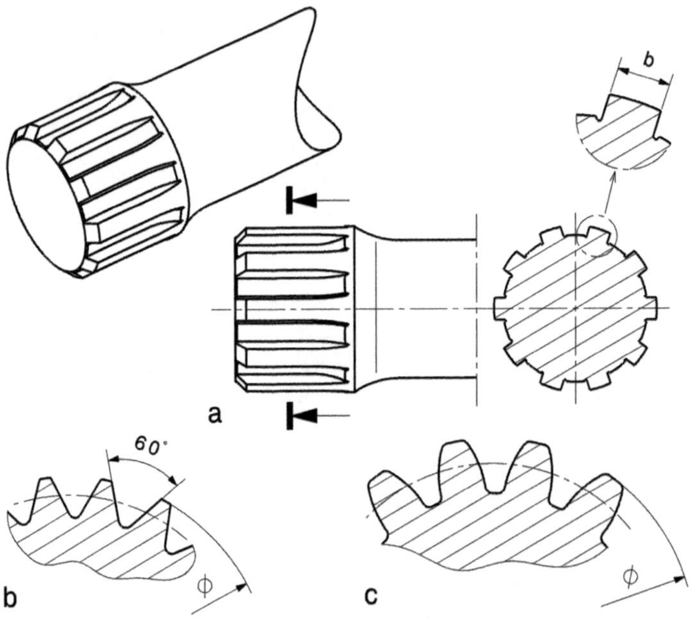

Fig. 5.110 Plug-in splines. (**a**) Splined shaft profile (DIN ISO 14, DIN 5464) . (**b**) Serrated *shaft* profile (DIN 5481). (**c**) involute tooth profile (DIN 5480 and 5482)

Axial securing of the hub on the shaft is ensured in various ways, Fig. 5.111. The greatest axial forces are transmitted by a rectangular ring made of spring steel (a), which also represents the safest connection. For assembly and disassembly, it is spread apart with a tool and pushed over the shaft. Snap rings (b to d) make assembly easier. For mounting and dismounting, access from one side is sufficient, but the transmittable axial force is considerably lower than that of the rectangular ring. The snap ring (b) springs into a groove in the hub during assembly. To facilitate dismounting, one flank of this groove has a chamfer. The solution (d) is basically the same, except that the hub is preloaded by a disk spring, thus compensating for the axial play. The snap ring (c) clamps against the bottom of the shaft groove. An inclined flank of this keyway facilitates the removal of the hub.

Connection Types of Drive Shafts

The sideshafts must be connected to the driven wheels and the axle drive. A number of principles have become established for this purpose. Which type is used depends on the design of the shaft and the adjacent components. In the case of purchased parts, the assembly is largely predetermined; in the case of in-house designs, there is every freedom for complete integration of the required parts. There are three main solutions for connecting the shaft to the wheel, Fig. 5.112. There are also variations of the basic variants shown.

Long trunnion (a) and short trunnion (b) connections are used on passenger cars. The wheel bearing (3) is preloaded by the trunnion bolt (1). In this respect, the short trunnion

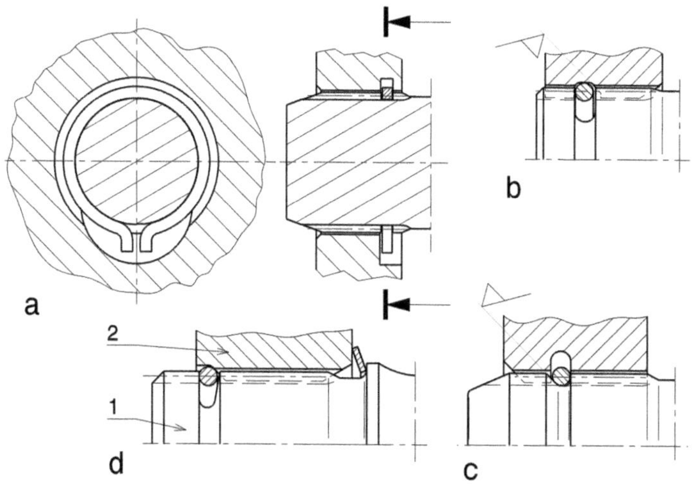

Fig. 5.111 Axial securing of plug connections. **1** Shaft. **2** hub. (**a**) Rectangular ring. (**b**) Snap ring in hub. (**c**) Snap ring in shaft. (**d**) Snap ring in hub with preload by disc spring (cup spring)

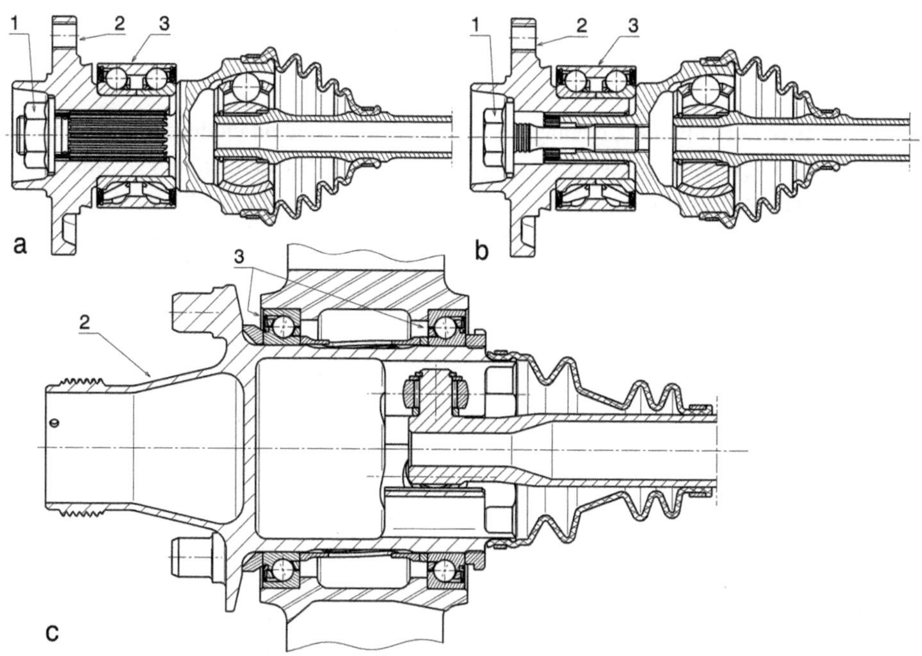

Fig. 5.112 Shaft connection to the wheel. (**a**) Long trunnion connection. (**b**) Short trunnion connection. (**c**) Integral connection. 1 Nut or bolt. 2 Wheel flange or hub. 3 Wheel bearing

solution is the better one because the expansion bolt (1) reacts much less sensitively to settlements than the short thread of the long trunnion solution, i.e. the preload force of the bearing screw connection remains at a higher level during operation. The short trunnion solution has the further advantage that the shaft does not have to be inserted as far into the wheel flange (2) during mounting. For cost reasons, the long journal solution is the standard variant for passenger cars. This also makes it easier to remove the shaft, because it may not be necessary to remove the entire wheel carrier. With the integral connection (c), no bolting is required because the raceway of the balls or tripods is machined directly into the wheel hub. This represents the most weight-efficient form. The shaft shown also saves mass by integrating the tripod spider into the shaft. The otherwise necessary shaft-hub connection, including the axial retaining rings, is thus eliminated. However, this also means that all parts are structurally matched to each other. The integral connection therefore forms a tailor-made solution and is therefore primarily found on racing vehicles.

The connection of the sideshafts to the gearbox is done in a similar way as to the wheel, except that sliding joints are used at this point, Fig. 5.108. The bolted flange solution (a) is generally applicable, but heavier because of the larger number of interfaces. The connection integrated into the flange allows the longest sideshaft for a given track width, but represents the greatest design effort.

5.6 All-Wheel Drive

All-wheel drive not only has advantages when starting off and on inclines, but also has an effect when cornering. Nevertheless, hardly any road vehicles were equipped with it. The additional expense did not seem worthwhile. Sporting success in the 1980s changed this fundamentally and many car manufacturers now offer all-wheel drive versions of their vehicles. In motorsport itself, the use of all-wheel drive systems is – as always – a matter of regulations.

5.6.1 Basics

Starting

Starting or accelerating on an incline with the same friction conditions at all four wheels can be analysed with a relatively simple vehicle model – see Fig. 5.113. Any lifting and pitching movements are neglected, as is the rotational acceleration of the wheels and the drive train, as well as the rolling resistance of the wheels. An expression $a*$ can then be found in which all the quantities governing traction – acceleration a_X, gradient α and drag F_L – can be combined. The factor ν determines the distribution of the drive torques between the two axles. For front-wheel drive, $\nu = 1$, for rear-wheel drive $\nu = 0$. The vehicle data show that the calculations are based on a slightly front-axle-heavy vehicle.

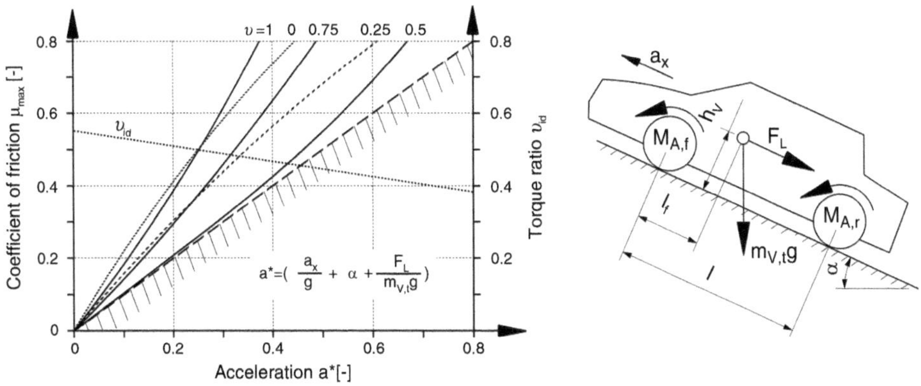

Fig. 5.113 Accelerated uphill drive (simulation), after [4]. Vehicle data: $m_{V,t} = 1360$ kg, $l = 2.6$ m, $l_f/l = 0.45$, $h_V/l = 0.2$, ν moment distribution: $M_{A,f} = \nu M_M$ and $M_{A,r} = (1 - \nu)M_M$

In the diagram of Fig. 5.113 it can now be read which maximum adhesion value μ_{max} must at least be required in order to achieve a certain value $a*$ *or* vice versa. For example, on a smooth road surface with $\mu_{max} = 0.2$, the single-axle drives (i.e. $\nu = 1$ or 0) can achieve accelerations of about 0.1 g, while with a torque split of 50:50 ($\nu = 0.5$) almost 0.2 g are possible. The dashed boundary line corresponds to a demand-dependent split of the drive torques with a rigid driveline via the slip at the front and rear axle. An equal splitting could only be achieved with a splitting factor ν_{id} that depends on $a*$. This optimal utilization of traction at all four wheels almost doubles the possible values $a*$ *of* the single-axle drives, especially in the range of road condition-related small maximum adhesion values, i.e., just for a driving condition where the maximum possible values of $a*$ can also be achieved on the engine side. On dry roads, a real gain in traction can only be exploited with a powerful engine.

Cornering

An advantage of all-wheel drive vehicles also occurs during accelerated cornering. The lateral force potential of all tyres is considerably greater for the same propulsive power because the propulsive power is distributed over four wheels instead of just two. The resulting force from circumferential and lateral force is limited by the friction conditions (see Sect. 4.1) and therefore a reduction in circumferential force allows an increase in lateral force. The behaviour during cornering determines driving safety to a greater extent than accelerated straight-ahead driving. Figure 5.114 represents the drivable lateral acceleration as a function of the longitudinal acceleration. The vehicle on which the calculations are based corresponds to that in Fig. 5.113. Aerodynamic downforce is not taken into account. The boundary curve describes a "rubberized" point mass. It reflects the known relationship between circumferential and lateral forces on a tyre. The chosen data of the base vehicle corresponds to a rear wheel driven one, therefore the potential with pure rear wheel drive ($\nu = 0$) is slightly higher than that of front wheel drive ($\nu = 1$). The curves

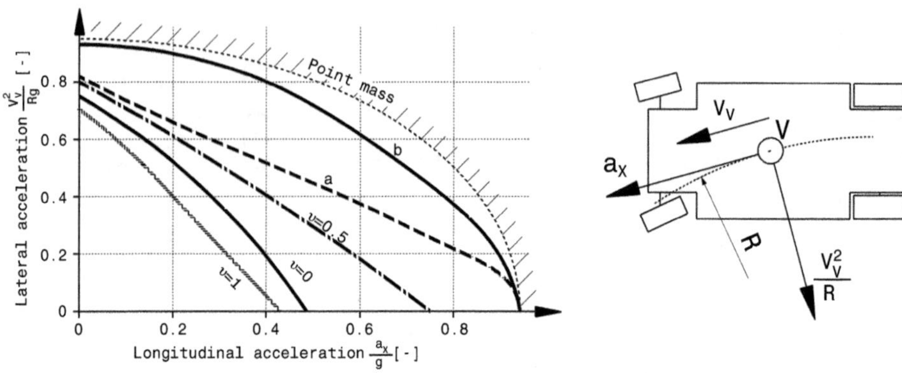

Fig. 5.114 Simulation of an accelerated circular motion (cornering), after [43]. $R = 100$ m, $\mu_{max} = 0.95$. Vehicle data and ν see Fig. 5.113. (**a**) Variable torque distribution between the two axles. (**b**) Variable torque distribution to all four wheels

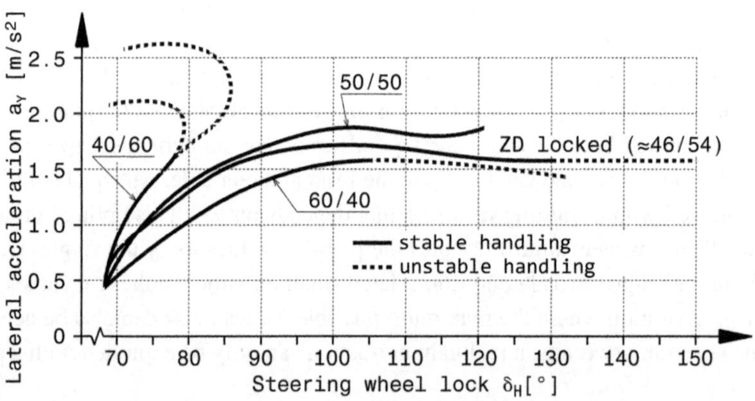

Fig. 5.115 Accelerated circular motion on low friction coefficient, according to [44]. Relationship between steering angle and lateral acceleration with different distributions of drive torque. 40/60 = 40% front axle/60% rear axle. 60/40 = 60% front axle/40% rear axle. ZD = Central differential. With the ZD locked, the power is distributed according to the dynamic axle loads

would be closer together with adjusted vehicle design. If the drive torque is distributed equally between the two axles by a rigid all-wheel drive ($\nu = 0.5$), much higher lateral accelerations can be achieved. Even better driving performance can be achieved if the distribution of the drive torque is not rigid but slip-optimal on the front and rear axles. The greatest possible approximation to the ideal of the point mass is possible with wheel-individual allocation of the drive torque. Due to the side-slip resistance of the tires, however, the limit curve is not reached.

The statements are supported by characteristic tests. Figure 5.115 shows the results for passenger cars when accelerating from a corner with low friction values. A vehicle related to rear-wheel drive with 40/60% drive torque distribution front to rear axle reacts in the known behaviour that initially high lateral accelerations are built up at relatively small steering angles. However, when the lateral acceleration is increased further, the rear end breaks away abruptly, whereby the yawing motion of the vehicle can only be intercepted with difficulty by countersteering (unstable handling). The vehicle with symmetrical power split (50/50%) can still be controlled at high lateral accelerations. For the normal driver, however, the vehicle with the 60/40% split is the easiest to control. It shows the stable, understeering behaviour. The steering wheel angle increases disproportionately with the lateral acceleration in the limit range and thus announces to the driver that the physical limit has been reached.

In summary, it can be stated that all-wheel drive systems have theoretical advantages over single-axle drives in terms of both traction and cornering. The prerequisites for their practical implementation are a high engine output and a dynamically controlled distribution of the engine output to the two axles or, even better, to each wheel individually.

5.6.2 *Racing*

One advantage of all-wheel drive is the possibility of achieving tyre protection with an ideal balance of the vehicle and thus being able to use softer tyres. However, the tuning of all-wheel drive vehicles is usually more complex than for vehicles with standard drive. For the ITC (*International Touring Car Championship*), a special racing ABS was developed that took into account the larger rotating masses and reduced the previously excessive tire wear. Ultimately, the same or even better deceleration values could be achieved than with standard-drive vehicles.

In some motorsport classes, a comparison is difficult because the regulations provide for weight disadvantages or similar for all-wheel drive vehicles. At present, the prevailing opinion is that all-wheel drive systems are too heavy and have poor efficiency in monoposti. They are also difficult to accommodate [33]. Next to the driver, the drive shaft has to pass and in front of the driver's feet, steering, final drive and suspension parts have to be accommodated. In Formula 1 there were experiments with four driven wheels. That was in the late 1960s, when tires didn't have the grip to deliver full engine torque to the road. Even so, the systems of the time didn't produce the desired results. When a Formula One car fully accelerates, the axle load transfer reduces the front wheel contact forces so much that the drive torque on the front tires causes understeer. Also, the usual reaction of drivers in such a situation – namely to provoke power oversteer with more throttle – causes the opposite, i.e. even more understeer [14]. Consequent improvements were therefore aimed at reducing the drive torque at the front, but this also virtually cancelled out the main advantage of this system. Today, tyres have evolved enormously, ground effect and wings increase the wheel contact force, and Formula 1 regulations now only permit two driven wheels.

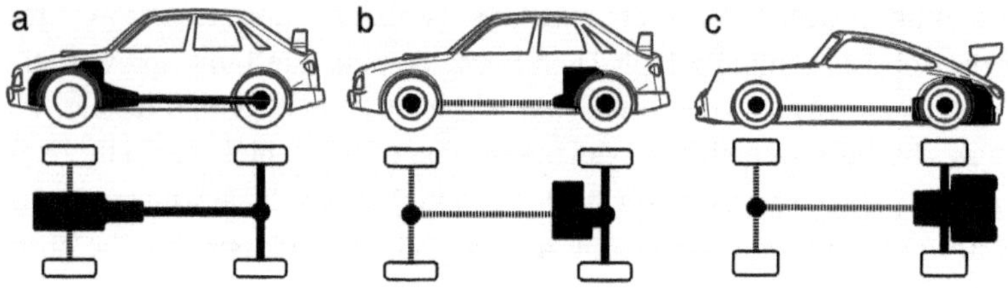

Fig. 5.116 All-wheel drive systems with different engine arrangements. (**a**) Front engine. (**b**) Mid-mounted engine. (**c**) Rear engine

In rallying, the gain in traction on loose tracks outweighs the disadvantages in terms of weight and driving stability. In touring cars, all-wheel-drive vehicles have shown superiority over powerful front-wheel-drive vehicles. Moreover, measurements on touring cars show that with optimally designed 4 × 4 systems, the sum of power losses to the road need not be greater than with rear-wheel-drive vehicles [17].

The all-wheel-drive touring cars definitely also show advantages over rear-wheel-drive ones. The lap time advantage can be up to 2 s. It is greatest at low friction coefficients and on race tracks with tight, undulating corners [17].

5.6.3 Types of Construction

All-wheel drive systems can be implemented with all common engine mounting locations, Fig. 5.116. Starting with an existing drive type (shown with solid lines), all-wheel drive is built by extending the drive to the non-driven axles. These extensions of the drive are shown with dashed lines.

A systematic classification of all-wheel drive systems can be made according to the type of power split. Four groups or generations of systems can be identified, Fig. 5.117.

The first group includes selectable four-wheel drives and permanent drive with central differential. The second generation mostly builds on these systems and expands them with a locking system of predetermined characteristics. From the third group onwards, the use of electronic control systems is required to regulate the power split. In these systems, the internal characteristic of the basic systems is superimposed with an external control. In the fourth group, power can be distributed in a controlled manner between the axles and/or between the wheels of an axle (*torque vectoring*).

The Audi Quattros of the first generation are famous representatives of permanent all-wheel drive with predetermined power distribution (first group in the systematics Fig. 5.117). However, if only one wheel loses contact with the ground, only that wheel will turn. The central differential of the following generation was a Torsen differential. A schematic of the first generation driveline is shown in Fig. 5.118. The engine torque, converted by the gearbox, acts on the centre differential (2). The output shaft of the gearbox

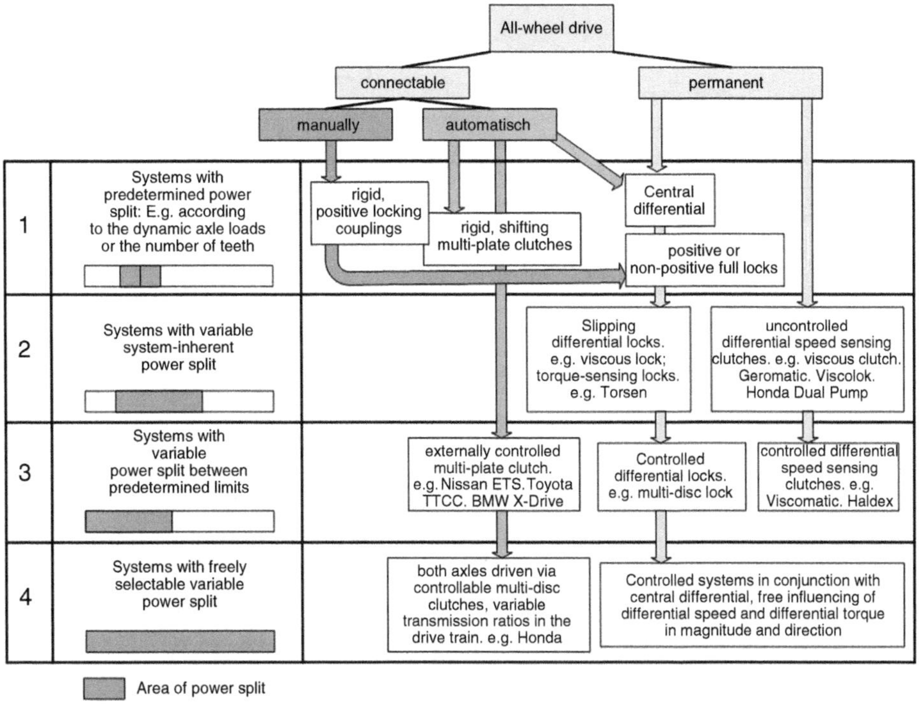

Fig. 5.117 Systematics of all-wheel drive systems [8]

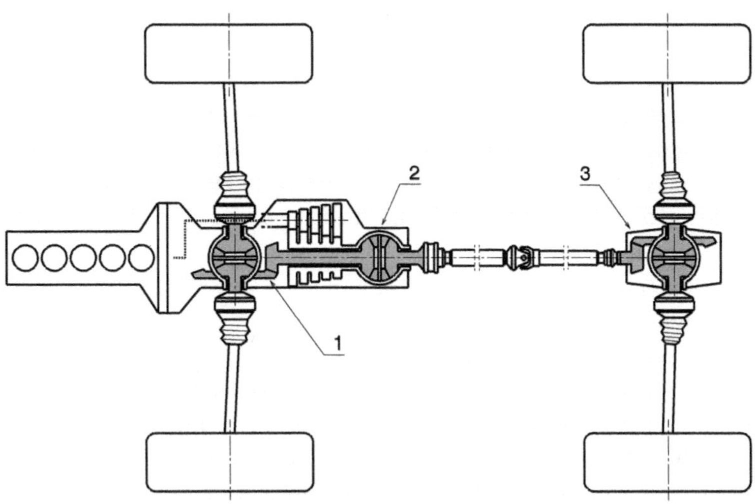

Fig. 5.118 Drive diagram Audi Quattro first generation. Permanent all-wheel drive with predetermined 50/50% power split. **1** Front axle drive. **2** Centre differential. **3** Rear axle drive

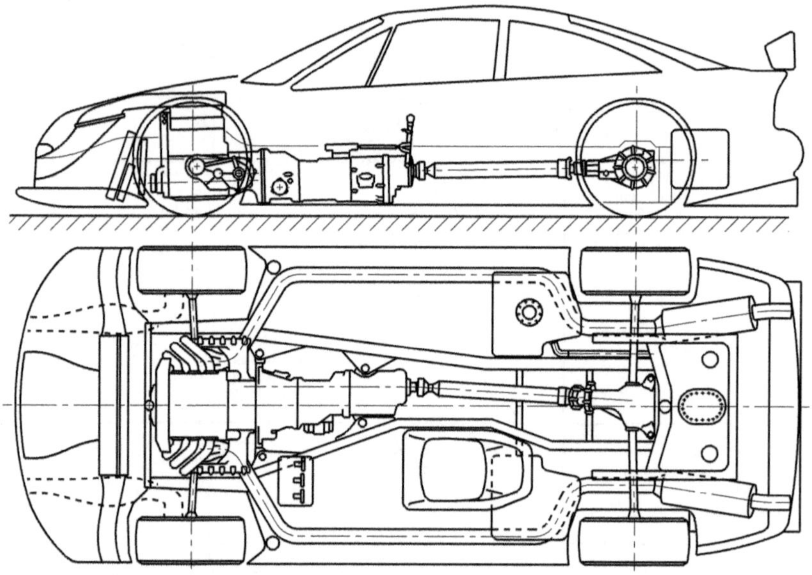

Fig. 5.119 Opel Calibra ITC season 1996 (*International Touring Car Championship*). Touring car with all-wheel drive. The front wheels are driven by two flat two-stage spur gears housed next to the engine. The axle load distribution was 50:50

is designed as a hollow shaft. The drive shaft runs through this to the front axle drive (1). A cardan shaft leads to the rear axle drive (3).

An example of a four-wheel drive touring car is shown in Fig. 5.119. The vehicle has a front engine to which the manual gearbox is flanged as in a standard drive. The transfer case with centre differential is flanged to the 6-speed manual gearbox. Three torque splits are selectable. The rear wheels are driven by a relatively short cardan shaft made of CFRP. The shaft is offset from the longitudinal center plane to bring the driver closer to it. The axle gearbox sits centrally in the vehicle so that the two sideshafts are of equal length. A shaft runs from the transfer case inside the transmission case to the front differential. The front wheels are driven by sideshafts from two flat three-stage spur gears mounted to the side of the engine.

All three differentials are lockable. The best solution for this vehicle turned out to be a hydraulic/electronic lock for the middle and rear differential, while a visco-lock with freewheel was preferred on the front axle [17].

There is another special feature on rally cars, which is due to the driving style of these competitions. In the area of the centre differential there is a *disconnecting device* (*centre clutch*) with which the driver can disconnect the rear axle from the drive. He uses this option when he wants to break out the rear of the vehicle with the handbrake acting on the rear wheels, i.e. when lining up the car for tight bends and skidding turns.

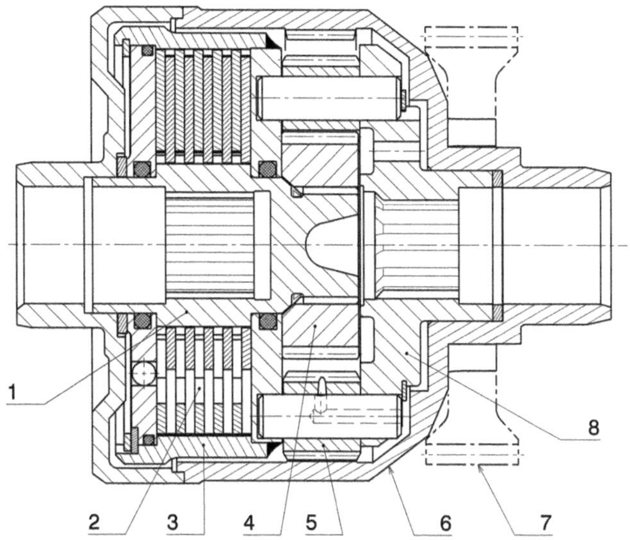

Fig. 5.120 Viscous centre differential with planetary gear, sectional view. 1 *hub* (output to front axle), 2 plates *viscous coupling unit*, 3 clutch housing (*clutch* basket), 4 sun gear, 5 planet gear, 6 differential housing (*differential case*), 7 drive gear, 8 planet carrier (*output to rear* axle). The drive is provided by the housing (6), in which the ring gear is integrated. The housing thus drives the outer planet gears (5), which distribute the torque to the sun gear (4) or arm (8)

A central element in an all-wheel drive powertrain is literally the *centre differential*. It is responsible for distributing the drive torque between the front and rear axles and thus has a significant influence on handling. Because all wheels are coupled together in one way, the centre differential has an effect on handling not only when driving but also when braking. Especially in rally vehicles, an electronically controllable centre differential is an essential element for increasing competitiveness. One possibility is a switchable clutch which, controlled by the on-board computer, bridges the differential depending on the driving condition and the driver's preference (asphalt, gravel, snow) by connecting the drive of the front and rear axles in a certain way. In WRC cars, such electronically controlled systems are now banned. The successor solution is that an automatic clutch has taken the place of the controlled one. Figures 5.120 or 5.121 shows an example of a planetary gear centre differential with a viscous coupling. This fluid coupling connects the hub (1) to the web of the planetary gear (8), thus connecting the drive to the front axle to that of the rear axle. The coupling consists of a sealed housing (3) which is filled to approx. 90% with silicone oil of high viscosity (toughness). In contrast to known oils, the viscosity of this special fluid actually increases above a certain temperature. Unlike multi-plate clutches, no external contact pressure is applied to the plate pack (2) here. The torque between the inner and outer plates is transmitted solely via the shear gradient in the working fluid that occurs during relative movement. The transmittable torque depends on the relative speed of

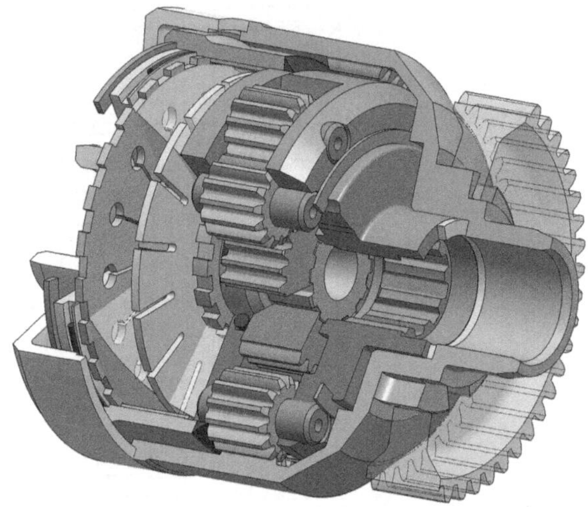

Fig. 5.121 Viscous centre differential, partially cut open. The diagram shows the differential of Fig. 5.120. For better visibility, only one inner and one outer plate are shown for the viscous coupling

opposing plates, the viscosity of the oil and the number of effective surfaces. The resulting torque has a braking effect on the faster rotating axle (slipping, cornering) and thus increases the torque for the other axle with the greater grip. If such a condition is maintained for a longer time, the clutch heats up due to the friction work and the pressure in the oil-filled space increases. This can go so far that the fluid is displaced from the plate-disc area, the plates now have mechanical contact after all and cause a 100% lock-up. This effect remains for a certain time even if the differential speed is reduced. In the differential speed torque curve, this phenomenon is represented as a *hump*, which is why it is also called the hump effect. However, the engagement of the viscous clutch is always soft and the driver is not surprised by an abrupt locking effect.

5.7 Examples

If the covers and bodywork parts are removed, the drive train is revealed on production cars and formula cars. In the case of touring cars and near-series cars, this assembly can only be fully seen as a unit – if at all – when the car is dismantled, Figs. 5.122, 5.123, 5.124, and 5.125.

Fig. 5.122 Drive train of a production sports car (Osella Honda). A massive clutch housing is flanged to the in-line engine and carries the spring damper unit. The gearbox sits at the end of the car and is constructed in such a way that the wheel sets lie behind the axle gearbox in the direction of travel. This makes it possible to change the transmission ratio on the stationary vehicle without major disassembly work. The gearbox housing carries the towing lugs and the rear wing at its end (not on the picture). Shifting is done manually via the linkage, which is guided through the exhaust manifold and wishbone

Fig. 5.123 Drive train of a production sports car. The gearbox carries the majority of the rear chassis, which is mounted on the housing. The axle gear is integrated in the gearbox. The clutch bell is removed, therefore the clutch shaft can be seen well

Fig. 5.124 Drive train of a formula car (Reynard Cosworth), direction of travel to the right. The gearbox is bolted directly to the V-engine. In order to reach the outside cylinder heads, additional brackets (right front in the picture) are used. The gearbox accommodates all chassis parts of the rear axle. The axle gearbox (left) forms the end of the housing. The mouth of the shift linkage of the manual gearshift can be seen in the lower, middle part of the picture

Fig. 5.125 Drive train of the electric racing vehicle TMG EV P001 [Toyota Motorsport GmbH, Cologne]. This vehicle holds the lap record for electric vehicles on the Nordschleife of the Nürburgring with road tires (Aug. 2011: 7 min 47.794 s for 20.8 km). It is based on a Radical SR 8 and is driven by two permanent magnet synchronous motors in axial flux design, cf. Fig. 2.19, which drive the rear wheels in tandem arrangement. The motors are liquid cooled. The associated heat

References

1. Lechner, G., Naunheimer, H.: Fahrzeuggetriebe, 2nd edn. Springer, Berlin (2007)
2. Roth, K.: Konstruieren mit Konstruktionskatalogen. Band 2 Konstruktionskataloge, 3rd edn. Springer, Berlin (2001)
3. Wagner, G.: Berechnung der Verlustleistung von Kfz-Vorgelegegetrieben. VDI-Berichte (1992), Nr. 977, S. 175–198
4. Trepte, S.: Einlaufverhalten technischer Reibbeläge. ATZ. **12**, 1168 (2001)
5. High-End Racing Technology (Sachs) in AutoTechnology 5/2002, S. 36 f
6. Segers, J.: Analysis Techniques for Racecar Data Acquisition, 1st edn. SAE International, Warrendale (2008)
7. VDI 2241 Blatt 1 und 2: Schaltbare fremdbetätigte Reibkupplungen und -bremsen. Begriffe, Bauarten, Kennwerte, Berechnungen. VDI Verein Deutscher Ingenieure e.V., Düsseldorf (1982)
8. Braess, H.-H., Seiffert, U. (eds.): Vieweg Handbuch Kraftfahrzeugtechnik, 4th edn. Vieweg, Wiesbaden (2005)
9. Katalog RCS Racing Clutch System von Sachs Race Engineering GmbH, Schweinfurt (2004)
10. Smith, C.: Prepare to Win, 1st edn. Aero Publishers, Inc, Fallbrook (1975)
11. van Basshuysen, R., Schäfer, F. (eds.): Handbuch Verbrennungsmotor, 7th edn. Springer Vieweg, Wiesbaden (2015)
12. www.springerprofessionals.de/neues-rennsportgetriebe-von-zf/6021470.html. Accessed 17 Nov 2015
13. Seeberger, M. et al.: Das Renngetriebe 8P45R von ZF. MTZ, 62–66 (2016)
14. Wright, P.: Formula 1 Technology, 1st edn. SAE, Warrendale (2001)
15. Kretschmer, J. et al.: Neues BMW Sechsgang-sequenzielles manuelles Schaltgetriebe (SMG). ATZ **7/8**, 648 ff. Vieweg, Wiesbaden (2003)
16. Nascimbene, M.: Composite body and carbon brakes for the ultimate Ferrari. AutoTechnol. **5**. Vieweg. Wiesbaden. (2002)
17. Indra, F.: Grande complication, der Opel Calibra der ITC-Saison 1996. Automobil Revue Nr. 50–1996
18. N.N.: Montage–/Einstellanleitung KLS-Schaltautomat, KLS Motorsport Januar (2002)
19. Smith, C.: Drive to Win, 1st edn. Carroll Smith Consulting, Palos Verdes Estates (1996)
20. Schmitz, H., Krauss, Ch., Leibbrandt, M.: Powershift Getriebe: Schalten ohne Pause. Beitrag zur Tagung Race.Tech München, 23.–24. Nov. (2006)
21. Newey, A.: How to Build a Car. HarperCollins Publishers, London (2017)
22. Dinner, H.: Formel-1: Schnelle Autos, schnelle Simulation. In: Konstruktion Sonderheft S 1/2004, S. 26–29. Springer-VDI, Düseldorf
23. Wemhöner, J., Bergrath, E.: Keramik in Getriebe- und Radlagern. Von der Formel 1 zum Serieneinsatz? ATZ **11**, 1080 ff (2001)
24. Haberhauer, H., Bodenstein, F.: Maschinenelemente. Springer, Berlin (1996)

Fig. 5.125 (continued) exchanger is clearly visible in the rear. It operates at a much lower temperature level than one for internal combustion engines. The lithium ceramic traction battery is located behind the cockpit, right where the fuel tank usually sits. The battery weighs 350 kg and stores 41.5 kWh of energy at a nominal voltage of 520 V. The two engines can thus achieve maximum power. The two motors can thus release a maximum torque of 800 Nm or a maximum output of 280 kW. The car, which weighs 970 kg, reaches 100 km/h from a standstill within 3.9 seconds. The top speed is 260 km/h

25. Haas, W.: Dichtung mit ewigem Leben. In: Konstruktion Heft 7/8-2007, S. 40–43. Springer-VDI, Düsseldorf
26. Incandela, S.: The Anatomy & Development of the Formula One Racing Car from 1975, p. 2. Haynes, Sparkford (1984)
27. Tremayne, D.: Formel 1, Technik unter der Lupe. Motorbuch, Stuttgart (2001)
28. Wagner, G.: Six-Speed Automatic Transmissions and CVT's to Support Fuel Economy. AutoTechnol. 4 (2002)
29. Adamis, P., Petersen, R., Hofmann, L.: Antriebskonzepte mit automatisierten Schaltgetrieben. VW AG Aggregateforschung, Wolfsburg (1999)
30. Schmelz, F., Seherr-Thoss Count, H.-C., Aucktor, E.: Universal Joints and Driveshafts, Analysis, Design, Applications. Springer, Berlin (1988)
31. Guiggiani, M.: The Science of Vehicle Dynamics. Handling, Braking, and Ride of Road and Race Cars, 1st edn. Springer, Dordrecht (s)
32. Ludvigsen, K.: Mercedes Benz Renn- und Sportwagen, 1st edn. Motorbuch, Stuttgart (1999)
33. McBeath, S.: Competition Car Preparation, 1st edn. Haynes, Sparkford (1999)
34. Smith, C.: Tune to Win. Aero Publishers, Fallbrook (1978)
35. Trommler, J.: Entwicklung von Leichtbau-Längswellen. In: Oetting, H. (Hrsg.) Leichtbau im Antriebsstrang, S. 260–281. expert, Renningen-Malmsheim (1996)
36. Pierburg, B., Amborn, P.: Gleichlaufgelenkwellen für Personenkraftfahrzeuge. Verlag Moderne Industrie, Landsberg/Lech (1998) (Die Bibliothek der Technik, Bd. 170)
37. Küçükay, F.: Gewichtsreduzierung im PKW-Antriebsstrang. In: Oetting, H. (Hrsg.) Leichtbau im Antriebsstrang, S. 225–247. expert, Renningen-Malmsheim (1996)
38. Muhs, D., et al.: Roloff/Matek Maschinenelemente, 17th edn. Vieweg, Wiesbaden (2005)
39. Klimach, M., Schürmann, H.: Beitrag zur konstruktiven Gestaltung einer Präzisionslagerung hochsteifer Faserverbund-Wellen. In: Konstruktion, Heft März 2015, S. 75–81
40. Amborn, P.: Entwicklung von Leichtbau-Seitenwellen. In: Oetting, H. (ed.) Leichtbau im Antriebsstrang, pp. 282–289. expert, Renningen-Malmsheim (1996)
41. Ross, B.: Investigating Mechanical Failures. The Metallurgists Approach. Chapman & Hall, London (1995)
42. Crampen, M., et al.: Verlustleistungen von Gelenkwellen und Radnaben einschließlich Bremse. Beitrag zur ATZ/MTZ-Konferenz Reibungsminimierung im Antriebstrang, Esslingen (2009)
43. Lugner, P.: Theoretische Grenzen und Möglichkeiten der Fahrdynamik von Pkw mit Allradantrieb, in VDI Reihe 12 Nr. 81. VDI, Düsseldorf (1986)
44. Stockmar, J.: Technische Lösungsmöglichkeiten des Allradantriebs beim Pkw, in VDI Reihe 12 Nr. 81. VDI, Düsseldorf (1986)

6 Fuel System

The energy for propulsion is transferred to the wheels by the much-noticed engine. But it too can only convert energy, not generate it. The fuel system stores energy and provides the engine with the basis of its function.

M. Trzesniowski, *Powertrain*,
https://doi.org/10.1007/978-3-658-39885-9_6

6.1 Requirements and Overview

The fuel system provides the actual energy for propulsion in the compact form of liquid fuel. To ensure that this energy is only used for propulsion, this assembly is very much restricted by law or regulations in both production and racing vehicles.

Figure 6.1 shows the essential parts of a fuel system, as they are necessary in every internal combustion engine – regardless of whether it is a petrol or diesel engine. A tank (1) which holds the fuel and prevents or dampens disruptive movements caused by internals (2). A collection pot (3) with check valves ensures that the pump (5) can draw in fuel at any driving condition and deliver it through a filter (6) to the fuel metering device (7), from which the fuel goes to the carburetor, into the intake manifold or directly into the combustion chamber. In the case of high-pressure systems (diesel engine, direct-injection petrol engine), a mechanical high-pressure pump additionally generates the required fuel pressure. In this case, the pump (5) serves only as a pre-feed pump. A pressure regulator (8) sets the desired fuel pressure and/or dampens pressure fluctuations. Excess fuel returns to the tank, in many designs to the collection pot, via a return line (9). During filling and when the engine consumes fuel, a vent (10) must be used to equalize pressure in the tank. If the pressure in the tank becomes too high due to extreme driving manoeuvres or heat input, an overflow tank (11) allows the pressure to be reduced without fuel escaping into the atmosphere.

The fuel system of a racing vehicle must ensure the continuous supply of fuel to the engine under all extreme driving conditions, right down to the proverbial last drop. Safety margins in the total amount of fuel would mean extra weight, which is to be avoided. This means the system must be able to completely empty the tank while driving. Furthermore, it must be a closed system, i.e. no fuel must leave the system except for engine operation.

Therefore, in the fuel tank there is a collecting pot, which receives the fuel during acceleration or braking by its inertia. The fuel feed pumps draw in from this pot. The

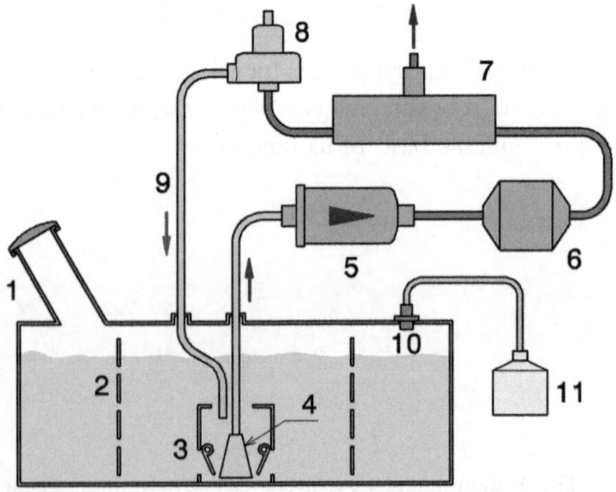

Fig. 6.1 Fuel system in general. 1 Fuel tank with filler neck. 2 *Baffle plates*, 3 Collector pot with flap check valves, 4 pick up funnel with coarse filter, 5 Fuel pump, 6 Fuel filter, 7 Metering unit, 8 Pressure regulator, 9 Return line, 10 Vent, 11 catch can (overflow tank)

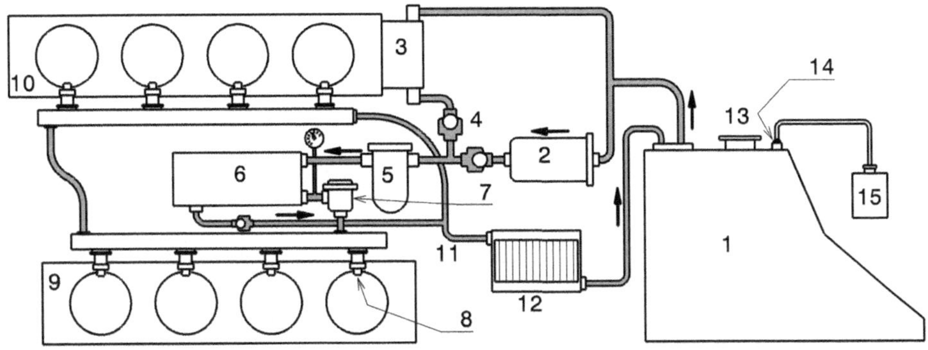

Fig. 6.2 Fuel system of a racing vehicle with intake manifold injection. **1** Fuel tank. **2** electric high pressure fuel pump. **3** Engine fuel pump. **4** Check valve (one-way valve). **5** Engine fuel filter. **6** Metering unit. **7** Pressure relief valve. **8** Injectors. **9** right hand bank manifold. **10** left hand bank manifold. **11** Return fuel line into collector pot. **12** Heat exchanger (cooler). **13** Quick *release fuel filler* valve. **14** tank breather (ventilation). **15** catch tank (overflow tank)

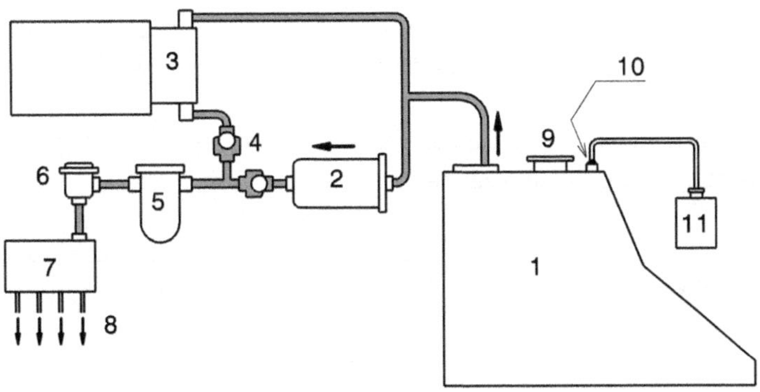

Fig. 6.3 Fuel system for carburetor engine. 1. Fuel tank. 2. Electric high pressure fuel pump. 3. Engine fuel pump. 4. Check valve (one way valve). 5. Engine fuel filter. 6. Pressure relief valve. 7. distributor block. 8. To carburettors. 9. Quick release fuel filler valve. 10. tank breather (ventilation). 11. catch tank (overflow tank)

electric fuel pump is only needed to start the engine. Basically, there is little difference between systems for fuel injection or carburetor engines, Figs. 6.2 and 6.3. In systems for fuel injectors, a certain differential pressure is set between the intake manifold and the *fuel rail* so that the metering amount of fuel can be controlled solely by the opening time of the injection valve. The excess fuel is returned to the tank. In the process, it can also be cooled and thus fed directly back into the collection tank.

Passenger car systems must also be closed systems. There are two ways of compensating for the increase in fuel volume due to heating. One variant has an external expansion tank, the other an internal expansion volume. Accordingly, the fuel system of racing vehicles is designed like the former variant. The minimum size of the overflow tank should

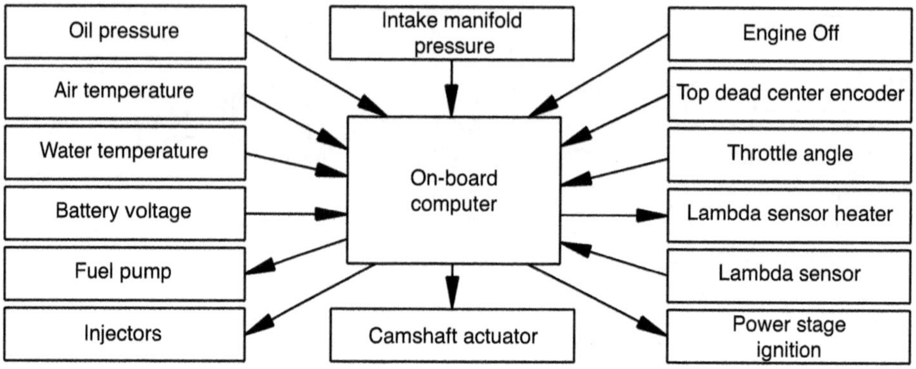

Fig. 6.4 Fuel metering system overview

be 3–5% of the tank volume. Mostly regulations specify a size. Without an overflow tank (catch can), at least 10% of its filling space must be kept free in the fuel tank to compensate for thermal expansion [1].

The metering of the fuel to the engine is nowadays done by an on-board computer in passenger cars and racing cars, Fig. 6.4. Only in a few racing classes carburettors are still used for mixture formation.

The on-board computer processes all relevant signals and controls the fuel injection quantity according to a programmed map.

6.2 Fuel Tank

6.2.1 Size

The capacity of the tank is determined by the required range or the range derived from a race strategy. In some regulations, the maximum tank size is staggered according to the engine displacement. In the case of production vehicles, if no data is available, the requirement is determined by means of a rough estimate of the fuel consumption.

For racing vehicles without refuelling possibility during the competition, the quantity is adjusted to the length of the race without safety allowance. Table 6.1 lists typical values of tanks. The maximum capacity of the tank is generally limited by the regulations for safety reasons. In the turbo era of Formula 1 (1984/85), the tank was allowed to hold 220 l. This was reduced to 195 l in 1986. This was reduced to 195 l in 1986, and today the volumes are between 110 and 150 l. The average fuel consumption of Formula 1 cars is around 70 l/100 km. Production cars consume around 8.5–10 l/100 km, depending on driving style and load. Another consideration for tank size is the additional mass that must be moved along in the form of fuel. In Formula 1, for example, one calculates an advantage of 0.3–0.4 s per lap

Table 6.1 Typical values of FIA FT-3 safety tanks, foam-filled

Contents, l	Dimensions $L \times W \times H$, mm	Mass, kg
15	254 × 240 × 260	1.5
30	510 × 241 × 260	2.3
60	635 × 390 × 260	3.8

and per 12 l less fuel.[1] If refuelling is permitted in the race, a smaller and thus lighter tank can be quite sensible. In most racing classes the fuel is prescribed, but the allowed range allows to influence the characteristics within certain limits. After all, a fuel consists of up to 250 chemical components. The FIA Formula 1 regulations limit the allowable density between 0.72 and 0.775 Kg/l at 15 °C [3]. Fuel manufacturers can thus provide the lightest possible (gravimetrically favourable) or the densest possible (volumetrically favourable) fuel for the same energy content. A 100 l tank filling with gravimetrically favorable fuel saves 5.5 kg of mass compared to the denser fuel. This gives a lap time advantage of about 0.2 s. In races with several refueling stops, the volumetrically more favorable fuel again brings advantages. The vehicle can be refuelled with the required amount of energy in a shorter time. The differences may be minimal, but in the highest leagues of racing, successful teams gain the decisive competitive advantage with such "little things".

If fine tuning of the tank contents is required, this is done, for example, with plastic volume compensation balls. Typical diameters are between 70 and 150 mm.

6.2.2 *Arrangement*

The arrangement of the tank in a passenger car is determined by the necessary crash protection measures. A characteristic feature is the placement in the crash-protected area in front of and in the area of the rear axle in front-engined vehicles. A conflict of objectives arises in the case of rear-axle-drive vehicles, where the tank would preferably be positioned behind the axle in favor of a higher rear-axle load. The usual arrangement of the tank in the crash-protected area today means that if the customer wants a through-loading option (e.g. for skis), separate tank variants with a smaller volume are offered [6].

On mid-engine and rear-engine vehicles, the tank is located in the crashworthy area behind the front axle.

In the case of racing vehicles, a low influence on the driving behaviour is particularly important, i.e. the tank should be positioned as low and centrally as possible and thus have little influence on the mass moments of inertia of the vehicle. The position behind the driver in the middle of the vehicle (Fig. 6.5), which is common for monoposti, has only little influence on the driving behaviour even with different filling levels. The tank should be

[1] Cf. Racing Car Technology Manual, Vol. 2 *Complete Vehicle*, Sect. 2.3.2.

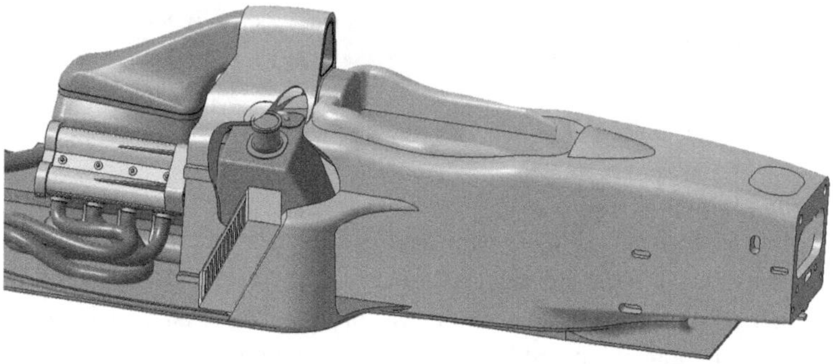

Fig. 6.5 Location of the fuel tank on a formula car. The tank is protected between the driver and the engine. This position, which is close to the centre of gravity, also has little influence on the driving behaviour

protected as far as possible by load-bearing structures. In the event of a rollover, no fuel should escape.

In many regulations, the position is prescribed in more detail: The fuel tank must be located in front of the rear axle and in an area around the vehicle's longitudinal centre plane.

In some vehicles, the fuel volume is divided among several tanks. This facilitates the accommodation of the tanks and also allows selective tank emptying, which can be controlled by the driver to influence the mass distribution according to his wishes [8]. An extreme in this respect was probably represented by the Ligier Formula 1 cars of the 1970s, which had seven tanks. Six of them were connected to each other with non-return valves at the side next to the driver and the last tank was located centrally behind the bulkhead to the cockpit. Due to the inertial forces acting during acceleration, the tanks emptied from the front to the rear [8].

In any case, a drainage facility must be provided in the tank room or in the structure surrounding the actual tank in case fuel leaks from the tank. Otherwise, a leak in the tank might be discovered much too late. In addition, fuel could leak into the cockpit. The easiest way to do this is to drill 3 mm holes around the edge of the floor.

6.2.3 Construction Types

The mechanical fuel pump usually does not deliver fuel directly from the tank, but from a collection pot to the engine, Fig. 6.6. The fuel enters this pot through usually four pipes with check valves. The four pipes extend into the four bottom corners of the tank. During braking and acceleration, inertia forces the fuel into the appropriate tubes and fills the collection pot. The check valves prevent the fuel from leaking into the opposite pipes.

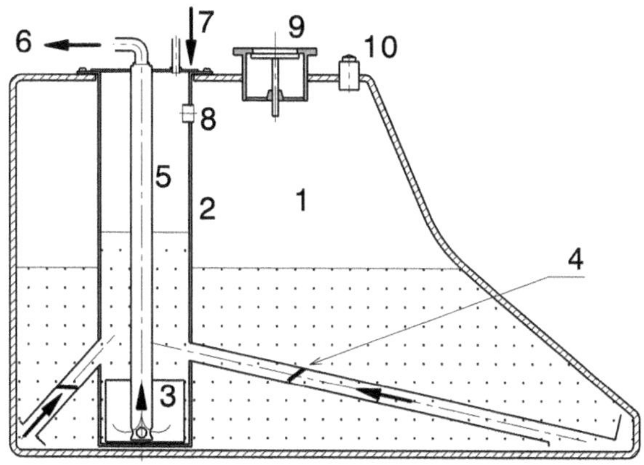

Fig. 6.6 Schematic of a fuel tank. 1. Fuel tank. 2. Collector pot. 3. Low pressure filter. 4. pick up line with one way valve. 5. *Feed line.* 6. To the fuel pump. 7. Return line. 8. Pot breather. 9. Quick release fuel filler valve. 10. Tank breather

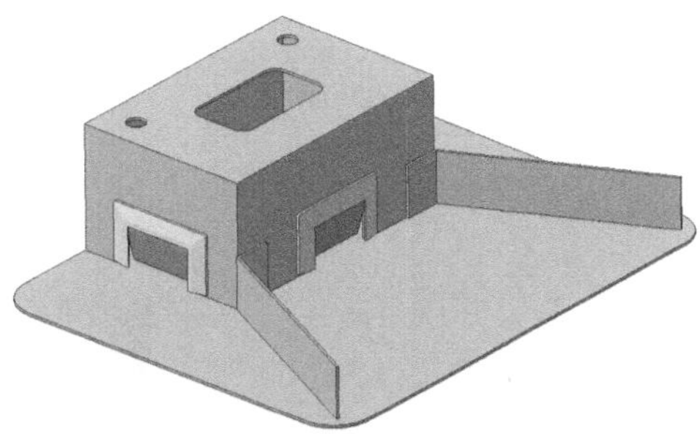

Fig. 6.7 Collecting pot

The collection pot is inserted into the tank from above and screwed on via a flange. This allows the filter to be easily reached even when the tank is assembled. Collection pots are also available which are placed in the tank and held in position by the foam, Fig. 6.7.

The collection pot is inserted into the fuel tank and held in place by the foam filling. The three flaps act as check valves and allow the fuel to flow only into the internal collection pot. The volume of the collection pot is 3 l.

Fuel tanks of racing vehicles can and should also be filled with foam, Fig. 6.8. The foam displaces approx. 3% volume per litre, acts as a vibration protection and increases the effectiveness of elastic safety tanks in the event of an accident. In addition, with an appropriate suction device, it enables the fuel to be sucked off almost completely. This principle also contributes to weight savings, as otherwise more fuel than necessary would have to be filled in order to avoid air suction. The foam must be cut out at places where ventilation valves, suction nozzles, etc. are located.

The foam block is inserted into an elastic tank and serves as a surge brake. It is cut out at areas of the suction port and vent valve.

Figure 6.9 illustrates the interaction of the above components in a complete tank for motorsport use.

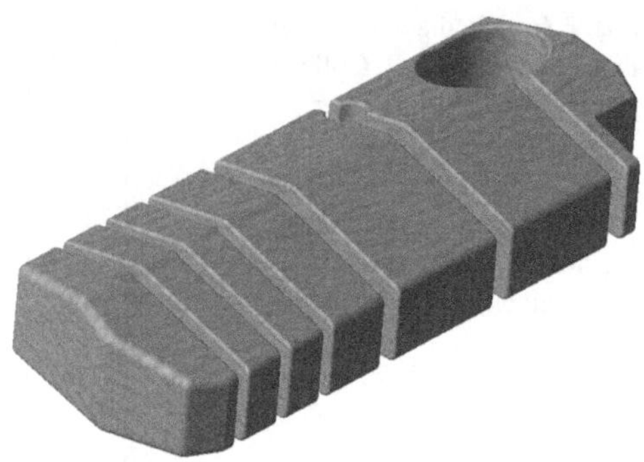

Fig. 6.8 Tank foam

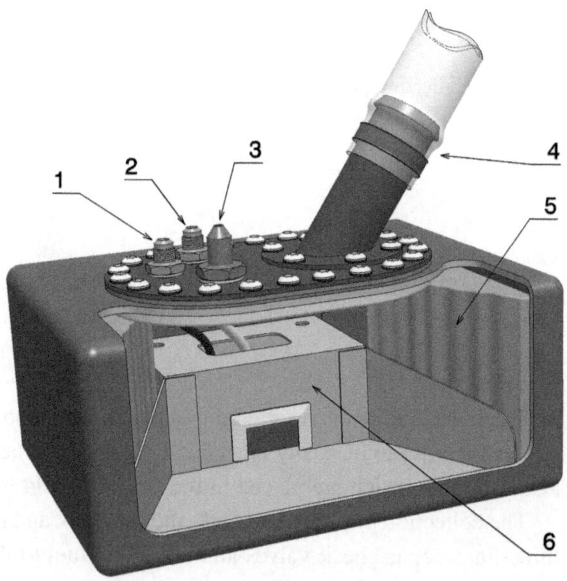

Fig. 6.9 Fuel tank complete, partially cut open (FIA FT3). 1 return line connection, 2 connection outlet, 3 Venting, 4 Filler neck with hose, 5 baffle foam, 6 Collector pot with one-way trap doors

All connections are located on a flange plate, which provides a large mounting opening and is tightly screwed to the tank. A hose is plugged onto the filler neck and secured with a clamp. It provides a flexible connection to the tank valve in the bodywork.

If the tank is installed in the vehicle interior without a bulkhead facing the driver, the tank must be housed in a fireproof and liquid-tight containment vessel. This is made of aluminium, for example, Fig. 6.10.

The actual fuel tank consists of elastic fibre-reinforced elastomer and is located in a box-shaped safety cell. In accordance with FIA specifications, the manufacturer and the expiry date of the validity for use, among other things, are indicated on the outside.

Resilient tanks in the trunk of production touring cars must be surrounded by a fireproof and liquid-tight structure, Fig. 6.11.

Fig. 6.10 Fuel tank for racing vehicle

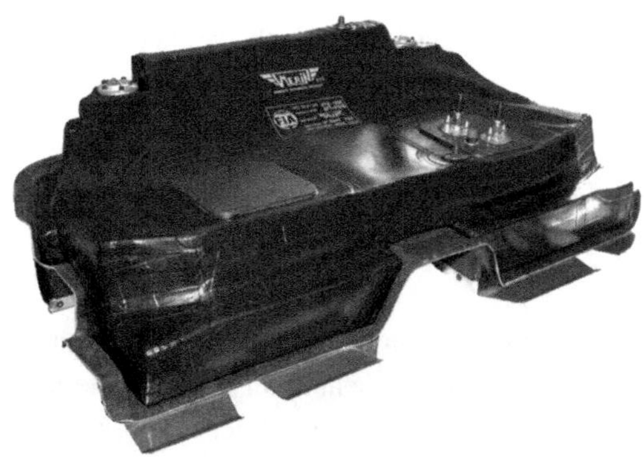

Fig. 6.11 Fuel tank for touring cars

The tank is saddle-shaped because it has to leave space in the middle for the propeller shaft tunnel. In the right flange plate you can see two fuel pumps.

Safety tanks consist of a fibre-reinforced elastomer, composite materials or a combination of both. The tank is first assembled from individual, cut-to-size parts and then vulcanized or bonded together in an autoclave to form a composite. The tank nestles against the supporting surrounding walls. During assembly, it can be folded and inserted through a relatively small opening into the receiving space. In the case of single-seaters, for example, it is inserted through an oval opening in the monocoque into the receiving area and then the flange plate for all connections is screwed on, Fig. 6.12.

This tank bladder is made of fiber reinforced elastomer. The filler necks, vents, etc. are not mounted. These are screwed to the tank by means of metal flanges.

The FIA requires a tank that comes from a manufacturer accredited by it. Among other things, the tank must have the name of the manufacturer, a serial number and the date of manufacture printed on it. Because the material properties of the tank suffer as a result of

Fig. 6.12 Elastic safety tank for racing vehicle

ageing, it must either be replaced after five years or inspected by the manufacturer. A positive inspection extends the service life of the fuel tank by a maximum of two years.

Tanks of series production vehicles are made of plastic (Lupolen, Hostalen, etc.) or metal (alloyed aluminium sheets, deep-drawing steels and high-alloy steels).

6.3 Connections

The fuel tank openings are closed with aluminium flanges. The openings are kept large, which allows easy cleaning of the tank interior. Flanges are either simple blind flanges for mere sealing or they carry filler necks, vent valves, collecting pots as well as tank fuel pumps, Fig. 6.13.

The filler neck is screwed to the tank by means of a flange plate. An interposed gasket seals the plate. The flange plate also carries four vent valves.

All connections shall be fire resistant and shall be located inside the vehicle. Self-closing breakaway valves shall be fitted to connecting lines to prevent fuel leakage in the event of accidental separation between tank and frame.

Filler Neck

When refuelling, the fuel is filled via a filling pipe. The end of this tube is sealed with a rotary cap (Fig. 6.14) or with a quick refuelling valve (Fig. 6.15). This valve allows quick refueling during a pit stop without having to open or close any caps, Fig. 6.17. In addition, this valve seals the filler opening even in the event of a rollover. This is not the case with a simple twist-off cap, and a separate check valve must be provided. A quick tank valve consists of a valve disc (2) which is pressed open by the hose valve during refuelling. Conversely, a collar in the flange (1) of the quick tank valve opens the hose valve and at the same time the inside of the flange seals the hose from the environment. When the hose is pulled off, both valves close spring-loaded.

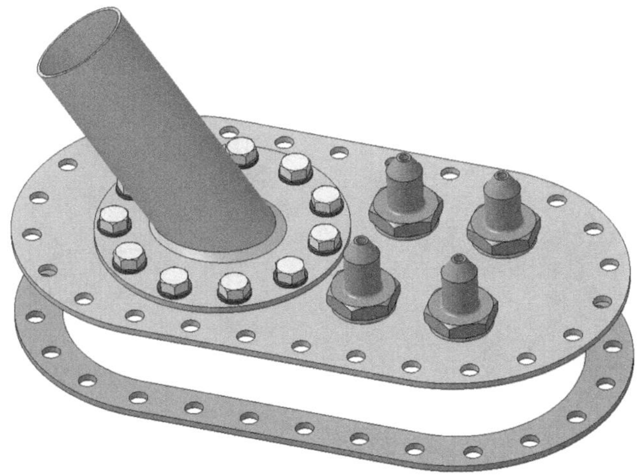

Fig. 6.13 Filler neck for racing vehicles

Fig. 6.14 Rear of a touring car

The tank filler neck is sealed with a screw cap. An overflow tray catches dripping fuel during refuelling. Also visible in the picture is an upright dry sump tank including the overflow vessel of the lubrication system and, behind it, a sealed battery box.

The valve can be installed directly into the filler neck of the fuel tank via its flange. When refuelling during the race, as is necessary e.g. in endurance races, a suitable hose valve (Fig. 6.16) is fitted, which presses on the disc (2).

During refueling, the valve is placed on the tank valve and depressed via the two grips, causing the hose valve disc to press on the tank valve disc and fuel to begin flowing. These valves are designed for high flow at low differential pressures.

The position and, above all, the orientation of the tank valve on the vehicle is worth some ergonomic consideration for accident-free quick refuelling. The hose together with

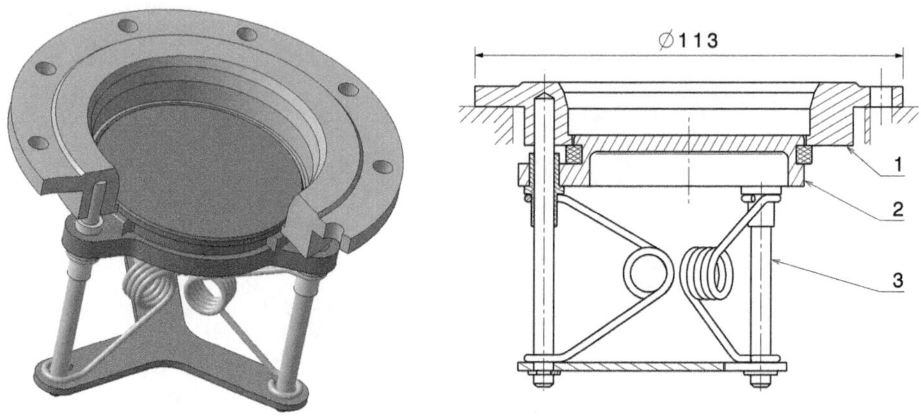

Fig. 6.15 Quick tank valve (*refuelling valve*). 1. Flange. 2. Valve discs. 3. Guide column

Fig. 6.16 Hose valve

the hose valve is relatively heavy and must be easy for the mechanics to put on and take off again, Fig. 6.18. A waist-high arrangement that allows the hose valve to be moved horizontally towards the vehicle proves to be favourable. Ferrari's Formula 1 team had to painfully realise this when the position of the tank valve on the F1–2000 was relocated for aerodynamic reasons. Unlike the previous car, the valve was almost vertical and hidden behind the cockpit fairing to boot. The mechanics now had to put the heavy hose over their shoulder and press it onto the tank nozzle from above. This ergonomically unfavourable position then also led to an incident during refuelling in the race [4].

Two valves are fitted for rapid refuelling. Both valves are pressed open during refuelling. The fuel enters through the lower one and the upper one is used for the air displaced from the fuel tank to escape.

When refuelling, the hose is held by the handle bar by a mechanic and pressed against the refuelling valve.

Fig. 6.17 Quick tank valve on a production sports car (Pro Sport 3000)

Fig. 6.18 Fuel hose for refuelling process on a Formula 1 car

When refuelling during a pit stop, up to 12 l/s flow through the fuelling nozzle. At a normal filling station, the throughput rate is comparatively about 40 l/min, i.e. approx. 0.7 l/s. Figure 6.19 shows a Formula 1 refuelling system. The vehicle is grounded during refueling so that no sparks can jump between the vehicle and the filler neck due to electrostatic discharge. The pit crew handling fuel should also wear antistatic clothing.

Check Valve

Check valves are installed in the filler neck between the fuel filler cap and the fuel tank. They prevent fuel from escaping when the vehicle rolls over, Fig. 6.20.

These check valves are installed in the filler neck. They allow the fuel to flow in only one direction and thus prevent fuel from leaking through the filler neck in the event of an accident.

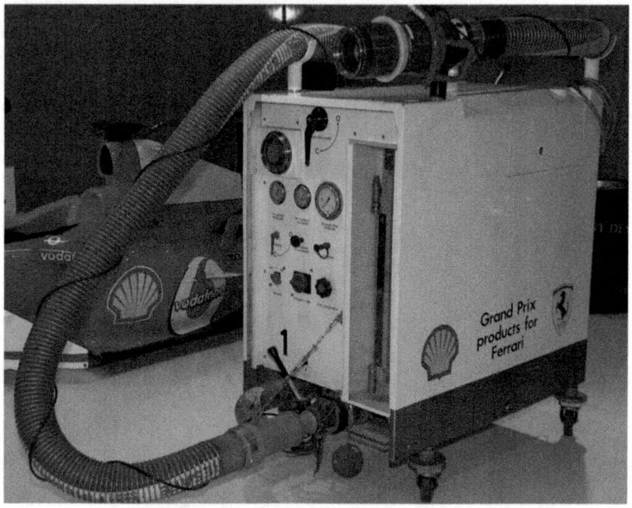

Fig. 6.19 Formula 1 refuelling system

Fig. 6.20 Check valves

Tank Breather Valve **(Vent Valve)**

Tank ventilation is necessary because the contents are emptied by the engine during operation and the corresponding amount of air must be able to flow in to equalize the pressure. When the temperature rises, pressure compensation must also be possible in the opposite direction due to the rising fuel vapour pressure. Both are ensured by a vent valve, Fig. 6.21. In addition, no fuel may escape into the atmosphere in the event of an accident. Accordingly, a non-return valve must also be integrated. Vent valves are located on the top of the tank or are installed in the vent line.

This valve is screwed into the top of the tank or vertically into the vent line to the overflow tank.

The valve opens by the own weight of the ball only from a certain overpressure to the tank ventilation. If the pressure in the tank drops, the ambient pressure pushes the disc

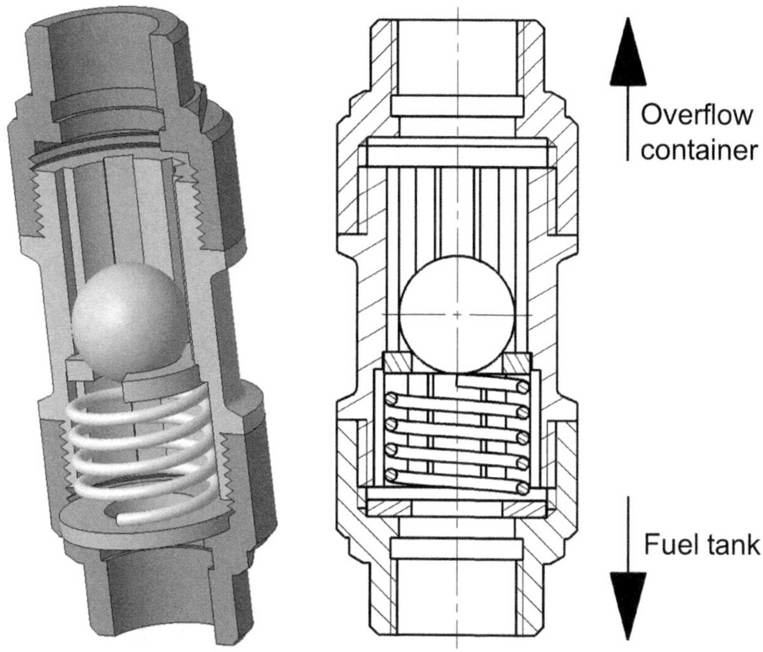

Fig. 6.21 Vent valve

downwards, allowing air to flow into the tank via the grooves on the circumference. If the valve tilts or the flow rate towards the overflow increases extremely (accident), the ball closes the valve completely.

Breakaway Valve

If there is a leak in a fuel pressure line (rupture of a line due to an accident, etc.), the valve closes by increasing the flow rate above a set value.

Quick-Release Coupling

Some events (e.g. rallies, Formula 3, Formula Renault) require fuel sampling valves. Technical commissioners take a fuel sample. For this purpose, a sampling nozzle with a self-closing coupling of a prescribed size must be fitted.

In summary, Fig. 6.22 shows a designed flange plate with typical tank connections.

The flange is attached to the top of a containment vessel. The container is located in the trunk of a touring car. The filler neck is located in the middle surrounded by connections. Two outlets (OUTLET) lead to two fuel pumps outside the tank. One line is the return line and the line in the foreground is the vent line (VENT) to the overflow tank.

Fig. 6.22 Tank connections

6.4 Fuel Lines

The position of cables in passenger cars is determined by production requirements and safety-relevant criteria. Crash-proof, crossover-free routing must ensure high production quality by means of safe and confusion-free quick connections. Minimizing the joints of fuel-carrying lines in conjunction with materials that allow the lowest possible fuel diffusion is necessary to reduce hydrocarbon emissions. For the same reason, activated carbon tanks (volume 1.5 l to approx. 5 l depending on tank volume and refueling venting system) are used for temporary storage of the gasoline vapors released during vehicle operation (e.g. tank heating, in the USA also for the refueling process). Fuel lines on racing cars are often stainless steel encased rubber or Teflon hoses. The connections are made by reusable fittings made of aluminium or stainless steel. The actual connection to the hose is ensured by a cutting ring or a sealing olive.[2]

In racing cars, fuel lines may also be routed through the cockpit. Such lines should be steel braided pressure lines and joined with screw connections [7].

If elastomer hose lines are used, special clamps must be used at the connection points, Fig. 6.23. These clamps completely enclose the line, have rolled-up edges and are tightened by means of a bolt and nut.

6.5 Fuel Pump

The delivery volume of the pump must be significantly higher than the maximum fuel requirement of the engine in systems with fuel return. On the one hand, the pressure regulator needs a certain volume flow for a stable control function, on the other hand, the

[2] See also e.g. Racing Car Technology Manual, Vol. 4 Suspension System, Chap. 6, Fig. 6.32.

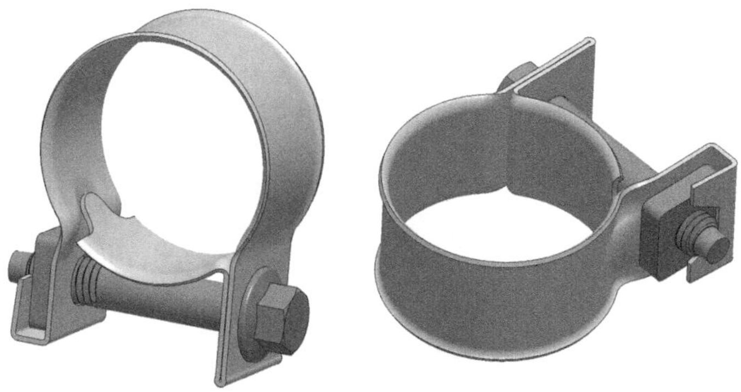

Fig. 6.23 Clamp for fuel line made of elastomer

Fig. 6.24 Electric fuel pump (high pressure pump)

Table 6.2 Data of electric fuel pumps [5]

Model	Voltage, V	Delivery pressure, bar	Delivery volume, l/h	Fuel connection
A	12	2.5	135	in[a]: M10 × 1.0; out[a]: M10 × 1.0
B	12	3	148–237	in: M14 × 1.5; out: M12 × 1.5
C	12	5	135–220	in: M18 × 1.5; out: M12 × 1.5

[a]in: pump inlet, suction side
out: Pump outlet, pressure side

hot delivery behaviour of the pump with up to 50% losses of the nominal delivery rate must be taken into account.

Electric and mechanical fuel pumps are used. Electric pumps can be installed in the tank or in the fuel supply line (inline pump). The inline pump must have a good suction performance because of the distance to the tank, but it is easy to check and repair, Fig. 6.24.

At the ends are the screw connections for suction and pressure connection. The two electrical connections can be seen on the right side.

Table 6.2 provides an overview of important data of typical fuel pumps.

Fig. 6.25 Fuel pumps

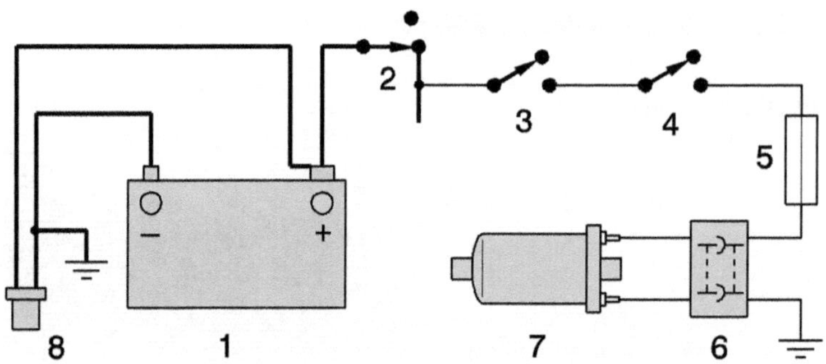

Fig. 6.26 Circuit diagram for electric fuel pumps. 1 Battery, 2 Main switch (*master switch)*, 3 *Ignition switch*, 4 Fuel pump switch, 5 Circuit breaker (*pop-out fuse)*, 6 Quick disconnect plug, 7 Fuel pump, 8 Auxiliary battery plug

An electric fuel pump is only needed in Formula 1 engines when starting and driving slowly. At higher speeds, the required fuel pressure comes from a mechanical pump driven by a camshaft. A purely electrical supply would consume too much power [5].

As engine speed increases, the time available for mixture formation becomes shorter and shorter. For this reason, fuel delivery pressures of 70–80 bar are built up in the high-revving Formula 1 engines. This allows the injection time to be reduced to 80 °CA for the same amount of fuel introduced [2].

The maximum flow velocity in fuel lines shall be 20 m/s. This allows the line cross-sections to be dimensioned for full-load operation.

The pump should be installed as low as possible in the vehicle. The centre of gravity remains low and the pump does not start empty. In general, pumps should have the shortest possible suction line, because the fuel is pumped to the pump by the pressure difference to the environment or gravity. So pumps can pump much better than suction. It should also be installed away from heat sources or at least be supplied with cooling air to avoid vapour bubbles. The electrical connections must be fixed, e.g. screwed, Fig. 6.25. For long-distance vehicles, an adequate quick-connect system can be advantageous when replacing the pump during the race (cf. Fig. 6.26, item 6).

On most race cars it must be possible to switch off the fuel pump from the driver's seat, e.g. with a toggle switch.

Figure 6.26 shows a corresponding electrical circuit diagram for a fuel pump. On some vehicles there is also another external electrical connection, which makes it possible to operate the pump with an external battery alone, e.g. to empty the tank.

The two fuel pumps are located in the trunk of a touring car and are connected in parallel.

References

1. Braess/Seiffert: Vieweg Handbuch Kraftfahrzeugtechnik, 4th edn. Vieweg, Wiesbaden (2005)
2. Piola, G.: Formula 1 Technical Analysis 2003/2004. Giorgio Nada Editore, Mailand (2004)
3. Internetseite der FIA.: http://www.fia.com/sport/Regulations/f1regs.html. Accessed 12 Dec 2005
4. Piola G.: Formel 1. Die Hightech-Geheimnisse der Rennställe. Copress, Münchend (2001)
5. Hack, I.: Formel 1 Motoren, 2nd edn. Motorbuch, Stuttgart (1997)
6. ATZ und MTZ Sonderheft 10/98: Die neue S-Klasse von DaimlerChrysler. Vieweg, Wiesbaden (1998)
7. McBeath, S.: Competition Car Preparation, 1st edn. Haynes, Sparkford (1999)
8. Incandela, S.: Aufl., Sparkford: Haynes 1984. The Anatomy & Development of the Formula One Racing Car from, 2 (1975)

7 Electrical System

The electrical system of a racing car was once limited to a generator, possibly a starter and the ignition system together with some cables. With the increase in electronic aids and measured value acquisition, the electrical system has also become so important in drives with internal combustion engines that its design and functional safeguarding occupy just as much space in development as that of other assemblies.

M. Trzesniowski, *Powertrain*,
https://doi.org/10.1007/978-3-658-39885-9_7

7.1 Wiring Overview

As with the drivetrain, there is little difference between vehicles with internal combustion engines and those with electric motors in terms of the power supply systems. Only the high-voltage range is added in the case of battery-electric drive systems, Fig. 7.1. Instead of a generator, a voltage converter (4) is used in a high-voltage system to charge the low-voltage battery (7). The ground of the high-voltage system is connected "centrally" to the frame ground (low-voltage system ground) (6). This is done using high-resistance resistors as voltage dividers. The central connection is made for safety reasons. In the event of contact, a maximum potential difference of only half the nominal voltage of the high-voltage system is present between the high-voltage and low-voltage sides [1]. In addition, an insulation monitoring device (5) constantly checks whether a dangerously high potential difference against frame ground is present due to short circuits (shunt fault). The consumers R1 to R1 stand for all electrically operated systems such as windscreen wipers, headlights or electric coolant pump.

The wiring harness is usually part of the vehicle concept, because it is relatively heavy and often the reason for reliability problems. Even in production vehicles, the electrics often cause problems, but here the fault is mainly to be found at the interfaces (= plugs, connectors and switches). In principle, the wiring is kept relatively simple. Figure 7.2 shows a simple electrical system. The more complex the wiring harness, the more complex the troubleshooting and the more likely a fault. In some vehicles, fuses are not even provided because the currents of the engine-side generators are relatively small [2]. However, if the vehicle is equipped with electronic driving aids and data acquisition systems, the complexity increases to the point where thoughtful integration of all systems is appropriate. In addition to the engine control unit, there are also transmission control units, other control units, sensors (speed, temperature, pressure,. . .) and actuators (injection valves, throttle valve, gearshift,. . .) on board. All of these have to communicate with each other, at least in part. This leads to the fact that also in racing cars a CAN bus (see also appendix) is used like in series vehicles, Fig. 7.3. Here a central two-core twisted cable serves for the transmission of all information. Control units, sensors and actuators send and receive serial messages, which must have a certain format, via this cable. This format includes the content of the message (e.g. engine speed) as well as its priority. Receivers only accept those messages that are included in their list of messages to be received. The system is monitored by the integrated control units (multi-master principle), which all have equal rights. If one participant fails, the structure still remains fully functional for the others. A CAN system saves line length because the alternative consists of many individual lines between all the addressed devices. As complexity increases, so does the advantage of fuses. In troubleshooting alone, the resulting restriction to a specific circuit can already be worthwhile. Further advantages are, in addition to real-time transmission, smaller control units due to fewer plug connections, smaller number of sensors (multiple use possible) and, thanks to standardization, no ties to one manufacturer; on the contrary, control units from different manufacturers can communicate with each other.

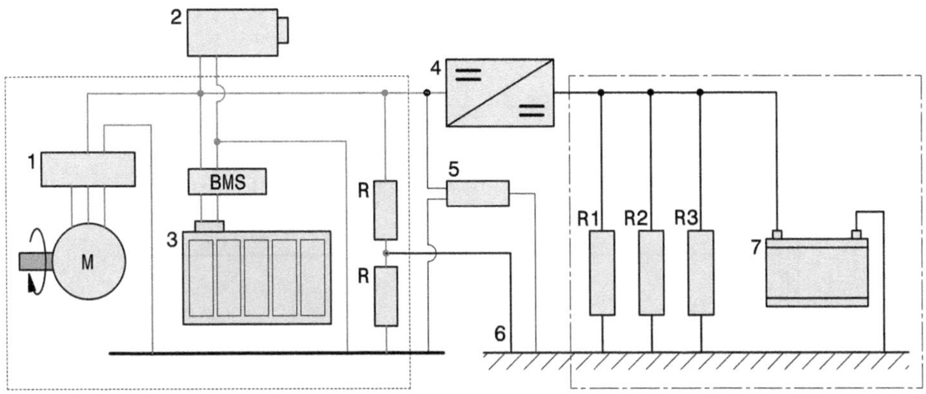

Fig. 7.1 Overview of electrical system architecture with electric drive. The right-hand side, the low-voltage range (dotted line), is found in every vehicle. For battery-electric and hybrid vehicles, the high-voltage range (dotted) is added. 1 inverter, 2 on-board charger, 3 high-voltage battery, 4 DC-DC converter, 5 insolation monitoring device, 6 ground connection, 7 low-voltage battery, M electric machine (motor-generator unit*)*, BMS battery management system, R1 to R3 consumer (loads)

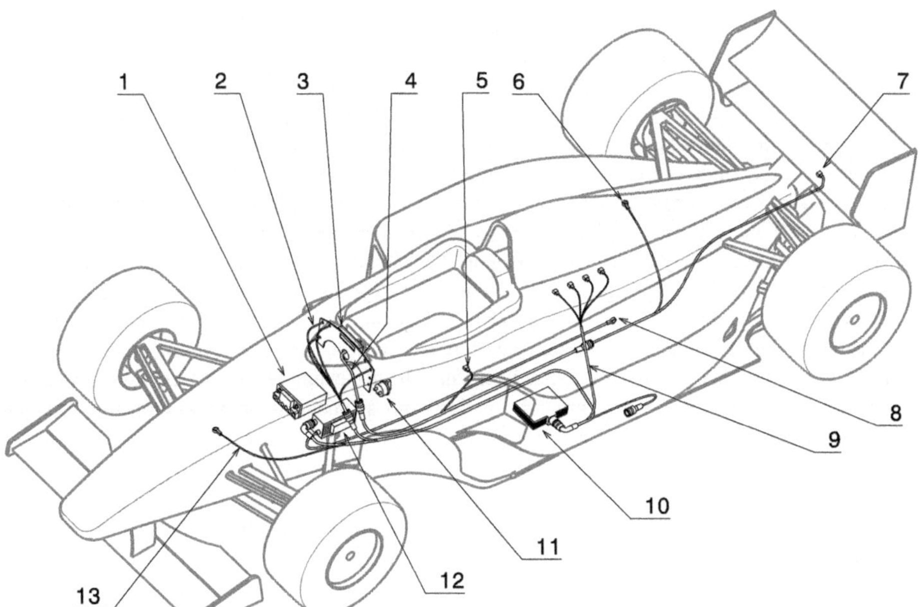

Fig. 7.2 Electrical system of a racing vehicle. **1** Battery. **2** Dashboard wiring. **3** Dashboard display. **4** Steering wheel wiring. **5** Electrical connection for fire extinguisher. **6** electric frame connection. **7** brake/tail light. **8** Starter connection (starter wiring). **9** Engine connection. **10** Control unit. **11** Battery *main switch*. **12** Switchgear. 13 Battery earth connection

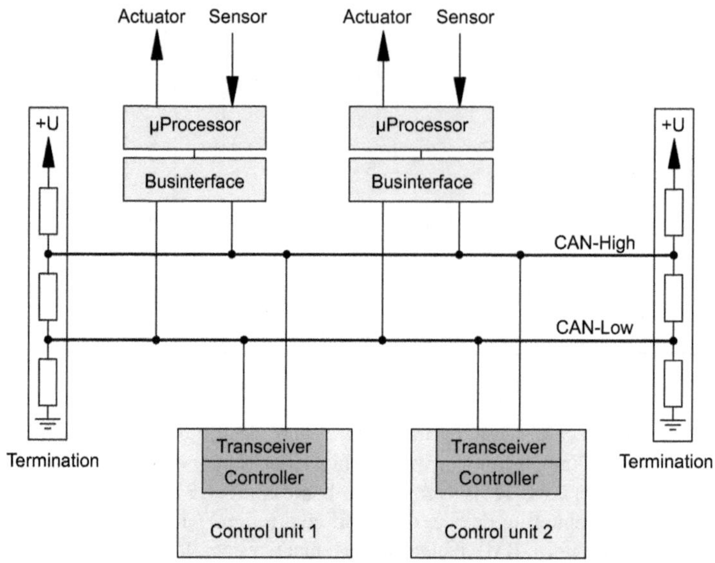

Fig. 7.3 Schematic structure of a CAN network. A two-wire line (CAN bus consisting of CAN high and CAN low line) transmits all information from sensors and actuators and connects all control units with each other. The ends of the two-wire line are terminated by resistor networks for reflection reduction. The data to be sent from an ECU is processed by the controller and transferred to the transceiver (amplifier). This is connected directly to the CAN bus and works as a transmitter and receiver. The information sent is received by all networked participants. If this information is important for them, it is processed further, otherwise it is ignored. A standardized data protocol indicates, among other things, the priority of the message. If several participants want to send at the same time, the highest priority is given priority and the others become recipients (arbitration). After the end of the message, they can try to send again. The signals, designed as square-wave pulses, are transmitted as electrical voltages (0 and 5 V). These are opposite in the two conductors, so that the electromagnetic radiation to the outside is zero and at the same time a control is given by the redundancy of the information. To inhibit interference from outside (transmitting stations, mobile phones,. . .) the two conductors are twisted together. In the event of a short circuit between the conductors or a short to positive or ground, the CAN bus switches to single-wire operation (emergency operation)

In the case of switches and plug connections, products from the aviation industry are often used because of their reliability.

7.2 Electronic Control Unit (ECU)

For engine control units and similar computing units that need to be cooled, the air ducts to the heat exchangers are ideal. In these, they can be located low down in the air intake area. This also ensures good accessibility, Fig. 7.4.

Fig. 7.4 Favourable arrangement of a control unit (Formula Renault, 2000). The vehicle shown has two heat exchangers symmetrically on both sides of the cockpit. The heat exchanger under consideration is located to the left of the cockpit. The engine control unit is located above the air intake. It is guided by two angles and fixed by an easily removable elastic band

Control units are housed in a dustproof and waterproof aluminium, magnesium or CFRP housing and have connections for plugs according to aviation standards or military standards, Fig. 7.5. Compared to series devices, the racing versions are sealed against moisture and dust by special measures. In addition, great importance is attached to resistance to vibrations and heat.

Control units for combustion engines are mostly derived from series applications, although the goals are partly opposite. This is due to the fact that the same actuators and sensors are required to fulfill the functions. For production vehicles, the focus is on comfort, safety, durability and exhaust emission limits. For racing vehicles, the focus is on short-term maximum performance. In both cases, the lambda sensor and knock sensor must supply signals to the control unit. Fuel consumption is also receiving increasing attention in racing vehicles. On the one hand, some regulations restrict the maximum fuel flow to the engine, on the other hand, lower fuel consumption means fewer fuel stops in endurance races or a smaller, lighter tank in sprint races. Exhaust catalysts are also already mandatory in some racing series – such as the German Touring Car Masters (DTM).

Racing engines rev much higher than series production units. For the control unit, this means a correspondingly higher clock frequency and possibly multiple computer systems. Of course, electronic components whose function is speed-dependent must also be able to keep up. For example, ignition coils for racing gasoline engines must be trimmed for shorter charging times.

Motorsport control units allow – as is usual in the chassis area – simple (e.g. by means of switches) changes to be made for tuning to the track and weather conditions. Parts of the software can also be changed by the user.

a b

Fig. 7.5 Engine control units. (**a**) Control unit for the DTM. (**b**) Control unit for racing series in the USA (Grand Am)

7.3 Battery

Batteries supply power to the electric fuel pump (start, slow speed), electronic ignition and fuel injection system including on-board computer. There are different considerations or regulations for engine starting, so that the battery does not have to provide the energy for the starter motor in every case. In some vehicles, starter motors are either not provided on the vehicle at all or are supplied by compressed air tanks. Other vehicles must be able to start without external assistance, but use external voltage sources for the initial start or for starting in the pit, which conserve or even charge the on-board battery.

Electrical protection systems (fire extinguisher, air bottle) must have their own power source for safety reasons.

Closed and maintenance-free batteries are used in racing vehicles, which can be installed in any position, Fig. 7.6. Nevertheless, completely tightly enclosing, non-conductive containers, e.g. made of GRP, are prescribed if the battery is accommodated in the vehicle interior.

The battery mounting must be able to withstand the accelerations of driving manoeuvres and the electrical connections must not come loose. Nevertheless, the battery should be easy and quick to remove for regular maintenance (charging).

The capacity of the batteries is chosen relatively small to keep the mass low. Table 7.1 shows examples of common batteries. Motorcycle batteries are usually sufficient. Even in Formula 1, two 6 V motorcycle batteries were still used in the 1980s when air starters were used for the engines [2].

In formula vehicles that do not have an on-board starter motor, some manufacturers use a supercap instead of a battery. The double-layer capacitor (see Chap. 3 *Hybrid drives*) saves weight compared to the battery. These weights are 1.3 kg for the 15-V capacitor and 13 kg for the replaced 12-V battery in a 3.5-l Formula Renault car of the 2012 season [3]. The electrical energy stored in the capacitor is used, among other things, to change gears.

Fig. 7.6 Battery for formula cars, such as Formula ADAC, Formula König, etc.

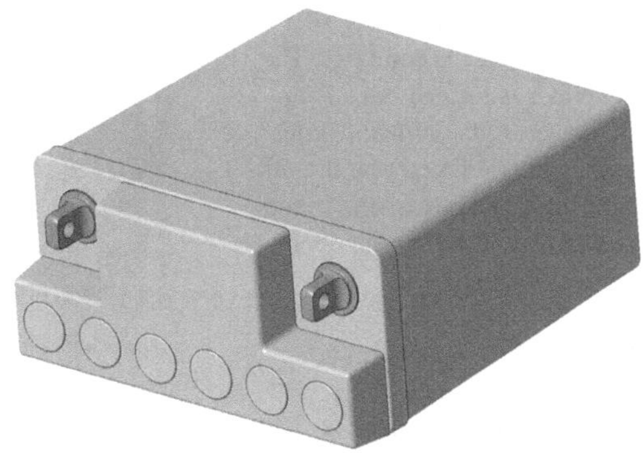

Table 7.1 Batteries for racing vehicles

Insert	Voltage, V	Capacity Ah	Dimensions mm	Mass kg	Type	Comment
Motorcycle	12	8	152 × 88 × 106	Approx. 3–4	YTX9-BS	Maintenance-free (gel technology): Any mounting position
Formula König, ADAC, ford	12	15	181 × 76 × 167	6.0	HP15-12 W	Leak-proof. Figure 7.6

7.4 Generator (*Alternator*)

A small alternator is driven directly by the engine and supplies various electrical pumps and the ignition system. In sprint competitions it should be checked whether a generator is needed at all or whether the capacity of a battery alone is sufficient to supply all consumers for the duration of the race (*total loss electrical system*). At longer competitions an on-board generator pays off in any case, even if it contributes to the power losses of the engine due to the efficiency of its drive and its operating principle. Batteries show a voltage drop when current is drawn, which can lead to sensitive consumers (engine control unit, fuel pump, . . .) no longer fulfilling their function properly. If the race lasts longer than expected (yellow flag, accident, . . .) or if you need more energy than planned (engine starts after spinning), this can lead to a failure if the voltage is supplied by the battery alone. For lights that are mandatory, such as brake lights and rain lights, the low current consumption of LEDs (light-emitting diodes) in a generatorless system proves to be advantageous.

In Formula 3, for example, there is no generator on the engine and drivers rely solely on the on-board battery to meet energy needs over a race distance [4].

Fig. 7.7 Generator on a V10 engine (Ferrari Formula 1). The generator is located outside at the lower, right crankcase end and is driven directly by the oil centrifuge. The hydraulic pump is attached to the visible open generator flange

Extremely small (e.g., diameter 62 mm by 98.5 mm length) motors are used for the extremely high-speed motors of Formula 1 [5]. They deliver a current of 41 A at 9000 min^{-1} and are speed-resistant up to over 12,500 min^{-1}, Fig. 7.7.

7.5 *Leads* and *Connectors*

In principle, as many elements of the wiring as possible should be arranged as low as possible in the vehicle because of their high mass. Each consumer should also be connected directly to the vehicle ground with a suitable cable.

Conductor Cross-Section

(Wire Size). Conductor cross-sections are primarily determined by the current to be conducted and the ambient temperature. One will not choose cross sections that are too large in order to save weight. However, too small a cross-section is also bad, as it increases the risk of failure or even fire. First, the current I of the consumer to be supplied is calculated:

$$I = \frac{P}{U_n} = \frac{U_n}{R} \tag{7.1}$$

I	Current of a consumer, A
P	Power requirement of a consumer, W
U_n	Rated voltage, V
R	Electrical resistance of a consumer, Ω

With the known current, the conductor cross-section q increases:

$$q = \frac{I\rho l}{\Delta U_+} \tag{7.2}$$

q	Conductor cross-section, mm^2
ρ	Specific electrical resistance, $\Omega mm^2/m$ For copper $\rho = 0.0185$, for aluminum $\rho = 0.0303$
l	Cable length, m
ΔU_+	Permissible voltage drop of the positive line, V

The permissible voltage drop of the positive lead at 12 V nominal voltage is 0.5 V for control leads, 0.4 V for charge leads (from generator to battery), 0.1 V for light leads and 0.5 V for the starter main lead. In Anglican-speaking countries, the conductor cross-section is specified in AWG (**A**merican **W**ire Gauge) codes. The smallest code number represents the largest cross-section.

The power dissipation of the conductor, which leads to heating, is calculated as follows:

$$P_{ls} = \frac{I^2 l\rho}{q} \tag{7.3}$$

P_{ls}	Power loss, W

This power loss corresponds to the heat flow to be dissipated. The temperature difference ΔT to the ambient temperature thus follows from the definition of the thermal resistance:

$$\Delta T = T - T_0 = P_{ls} R_{th} \tag{7.4}$$

T	Temperature of the conductor, K
T_0	Ambient temperature, K
R_{th}	Thermal resistance, K/W

Lines

(**Wires**). Wires must be neatly laid and fastened. Avoid chafing or contact with moving parts at all costs. Proximity to hot parts (exhaust pipe, brake, …) is equally delicate and must therefore be avoided at all costs. Cables leading to moving parts can be kept protected in corrugated tubes. They must also be long enough to accommodate all movements without stretching. In extreme cases, helical cable sections that contract again when moving back can help.

If cables pass through bulkheads, they can be effectively protected with rubber or plastic grommets.

Certain colour codes are prescribed for some cables. For example, the battery ground cable must be yellow. But codes are also advantageous for other cables. They facilitate troubleshooting, which often has to be carried out quickly and in uncomfortable surroundings (outdoors, . . .), especially in the case of racing vehicles.

Electrical cables contribute to the mass of the vehicle – although not to the same extent as in a passenger car. Therefore, the idea of using light metal instead of the usual copper for cables is obvious. Although aluminium appears to be a simple solution at first glance, it cannot replace copper without additional measures. Aluminium has a poorer electrical conductivity than copper, so the aluminium cable cross-sections must be dimensioned approx. 60% larger when replacing copper. Aluminum also exhibits what is known as creep, which is an increase in permanent deformation over time. This means that special plugs are required to maintain the connection between the cable and the contact spring over a longer period of time. The contact force is therefore not ensured by the contact element itself but by a spring. Material pairings in direct contact with different potentials in the voltage series should be avoided so that no corrosion-promoting local elements occur. Coatings with more noble metals offer a remedy [6].

The ratio of electrical conductivity and density (so-called Performance Index P.I.) can be used for the final decision in the choice of material in order to form the best compromise between these two properties. Accordingly, aluminum is the most suitable (P.I. = 14), followed by magnesium (P.I. = 13) and copper (P.I. = 6.6) [7].

Crimping (pressing) is more suitable than soldering for connecting electrical contact partners. Solder joints have proven to be sensitive to vibration stress. Ninety percent of the reliability problems of engine control units are due to poor contacting [8].

In general, cables made of fine-stranded cores should be used in vehicles. Cables with rigid cores tend to break due to the inevitable driving movements.

In motorsport, cables of the Raychem Spec 44 specification have proven their worth. These have a high current carrying capacity, are mechanically more robust despite thinner insulation, but are more expensive than comparable vehicle cables. It is also advisable to connect these cables with MIL plugs, because the tightness with standard plugs can only be achieved with effort. Another option are cables whose insulation material is Tefzel (MIL Spec 22759), which can be used up to 150 °C.

The CAN bus consists of a 2-core stranded cable with shielding and filling lines and a defined characteristic impedance of 120 Ω. Shielded cables are used when sources of interference (ignition coils, radio link,. . .) could affect important signals (TDC signal cylinder 1, oil pressure, water temperature,. . .).

Connectors

(Plug and Socket Connector). Mechanical connections in electrical systems are one of the main causes of failure, even in production vehicles. The flat and round connectors commonly used in automotive construction are attractive due to their simplicity and low

Fig. 7.8 Multi-pole connector on an engine (Ferrari, Formula 1). The plug is mechanically coded (only one plug-in position possible) and fitted with a bayonet lock so that it is also held positively. In addition, the plug in the joined state seals the housing to the outside

price, but are not very suitable for racing vehicles. The planned reliability is limited to about 10 times release. In racing, connectors from the aviation industry and according to military standards have proven themselves, Fig. 7.8. These seal the electrical contacts against moisture and also ensure a reliable mechanical connection via union nuts or similar. If there is play between the electrical contact partners, this leads to the formation of fretting corrosion, especially in the case of "flying" plug connections, and thus inevitably to electrical problems due to changed contact resistance. Multiple plugs can be opened easily and, above all, often without affecting the electrical contact. In addition to taking up less space, multiple connectors also have advantages when installing and removing parts, because many connections are made in a single plugging process. This saves time and avoids forgetting a connection.

Electronic devices are sensitive to heat and vibration. They should therefore be stored in a vibration-isolated manner and in a place where the heat can be dissipated by air. Simple vibration decouplers are usually made of elastomer and are not conductive for this reason. These devices must therefore be connected to the vehicle ground with a separate cable. It is also advisable to mount the dashboard on the vehicle in a vibration-insulated manner.

Wiring Harness

Individual cables and wiring harnesses are combined to form a cable harness. For the design of a wiring harness, a mixed approach of constructional and design activities is recommended. The rough concept of the wiring harness is created directly on site in the vehicle (or, depending on the construction status, with a dummy engine in the frame/ chassis). The position of the control units, sensors, switches, etc. must already be known. Start with a straight section at least 100 mm long near the main connector to allow space for any terminals. From there, all devices and sensors are supplied with one outlet each. This makes it easy to determine the required cable lengths by laying a "measuring" cable. This concept, including special features – such as points where chafing protection is required or angled connectors – is recorded in a drawing. Cable cross-sections and cable type are calculated or determined and noted in the drawing, as are cross-sections for outgoing cables at thick points. Figure 7.9 shows an example of a drawing of a cable harness as a

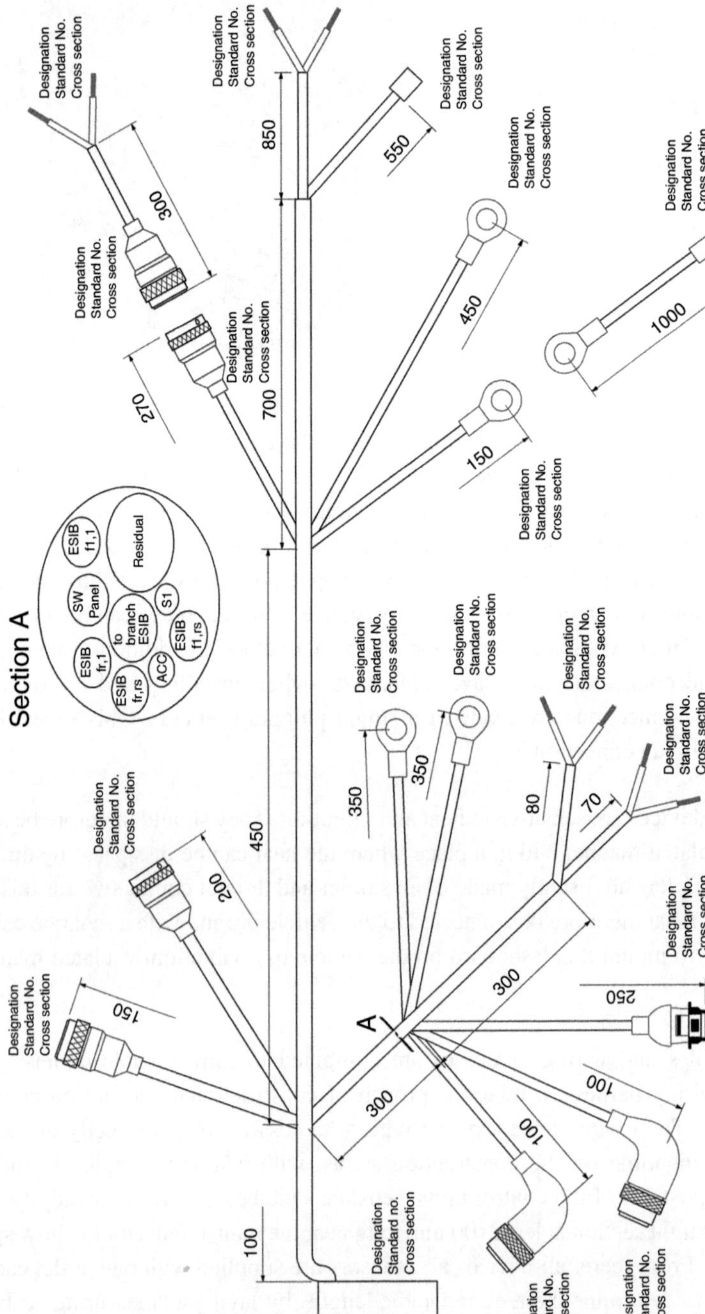

Fig. 7.9 Drawing of a wiring harness. This so-called nail board representation is needed for the production

manufacturing document. There are various types of sheathing (*looming, jacketing*) for bundling cable harnesses. The range extends from simple cable ties to spiral tubing, braided tubing to shrink tubing and, as additional abrasion and temperature protection (up to 800 ° C), to Kevlar temperature protection tubing. Cable ties are used for test setups. Spiral hoses work over a greater length and are similarly repair-friendly to rewinding. A more robust type of wrapping is represented by braided sleeving, which is only surpassed in mechanical properties by slotted tubular sheathing. Shrink tubing can withstand high temperatures and seals the wiring harness when the outlets are glued. However, the production is more complex compared to the aforementioned methods and subsequent modifications are no longer possible. Professional teams prefer this type of wiring harness production.

7.6 Fuses

Fuses have the task of protecting cables from excessive currents by interrupting the current flow. The conductor cross-sections do not have to be oversized and cable fires or destruction of consumers is effectively prevented. Fuses should be grouped together in an easily accessible location in a structural unit. This facilitates maintenance and troubleshooting. *Blade-type fuses*, as are also commonly used in passenger cars, have proved to be the best choice. Glass tube versions and their holders are too sensitive to vibration. A disadvantage of fuses results from their operating principle. If an excessive current flows for a certain time, this circuit is permanently interrupted, i.e. until the fuse is replaced. If this circuit supplies an essential component (fuel pump, coolant pump, ...), the race is over prematurely. Programmable *power distribution modules* offer an elegant method of protecting cables and consumers in a more targeted manner. A control unit monitors the currents occurring in the individual circuits and switches off the current briefly in the event of defined overcurrents. However, the circuit is then supplied with power again and continues to be monitored. If the cause was not a dynamic effect, but e.g. a mechanical problem (blocked pump, clogged line, ...) and thus the current remains permanently too high, the control unit switches off the circuit completely or it can switch the system to a substitute system (if available!) according to a predefined algorithm and thus maintain the function. Via the CAN bus the occurred error and the selected substitute function can be displayed in the cockpit, which makes diagnosis and repair much easier.

This also allows specific faults such as a cable break to be detected. If you know the interval in which the voltage moves for a conductor in normal operation, for example the state "on" = 8 V, "off" = 4 V, a voltage of 0 V means that the cable is broken. This is also used in production vehicles. Here, there is practically no wire that is not diagnosed.

Figure 7.10 shows an example of a power distribution system that replaces not only fuses but also relays and switches. This allows the wiring harness to be simplified by eliminating wires.

Fig. 7.10 PowerBox PBX 190, Bosch Motorsport, [9]. The PowerBox measures 245 × 183 × 37 mm and has 18 analog and 10 digital inputs. Via a programmable software 52 outputs can be addressed. The total current through the PowerBox can be 250 A (peak 310 A) continuously. Communication takes place via CAN bus, LIN or Ethernet

7.7 Switches

A main switch must be capable of interrupting all electrical loads, except possibly an existing electrical fire extinguisher. Its method of operation shall be easy to detect so that it can be quickly recognised by anyone in the event of an emergency, Fig. 7.11.

For safety reasons, it must be possible to switch off the ignition from the driver's seat, as well as the fuel pump.

An example of a central arrangement of switches and connectors in the cockpit of a racing vehicle is presented in Fig. 7.12.

7.8 Circuit Diagram

The circuit diagrams are kept as simple as possible. Fuses are hardly ever included. A basic circuit diagram with a prescribed main switch and a starter battery on board is shown in Fig. 7.13.

In addition, this common circuit of racing vehicles allows the starter motor to be operated independently of the ignition. This allows, among other things, to build up a pre-pressure in the lubrication system, to fill the dry sump system or, in the case of mechanical fuel pumps, to vent the fuel system.

If the regulations allow it, external starter batteries are to be preferred. They protect the on-board battery (at least if their voltage is higher than that of the on-board battery, in the other case they discharge the battery) and can be much heavier. Internal combustion engines also need a certain starter speed for safe starting, for this reason alone a starter

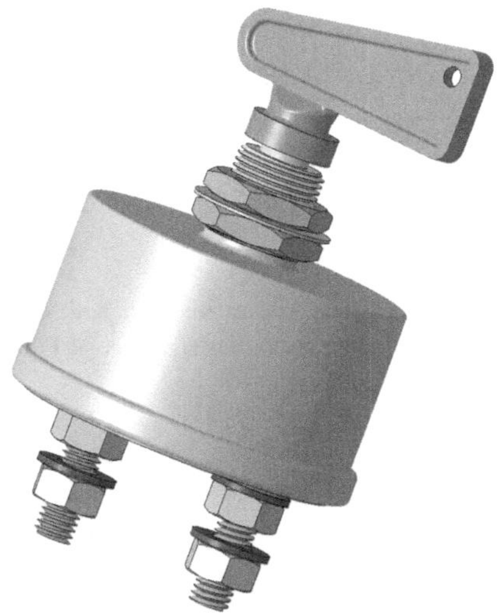

Fig. 7.11 Main switch. The switch housing is screwed onto a wall. It is operated with a red handle by a quarter turn. The handle of this switch can be pulled off similar to an ignition key

Fig. 7.12 Switches and connectors in an open cockpit of a Le Mans prototype (Audi R8S). This CFRP cover is mounted on the cockpit wall. Thus it is accessible from the driver and from the outside

battery with a large capacity is advantageous. The connection between the external battery and the vehicle electrical system is made via a large plug which is suitable for large currents and cannot be reversed. If the plug is in the driver's field of vision, the annoying case of the driver driving off with the starter battery plugged in can be easily avoided. A circuit diagram for this is provided in Fig. 7.14.

While driving, the driver can hardly process all the signals offered to him on the track and on the dashboard. In addition to the engine speed, the oil pressure is also "vital" for the engine. A simple circuit activates a warning light when the engine is running, as soon as the oil pressure drops below a certain value, Fig. 7.15.

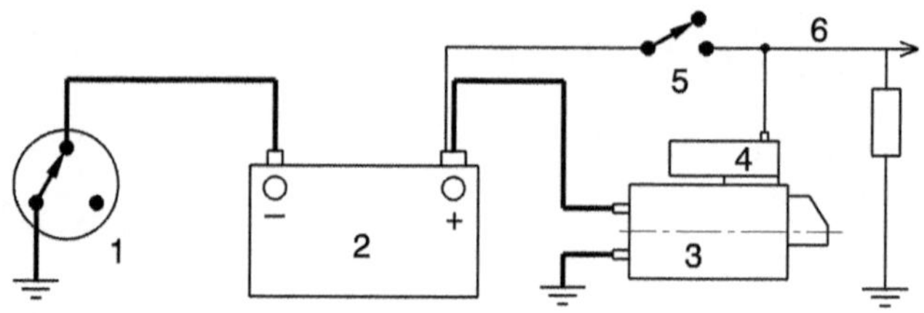

Fig. 7.13 Basic circuit diagram with on-board battery, according to [8]. **1** Main switch (master switch). **2** Battery. **3** Starter motor. **4** Pull-in solenoid. **5** spring loaded starter button. **6** to further consumers

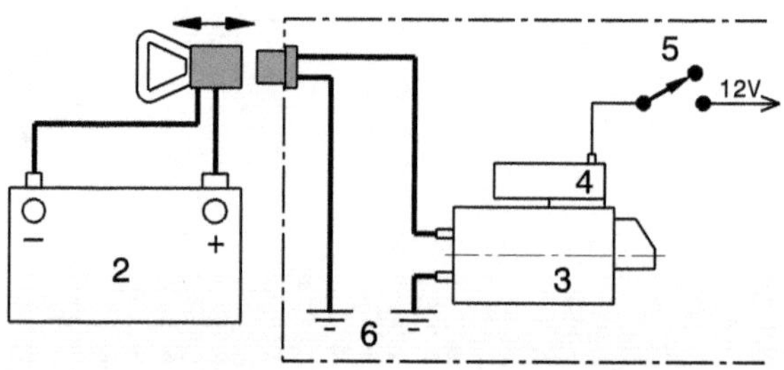

Fig. 7.14 Basic circuit diagram with external starter battery, according to [8]. **1** Connection plug (jump plug). **2** external starting battery. **3** Starter motor. **4** pull-in solenoid. **5** spring loaded starter button. **6** ground (earth) to engine/chassis

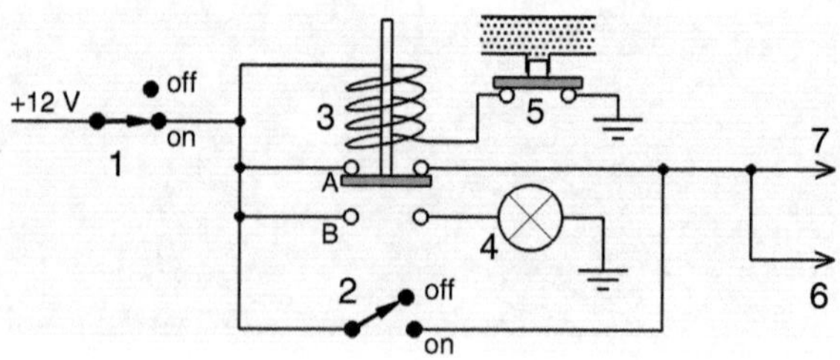

Fig. 7.15 Engine protection circuit for running engine, according to [8]. **1** Ignition switch. **2** spring loaded starter switch. **3** relay. **4** Warning light. **5** oil pressure switch 2,4 bar *(35 psi)*. **6** ignition system. **7 Fuel** pump(s)

When the relay (3) is energized, contacts A are connected and thus the fuel pump (7) and the ignition system (6) are energized. If the engine is switched off or the oil pressure drops below 2.4 bar, the relay switches to contacts B and the warning light (4) lights up.

To start, the ignition switch (1) is set to the "on" position, the warning light lights up and the start button (2) can be pressed. When the oil pressure has reached 2.4 bar, the warning light goes out and the starter button is released.

A circuit diagram of the circuit of an electric fuel pump can be found in Chap. 6 Fuel system.

References

1. Braess, H.-H., Seiffert, U. (eds.): Vieweg Handbuch Kraftfahrzeugtechnik, 7th edn. Springer Vieweg, Wiesbaden (2013)
2. Incandela, S.: The Anatomy & Development of the Formula One Racing Car from 1975, 2nd edn. Haynes, Sparkford (1984)
3. Ward, W.: Kers – Supercapacitors. www.ret-monitor.com. Accessed 30 Nov 2011
4. Indra, F., Grebe, D.: Der Formel-3-Rennmotor von Opel. MTZ 54/11, 576 ff (1993)
5. Wright, P.: Ferrari Formula 1. Under the Skin of the Championship-Winning F1–2000, 1st edn. David Bull Publishing, Phoenix (2003)
6. Pudenz, K.: Leitermaterial im Bordnetz: Aluminium statt Kupfer. www.atzonline.de/Aktuell/Nachrichten/. Accessed 08 Feb 2011
7. Arnold, B.: Werkstofftechnik für Wirtschaftsingenieure, 1st edn. Springer, Berlin/Heidelberg (2013)
8. McBeath, S.: Competition Car Preparation, 1st edn. Haynes, Sparkford (1999)
9. http://www.bosch-motorsport.de/de/de/produkte/catalog_products_1_796661.php. Accessed 07 June 2018

References

8 Electronic Driver Aids

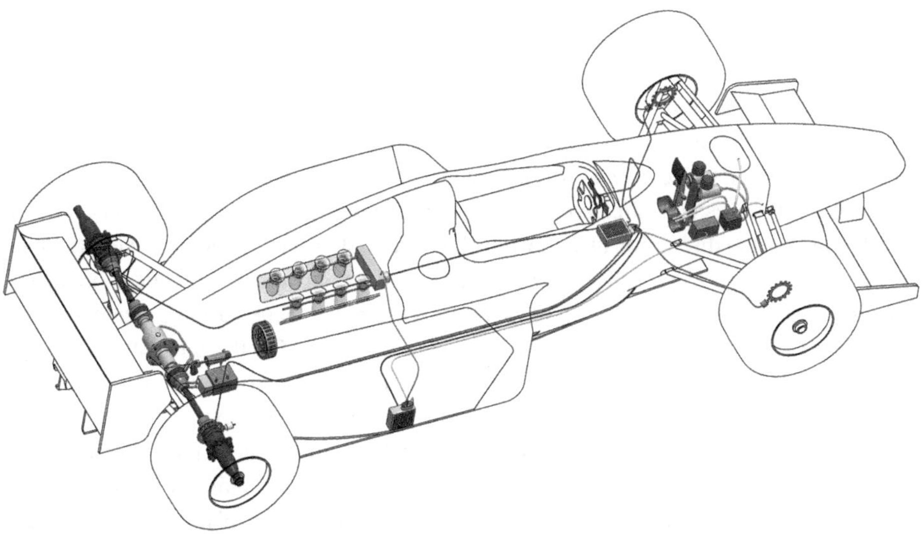

Electronic driving aids are not permitted in every racing class. Where they are permitted, they make the driver's job much easier and he can concentrate better on the other processes while driving and tires less quickly. Where the regulations prohibit them, it is usually so that the driver's influence remains in the foreground during the competition.

M. Trzesniowski, *Powertrain*,
https://doi.org/10.1007/978-3-658-39885-9_8

8.1 Introduction

In the case of passenger cars, numerous assistance systems (ABS, ESP, lane departure warning, etc.) and comfort-enhancing features (level control, air suspension, variable power steering, etc.) are now standard even in the small and mid-size range. In racing cars, on the other hand, comfort plays a role at most when it comes to long-distance races, and the additional expense in terms of space, energy and mass of assistance systems in the vehicle only makes sense if the (lap) time decreases.

In general, driver aids occur at three levels: Stability (ABS, ESP, ARP,[1] active suspension), path guidance (lane keeping, collision avoidance, parking/parking), and navigate (follow route). The required control frequency decreases in the listed order. The support measures for path guidance are summarized under the acronym ADAS (Advanced Driver Assistance Systems). Current developments are aimed at fully autonomous driving, i.e. the driver is to be completely relieved of the driving task. The trend in passenger cars towards autonomous driving is leading to a move away from muscle-actuated and only servo-assisted control systems towards by-wire systems, i.e. those in which the human no longer has a direct mechanical effect on the steering, brakes, throttle, etc., but only via electrical signals.

A further distinction is made between active and passive systems in the case of assistance devices which support the driver in driving the vehicle. While passive systems merely alert the driver to critical driving conditions by means of a signal (display, noise, vibration), active systems effectively intervene in the driving dynamics. It is precisely this fact that ignites the discussion as to how useful such systems are in racing, where the focus is on sporting competition. The driver's influence on the result is reduced in driving situations in which assistance systems at least help to shape driving manoeuvres in fractions of a second – i.e. faster than a human driver can. On the other hand, in many races not only drivers but also designers and manufacturers compare themselves. In this respect, a technically superior vehicle is a worthy winner. It is therefore ultimately up to the writers of the regulations to determine which competition should be in the foreground. This is not an easy task, because on the one hand racing is supposed to embody the spearhead of technical development and on the other hand some technical details of racing cars seem outdated compared to contemporary passenger cars.

An (as yet) unusual solution is represented by races with self-driving vehicles, in which the competition is primarily based on the previously created control software [1].

[1] Active Rollover Protection.

8.2 Active Systems

Active systems can be differentiated between automatic and manual systems. Automatic systems work completely without the driver's intervention, while manual systems are consciously requested by the driver.

8.2.1 Manual Systems

Throttle-by-Wire

The electronic accelerator system was the first by-wire system in passenger cars. The driver's wish is recorded via a potentiometer in the accelerator pedal and ultimately passed on to the throttle valve or the injectors. In between, the on-board computer calculates the appropriate position of the throttle valve in the case of the gasoline engine or the injector opening time in the case of the diesel engine, depending on the driving condition, gear engaged, engine temperature, etc. The throttle valve is set by a small geared motor. This separation of the accelerator pedal and the engine control unit allows any transmission characteristic that can be changed at any time. This makes ASR (acceleration slip regulation, traction control), ESP (electronic stability program), MSR (engine overrun torque control), cruise control and the control of hybrid vehicles possible without additional effort.

***Launch Control* (Starting Aid, Automatic Starter)**

Clutch actuation is not easy when starting from a standstill, especially with high-bred engines, and false starts or stalling of the engine occur time and again. Therefore, there have been and still are attempts to automate the clutch operation. Figure 8.1 shows an overview of the principle. The clutch (2) itself is the same as that operated by the driver with the foot pedal, except that in this case it is operated by an actuator (6) (electro-hydraulic, pneumatic, . . .). The driver's wish is entered via a switch on the steering wheel. A control unit (7) processes the input data from sensors that detect the operating range of the engine (4) and the vehicle (5) and, of course, the clutch function. For this purpose, the clutch actuator has an integrated displacement sensor. Recent developments of such actuators operate hydraulically as two-piston slave cylinders. The two pistons are arranged in series. The second piston enables fully automatic venting on the one hand and movement to "stop" on the other. This measure reduces the control time of the clutch in the open state [2].

At the start, a certain amount of tyre slip of the drive wheels is regulated, analogous to the traction control, so that the greatest possible circumferential force can be built up. During control, all the necessary information is reprocessed by the control unit in 10-ms cycles. The tire slip results from the difference in speed between the front and rear wheels. The non-driven front wheels provide an easy-to-measure approximate value for the vehicle speed. If the slip of the drive wheels exceeds the specified value, the clutch lever is moved in the "open" direction and the excess engine torque is dissipated in the clutch. As a

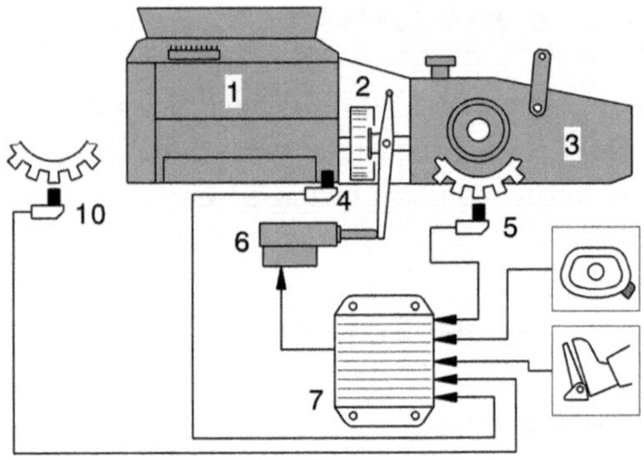

Fig. 8.1 Functional principle of automatic clutch. **1** Engine. **2** Clutch. **3** Gearbox (transmission). **4** Engine speed. **5** Wheel speed driving wheels. **6** Clutch actuator. **7** Control unit. **8** Clutch switch. **9** Accelerator pedal. **10** Wheel speed front wheels

result, the drive wheels decelerate. The control intervenes again as soon as the rear wheels have reached a preset differential speed to the front wheels. The clutch is then actuated in the direction of "closing". This control cycle can repeat up to eight times per second, Fig. 8.2. The engine power remains continuously at the maximum value during this.

Shift-by-Wire

Analogous to the electronic-accelerator system, the mechanical connection between the gearshift lever and the transmission can also be dissolved and replaced by an electric system. Here, too, the greatest advantages – in addition to greater freedom in the arrangement of the actuating device in the cockpit – are the freely selectable transmission behavior and the possibility of networking with other systems. In addition, the on-board computer can carry out plausibility checks and thus, for example, downshift only when the driving speed does not cause the engine to overrev. If the transmission including the drive train is protected by such systems during operation, these assemblies can be made lighter because no overdimensioning is required to protect against misuse.

Semi-Automatic Transmission

The gearshift is initiated by the driver using buttons or paddles on the steering wheel. The on-board computer controls everything else and implements the necessary actions via mostly electrohydraulic actuators. For example, the clutch is automatically opened and closed again after the gearshift. The tractive force interruption of the engine, e.g. via ignition interruption, is also carried out by the electronic control unit, see also Sect. 5.3 (Fig. 5.28).

***Power Steering*[2]**

The steering is one of the most essential actuators in a vehicle. Particularly in a racing vehicle, the driver should not have to reach around when driving and should be able to quickly change the position of the front wheels. This leads to a small steering ratio (direct

[2]For more details see Racing Car Technology Manual, Vol. 4 Suspension System, Chap. 5 *steering*.

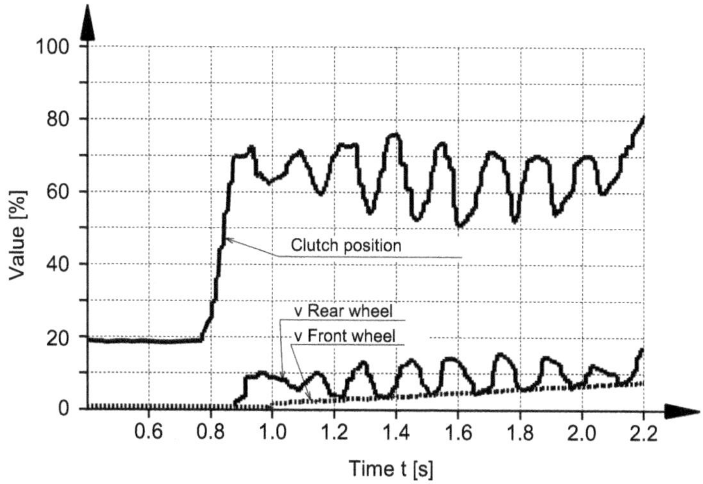

Fig. 8.2 Launch control by clutch control [2]. The clutch is set by the control unit via an actuator in such a way that the slip of the drive wheels remains within a certain range at constant maximum engine power. v Rear wheel Peripheral speed of a rear wheel (driven). v Front wheel Peripheral speed of a front wheel (not driven)

steering). On the other hand, the steering wheel moments should remain small for ergonomic reasons and still provide sufficient feedback on the grip level of the front tires. With increasing aerodynamic downforce, this situation is exacerbated. This argues for a large steering ratio, which ideally increases further with increasing road speed. Steering assist can resolve this inevitable trade-off. Power steering devices are allowed in some racing series. However, as with road cars, only in such a way that the wheels can still be steered by hand even if the assistance fails. The development of passenger car steering systems is primarily driven by comfort. Racing vehicles, on the other hand, should be able to be operated specifically at the limits. Even if a steering aid makes the driver's work easier in the cramped cockpit of a single-seater, in Formula 1 vehicles approx. 30% of the power of a servo assistance proves to be the upper limit, so that the necessary feeling for the road is maintained for the driver [3].

Not all types of power-assisted steering are suitable for racing vehicles either. In rally vehicles, electromechanical power-assisted steering systems (EPS), which are currently displacing hydraulically assisted steering systems (HPS) in passenger cars, are proving to be unsuitable. Average steering angle speeds of 700 °/s and peak values of 1500 °/s cannot be achieved with such systems [4].

Another principle-related disadvantage of EPS steering systems is that they are heavier than HPS systems, and the more mechanical power is required. In the small car segment, both systems still have about the same mass; in luxury class vehicles, the difference is about 1.4 kg [5].

Power-Assisted Braking

As with passenger cars, there are also additional devices for racing vehicles that support the driver's foot force during braking. This reduces driver fatigue and also makes it easier to modulate the braking force. The energy for this must of course be provided by an on-board system. In the case of gasoline engines, the pressure difference generated by the throttle valve in the intake tract lends itself to pneumatic assistance. With diesel and electric engines, the power must be provided by a hydraulic or vacuum pump. In either case, the constructional and control engineering effort increases. The mass of the vehicle is increased and the fuel or energy consumption increases.

Like many other auxiliary systems commonly found on passenger cars, power braking is banned in some racing series.

8.2.2 Automatic Systems

Traction *Control System (TCS)*

This device prevents the drive wheels from slipping too much (therefore also referred to as ASR – acceleration slip regulation). If the slip becomes too great, less circumferential force can be transmitted (this also unnecessarily increases tread wear) and the cornering potential of the tyres is greatly reduced. Both should be prevented for maximum acceleration and to maintain driving stability (directional stability), Fig. 8.3. The control range of the tyre slip is not constant but depends on the tyre, the road surface and the speed, among other things, and is in the order of 8–15%.

Good drivers react to excessive tire slip within a fifth of a second with their throttle foot. An electronic control system reacts much faster and also relieves the driver of this task. The driver makes his wish known via the accelerator pedal, but the control unit of the control system decides whether or how this is implemented. This requires, among other things, the use of an e-throttle. There is no mechanical connection between the accelerator pedal and the throttle valve, Fig. 8.4. The driver's wish made known via the accelerator pedal (1) is recognised, but is only converted into drive torque by the e-throttle control unit (2) up to the point where the drive wheels threaten to spin. Sensors measure the speeds of the front (7) and rear wheels (8) for this purpose. These values are compared with each other by the ASR control unit (3). The slip of a wheel results from the measured values to:

$$S_l = 1 - \frac{v_{W,l}}{v_V} \tag{8.1}$$

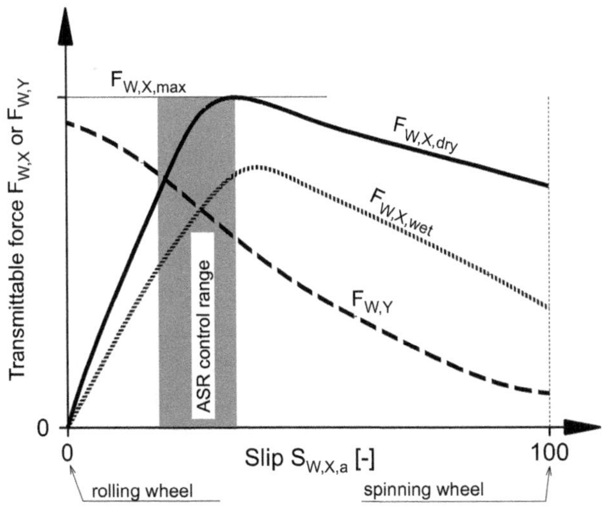

Fig. 8.3 Control range of a traction control system. A purely rolling wheel does not transmit any circumferential force $F_{W,X}$, but can build up the greatest lateral force $F_{W,Y}$. With increasing slip, the transmittable circumferential force initially increases sharply and reaches its maximum at relatively low slip values. For the greatest possible drive forces, the slip at no drive wheel may exceed this value. At the same time, the transmittable lateral force decreases, so that the control range ends at the maximum, so that not too much lateral force potential is lost. The characteristic curves depend, among other things, on the tyre, the road surface and the speed

Fig. 8.4 Traction control scheme. **1** Accelerator pedal. **2** drive-by-wire control unit. **3** *traction control unit.* **4** Engine control unit. **5** Throttle actuator. **6** Throttle potentiometer (throttle position sensor). **7** front wheel speed sensor. **8** rear wheel speed sensor. **9** Engine and transmission

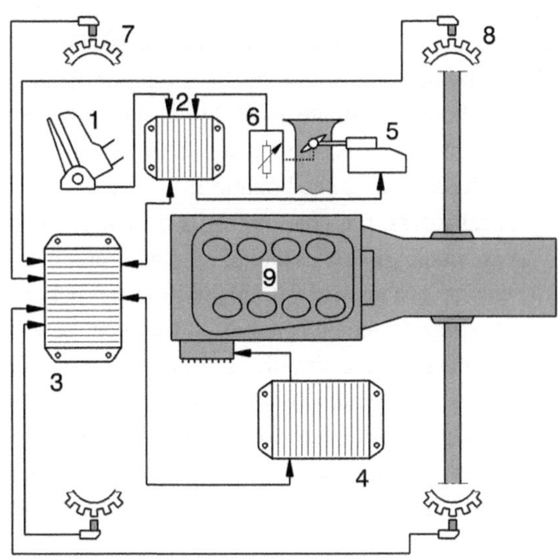

S_l	Slip of the left Tyre, –.[3] the slip of the right Tyre follows analogously from $v_{W,rs}$
$v_{W,l}$	Peripheral speed of the left Tyre, m/s. $v_{W,l} = r_{dyn}\ \omega_{W,1}$
v_V	Vehicle speed, m/s. the value is determined from the averaged front wheel speeds

If necessary, i.e. if the drive wheels slip too much, the engine output is reduced. This is done via the throttle valve, injection quantity and ignition angle in a coordinated sequence. These measures are implemented by the engine control unit (4). The ignition of individual cylinders is modulated with 15–30 Hz, while still switching cyclically over all cylinders so that the individual cylinders are relieved. It is not possible to control the throttle valve alone, this would be too slow, Fig. 8.5. A combination of cylinder deactivation (ZAS) and throttle valve control is best. The throttle is used for coarse pilot control, and ZAS is used for fine control. Other possibilities of traction control result from varying the air number λ (see appendix) or the ignition angle. A disadvantage of the latter method is the increase in exhaust gas temperature at small ignition angles, because combustion may not yet be complete when the exhaust valves are opened and may extend into the exhaust tract.

In principle, a torque curve that decreases above the engine speed also acts as a traction control. As the engine speed increases due to slipping drive wheels, the engine torque decreases and the wheels reach the adhesion limit again (dynamically stable behavior).

There are also systems that, in addition to engine control, also actuate the wheel brakes individually. This makes it possible, on the one hand, to emulate the function of a limited-slip differential by braking the wheel that is spinning and, on the other hand, to maintain driving stability by intercepting a yawing movement through targeted braking on one side. To do this, however, the control unit needs further inputs so that it can recognize the driving condition and calculate countermeasures. For example, it also receives data about the steering wheel angle and the car's rotation from a yaw rate sensor. If the car understeers too much, the brakes are applied primarily to the inside rear wheel, while if it oversteers, the brakes are applied to the outside front wheel. The otherwise disadvantageous decrease in lateral forces due to the braking forces generated has a supporting effect in this case.

In rally cars, traction control is difficult to implement because the ground conditions vary so much (gravel, asphalt, . . .). Therefore, many teams are content with a speed limit of the engine as a control variable to limit the tire slip.

For some drivers, it is helpful if they can "save" themselves in certain driving situations with a burst of the throttle, in which they cause the rear end to break away. In such situations, the traction control must be switched off, otherwise it has a counterproductive effect. This is achieved, for example, by also processing the gradient of the accelerator pedal movement by the control computer. If the rate of change of the accelerator pedal angle over time is extremely high, the computer will allow the wheels to spin.

[3] See Racing Car Technology Manual, Vol. 4 Suspension System, Sect. 1.2.3.

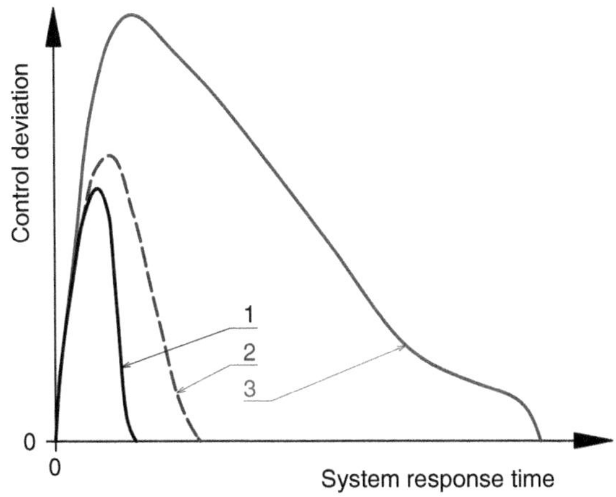

Fig. 8.5 Response times for different control interventions [6]. A control deviation is the difference between the setpoint and actual value. In the present case, this is the difference between the desired tire slip and the actual. Traction control is comparatively performed with different intervention combinations: 1 throttle/wheel brake intervention, 2 throttle/ignition intervention, 3 throttle intervention

Anti-Lock *Braking System* (ABS)

As with starting, there is also a desire to exploit the tyre potential to the maximum when braking. In contrast to a traction control system, however, the focus of the ABS design is on the requirement that the wheel never locks. A locking tire builds up less braking force (sliding friction!), the braking energy is primarily only converted into heat in the contact patch (and not in the brake disc) and lateral control is lost. However, an ABS function is a compromise between drivability and braking effect. On production vehicles, the greatest emphasis is placed on drivability: The vehicle must remain maneuverable under all circumstances. With racing ABS systems, the braking effect is paramount. Experienced pilots can still control a slightly unstable vehicle [7], Fig. 8.6.

To illustrate an ABS control process, let us first consider just one wheel, Fig. 8.7. When braking, the hydraulic pressure in the brake system increases and the wheel begins to decelerate more than the vehicle. The braking slip[4] thus increases starting from zero until a maximum value of the friction force is reached at a tire and road dependent value. A further increase in brake pressure leads to a decrease in braking force, with the risk that the wheel will lock in a short time because the wheel brake torque remains the same or increases while the possible reaction torque of the tire decreases (resulting in excess wheel brake torque). The control system must therefore keep the braking slip within the range of the maximum frictional force of the tire. Unlike a stock ABS, where different tires must be

[4]See Racing Car Technology Manual, Vol. 4 Suspension System, Sect. 1.2.3 *Influence on handling*.

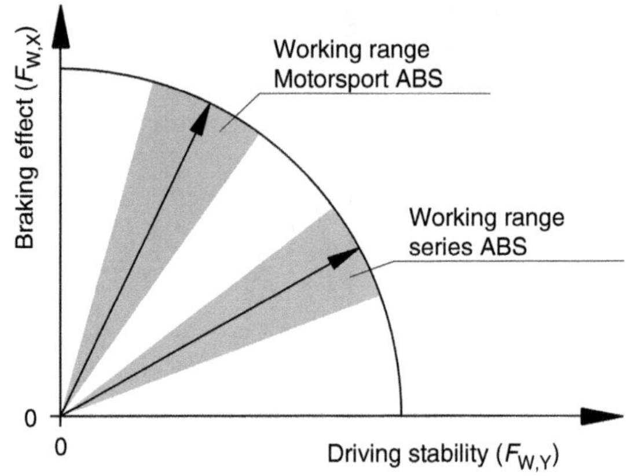

Fig. 8.6 Design differences of ABS systems, according to [7]. With the aid of the Kamm circle, the different design objectives of series and motorsport applications can be illustrated memorably. The circumferential force $F_{W,X}$ is responsible for the braking effect, while the cornering force $F_{W,Y}$ is decisive for stability

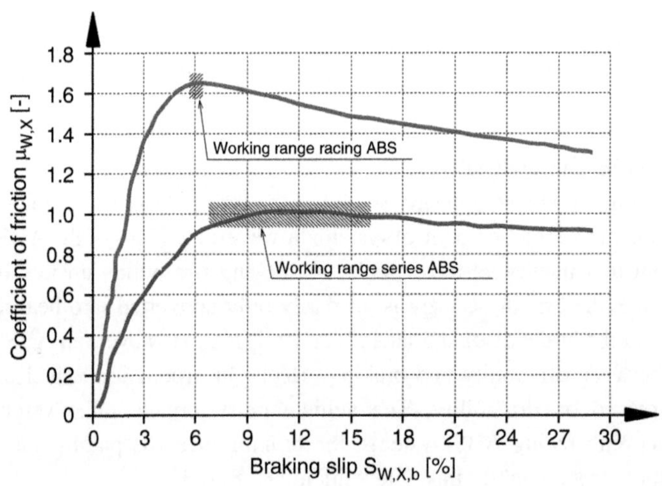

Fig. 8.7 Operating ranges of ABS systems, according to [8]. Production tire (blue) and race tire (red) at the same tire temperatures. Racing tyre: Dunlop DTM Slick 265/660–18. The operating range of the racing ABS is matched to the tyre in a much narrower window than the series application, which has to detect different tyres

covered in a wide slip range, the race version controller can be designed in a narrower range to match the behavior of the tire being used. The controller detects the wheel speed change (angular acceleration) and calculates a vehicle reference speed from data from other sensors (e.g. vehicle acceleration sensor). This enables the computer to determine the current braking slip and adjust it to the desired value by modulating the brake pressure. In the case of a so-called 4-channel system, all 4 wheels of a vehicle can be controlled individually. At least when it comes to braking, this approaches the ideal of taking into account wheel-specific, different wheel loads. Apart from the fact that such a thing is not possible

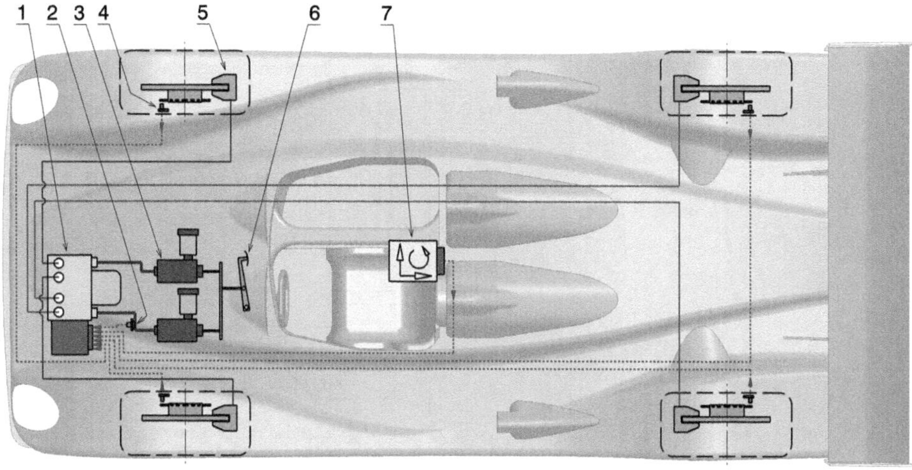

Fig. 8.8 ABS system overview, Bosch Motorsport M4. 1 Hydraulic unit with attached ECU, 2 Brake pressure sensor, 3 Master brake cylinder, 4 Wheel speed sensor (here: front right), 5 Brake caliper, 6 Brake pedal with balance bar, 7 Yaw rate/acceleration sensor (*gyro/acceleration sensor)*. Not shown: ABS function switch (map switch), malfunction indication lamp, brake light switch. All communication between the devices takes place via the CAN bus

with only one brake pedal anyway, a (human) driver could never manage this, especially when cornering.

A system overview with the functionally decisive components and their switching is provided in Fig. 8.8. The distribution of the force from the brake pedal (6) to the front and rear axles is done via a usual balance beam.[5] The brake lines of the two master brake cylinders (3) are connected to the corresponding inputs of the hydraulic unit (1). This influences the brake pressures to the individual brake calipers (5). The control unit, which is located directly on the hydraulic unit, receives the necessary information from the wheel speed sensors (4), from the brake pressure sensor (2) in the front axle brake circuit and from the yaw rate/acceleration sensor (7), which is located approximately centrally in the vehicle. Necessary vehicle parameters – such as wheelbase, track width, tire radius, vehicle mass, drive type, etc. – must be stored in the control unit. Current changes (road condition,. . .) can be set via a selector switch in the cockpit. With this switch, the ABS can also be completely deactivated, which is helpful, for example, during tuning drives for brake balance.

The central mechatronic element of an ABS system, the hydraulic unit, can be seen in Fig. 8.9. The unit is bolted to the car floor in a vibration-insulated manner via silent blocks and mounting plate. The brake pressures to the wheel brakes are modulated by two electrically driven piston pumps. The electrical power consumption in operation reaches

[5] See Racing Car Technology Manual, Vol. 4 Suspension System, Chap. 6 *Brake System*.

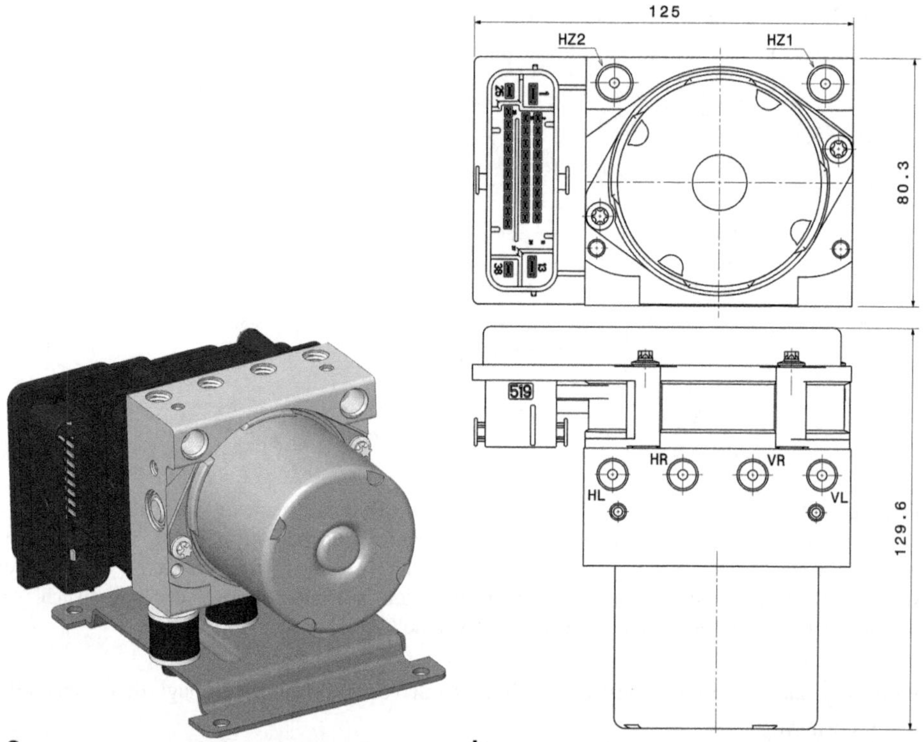

a b

Fig. 8.9 ABS hydraulic unit with attached ECU [Bosch Motosport ABS M4]. (**a**) Axonometric diagram with mounting plate, (**b**) Drawing with dimensions and connections. The connections are: HZ1 Front brake master cylinder, HZ2 Rear brake master cylinder, VR, VL Front brake calliper right or left, HR, HL Rear brake calliper right or rear brake calliper left

230 W. The mass of the unit is approx. 1850 g. To ensure that the hydraulic unit can actually oscillate, the brake lines must be at least 200 mm free and must not be fixed. Rigid brake lines are nevertheless recommended because they cause less control effort than elastic Stahlflex lines with inevitably greater volume fluctuations. The same applies, of course, to the brake calipers. Stiffer designs prove to be much better in terms of response and control quality. When ABS is used, brake piston seals are subjected to greater stress than in conventional brake systems due to the pressure pulsations resulting from the control process. The sealing rings must therefore be checked or replaced more frequently.

The acceleration sensor, Fig. 8.10, is mounted over a damper plate on the car floor near the centre of gravity of the vehicle. It is aligned longitudinally and the plug connection must point to the rear. The inertial forces of a small seismic mass, which is elastically supported, are used to detect the accelerations of the car and convert them into an electrical signal. The longitudinal and lateral acceleration as well as the yaw rate of the car are measured in this way.

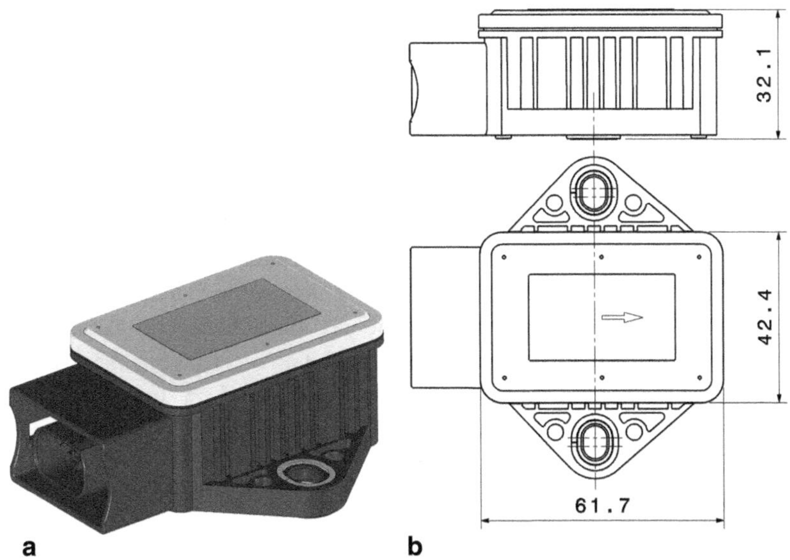

Fig. 8.10 Yaw rate/acceleration sensor *(gyro/acceleration sensor)* [Bosch Motosport ABS M4]. (**a**) Axonometric representation, (**b**) Drawing with dimensions

Anti-Stall System

A further or additional device that relieves the driver, especially in hectic situations – such as those that occur during a spin or emergency braking – is the stall protection. The on-board computer opens the clutch as soon as the engine drops below a certain speed at low driving speed, thus preventing the unit from stalling. The engine continues to run and the driver can ideally continue the journey without delay.

A centrifugal clutch (see Sect. 5.2.3), which engages or disengages only at a certain speed, fulfils the same purpose purely mechanically. Such a centrifugal clutch is also an enormous help at the race start: The clutch pedal does not have to be operated by the driver and the transmitted torque can be controlled purely via the accelerator pedal during the slipping period.

Active *Centre Differential*

Active centre differentials in an all-wheel drive train, which are controlled via a computer unit, are also an enormous help for rally drivers. They intervene in the torque distribution between the front and rear axles according to the driving situation and thus ensure the desired self-steering behaviour of the car on asphalt, gravel and snow. A centre differential like this, paired with a controllable axle differential at the front and rear, represents the pinnacle of what is possible in the drive system; however, it also consumes a great deal of money in development and tuning and, not least for this reason, was banned by the FIA for WRC vehicles in 2010.

Active Axle Differential

An axle differential distributes the drive torque equally to the left and right drive wheels when driving straight ahead and under the same road conditions. When cornering, it enables the kinematically required relative rotation between the two drive wheels. At high lateral accelerations with the associated lateral wheel load shifts and/or with different road friction, open differentials reach their limits. However, passive differentials also only work as desired under certain conditions, so that the idea of actively distributing the torque (*torque vectoring*) comes to mind. Section 5.4.2 Torque distributing differential describes how such differentials influence the drive torque of the two axle halves. Different torques on the tires result in different circumferential forces, which have a beneficial effect on the yawing motion of the car. This affects not only the acceleration phase but also the deceleration phase and the cornering itself.

8.3 Passive Systems

Shift Light

In vehicles with manual gearshift, you will often find shift lights that indicate to the driver that the engine speed for upshifting has been reached. This indicator can be programmed for individual gears, for example to prevent the maximum torque from being reached in the lower gears, which the tyres cannot transmit anyway.

Lock-Up Light

In vehicles with braking systems without ABS, a light on the indicator panel or on the steering wheel can signal to the driver that a wheel on the front axle is locked or is not braking enough. Such devices are naturally preferred in touring cars where the driver cannot see the front wheels.

References

1. www.roborace.com. Accessed on 05.04.2016
2. Eder, J. et al.: Sequenzielles M-Getriebe der zweiten Generation mit Drivelogic. ATZ **11**, 1024–1035 (2001) und **2**, 154–163 (2002)
3. Tremayne, D.: Formel 1, Technik unter der Lupe. Motorbuch, Stuttgart (2001)
4. Scoltock, J. et al.: Automotive Eng. **6** (Juli–August), 14–16 (2011)
5. Pfeffer, P., Harrer, M. (eds.): Lenkungshandbuch, 2nd edn. Springer Vieweg, Wiesbaden (2013)
6. Bauer, H., et al.: Bosch, Kraftfahrtechnisches Taschenbuch, 25th edn. Springer Fachmedien, Wiesbaden (2003)
7. N.N.: Handbuch, Das ABS M4-Paket. Bosch Engineering GmbH Motorsport, Abstatt (2011)
8. Henn, R., et al.: Technology testbed. Professional Motorsport World Magazine Sonderheft Annual Showcase. **2010**, 22–26 (2010)

9 Comparison Series: Racing

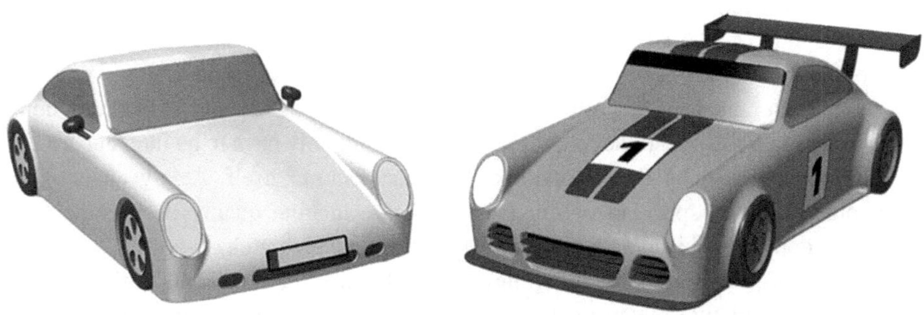

Comparisons between production and racing vehicles are made throughout the book to increase clarity, but here the similarities and differences are to be directly contrasted in a compact summary. This says a lot about both industries, their working methods and products in just a few pages.

9.1 Introduction

Developments in motorsport have always gone much faster from idea to implementation than series developments. This is also understandable. Once an idea has been born, it has to be turned into an advantage on the race track as quickly as possible, otherwise the opposition might take care of it. The results of the developments can be described in general terms as increased performance, increased efficiency, weight reduction, increased reliability and simplified handling. In other words, all achievements that are also quite

M. Trzesniowski, *Powertrain*,
https://doi.org/10.1007/978-3-658-39885-9_9

welcome in production vehicles. Numerous innovations in modern passenger cars also originate from racing. The development results flowed from the sport into the series vehicles (disc brake, four-valve engines, direct injection, light alloy wheels, double clutch transmission, composite materials, . . .). In the meantime, however, the roles have changed to some extent. In the meantime, the race track serves the development departments of large companies as an unbureaucratic test laboratory for pre-development. For example, when it comes to the practical testing of new materials or extreme designs of components. And if the development is successful, it is duly marketed. From the point of view of the car manufacturers, nothing has changed: Racing success brings sales success.

9.2 Development *Process*

The design and construction of an F1 car takes about 5 months. Depending on the team, around 300 engineers are employed. Thousands work on a production car for about 4–5 years. This may seem astonishing at first glance. But in fact there are significant differences in the goals. The racing car has to comply with the regulations and be ready in time for the start of the season. A passenger car also has a completion date. However, this is self-set and can (and will) be postponed if necessary. There are around 15,000 individual requirements in a car's customer requirements, which not only have to be met, but are also meticulously checked by the legislator or the manufacturer itself. A large part of the development time is spent on optimizing production in high quality despite the large number of units. This is the only way to ensure an affordable purchase price.

The competition car has ONE overriding goal, to be superior to the competition and win races. Cars should find buyers in large numbers so that the immense investments are worthwhile. They must therefore be adapted to the needs of the market and meet the (future) taste of the public.

A direct, reliable comparison between series and motorsport development is offered by companies that carry out both. As an example, consider the chronological sequence of engine development for Ferrari's production cars and Formula 1 cars, Fig. 9.1. While in production an engine matures from concept to producible unit in 42 months, the racing department goes through three full racing seasons. It not only designs, builds and develops new engines for each season, but also constantly improves them during the season in order to keep pace with its opponents or, ideally, to outdo them.

Even if the development processes are therefore different, one circumstance is (meanwhile) the same: the design of the vehicles takes place from the outside in. Based on market analyses, trend research results, strategic product specifications, etc., the design department specifies the size, external shape and appearance of the new car. The size, external shape and appearance of the new car. It is now the task of the design department to accommodate all the necessary and desired assemblies and components. In the case of racing cars, the

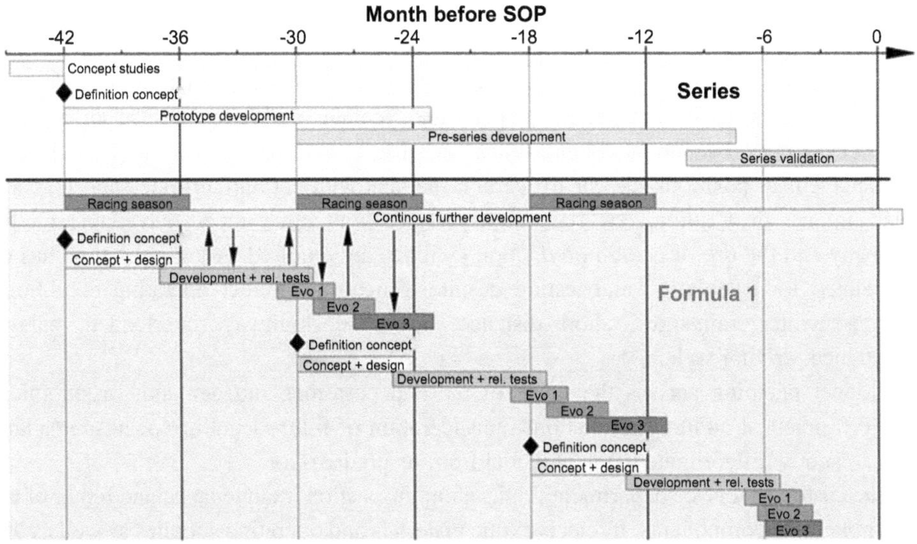

Fig. 9.1 Comparison of the flow charts of Ferrari's engine development, after [1]. Timing of engine development in series production (top) and in the Formula 1 team (bottom). The starting point is the start of production (SOP) of the series engine. Abbreviations: Gen.1: Generation 1, rel. tests: reliability tests, Evo 1: Evolutionary stage 1

process is exactly the same in that the aerodynamics department specifies the shape and thus the desired airflow around and through the car based on CFD calculations[1] and preliminary wind tunnel model tests, and all other development units must subordinate themselves to these specifications.

Often, in the course of a debate about profiting from a racing commitment, the question is asked as to which parts or assemblies have been incorporated from a racing vehicle into a production vehicle. The answer to this question does not have to be about components. Development tools and test methods are also advantageously adopted, as reported in [2], for example.

9.3 Development *Goals*

As mentioned, significant differences can be found in the development goals. Objectives are derived from requirements. In the case of passenger cars, these mainly come from the legislator (sometimes contradictorily from different states), from management, from product planning and from customer service.

[1]Computational Fluid Dynamics. Numerical flow simulations on a 2D or 3D computer model.

Legislation limits emissions (exhaust, evaporation, noise), specifies maximum dimensions and maximum masses, lays down safety standards to prevent accidents and reduce the consequences of accidents, restricts the use of substances that are hazardous to health, regulates the clear labelling of parts and vehicles and takes care of an orderly, resource-saving procedure at the end of the car's life.

Management positions the car for certain markets with certain attributes and planned sales figures. In addition, cost-reducing synergies with other models available in the company and the use of certain production facilities are required. A target lifetime has to be defined for calculation and testing despite difficult to predict operating conditions (summer/winter, rain/snow, short distance, bad road, highway, overloaded, garage maintained, driving style,. . .).

Product planning acts as the voice of the end customer and demands ergonomics, comfort, practical additional functions, consideration of future legal or social trends and aims to subtly differentiate the new model from its predecessor.

The customer service department thinks about inspection, maintenance and repair of all assemblies and components. It rejects exotic materials and operating supplies as well as the introduction of new special tools.

Racing vehicles must only comply with the regulations of the relevant sports authority and the specifications of the management. In individual cases, the main sponsor or purchaser will also make his demands and wishes known. The choice of material is only limited by the regulations and the time availability.

In general, it can be stated that the goals and methods of series and racing developments converge when the planned production quantity and the period of use of racing vehicles increase. Formula 3 cars are a good example of this. The cars are designed for a specific series but for unknown customers and are built for at least 3 years. Customer taste, cost and spare parts supply thus become significant issues in the design process.

9.4 Research and Development (R&D)

Continuous further developments are necessary – even if not for the same reasons – in both camps under consideration. The financial resources required for this must be generated. Twenty-five percent of the turnover of a racing team flows back into R&D expenditure. The automotive industry spends only 6% [3]. In this sense, a racing vehicle is never finished, but is subject to constant improvements, which are nourished by own research or ideas "inspired" by the opponent's solutions. Compared to the car industry, a competition vehicle is a permanent prototype.

Traditionally, racing has been extremely stingy when it comes to publishing research results or key findings. On the contrary, they even try to keep solutions they have found secret for as long as possible. After all, they promise an advantage over the opponents in the race. It is not uncommon for ingenious innovations to have decided entire world championships. Lotus, for example, was the lucky discoverer of winged cars using the ground effect. Williams developed an active suspension that made the aerodynamic effects work even more comprehensively. McLaren benefited from the fiber composite monocoque. Ferrari dominated races with a semi-automatic gearbox. Renault won the title with mass dampers on both axles. The double diffuser was the clever exploitation of a loophole in the regulations that brought Brawn GP laurels. And that's just a chronological list of a few cases.

However, history also offers reverse cases, i.e. those in which a team was able to successfully use an alien innovation. The most famous example is the turbocharger, which Renault introduced and developed for years, but in the end it was Honda's turbocharged engines that won multiple world championship titles.

The fear that one's own ideas can be implemented faster or even more successfully by others is therefore not unjustified. Secrecy, protection against industrial espionage and compartmentalization of the development departments are important strategic measures that shape the everyday lives of those involved. There are no major differences in mindset, regulations, contracts and delivery terms between racing and production development. The goals are basically the same: No one should know about groundbreaking creations before they go into the race or are offered in the new model, certainly not the direct opponents or the competition.

This omnipresent fear sometimes produces strange blossoms and makes cooperation with development service providers, suppliers and external manufacturers more difficult. Especially the change of drivers and employees is treated with great suspicion. One of the ways this is solved is that individuals only have access to very specific knowledge and do not have an overview of the big picture. Gone are the days when a handful of engineers designed and developed a complete car, complete with engine. The five-man engineering group of the Formula 1 teams of the 1980s has grown to over 200 engineers today. The developer of the brake caliper has no idea what his colleagues are doing to improve the steering. And drivers aren't privy to the technical details "under the hood" anyway. What's the point? They need to know which switch activates which function and concentrate on the track, literally.

Racing teams have to accept one major disadvantage in this context. Just as successes can be celebrated with great publicity, failures are also not hidden from the public. However, the press has never seen some of the mistakes made by car manufacturers, and for good reason.

9.5 Costs

Even if costs are at the top of the priority list for passenger cars, this does not mean that they do not play a role in motorsport. On the contrary: in numerous racing series, costs became an insurmountable financial problem in the course of the "arms race" between the participants and led to the end of this series of events. Even the top leagues in motorsport are by no means immune to this phenomenon. It is therefore with good reason that rule writers also look at the cost side when regularly adjusting the rule book. Particularly cost-driven racing series are GT3 and GT4, for example. GT3 cars are expected to be able to race for about 6000 km without having to change parts. In GT4, this figure is as high as 10,000 km or one year. Production cars drive around 250,000 km in 10–15 years (some wear parts such as tyres, brake pads, brake discs, exhaust systems, timing belts, etc. are of course replaced in the process).

However, costs are also becoming a problem in professional racing due to increasing demands, more complex development and internationalisation of the series. History bears witness to several prominent examples where precisely such a development has forced changes in regulations or even meant the end of the series.

Customers' expectations are rising. What was a sensation in the luxury class yesterday is tacitly expected in the small car segment today. Cars must be low-maintenance with long service intervals. Long gone are the days when the valve clearance had to be adjusted after 5000 km, or in winter a radiator grille with hinged elements was used to partially seal off the cooling air intake so that the engines warmed up (faster). Even punctures are hardly accepted and the average customer wants to be able to drive to the next garage without getting his hands dirty. Sensors that can continuously report the internal pressure of the tyres to the on-board computer while driving have proven their worth both in series production and in racing. Admittedly, pressure from the legislator was required for widespread use in passenger cars. For the race engineer, the current values of tire pressure and temperature are valuable information that can be used to determine the tire behavior and thus the driving performance of the vehicle. Over the course of the race, the development of the tires crystallizes and measures can be derived to influence traction and durability. Sensors located directly on the tire inner liner instead of at the usual rim position turned out to be ideal. They indicate the temperature close to the tread and provide an approximation of the actual internal pressure. The values of the pressure transducers on the rim have to be compensated with its temperature. Because the temperature of the rim rises much faster than that of the air in the tire during hard braking, overcompensation sometimes occurs and as a result annoying false alarms occur as the isochore-compensated cold pressure falls below the warning threshold [4].

In series vehicle production, the aim is to have the same (spare) parts for different versions of a type for reasons of cost and logistics. This can lead to parts being undersized for the strongest variant and oversized for the weakest variant. Rally teams are familiar with a similar dilemma, as they are used to maintaining their vehicles during competitions (under the most adverse conditions) and getting them back on the road again after damage. The fewer parts that have to be taken along, the better. As a concrete example, consider the wheel carriers. They are designed in such a way that one version can be fitted to all four wheels. The same applies to special tools. Incidentally, this is also a request made by customer service to the series designers. It should be possible to adjust and mount spare parts without special tools. During series assembly in the factory, robots carry out many work steps and assembly locations must be correspondingly accessible and assembly parts must be correspondingly clearly designed. This is where racing designs differ. The desired uncompromising fulfilment of function takes precedence. In any case, assembly is done by hand. Parts that work together – such as bearing journals and bearing shells, pistons and bushings, meshing gears, axial washers and housings, inner seal and inner channel contours – are measured, classified and meticulously mated beforehand. Not only dimensional tolerances, but also mass and surface tolerances are taken into account. In Formula 1, even screws with thread pitch matched to the elongation[2] are used.

9.6 Environmental Protection

In one area in particular, which is used by many sides as the main argument against motorsport, fundamental changes have taken place in recent decades. Thanks to environmental management, motorsport, like environmental technology, is regarded as a pacemaker for technical progress. This applies in particular to elementary areas such as energy efficiency, avoidance of pollutants, choice of materials and handling of hazardous substances [5]. Based on the ADAC Environmental Plan 2000+, environmental strategies are successfully implemented in 4 central fields of action: organization, technology, infrastructure, and research and science [5]. Lightweight design, tyre technology, occupant protection, alternative drive systems, increased efficiency and wear minimisation can be listed under the area of "technology" which is of interest here [5]. "Green racing" or "clean racing" are the buzzwords under which manufacturers, teams and sponsors visibly (and hopefully pioneeringly) pursue sustainable racing.

In the case of series-produced vehicles, the idea of environmental protection has been imposed much earlier. With mass motorization in the middle of the last century, vehicles not only became affordable for everyone, but also became a burden due to the large number of vehicles. Traffic regulations had to be set up, traffic guidance systems became necessary and ultimately exhaust and noise emissions had to be restricted. However, this does not only concern the operation of the vehicles themselves, but also their production and disposal. In the case of racing vehicles, exhaust emissions are not (yet) the focus of the regulation writers. And for good reason. The magnitude of the effects compared to production vehicles is completely different: At major DMSB events[3], around 95% of the total vehicle kilometres travelled are attributable to spectators travelling to and from the event, who thus contribute around 93% of CO_2 emissions [5]. The following estimate illustrates even more clearly the impact of millions of vehicles compared to a few hundred. The total fuel consumption in German motor sports is less than 3% of the evaporative losses during parking and refuelling of passenger car road traffic [5]. To this should be

[2] In common bolted joints, due to the elongation of the bolt and tolerances of the bolt and nut threads during pretensioning, the first thread generally carries approx. 25–35% of the load.

[3] Deutscher *Motorsport* Bund e. V., umbrella organization responsible for motor sports in Germany. Exercises national sporting authority for automobile and motorcycle sport.

added that the total vehicle evaporative emission of vehicles with gasoline engines at standstill is regulated by law and checked in the so-called SHED[4] test. According to the US typing regulations, hydrocarbon emissions are also limited and monitored during the refuelling process.[5]

CO_2 emissions are directly (and linearly) related to fuel consumption with hydrocarbon fuel. A legal restriction on the fuel consumption of production vehicles and thus on climate-damaging CO_2 emissions therefore makes sense. In the case of racing cars, another consideration has led the teams themselves to use liquid energy storage as sparingly as possible. The less fuel a car needs for the targeted distance, the smaller and lighter the fuel tank can be. As a result, the car has lower driving resistance (mass, cross-sectional area) and benefits from higher performance and greater range. Conversely, higher efficiency means lower losses. These become noticeable through the need for heat dissipation and wear. Increased efficiency pays off directly through smaller heat exchangers as well as shrunken cooling air ducts and less coolant required, which in turn reduces drag and helps reduce mass. Reduced wear lowers the need for lubricating oil and its volume, which must neutralize or suspend abrasive particles with special additives (detergents and dispersants). In addition, the component with the wearing surface can be made thinner. It is no longer necessary to maintain so much wear volume so that the residual load-bearing capacity of the affected component wall remains large enough.

9.7 Technology

9.7.1 Frame and Body

Early in the history of development, the paths of production and racing vehicles diverged when it came to frame construction. A passenger car is supposed to accommodate passengers and luggage, protect them from wind and weather... A racing vehicle is, to put it exaggeratedly, an engine on wheels that can be manoeuvred on board by one person, or in exceptional cases two people. Initially, the ladder frame was standard for all vehicles, but for racing cars it was replaced by tubular space frames, box frames and ultimately spatial shell structures made of fibre-reinforced composites. In passenger cars, these solutions are only found in very small numbers in sports cars, i.e. a preliminary stage to the pure-bred, purpose-oriented racing car. The self-supporting steel body has become established for passenger cars. Strictly speaking also a shell structure. This is no different for apparently production-based vehicles such as rally and touring cars (DTM, NASCAR). They may have a similar shape, which is crucial for recognition value, but under the outer skin there is sometimes completely different technology. Naturally, the difference between

[4] Sealed Housing for Evaporative Determination.

[5] ORVR: On-Board Refuelling *Vapour* Recovery, *vapour recovery* system during refuelling.

the road car and its track counterpart is much smaller for GT3 and GT4 cars. As is so often the case, the degree of specialisation to become a race car is, after all, a question of money. The more cost-effective a racing series is to turn out for the participants, the smaller the extent of the permitted or made conversions may be. This is where the important role of the regulations comes into play, to maintain the orientation of the race series and equal opportunities.

The numerous tasks of the bodywork or body also include the ventilation of vehicle areas (passenger compartment or cockpit, engine compartment, brakes, etc.) and thus, in the broadest sense, the aerodynamic behavior. While air resistance, noise generation, soiling of windows and lights are the most important issues in passenger cars, downforce dictates the development of high-performance vehicles in particular. The extreme in this respect is Formula 1, which even subordinates chassis designs to this issue. For passenger cars, it is sufficient if the shape does not generate too much lift; sports cars should at least achieve slight downforce. Racing car designers also have an easier time of it; they don't have to take into account fields of vision, pedestrian protection, parking pile-ups or tyre covers, but can place wings, chimneys, deflector plates, splinters, spoilers, vortex generators, etc. on the vehicle according to purely technical aspects, subject to the regulations.

9.7.2 Engine

The main differences in the internal combustion engine result from the list of requirements. Car engines should be easy to start in all seasons, regardless of previous operation, operate quietly, have low pollutant emissions, be fuel-efficient over a wide speed and load range, and function as intended for a long time with long service intervals. A racing engine must also last, but in the crassest ideal case only until the finish line. By then, however, it should have converted maximum power or torque from the energy provided in the fuel. The noise released in the process is perceived favourably, at least by the public, and is not perceived as a noise nuisance.[6]

If there is no need to compromise on suitability for everyday use, component designs can be precisely aligned with the desired objective. In extreme cases, this goes so far that racing engines may only be transported in a certain position and with pressurized valve cups. That they must be preheated before they can be started because the bearing and piston clearances are only suitable at operating temperature. That separate spark plugs are screwed in for the warm-up phase.

The engine is a heavy subsystem and should therefore be installed as low as possible in the vehicle. The flywheel diameter becomes the determining factor in this context. Not least

[6]This is a phenomenon that can also be observed at major musical events.

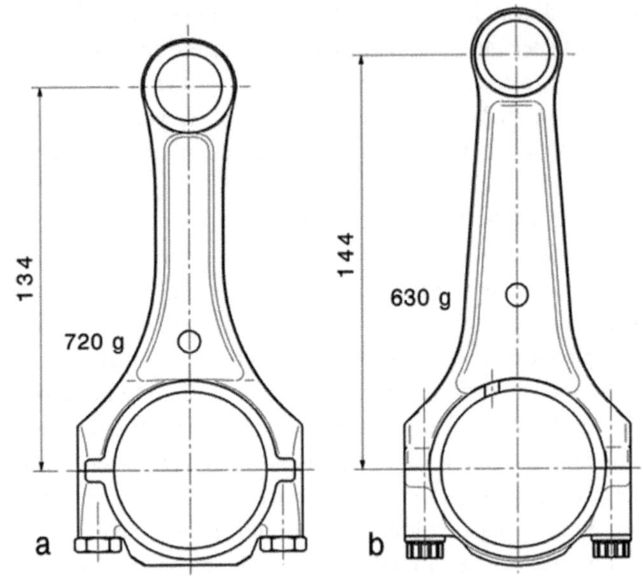

Fig. 9.2 Comparison of connecting rods of a 3.0-l petrol engine, according to [6]. A racing engine was derived from an in-line six-cylinder. With the height of the crankcase remaining the same, the longer connecting rod of the racing engine resulted in a shortened compression height of the piston. Both connecting rods were forged, but the material of the racing connecting rod was of higher quality. This was therefore lighter despite the larger diameter. **a** Series connecting rod. **b** Racing connecting rod

for this reason, racing engines have small flywheels or none at all. High idle speeds and low elasticity only interfere with passenger cars.

Again, a direct comparison should make us aware of the differences that characterise this chapter. A DTM engine is derived from a series engine. In the example considered, Fig. 9.2, the racing connecting rod is longer and more flexurally rigid in the transverse direction, yet its total mass is lower. However, it must be mentioned that the material of the racing connecting rod is of higher quality.

Car engines have to cope with different fuel qualities, which, thanks to electronic control and corresponding sensors, involves fewer compromises nowadays than it did a few decades ago. Nevertheless, racing engines have an advantage here because they only have to be designed for a specific fuel and this fuel is also delivered with a much narrower tolerance of its composition.

Exhaust systems have the task of removing the combustion gases in a targeted manner, lowering the noise emission and reducing the proportion of harmful components. At the same time, the design of the pipe/reservoir system enables the gas dynamic behaviour to be specifically tuned for certain speed ranges. The positive scavenging gradient from the intake to the exhaust side is supported and the residual gas quantity (performance!) or the scavenging losses (fuel consumption!) are kept small. In the production vehicle, a compromise is made in the direction of low emissions (noise, pollutants). In the case of a racing vehicle, the focus is on performance. A power-demonstrating soundscape is even enjoyed by the public to a certain extent. Other uses of exhaust gas in racing vehicles are as an energy source in *Energy Recovery Systems* (ERS) and (at least in large-volume, high-revving engines) in supporting aerodynamic elements such as the rear diffuser and rear wing. In production vehicles, ERS will be used in hybrid powertrains. This allows the energy present in the fuel to be used more fully, or at least stored for later use.

Basically, the same applies to electric motors, even if in this case the many years of field experience are still lacking on both sides. Electric vehicles eke out a niche existence in road traffic and on the racetrack. Nevertheless, a lot has already been done in the comparatively short development time for racing vehicles and they often attract attention by setting a new record. The higher electrical currents caused by the greater power output in racing vehicles drive batteries with high internal resistance to the limits of their thermal load capacity and these thus become the bottleneck in the increase in power. And this is true in both directions, i.e. when driving and when (regenerative) braking. Other energy storage devices – such as supercaps – are more advantageous in this context. Improvements have also been made to current-carrying parts in the motor. For example, conductor cross-sections have been optimized (trapezoidal shape, etc.). Cooling types and media and magnetic flux directions are varied with the aim of further increasing the power density. Wheel hub motors or motors close to the wheel make it possible to come remarkably close to the driving dynamics goal of allocating torque to individual wheels (torque vectoring).

9.7.3 Power Train

The differences in the powertrain are similar to those in the engines. Service life, operability and comfort characterize the development and results of series-produced vehicles. Low-noise, easy-to-change gears or transmission systems that automatically change gear ratios to save fuel are in demand for passenger cars.

Lightweight transmissions that transmit engine power to the wheels with low losses and are also a load-bearing part of the vehicle are preferred for racing vehicles. Damping elements, synchronising devices and additional masses on the outer gearshift to support the gearshift movement are not to be found on a racing gearbox. These parts increase mass and create losses. The disadvantage of lack of comfort is not an issue on race cars because it is not a competitive factor. Much more important is the ease of adapting gear ratios to engine and track conditions. In production transmissions, oil is generally not even changed in the course of a vehicle's life. Gear geometry in racing transmissions is designed with shaft deflections in mind for the highest transmission efficiencies. Series gearboxes must be able to be manufactured and tested in large numbers using standard tools, and gear meshing must be quiet. This is one of the reasons why hypoid gearing is often used for axle drives, although it has disadvantages in terms of transmission efficiency due to a high degree of longitudinal sliding in the rolling motion.

9.7.4 Suspension

The biggest differences between production and racing vehicles are probably in the area of suspension. This is not surprising, since the suspension has the greatest influence on handling. Cars are designed with comfort and safety in mind. Great compromises have to be made in the process. The vehicles are operated without modification – apart from summer and winter tyres – all year round in all weather conditions with different load conditions, with and without roof superstructures, with and without trailers on different road surfaces. Furthermore, very few drivers check the inflation pressure of the tyres or even their settings such as toe-in or camber before setting off. In the case of racing vehicles, these are routine activities that can also make all the difference on the results list. Racing suspensions must therefore be easy to adjust and within the required range. Production tyres should deliver long mileage with low rolling resistance and excellent wet grip (a contradiction that constantly challenges tyre developers). Noise emissions and rolling comfort are also considered and evaluated. Racing tires seem to have it easier in this regard. They are supposed to have consistently high grip in circumferential and lateral directions and not change their behaviour. The short service life is accepted and sometimes even deliberately used by the organisers of racing series to increase the excitement for the spectators. Compared to their production counterparts, racing tyres are only partially vulcanised and only cure during operation. The rim width of production wheels is made as narrow as possible. Racing tyres, on the other hand, are mounted on wheels whose rim width is 1–2 inches wider than the tyre. This noticeably increases tire volume and lateral stiffness. The offsets of the standard wheels primarily take into account the balanced load on the wheel bearing pairs when driving straight ahead. In the case of racing vehicles, camber stiffness is decisive in this context. Figure 9.3 illustrates how corresponding compromises are made in the suspension designs. If – as in many passenger cars – the focus is on high tyre mileage, the static camber of 0° is selected and the tyres are fully utilised when driving straight ahead (top row). When cornering, the low camber is disadvantageous; the tyre cannot transmit the maximum possible lateral force. For sportier cars and sports cars, the compromise can be shifted in favour of cornering (middle row). For racing cars, the compromise is to the detriment of straight-line driving (bottom row). Tyre mileage is not an issue and for most tracks the importance of lateral dynamics far outweighs longitudinal dynamics (ratio up to 4:1). Tire utilization at maximum lateral acceleration is paramount.

Even though driving safety is the primary concern, there is still a classic conflict of objectives when it comes to spring/damper tuning in production vehicles. On the one hand, ride comfort should be high (soft spring, low damper effect) and on the other hand, wheel load fluctuation should be as small as possible (high damper force), Fig. 9.4. A solution to this conflict of objectives is provided by (technically complex) variable dampers.

To increase driving safety, the suspension joints on passenger cars are deliberately designed to be flexible. When lateral or circumferential forces are applied, the corresponding wheel "steers" in such a way that understeering behaviour results. Not

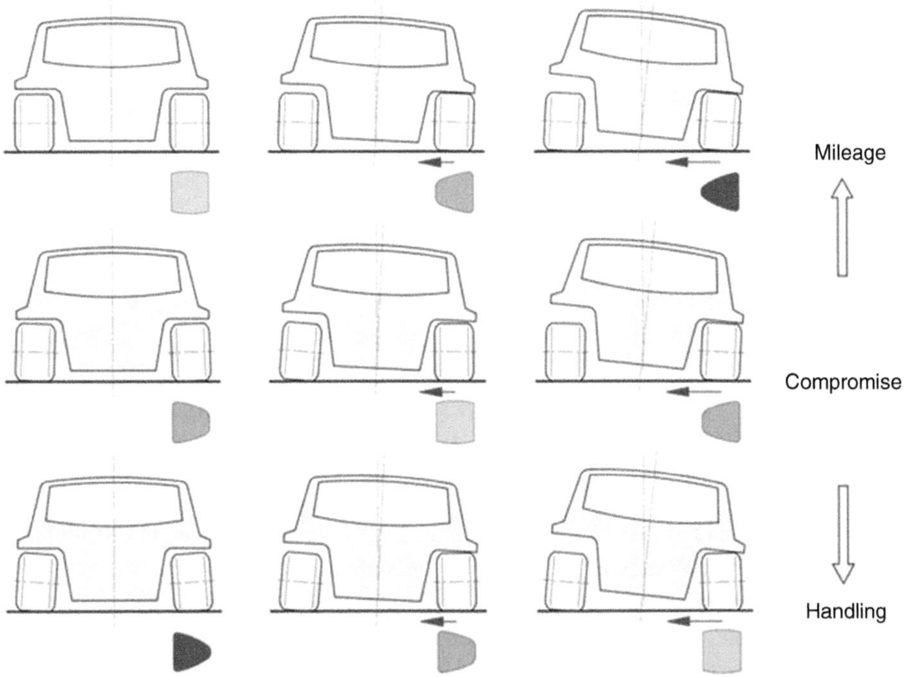

Fig. 9.3 Types of chassis designs, according to [7]. Three designs each are shown as a view of a wagon from behind when driving straight ahead (left), cornering (centre) and extreme cornering (right). In addition, the tyre contact patch shape is shown under the right-hand wheel or the wheel on the outside of the corner. The ideal contact patch utilization is shown in green. In this case the maximum tyre surface is evenly loaded on the road. In combination with lateral slip and missing or unfavourable camber angle, the worst utilisation of the tyre results, the tyre contact patch is strongly trapezoidal (red). Between these extremes you find the middle contact patch utilization (yellow). Depending on the primary goal (mileage or handling), different camber angles are accepted or aimed for, and thus different utilization or wear of the tire

only can racing drivers not use this elastokinematic wheel suspension, it also no longer functions at all in the saturation range[7] of the tires. Racing suspensions are therefore designed with almost play-free, low-friction joints. The racing driver assumes that the vehicle does what he expects through his steering inputs. The essential difference between production and racing suspensions becomes apparent when a sports car is converted to GT3 use. In [9] it is reported that the suspension of the sports car has been simplified for racing use. Not least so that problems on the track could be dealt with more quickly. For similar reasons, the adaptive dampers of the production car were replaced, but the control arms

[7] If the lateral force no longer changes via the slip angle, a steering movement of this tyre also no longer brings about a change in the lateral force.

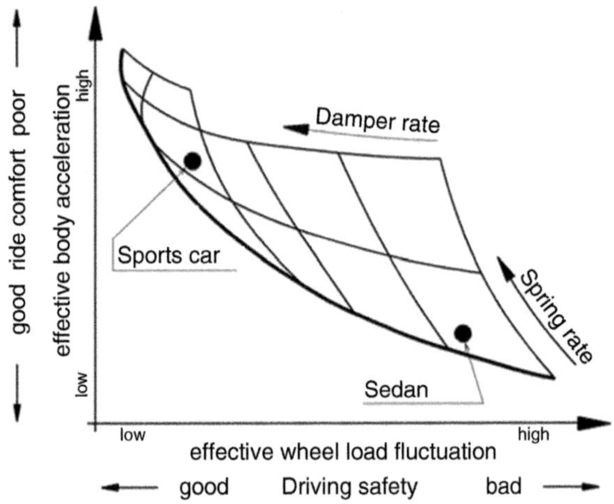

Fig. 9.4 Influence of spring stiffness and damper rate on the driving behaviour, after [8]. A stiff body spring and a hard damper increase driving safety enormously, but reduce driving comfort due to the high body accelerations caused. For racing vehicles, the choice of spring and damper rate is easier in this respect. For them, only driving safety counts

remained the same. On the GT3 car, the front tires are wider and the weight distribution is more balanced than on the base car. The race car is about 100 kg lighter.

In general, racing vehicles are tuned more neutrally than production vehicles, for which (dynamically stable) understeering behaviour is recommended for safety reasons.

The disturbance force lever arm – the normal distance of the wheel center or the contact point from the steering axis – together with circumferential forces on the front tire (changes in rolling resistance, fluctuations in braking force,. . .) cause disturbance torques around the steering axis, which the driver perceives at the steering wheel. In comfort-oriented passenger cars, this disturbance information is kept as small as possible, even if it does contain a certain amount of useful information. In sporty vehicles and racing cars, this information should be more pronounced.

References

1. Jenkins, M., et al.: Performance at the Limit. Business Lessons from Formula 1 Motor Racing. Cambridge University Press, Cambridge (2005)
2. Steiner, M.: Serienmodelle profitieren vom Rennsport-Know-How. ATZ Jubiläumsausgabe 120 Jahre ATZ 03, pp. 138–142. Springer, Berlin (2018)
3. Reuter, B. (ed.): Motorsport-Management. Grundlagen – Prozesse – Visionen. Springer Gabler, Berlin (2018)
4. Kunzmann, S.: Elektronische Reifendrucküberwachung mittels batterieloser Transpondertechnologie. In: Krappel, A. (ed.) Rennsport und Serie – Gemeinsamkeiten und gegenseitige Beeinflussung, pp. 183–197. expert, Renningen (2003)
5. Ziegahn, K.-F.: Umweltschutz und Nachhaltigkeit im Motorsport. In: Reuter, B. (ed.) Motorsport-Management. Grundlagen – Prozesse – Visionen, pp. 311–333. Springer Gabler, Berlin (2018)
6. Indra, F., Tholl, M.: Der 3,0-l-Opel-Rennmotor für die Internationale Deutsche Tourenwagenmeisterschaft. MTZ. **52**(9), 454 ff (1991)

7. Serra, L., Andre, F.: Suspension systems: Optimising the Tyre contact patch. Auto Technol. **4**, 66–68 (2001)
8. Krimmel, H., et al.: Elektronische Vernetzung von Antriebsstrang und Fahrwerk. ATZ. **5**, 368–375 (2006)
9. Scoltock, J.: McLaren MP4-12C GT3. Autom. Eng. **Jul.–Aug**, 8 f. Caspian Media, London (2011)

Appendix

Glossary

1D simulation: One-dimensional charge exchange calculation for pre-optimization of pipe lengths, tank volumes and valve timing of an internal combustion engine. Based on one-dimensional unsteady, compressible filament flow (acoustic theory), the engine is simulated as a system of pipes and tanks and wave travel times are determined. The torque curve over the rotational speed can be determined via the cylinder filling that occurs. With this method, cam profiles, valve timing, intake manifold lengths, distributor volumes, duct geometries, exhaust pipe lengths and mufflser designs can be pre-optimized without an existing test bench engine. Well-known software tools are available from AVL, Gamma Technologies, LMS, Lotus or Ricardo, among others.

ABS: *Anti-lock braking system.* A control system in the hydraulic circuit of brake systems reduces the pressure in the brake line applied by the driver via the brake pedal as soon as a wheel threatens to lock. This requires, among other things, sensors that record the wheel speeds and compare them with a setpoint calculated from the deceleration. The main function of an ABS is to maintain the steerability of a vehicle. Locking wheels cannot build up usable lateral forces, which can lead to a loss of stability, especially on the rear axle.

If different friction values occur on the left and right (μ-split), the driver must countersteer when braking. ABS can support the driver in this by building up braking force more slowly on the front wheel with more grip (yaw moment control). The yaw moment acting on the vehicle thus also builds up more slowly and there is

M. Trzesniowski, *Powertrain*,
https://doi.org/10.1007/978-3-658-39885-9

more time to countersteer. This inevitably increases the braking distance. In addition, the rear axle is controlled according to the wheel on the low friction value side (select low) [1].

An extension of the ABS control system is ABSplus or CBC (Cornering Brake Control). Here, the system detects the driving situation – in particular cornering – by means of the wheel speeds and regulates the braking forces on the individual wheels accordingly in order to keep the vehicle on track.

Acceleration: Is the rate of change of velocity over time. In purely physical terms, it can be positive or negative, i.e. the speed increases or decreases. In the case of vehicles, however, we generally speak of acceleration and deceleration.

ACO (Automobile Club de L'Ouest): Automobile club that has been organizing the 24 hours of Le Mans since 1923 and issues the regulations for the vehicles eligible to start. It also sets the regulations for the former European Le Mans Series (ELMS) and the American Le Mans Series (ALMS).

Actuated transmission *(shift by wire)*: Manually operated manual transmissions have a mechanical connection (linkage, cables) between the shift lever and the actual actuating device on the transmission housing. If the actual shifting process is carried out via actuators (pneumatic or hydraulic cylinders, electric motors, ...), shifting can be initiated by the driver at the push of a button or by the on-board computer (automated transmission).

ALMS: Abbreviation for American Le Mans Series. In this American racing series, the same regulations apply as in the famous 24-hour Le Mans race. However, the races are shorter and last between 2:45 and 12 hours.

Anisotropy: Directional dependence of certain material properties, e.g. modulus of elasticity, strength. The opposite behaviour is called isotropic.

Boundary layer: If air flows around a stationary body with a favourable flow, the air follows the contour of this body the closer to the surface the considered air layer is. Due to friction effects, an air flow slows down the closer it gets to the surface of a stationary body. Thus a static to slow flow is formed at the surface of the body, the thickness of which increases towards the end of the body, the so-called boundary layer. Depending on the shape of the body and pressure conditions, this B. detaches from the surface of the body with increasing thickness and turbulence after a certain distance of flow along it. Outside this B. the friction can be neglected and the velocity of the particles increases with the wall distance.

Buckling: Type of failure of slender, bar-shaped components that transmit compressive forces. Compared to an ideal load that only presses the bar, imperfections occur in reality that lead to additional bending of the component. If the compressive force becomes too great, the bar deflects laterally in the middle and fails due to the excessive bending stress.

CAD: Abbreviation for Computer Aided Design. Components and their assemblies are designed three-dimensionally with the aid of suitable software. Clearances and movement spaces can be controlled more easily than on the drawing board, and numerical simulations (strength, flow investigations, . . .) can also be carried out. Some of the data can be used directly for the production of real components. *See also*: Rapid Prototyping.

CAN: Abbreviation for Controller Area Network. A two-core cable harness used instead of many lines to transmit signals in vehicles. It is a serial bus system in which messages from all participants (ABS control unit, engine control unit, sensors, actuators, . . .) can be sent or received one after the other. The CAN controller controls this sequence and sets the priorities if several signals are to be sent simultaneously. The wiring harness in a vehicle with CAN is much shorter than in a conventional system and the number of plug connections is halved.

CART: Abbreviation for Championship Auto Racing Teams. American formula series that was contested in oval stadiums and on road courses. The 2.6-liter V8 engines were powered by methanol and accelerated the single-seaters to 400 km/h. 2003 bankruptcy. Afterwards new start as ChampCar. Champ Car has since (early 2008) initially merged with IRL to form a formula series for financial reasons, and shortly afterwards became officially insolvent.

CFD *(computational fluid dynamics)*: Similar to the finite element method (FEM), the geometry to be investigated is broken down into small areas (“grids”) for which the equations describing the flow are solved numerically. Depending on the equation used (potential equation, Euler equation or Navier-Stokes equation) and computer performance, even hydrodynamic boundary layers, turbulence and flow separation can be determined.

CFRP: Carbon-fibre-reinforced plastic. A composite material in which carbon fibre fabrics are impregnated with reactive resins and processed in several superimposed layers to form moulded parts or with internal honeycomb cores to form sandwich constructions. The targeted arrangement of the aligned fibres makes it possible to influence the mechanical component behaviour in the desired way.

Chassis: This term is widely used from chassis to body. Probably not least because a clear separation into different assemblies cannot be made in every case. In this book, it is intended to mean the actual, load-bearing structure of a vehicle, to which wheel suspension, powertrain and bodywork components are attached. Accordingly, another term for C. is frame. In most passenger cars, the body is of self-supporting design and thus the bodywork, frame and floor panel are combined to form a structural unit. A clear assignment of the terms to one component each is therefore not possible in this case.

Coefficient of friction μ: Value determined by tests to calculate the frictional force between two bodies. The C. depends, among other things, on the material pairing.

Compression ratio ε: The C. of an engine is the ratio of maximum and minimum cylinder volume. The largest volume results when the piston is at bottom dead center. This volume is therefore the displacement of a cylinder plus the so-called compression volume. The smallest volume is enclosed by the piston at top dead centre. This volume represents the compression volume. The compression volume is made up of the combustion chamber volume and other components that result from the piston crown shape.

Concept: First phase in a design process. In this phase, possible solutions for sub-functions of the overall system are sought and assembled into an effective structure. This phase is followed by the design phase.

Coordinate system: Of the common, vehicle-fixed coordinate systems, the following is used in this book in accordance with DIN ISO 8855 (was DIN 70000) and ISO 4130: The coordinate origin is the intersection of the vehicle longitudinal center plane with the front axle. The trihedron is aligned to this as follows. The positive x-axis points in the direction of travel, transversely the y-axis points to the left and the z-axis points upwards.

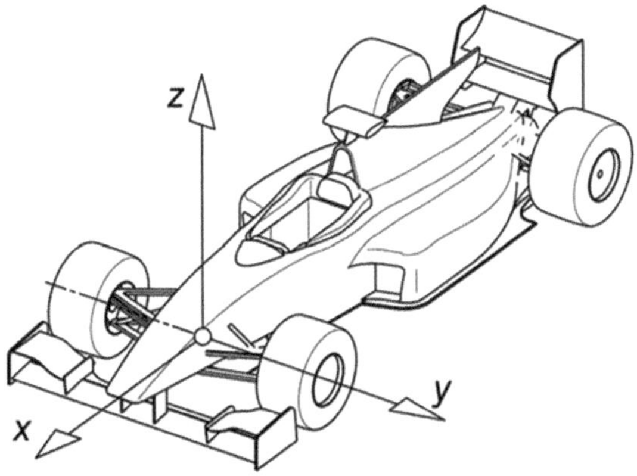

Degree of freedom (DOF): An DOF. is a defined change of position of a rigid body according to a unique and reproducible function. A body has six DOF. in space (three translations and three rotations). The machine elements that enable such DOF. are called joints. A spherical bearing offers three (rotational) DOF. as a ball joint. All displacements (the possible three translations) are locked. The piston rod of a damper leg is a rotational joint. It has two DOF: A translation (compression/deflection) and a rotation (rotation around the piston rod axis).

Design position *(reference standing height)*: Certain position of a vehicle in relation to the roadway, which is used as a basis for the design of running gears. Usually, the car is ready to drive with a half-full tank and the driver on board. Based on this position, the car can bounce in and out or pitch and roll. All nominal dimensions, e.g. for ground clearance, ground distance, king-pin inclination, caster, etc., are thus obtained in D.

Differential construction: Design principle in which a functional carrier (component) is broken down into several parts. Each part can then be optimised for its partial function, e.g. multi-part wheels. The opposite is represented by the integral design.

Draft *(embodiment design)*: Phase of design activity in which the proposed solutions literally take shape. The search for solutions before the layout is the concept (conceiving) phase.

Drag: Force acting on moving bodies due to the fact that they displace air and that the air rubs against the surface of the body.

Drivability: For the human driver, a linear, predictable response of a system to his input is best. This is also the case with the accelerator pedal: A good D. means that the engine delivers as much torque as the driver expects based on his foot movement when he accelerates. Special attention is paid to the breakaway behaviour (*tip-in*), i.e. the opening of the closed throttle. Here the engine should increase its torque gently and not move the vehicle forward abruptly. A good D. supports the driver during acceleration maneuvers, especially in overpowered, traction-limited vehicles.

Driving *(operating behaviour)*: In the picture (according to [2]) the trajectories of three vehicles are shown, which drive a turn with constant steering angle. The only difference between the vehicles is the position of the centre of gravity. In the understeering car the center of gravity is more forward, in the oversteering car it is more backward compared to the neutral vehicle. All vehicles require a run-in period where slip angles of the front wheels are first established, followed by a slip of the rear wheels. The vehicle begins to yaw and deviates from the original straight line. Only then comes the phase of constant cornering. In the case of a neutral vehicle, the slip angles of both axles are the same.

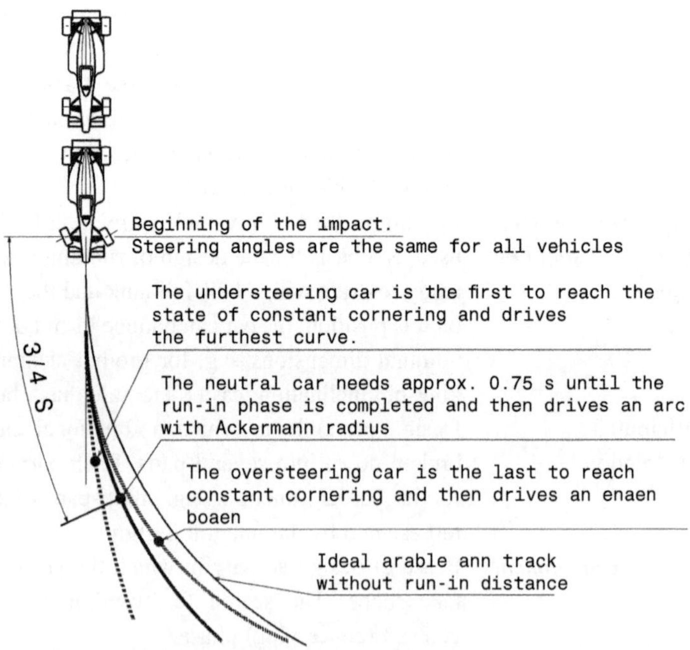

DTM: Abbreviation for Deutsche Tourenwagen Masters (*German Touringcar Championship*). Touring car series whose vehicles must be based on production cars with at least four seats. The engines must be four-stroke gasoline engines with eight cylinders in a V arrangement with 90°. The engine capacity is limited to 4 litres.

Dynamic rolling radius: The distance between the centre of the wheel and the contact patch surface of a stationary wheel is smaller than that of a rolling wheel (static tyre radius). Depending on the tyre design and the wheel speed, the distance increases with increasing speed. The dyn. R. as measured value is calculated from the measured rolling circumference of a tire at 60 km/h.

Elongation at rupture: Relative elongation of a test bar at which fracture occurs. The E. is thus a measure of the toughness of a material. The higher the E., the more favourable is the fracture behaviour of a material, because failure is announced slowly.

ESP: *Electronic stability programme.* Control system that influences driving stability. Sensors record the driving condition of the vehicle, in particular the yaw movement and the steering reaction of the driver. If the state of the car deviates from the calculated target state, the system intervenes by applying brakes to individual wheels or influencing the engine management system. ESP has a stabilising effect, for example, in the event of panic evasive manoeuvres, corners taken too fast or tyre blow-outs.

FIA: Federation Internationale de l'Automobile. World automobile federation with headquarters in Paris. Issues the international sporting code and is thus also the supreme motorsport authority.

Filling pressure *(inflation pressure)*: The pressure difference between a tyre and the ambient pressure. The F. is usually measured on a cold tyre. For example, if the air pressure is 1 bar[1] and the absolute pressure in the tyre is 2.5 bar, the inflation pressure is 1.5 bar. This is also referred to as overpressure.

Finite element method (FEM): Stress calculation of components with numerical methods by a computer. In this process, the component is broken down into (thousands!) finite elements and each element is calculated according to the laws of mechanics. These approximation methods also allow the stress calculation of parts of complex geometry and load, which cannot be calculated with formulas.

GFRP *(glass-fibre-reinforced plastics)*: Glass fibre reinforced plastic. Plastics reinforced with glass fibres in the form of mats, fabrics and strands of parallel threads to increase strength. GFRP parts are used as bodywork parts, wings, moulded parts.

Glass transition temperature In plastics, a characteristic change in behaviour occurs when a certain temperature is reached. Below this so-called G., the oscillatory movements of the macromolecules come to a standstill and the materials become brittle. On further cooling, they reach a

[1] 1 bar = 100 kPa. Although the valid SI unit for pressure is Pascal (Pa), the book uses the unit bar, which is more "handy" in practice.

	glassy-hard state. In the case of tires, the greater the difference between the G. of the rubber compound and the operating temperature, the softer the rubber becomes and the more friction it builds up.
Ground clearance:	Distance between the vehicle underbody and the road surface. A distinction must be made between this and the ride height.
Haptics:	H. is the study of haptic perception. Haptic perception is the active sensing of size, contours, surface texture, weight, etc. of an object through the sense of touch.
Heat exchanger:	Structure in which heat is transferred from one liquid or solid substance with a higher temperature to another with a lower inlet temperature without the two substances being mixed with each other. Depending on the media involved, a distinction is made between, for example, water/air or air/air heat exchangers for the charge air cooling of a turbocharged engine.
High voltage (HV):	Electrical voltages greater than 60 V DC or 25 V AC are referred to as high voltage in the vehicle sector. This term is used to distinguish this range from "high voltage" in industrial standardization, which also has completely different numerical values. HV cables and connectors are identified by orange insulation.
IMSA:	International Motor Sports Association. International motorsport authority that runs the American Le Mans races, for example.
IndyCar Series:	Formerly IRL (Indy Racing League): Organizer of the Indianapolis 500 miles (Indy 500 on Memorial Day, May 30) and other races under the same rules on oval tracks. The cars are single-seaters and were powered by methanol-fuelled V8 engines with a displacement of 3.5 litres. Ethanol is now used as fuel and the engines are 2.2-liter V6 biturbo engines. The cost of the vehicles is limited by the regulations.
Integral construction:	Design principle in which an attempt is made to accommodate all the functions that a component must fulfil in one component. This eliminates the need for weight-increasing and strength-reducing joints. One example of this is sideshafts made from one piece with integrated tripod journals. The opposite is the differential design.
Isotropic:	The material properties are the same in all directions. The opposite behaviour is called anisotropic.
Knocking:	In a gasoline engine, a limit to the increase in compression results from (partially audible) knock at full load. Knocking is an uncontrolled sequence of a combustion initiated by the spark plug. Especially towards the end of a knocking combustion, high pressure peaks occur which propagate at the speed of sound in the

combustion chamber and damage the piston crown, seal surrounds and cylinder head. Therefore, continuous knocking must be avoided at all costs. This is achieved, among other things, by fuel additives, setting a rich fuel-air mixture, reducing the ignition angle, reducing the boost pressure, cooling the intake air, shaping the combustion chamber and targeted cooling of problematic combustion chamber areas (spark plug seat, exhaust valve seat rings).

Labisator, balance spring, Z-bar: Z-shaped connecting rod of the wheels of an axle. In contrast to a stabilizer bar, this arrangement reduces the wheel load differences and thus increases the grip on this axle. The self-steering behavior is thus influenced in exactly the opposite direction compared to the stabilizer.

Laminar flow: The flow runs in superimposed layers that do not mix. This means that there are no cross flows (turbulence).

LMS Le Mans Series: Is a racing series held according to the rules of the famous 24-hour endurance race at Le Mans. The races are usually held over 1000 km. Several drivers are entered per vehicle due to the duration of the race.

Load collective In general, the load on a component is not constant over time, but changes irregularly. A drive shaft, for example, is subjected to extremely high loads during start-up and after a gear change, but almost no loads at all when braking and driving through a turn. However, simplified representations of loads (forces, moments) are required for the design of components. In test series (e.g. driving through a certain course), loads are therefore recorded and evaluated over time. In such evaluations, among other things, the load heights that occurred and their frequency (temporal proportion, number of load changes) are determined. The figure shows how a load collective is created from a load course.

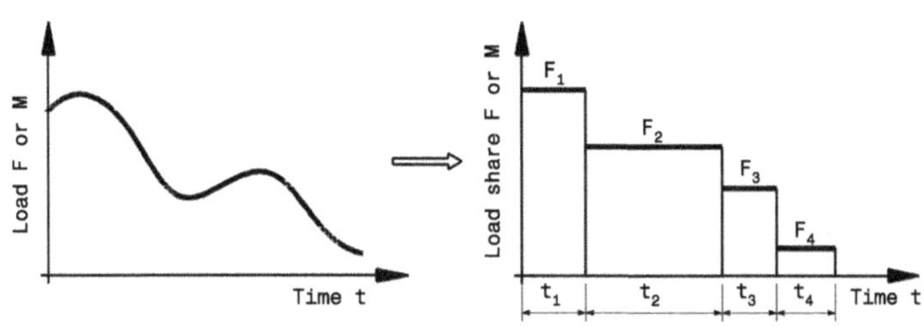

Mass inertia (first Newtonian axiom): For a body to change its direction of motion or speed, a force must act on it. This force is proportional to the acceleration and the mass (Newton's second axiom), $F = m \cdot a$.

Mean effective pressure $p_{m,e}$: During an operating cycle of an internal combustion engine, the pressure in the combustion chamber changes. The mean pressure is a calculated comparative variable. It is an imaginary constant pressure that would perform the same work on the crankshaft as the actual periodically changing pressure in the course of an operating cycle.

Mixture formation: The task of mixture formation in an engine is to produce an ignitable and combustible air-fuel mixture under all operating conditions. These mixtures only burn satisfactorily in a narrow mixture range. If the air content increases (lean mixture), fuel consumption decreases until combustion misfires increase and the running limit is reached. If the fuel content increases (rich mixture), the engine power increases until the fuel can no longer be completely burned due to a lack of oxygen.

Moment of inertia J_{polar}: The M. in a rotation is a measure of the resistance to changes in angular velocity and is thus comparable to the mass in a translation. The M. depends on the distribution of the mass in relation to the axis of rotation. The further away mass components are from the axis of rotation, the greater the M.

Momentary pole *(instanteneous centre)*: Every movement between two rigid bodies can be described by a rotation around an instantaneous (= momentary) axis of rotation (= instantaneous pole). The location of the M. is therefore also the location at which no velocity exists between the bodies under consideration. The specification of the M. in coupling gears is done by the combination of the related links. In the picture a four-link gear is shown. If links are mounted in the (fixed) frame 1, the bearing point is taken as the M., i.e. in the example joints 12 and 14 for links 2 and 4. If links under consideration are not directly coupled, the M. can be determined by knowing two velocity vectors belonging to the rigid body. So here the pole for the links 1 and 3.

If forces act between two links, the position of the line of action of the force in relation to the M. of these links determines which kinematic state occurs. In the example, the force F_{31} (force on link 3 of link 1) causes a clockwise rotation. If the M. 13 would lie on the line of action of F_{31}, the gear would remain in equilibrium. If the line of action is below the M. 13, link 3 rotates counterclockwise [3].

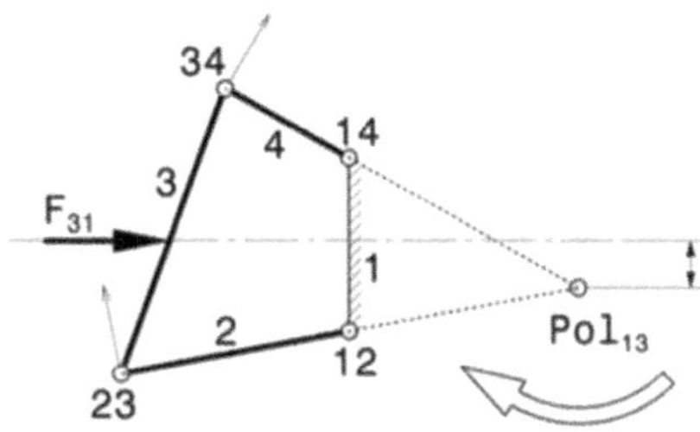

Monoposto *(single-seater)*: Single-seater racing car in which the driver's seat is located on the longitudinal centre plane of the vehicle.

NACA air inlet *(NACA air duct)*: (National Advisory Council for Aeronautics). Design of an air shaft according to the recommendations of the NACA.

NASCAR: Abbreviation of National Association for Stock Car Auto Racing Inc. This is the rules authority for the NASCAR Sprint Cup Series (was 2004–2007 Nextel Cup Series, before that Winston Cup), a popular racing series in North America, the majority of which is run on oval tracks in stadiums. It represents the highest level racing series in the United States. The cars look like production cars on the outside, but consist of a tubular space frame and until 2011 were powered by carburetor engines driving a rigid axle on trailing arms via a cardan shaft. Since 2012, gasoline injection (Multipoint Fuel Injection MPFI) has been used.

Natural frequency: An oscillating structure performs an oscillation (= a periodic movement around the rest position) after a single impulse left to itself. The frequency occurring in this process is the natural frequency. If such a structure is excited with a frequency equal or nearly equal to the natural frequency, the oscillation amplitudes become maximum (resonance).

Notch factor: The stress on a component at a point is determined by calculating the mechanical stresses (bending stresses σ, torsion stresses τ etc.). In conventional calculation methods, the so-called nominal stresses are first determined, which result from the cross-section at the notch bottom of the unnotched component and the load. (In contrast, numerical methods exist that allow the approximate calculation of the stress curve, see Finite Element Method). The component is subjected to higher stresses at notch locations. The local stresses at the notch base are considerably greater than the nominal stresses. The notch factorK_f indicates by how much the maximum stresses become greater than the nominal stresses under dynamic, i.e. time-varying, loading. For bending applies:

$$\sigma_{b,\,\max} = K_{f,\,b} \cdot \sigma_{b,\,n}$$

and for torsion:

$\tau_{ts,\,\max} = K_{f,\,ts} \cdot \tau_{ts,\,n}.$

Where the index b stands for bending, ts for torsion and n for nominal. A value of $K_f = 1$ therefore means that the component is completely notch insensitive.

Trace of axial stresses in a notched tension bar:

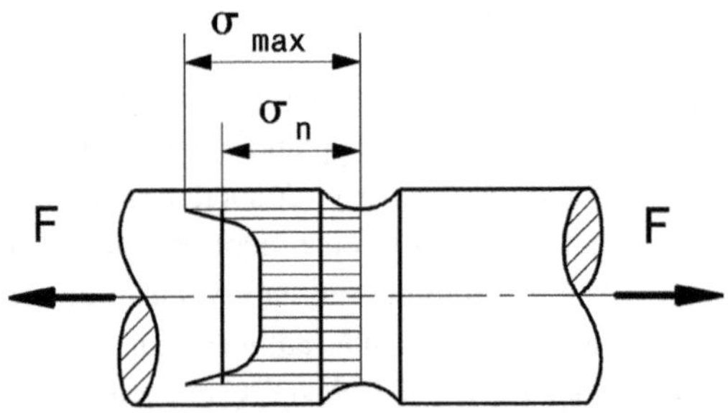

O-arrangement *(back to back a.)*: Two angular contact ball bearings or tapered roller bearings can be arranged in mirror image. If two bearings are fitted in such a way that the pressure lines point outwards (i.e. as in "O"), this is referred to as an O arrangement of bearings. If the pressure lines point towards each other, this is known as an X arrangement.

Octane number: A parameter for the anti-knock properties of a fuel. The higher the octane number, the more resistant the fuel is to knocking. Two different methods are used to determine the octane number: The engine method (engine octane number MON) and the research method (research octane number RON).

Oversteer, *AE: loose*: see driving behaviour

Percentile: Division of a population (normal distribution) into 100 sections. Here statistical division of the dimensions of the human body. This is used to design cockpits and passenger compartments that are suitable for a large proportion of the population. Thus, in car design, the 95% man and the 5% woman cover 90% of the total population. That means only 5% of men are taller and only 5% of women are shorter than the percentiles used.

Pitching: see vehicle movements

Planar moment of inertia: Mathematical quantity that follows from the geometry of a cross-section. The MoI. is required in the strength calculation for bending stress on components.

Pressure angle: At this angle, the force is transmitted from the outer ring and inner ring in a rolling bearing. The greatest load carrying capacity for a bearing is obtained when the contact angle coincides with the angle of the external bearing force.

Prototype:	Racing car of a certain category and group, which is only produced in small numbers or as a one-off.
Quality control:	The control of the load (and thus, in the case of constant load, of the speed) takes place in diesel engines by controlling the fuel supply to the combustion chamber. The engine draws in the combustion air unthrottled. This results in the desired air-fuel mass ratio in the combustion chamber solely by changing the fuel quantity.
Quantity control:	In gasoline engines with conventional mixture formation (carburetor, intake manifold injection), the load (and thus the speed at constant load) is controlled via throttle elements (throttle valve, slide valve). In the partial load range, the air or mixture quantity supply to the engine is changed by throttling the intake cross-section. At full load, the entire cross-section is released.
Raid, Rally Raid:	This generic term covers endurance races that are held cross-country in open terrain, primarily in desert regions. The basic course of the competition is similar to that of a rally, i.e. the vehicles drive from a starting point to a specific destination.
Rally:	These are competitions that are held on sections of road that are closed off for the duration of the competition. The road surface can be asphalt or similar, gravel, but also snow and ice. Each vehicle usually drives the route alone. A characteristic feature of R. is that a co-driver announces the course to the driver.
Rapid prototyping (3D printing, additive manufacturing):	This includes all processes with the help of which real models can be created directly from 3D CAD information. Some of these processes work like a printer that prints out three-dimensional plastic bodies. Depending on the process and the purpose, these models can be demonstration objects, test parts, casting models or molds. The aim is to quickly arrive at a functional (prototype) part (name!) based on CAD data.
Reynolds number *Re*:	Is a dimensionless similarity ratio in fluid mechanics. It compares the inertia forces with the friction forces in a fluid. In a wind tunnel test with a scaled-down vehicle model, the R. values of the model and the original must be the same in order to obtain comparable flow fields and thus useful measurement results.
Rockwell hardness:	Indication of the hardness of a material. Is determined via the permanent penetration depth of an indenter (cone, ball) into the workpiece.
Roll:	see vehicle movements
Ride height:	Is the distance of any point fixed to the vehicle from the road. During set-up, a certain ground clearance is assumed as a reference

value and the car is set higher or lower. The ground clearance is therefore only a metrological simplification for determining the ground clearance.

Rubber: Collective term for rubber-based elastomers (a subgroup of plastics). The actual rubber is obtained from the thickened sap (latex) of the rubber tree by sulphur treatment (so-called vulcanisation, leading to wide-meshed cross-linking of the molecules). In addition to this natural rubber, there is also synthetically produced rubber. The best known representative is Buna, which is produced by polymerizing butadiene.

For elastomers, the service temperature is higher than the glass transition temperature. For the other plastics (thermoplastics and thermosets) it is exactly the opposite.

Self-steering properties: (See also Driving Behaviour.) In the limit range of the drivable lateral acceleration, the vehicle rotates about its vertical axis in a different way than corresponds to the steering angle during pure rolling of the tyre. The lateral forces increase differently at the front and rear axle (more precisely at each individual wheel). However, lateral forces are only transmitted to the rubber-tyred wheel if it rolls at an angle to its plane (slip). If the slip angle on the front axle of a vehicle increases faster than on the rear axle, the car "pushes" out of the curve via the front wheels. The driver has to turn in more than he would have to if the car was just rolling (understeering S.). The opposite behaviour is called oversteer. The behaviour of a vehicle with (approximately) equally increasing slip angles at all wheels is called neutral. However, a given vehicle need not exhibit the same self-steering behavior over the entire drivable limit range. In addition to vehicles that exhibit constant behavior, there are also those that understeer at low lateral accelerations, but switch to oversteering behavior at higher lateral accelerations, and vice versa. In addition, there is the not inconsiderable influence of circumferential forces on the drive wheels, especially at high engine outputs. For example, a rear-wheel drive vehicle that is neutral when rolling will oversteer when accelerating strongly because the drive forces cause the tires to become laterally softer.

Sequential shifting: A type of gear change in a manual transmission in which the individual gears are only engaged one after the other (sequentially). The driver merely has to make a simple movement. Motorcycle transmissions are an example of this. In contrast, common passenger car manual transmissions have an H-shift, where any gear can be engaged with a compound movement.

Shear modulus (slip modulus) G_{shear}: Material constant determined by shear tests on test specimens. For many materials, the ratio between shear stress and angular distortion remains the same under shear loading. This ratio is the S.

Simulation: Simulations are used to calculate the effects of complex physical relationships, usually over time. For this purpose, the system to be investigated is first represented in simplified form by a model. This model is then described mathematically by a system of equations. With the help of a computer, this system of equations is solved (usually by numerical approximation methods). The results are then displayed (visualized) as graphics or animations. Simulations allow many changes to be made to the system under investigation in a short time, which would either not be possible or too expensive on the real object in isolation. Among other things, the driving behaviour of a car with different tyres, axle loads, centre of gravity heights, downforce, etc. on different routes (which must of course have been recorded three-dimensionally for this purpose) is simulated. Because of the simplifications made, a simulation does not represent reality exactly, but it does provide qualitative information about the factors influencing the system under investigation. By comparison with measured test results, models are tested for their usefulness and subsequently improved.

Spring rate: Specification of the spring stiffness. If the behaviour of a spring is plotted in a force/displacement diagram, the spring characteristic curve is obtained. The slope of the characteristic curve is the spring rate c_{Sp}. The spring rate does not have to be constant, but can change during compression. If the spring becomes stiffer during compression (the line becomes steeper), this is called progressive behaviour. The opposite behaviour is called degressive. The characteristic curve flattens out and the spring becomes increasingly softer when loaded.

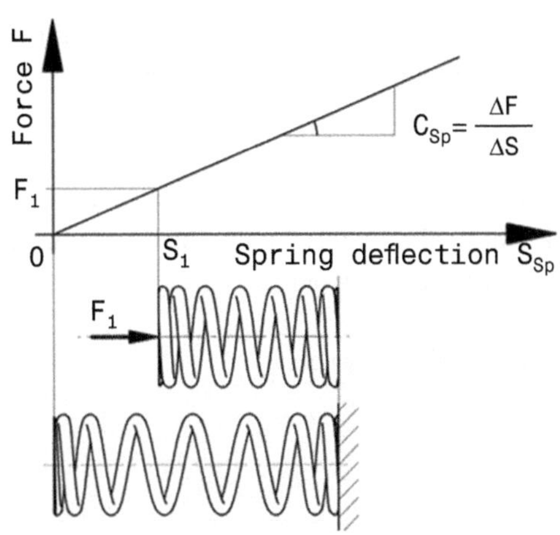

Stress: An external load (force, moment, torque) causes a stress state in the material structure of a component. This stress state is the stress. It is recorded by (technical) stresses (tensile stress, compressive stress, shear stress, . . .).

Stroke/bore ratio: The ratio of the piston stroke s to the cylinder bore B in a reciprocating engine. Based on the appearance of a cylinder from the side, a distinction is made between square (stroke = bore), undersquare or long-stroke (stroke > bore) and oversquare or short-stroke (stroke < bore) engine designs. The figure shows schematically a short-stroke (a) and a long-stroke (b) design of a crank mechanism.

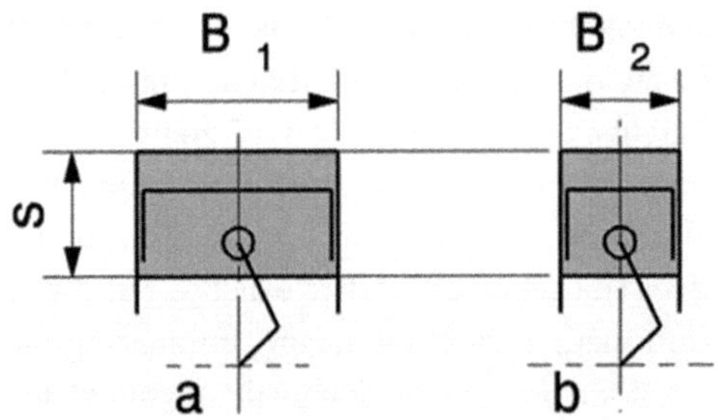

Tensile strength R_m: Material characteristic value determined in a tensile test. It results from the quotient of the maximum force during the test and the cross-section of the test bar before the test. The T. is included in many material abbreviations.

Tension stress: If a component is loaded by external forces and/or moments or if it is hindered in its thermal expansion, a stress occurs in the interior. This stress is recorded mathematically by mechanical stresses, e.g. in N/mm^2= MPa. If the stress at a point in the component exceeds a material-dependent characteristic value, failure (cracking, flow, . . .) occurs at this point.

Tribology: The study of the interaction of friction, lubrication and wear. If relative motion occurs between bodies, this leads to loss of energy (friction) and material removal (wear).

Turbulent flow: Is a flow form in which cross flows and turbulence occur in different sizes and directions.

Tyre contact patch: The contact area of a tyre. All forces between the tyre, and therefore the vehicle and the road, are transmitted via this surface.

Understeer, AE: push: see driving behaviour

Vehicle coordinate system *(axis system)*: see coordinate system

Vehicle level: see ground clearance

Vehicle motion: A vehicle – like any rigid body – has six degrees of freedom in space. The possible individual movements (displacements and rotations) about the three main axes are designated as follows:

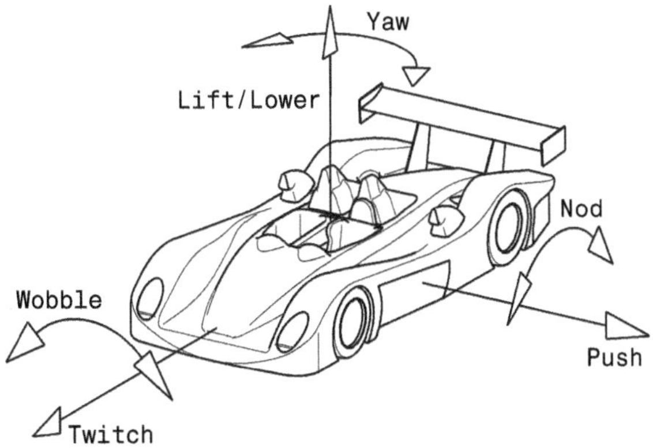

Displacements (translations): Along the longitudinal axis: twitch *(jerk)*.

Along the transverse axis: Push *(drift)*.

Along the vertical axis: lifting or lowering *(heave)*.

Turns (rotations): Around the longitudinal axis: roll (tilt).

Around the transverse axis: pitch.

Around the vertical axis: *yaw*.

If a vehicle drives on a roadway, the movements are a combination of the possible individual movements and result from the given movements of the roadway and the driver's influence by steering.

Wheel frequency: Natural frequency of an oscillating wheel connected to the car body by spring and movable links.

WRC - *World Rally Car*: Rally car based on a generous set of regulations that do not stipulate a minimum number of cars built. The minimum weight is 1230 kg. The number of cylinders in the engines is limited to eight. The displacement depends on the number of valves and the supercharging method. Other rally cars belong to group A and N. For these cars it is required that 2500 basic models are built within one year. To Group A we owe such road cars as the Lancia Delta Integrale, Mitsubishi Lancer Evo and Ford Escort RS-Cosworth.

Yawing: see vehicle movements

Yield strength R_e: Material characteristic value determined in a tensile test. If a bar is pulled with increasing force, it remains elastic until the yield point is reached, i.e. it returns to its original length when the load is removed. For materials without a distinct yield point, a substitute value is determined, the proportional limit$R_{p0,2}$.

Young's modulus: Material constant determined by elongation tests on test specimens. For many materials, the ratio between the stress (load) and the strain obtained (elongation) remains the same. This ratio is the modulus of elasticity. The modulus of elasticity can also be seen as the (of course only theoretical) stress at which the elongation of a bar is 100%, i.e. the bar has reached twice its original length.

Charging efficiency λ_a: In an internal combustion engine, the C. is the ratio of the fresh charge supplied (this is everything that flows through the air filter) to the charge mass theoretically possible in the cylinder. Thus, the C. is not equal to the degree of delivery. Due to scavenging losses in the charge exchange top dead center, for example, fresh charge can be lost via the exhaust tract. This loss is taken into account in the C., but not in the degree of delivery. In this example, the C. would be greater than the degree of delivery if the mass supplied is greater than the theoretically possible mass. The C. is easier to measure than the degree of delivery.

Volumetric efficiency λ_l: In an internal combustion engine, the V. is the ratio of the charge mass actually in the cylinder after completion of the charge exchange compared to the charge mass theoretically possible in the cylinder (= swept volume times air density). In naturally aspirated engines, the V. is less than 1. As the flow velocity (speed) increases, the losses increase due to throttling in the lines and valves. This is partly compensated or even overcompensated by gas dynamic effects at certain speeds.

Air-fuel ratio λ: The air-fuel mixture in the engine ignites and burns satisfactorily only within a certain mixture range. For gasoline, this ratio is about 14.7:1, i.e. 14.7 kg of air are required for complete combustion of 1 kg of fuel (stoichiometric mixture).

The air number λ compares this theoretical demand with the actual mixture present.

$$\lambda \frac{\text{existing mixture}}{\text{stoichiometric mixture}}$$

$\lambda = 1$ means that there is a stoichiometric mixture in the combustion chamber. $\lambda < 1$ means there is a lack of air (rich mixture). $\lambda > 1$ means there is excess air (lean mixture).

Listed below are the differences between corresponding American (AE) and British terms (BE) for some common parts:

Component	American	British
Side shaft	Axle shaft	Half shaft
Drive shaft	Driveshaft	Prop shaft
Wheelhouse	Fender	Wheel arch
(Engine) hood	Hood	Bonnet
Oversteer	Loose	Oversteer
Bevel gearbox	Ring & pinion	Crown wheel & pinion
Understeer	Tight (push)	Understeer
Trunk	Trunk	Boot
Shock absorber	Shock absorber	Damper
Torsion stabilizer	Sway bar	Anti roll bar
Gurney bar	Wicker	Gurney
Windscreen	Windshield	Windscreen

Different racing classes also use different names for what is essentially the same component:

- Wishbone: A-arm/wishbone, control arm
- Wheel carrier: spindle, knuckle (touring car)/upright (monoposto)
- tie rod: tie rod/toe link

References

1. Breuer, B., Bill, K.-H. (eds.): Bremsenhandbuch, 1st edn. GWV Fach/Vieweg, Wiesbaden (2003)
2. Milliken, W.F.: Chassis Design: Principles and Analysis. Society of Automotive Engineers, Warrendale (2002)
3. Neumann, R., Hanke, U.: Eliminierung unerwünschter Bewegungen mittels geeigneter Momentanpolkonfiguration. Konstruktion, Heft 4, pp. 75–77. Springer, Berlin (2005)

Index

M. Trzesniowski, *Powertrain*,
https://doi.org/10.1007/978-3-658-39885-9

GPSR Compliance

The European Union's (EU) General Product Safety Regulation (GPSR) is a set of rules that requires consumer products to be safe and our obligations to ensure this.

If you have any concerns about our products, you can contact us on ProductSafety@springernature.com

In case Publisher is established outside the EU, the EU authorized representative is:

Springer Nature Customer Service Center GmbH
Europaplatz 3
69115 Heidelberg, Germany

Batch number: 08985863

Printed by Printforce, the Netherlands